Advances in Postharvest Fruit and Vegetable Technology

Contemporary Food Engineering

Series Editor

Professor Da-Wen Sun, Director

Food Refrigeration & Computerized Food Technology
National University of Ireland, Dublin
(University College Dublin)
Dublin, Ireland
http://www.ucd.ie/sun/

Advances in Postharvest Fruit and Vegetable Technology

EDITED BY

Ron B.H. Wills • John Golding

CRC Press
Taylor & Francis Group
Boca Raton London New York

CRC Press is an imprint of the
Taylor & Francis Group, an **informa** business

CRC Press
Taylor & Francis Group
6000 Broken Sound Parkway NW, Suite 300
Boca Raton, FL 33487-2742

First issued in paperback 2018

© 2015 by Taylor & Francis Group, LLC
CRC Press is an imprint of Taylor & Francis Group, an Informa business

No claim to original U.S. Government works

ISBN 13: 978-1-138-89405-1 (pbk)
ISBN 13: 978-1-4822-1696-7 (hbk)

Visit the Taylor & Francis Web site at
http://www.taylorandfrancis.com

and the CRC Press Web site at
http://www.crcpress.com

Contents

Series Preface

CONTEMPORARY FOOD ENGINEERING

Food engineering is the multidisciplinary field of applied physical sciences combined with the knowledge of product properties. Food engineers provide the technological knowledge transfer essential to the cost-effective production and commercialization of food products and services. In particular, food engineers develop and design processes and equipment to convert raw agricultural materials and ingredients into safe, convenient, and nutritious consumer food products. However, food engineering topics are continuously undergoing changes to meet diverse consumer demands, and the subject is being rapidly developed to reflect market needs.

In the development of food engineering, one of the many challenges is to employ modern tools and knowledge, such as computational materials science and nanotechnology, to develop new products and processes. Simultaneously, improving food quality, safety, and security continues to be critical issues in food engineering studies. New packaging materials and techniques are being developed to provide more protection to foods, and novel preservation technologies are emerging to enhance food security and defense. Additionally, process control and automation regularly appear among the top priorities identified in food engineering. Advanced monitoring and control systems are developed to facilitate automation and flexible food manufacturing processes. Furthermore, energy-saving and minimization of environmental problems continue to be important food engineering issues, and significant progress is being made in waste management, efficient utilization of energy, and reduction of effluents and emissions in food production.

The *Contemporary Food Engineering Series*, consisting of edited books, attempts to address some of the recent developments in food engineering. The series covers advances in classical unit operations in engineering applied to food manufacturing as well as topics such as progress in the transport and storage of liquid and solid foods; heating, chilling, and freezing of foods; mass transfer in foods; chemical and biochemical aspects of food engineering and the use of kinetic analysis; dehydration, thermal processing, non-thermal processing, extrusion, liquid food concentration, membrane processes, and applications of membranes in food processing; shelf-life and electronic indicators in inventory management; sustainable technologies in food processing; and packaging, cleaning, and sanitation. These books are aimed at professional food scientists, academics researching food engineering problems, and graduate-level students.

The editors of these books are leading engineers and scientists from different parts of the world. All the editors were asked to present their books to address the market's needs and pinpoint cutting-edge technologies in food engineering.

All contributions are written by internationally renowned experts who have both academic and professional credentials. All authors have attempted to provide critical, comprehensive, and readily accessible information on the art and science of a relevant topic in each chapter, with reference lists for further information. Therefore, each book can serve as an essential reference source to students and researchers in universities and research institutions.

Da-Wen Sun
Series Editor

Series Editor

Prof. Da-Wen Sun, born in southern China, is a global authority in food engineering research and education; he is a member of the Royal Irish Academy (RIA), which is the highest academic honor in Ireland; he is also a member of Academia Europaea (The Academy of Europe) and a fellow of the International Academy of Food Science and Technology. He has contributed significantly to the field of food engineering as a researcher, as an academic authority, and as an educator.

His main research activities include cooling, drying, and refrigeration processes and systems, quality and safety of food products, bioprocess simulation and optimization, and computer vision/image processing and hyperspectral imaging technologies. His many scholarly works have become standard reference materials for researchers, especially in the areas of computer vision, computational fluid dynamics modeling, vacuum cooling, and related subjects. Results of his work have been published in over 800 papers, including more than 390 peer-reviewed journal-papers (Web of Science h-index = 62). He has also edited 14 authoritative books. According to Thomson Scientific's *Essential Science Indicators*[SM], based on data derived over a period of 10 years from Web of Science, there are about 4,500 scientists who are among the top 1% of the most cited scientists in the category of agriculture sciences; for many years, Professor Sun has consistently been ranked among the top 50 scientists in the world (he was at 25th position in March 2015, and at 1st position if ranking is based on "Hot Papers," and in 2nd position if ranking is based on "Top Papers" or "Highly Cited Papers").

He received a first class BSc honors and MSc in mechanical engineering, and a PhD in chemical engineering in China before working in various universities in Europe. He became the first Chinese national to be permanently employed in an Irish university when he was appointed college lecturer at the National University of Ireland, Dublin (University College Dublin, UCD), in 1995, and was then progressively promoted in the shortest possible time to senior lecturer, associate professor, and full professor. Dr. Sun is now a professor of food and biosystems engineering and the director of the UCD Food Refrigeration and Computerized Food Technology Research Group.

As a leading educator in food engineering, Professor Sun has trained many PhD students, who have made their own contributions to the industry and academia. He has also frequently delivered lectures on advances in food engineering at academic institutions worldwide, and delivered keynote speeches at international conferences. As a recognized authority in food engineering, he has been conferred adjunct/ visiting/consulting professorships from 10 top universities in China, including

Zhejiang University, Shanghai Jiaotong University, Harbin Institute of Technology, China Agricultural University, South China University of Technology, and Jiangnan University. In recognition of his significant contribution to food engineering worldwide and for his outstanding leadership in the field, the International Commission of Agricultural and Biosystems Engineering (CIGR) awarded him the "CIGR Merit Award" in 2000, and again in 2006, the Institution of Mechanical Engineers based in the United Kingdom named him "Food Engineer of the Year 2004." In 2008, he was awarded the "CIGR Recognition Award" in honor of his distinguished achievements as one of the top 1% among agricultural engineering scientists in the world. In 2007, he was presented with the only "AFST(I) Fellow Award" given in that year by the Association of Food Scientists and Technologists (India), and in 2010, he was presented with the "CIGR Fellow Award"; the title of Fellow is the highest honor at CIGR and is conferred to individuals who have made sustained, outstanding contributions worldwide. In March 2013, he was presented with the "You Bring Charm to the World" Award by Hong Kong-based Phoenix Satellite Television with other award recipients including the 2012 Nobel Laureate in Literature and the Chinese Astronaut Team for Shenzhou IX Spaceship. In July 2013, he received the "Frozen Food Foundation Freezing Research Award" from the International Association for Food Protection (IAFP) for his significant contributions to enhancing the field of food-freezing technologies. This is the first time that this prestigious award was presented to a scientist outside the United States.

He is a fellow of the Institution of Agricultural Engineers and a fellow of Engineers Ireland (the Institution of Engineers of Ireland). He is editor-in-chief of *Food and Bioprocess Technology—An International Journal* (2012 Impact Factor = 4.115), former editor of *Journal of Food Engineering* (Elsevier), and a member of the editorial boards for a number of international journals, including the *Journal of Food Process Engineering, Journal of Food Measurement and Characterization,* and *Polish Journal of Food and Nutritional Sciences.* He is also a chartered engineer.

On May 28, 2010, he was awarded membership in the RIA, which is the highest honor that can be attained by scholars and scientists working in Ireland. At the 51st CIGR General Assembly held during the CIGR World Congress in Quebec City, Canada, on June 13–17, 2010, he was elected incoming president of CIGR, became CIGR president in 2013–2014, and is now CIGR past president. On September 20, 2011, he was elected to Academia Europaea (The Academy of Europe), which is functioning as the European Academy of Humanities, Letters and Sciences, and is one of the most prestigious academies in the world; election to the Academia Europaea represents the highest academic distinction.

Preface

The aim of postharvest technology is to transfer fruit and vegetables from the farm to consumers without loss of quantity and quality. Great advances have occurred since the 1950s, when there was a seasonal availability of most produce with consumption close to production areas and where quality was secondary to availability. In the intervening years, the horticulture industry has advanced to a situation where there is now year-round availability for many commodities, which is supported by a substantial international trade. In addition, consumer expectations of fresh produce and quality have substantially changed in the last few decades. A greater diversity and quantity of produce are marketed but high quality is paramount, although there are various concepts of what constitutes quality. This development was made possible by the introduction of a range of technologies but the major technological impacts can be ascribed to

- Cool-chain management, where a controlled temperature environment is maintained from farm storage to domestic consumption to inhibit ripening and senescence.
- Availability of synthetic chemicals, mainly to inhibit the growth of diseases and pests.

However, changes in the use of these technologies are being driven by evolving consumer expectations, community attitudes, and economic pressures on the industry. Important among these are the

- Desire by the international community to minimize environmental degradation; this has led to a range of government policy initiatives to stop the use of ozone depleting substances and reduce greenhouse gas emissions.
- Growing consumer aversion to the presence of synthetic chemicals in foods and a desire for what is perceived to be more "natural" and "safe" foods, which effectively translates into reduced usage of synthetic chemicals.
- Use of smarter, more efficient technologies to assess and manage quality.

This volume examines a range of recently developed technologies and systems that will assist the horticulture industry in being more environmentally sustainable and economically competitive, in minimizing postharvest quality loss, and in the use of biomolecular tools in research and applications.

Acknowledgment

The editors wish to acknowledge the mentoring role of Dr. Barry McGlasson, first as the supervisor of their doctoral programs, but also as the source of wise counseling and friendship over many years. Dr. McGlasson's significant scientific advances to knowledge during his 50-year career have profoundly influenced the postharvest community around the world. He has an infectious enthusiasm for grasping new scientific tools that lead to a better understanding of the physiology and biochemistry of postharvest systems, and then finding innovative ways of translating this knowledge into practical outcomes. Dr. McGlasson has a unique ability to inspire his students and colleagues to be creative thinkers not only in postharvest but in life generally. He is passionate about the importance of horticulture and has been a strong and enthusiastic advocate for horticultural science, and the future of horticultural education. Dr. McGlasson's contributions to horticulture have been recognized in both Australia and internationally. He is a fellow of the International Society of Horticultural Science. His continuing support and lifelong friendship are greatly valued.

Editors

Dr. Ron B.H. Wills is an emeritus professor in the School of Environmental and Life Sciences at the University of Newcastle, Australia. He has 50 years experience in many aspects of postharvest horticulture and has published 300 research papers. He has held government, university, and industry positions in Australia and New Zealand, and has consulted for government and international agencies on postharvest development projects throughout Asia and the South Pacific.

Dr. John Golding works for the New South Wales Department of Primary Industries (Australia), where he conducts applied postharvest and market access research for a range of horticulture industries. He has had extensive experience in applied food science, postharvest physiology, and supply chain work. Dr. Golding also holds a conjoint position within the School of Environmental and Life Sciences at the University of Newcastle in Australia.

Contributors

Sajid Ali
Institute of Horticultural Sciences
University of Agriculture
Punjab, Pakistan

Salvador Castillo
Department of Food Technology
Miguel Hernández University
Alicante, Spain

Filippo De Curtis
Department of Agricultural
    Environmental and Food Sciences
University of Molise
Campobasso, Italy

Huertas M. Díaz-Mula
Department of Food Technology
Miguel Hernández University
Alicante, Spain

Charles F. Forney
Agriculture and Agri-Food Canada
Atlantic Food and Horticulture
    Research Centre
Kentville, Nova Scotia, Canada

John Golding
New South Wales Department of
    Primary Industries
New South Wales, Australia

Fabián Guillén
Department of Food Technology
Miguel Hernández University
Alicante, Spain

Maarten L.A.T.M. Hertog
BIOSYST-MeBioS, KU Leuven
Leuven, Belgium

Antonio Ippolito
Department of Soil, Plant and Food
    Sciences
University of Bari Aldo Moro
Bari, Italy

Pongphen Jitareerat
Division of Postharvest Technology
School of Bioresouces and Technology
King Mongkut's University of
    Technology Thonburi
Bangkok, Thailand

Ahmad Sattar Khan
Institute of Horticultural Sciences
University of Agriculture
Punjab, Pakistan

Giuseppe Lima
Department of Agricultural
    Environmental and Food Sciences
University of Molise
Campobasso, Italy

Babak Madani
Department of Crop Science
Universiti Putra Malaysia
Selangor, Malaysia

Marta Mari
CRIOF, DipSA
University of Bologna
Bologna, Italy

Domingo Martínez-Romero
Department of Food Technology
Miguel Hernández University
Alicante, Spain

Fiorella Neri
CRIOF, DipSA
University of Bologna
Bologna, Italy

Bart M. Nicolaï
BIOSYST-MeBioS, KU Leuven
and
VCBT
Leuven, Belgium

Simona Marianna Sanzani
Department of Soil, Plant and Food
    Sciences
University of Bari Aldo Moro
Bari, Italy

María Serrano
Department of Applied Biology
Miguel Hernández University
Alicante, Spain

Sukhvinder Pal Singh
New South Wales Department of
    Primary Industries
New South Wales, Australia

Zora Singh
Department of Environment and
    Agriculture
Curtin University
Western Australia, Australia

Jun Song
Agriculture and Agri-Food Canada
Atlantic Food and Horticulture
    Research Centre
Kentville, Nova Scotia, Canada

Alice Spadoni
CRIOF, DipSA
University of Bologna
Bologna, Italy

Lorraine Spohr
New South Wales Department of
    Primary Industries
New South Wales, Australia

Apiradee Uthairatanakij
Division of Postharvest Technology
School of Bioresouces and
    Technology
King Mongkut's University of
    Technology Thonburi
Bangkok, Thailand

Daniel Valero
Department of Food Technology
Miguel Hernández University
Alicante, Spain

Juan M. Valverde
Department of Food Technology
Miguel Hernández University
Alicante, Spain

Kerry Walsh
Central Queensland University
Queensland, Australia

Chris B. Watkins
Department of Horticulture
Cornell University
Ithaca, New York

Ron B.H. Wills
School of Environmental and Life
    Sciences
University of Newcastle
New South Wales, Australia

Pedro J. Zapata
Department of Food Technology
Miguel Hernández University
Alicante, Spain

1 Postharvest Technology Experimentation
Solutions to Common Problems

John Golding and Lorraine Spohr

CONTENTS

1.1 INTRODUCTION

Postharvest research aims to improve the availability and quality of horticultural produce by developing new ideas and assessing these ideas by conducting experiments and following the scientific method. The scientific method requires a well-defined and reported experiment, which then ensures the validity and defines the scope of any subsequent conclusions.

Sound statistical design and analysis of postharvest experiments are essential to the interpretation of results and inferences that can be drawn. Many of the tools that postharvest researchers use to design, conduct, and analyze experiments are sometimes misused, and consequently find their way into the literature. Poorly designed, and analysed experiments often lead to biased results and conclusions. This chapter is a collection of common problems encountered in postharvest technology research and suggested solutions to these problems.

After reviewing papers from three international postharvest journals published in 2014, 50 papers were studied in detail. We found 56% of these papers had no or poor descriptions of the experimental design. This poor description of the design makes it difficult to know how the experiment was setup and conducted. While minute details are often provided, describing the manufacturer of reagents and equipment, comparatively less detail is provided in describing the experimental design.

In this survey of recently published postharvest literature, we also found that the majority of papers surveyed had poor statistical analysis descriptions and presentation of the results. We found that 62% of the papers had either no or poor descriptions of the statistical methods, while 56% had poor presentation of the results. From this survey, we identified twelve common problems in postharvest technology research:

1. Acknowledging the literature
2. Ignoring variability
3. Unknown history of experimental material
4. Failing to acknowledge scope of inference
5. Failing to use a good experimental design
6. Poor description of experimental design
7. Understanding replication
8. Pseudoreplication
9. Sub-sampling
10. Incorrect statistical analysis, for example, using ANOVA (analysis of variance), regardless of data type and not checking the assumptions
11. Reporting of statistical analysis methods and results
12. Statistical significance versus biological significance of results

1.2   SOLUTIONS FOR COMMON PROBLEMS IN POSTHARVEST TECHNOLOGY EXPERIMENTATION

This section discusses the issues and problems that were identified in the literature survey (Section 1.1) and provides potential solutions. In addition, Chapter 16 further explores some of these issues in the context of modeling.

1.2.1   Acknowledging the Literature

As a first step in planning a postharvest experiment, it is essential to adequately acknowledge previously published results by thoroughly exploring the literature. This sounds like a menial task, but there have been many well-meaning experiments that have failed to access the literature. To conduct a thorough literature review, begin an online keyword search with a library database search engine, searching over all possible years. Repeat this search with at least one other search engine. Read the relevant papers and search the references from those papers. Attempts should be made to access book chapters, reviews, and conference proceedings in your literature review. These sources may not be easily located in the searchable electronic journals. This suggestion may appear trite to experienced researchers; however, there are many relevant older studies that could easily be overlooked if only databases are used.

1.2.2 IGNORING VARIABILITY

Variation is a natural feature of biological systems. Fundamental to the design and analysis of experiments is the appreciation and management of variability. The variation in produce quality parameters, such as soluble solids content (SSC) levels, is an inherent part of horticulture and postharvest science. Acknowledgment and management of produce variation, such as where it comes from, why it occurs and how to manage it, is crucial to good postharvest science. Identifying the sources of variability and taking them into account when designing postharvest experiments is essential, as undetected bias can contribute to spurious treatment effects.

1.2.3 UNKNOWN HISTORY OF EXPERIMENTAL MATERIAL

Finding out as much as possible about the history of your experimental material will assist in experimentation. Produce variation comes from a wide range of interacting pre- and postharvest factors. Preharvest factors such as cultivar, rootstock, soil type, irrigation and fertilizer regime, and fruit maturity at harvest, can affect postharvest storage behavior.

For example rootstocks can significantly affect maturity and SSC accumulation (Stenzel et al., 2006). It is easy to ignore fruit maturity and assume all fruit are "commercially mature," based on background color or fruit blush since color may not necessarily reflect physiological maturity. Relying on "commercial maturity" is a common practice in postharvest experimentation but can contribute to the often contradictory results sometimes found in the literature. For example, the effect of 1-methylcyclopropene (1-MCP) application to peaches has been reported to have varying effects on the development of chilling-related injury. Many peach studies with 1-MCP have been conducted using fruit with poorly defined maturity at harvest. However, in a clearly defined fruit-maturity study, 1-MCP treatment increased the severity of chilling-related injury in early maturity peach fruit but had either no effect on commercially mature fruit or even slightly delayed chilling-related injury (Jajo et al., 2012). This illustrates the importance of knowing fruit maturity and adequately defining the physiology of the fruit before and after storage. The concept of biological variation is further explored in Chapter 16.

Ripening and climacteric behaviors, chilling-sensitivities, and storage potential, can all vary between different cultivars of the same produce type. It is therefore important to adequately report the cultivar (or variety) used in an experiment.

Sourcing of fruit from the general market is a convenient method to obtain produce for postharvest experiments but is not recommended, unless the aim of the research is specifically to describe variability in the marketplace. Produce purchased from the market floor may not have enough relevant history. For example, fruit sourced from the market is of unknown biological maturity and origin and may have been harvested several days or even weeks previously. In addition, the storage conditions during the time spent in the market chain are unknown. The produce has been exposed to unknown temperature conditions and ethylene levels. To have confidence in the experimental results, it is important to have a thorough knowledge of the background of the produce to be used in the experiment.

In addition, the application of plant growth regulators in the orchard can have a significant impact on ripening behavior during postharvest storage. For example, it is essential to know if your produce had been treated with orchard treatments such as aminoethoxyvinylglycine (AVG). This compound is known to competitively inhibit the activity of the enzyme 1-aminocyclopropane-1-carboxylic acid synthase which is the rate limiting enzyme in the ethylene biosynthetic pathway. AVG is generally applied 28 days before harvest and is used to extend the harvest period and increase the storage-life of the apples. Knowing if your produce has been treated with such chemicals is essential to properly conduct and interpret postharvest storage experiments.

Produce variation occurs at many different levels:

- *Within a single fruit*: Physiological differences may be observed within a single fruit. An example of this is the red blush (anthocyanin) associated with the sun-exposed side of apple fruit. In addition, temperature differences across the fruit, from exposed to unexposed sides, can exceed 10°C (Woolf et al., 1999). This means that the sun-exposed side of the fruit may have a different physiology at certain times of the day and this may have consequences for the fruit's postharvest behavior. There may also be significant gradients in SSC levels within some fruits. In extreme cases, the SSC may vary by up to 4% Brix levels within a single peach fruit (Golding et al., 2006). This variability has important consequences when sampling and measuring a fruit SSC.
- *Within a tree*: There can be large differences in produce quality from the same tree. Variation in fruit-size and dry matter content, within a tree and within individual branches, has been related to heterogeneity in light interception, the fruit's proximity to carbohydrate sources, fruit-load, and inter-fruit competition (Lopresti et al., 2014). The SSC levels within a single nectarine tree at harvest are presented in Figure 1.1. Although the average SSC level of all fruit was 14.2% Brix, the range of fruit SSC was 10%–21% Brix of fruit on the same tree at "commercial" harvest. Harvesting and using all the fruit from this tree in a postharvest experiment to investigate the effect of postharvest treatments on SSC would be challenging. In general, the top half of the tree had higher levels of fruit SSC compared to the bottom half. Even within the top of the trees, the average SSC of

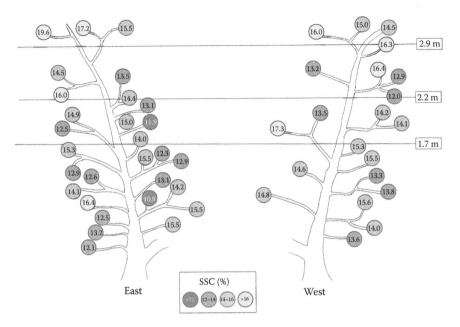

FIGURE 1.1 Schematic version of Arctic Snow nectarine tree showing average SSC% on each fruiting lateral. (From Golding JB. 2008. Scoping study to identify and quantify factors to improve summer-fruit quality and consistency. Horticulture Australia Project SF 06013. ISBN 0 7241 1909 7. p 161. With permission.)

some fruiting laterals was lower and not consistent with their position on the branches. Not only can fruit SSC vary within a single tree, numerous other physiological and biological factors can vary and affect postharvest behavior. For example, Magwaza et al. (2013) showed that variations in microclimatic conditions inside a citrus-tree canopy during the growing season, affect the biochemical profile of the fruit rind, which, in turn, influences fruit response to postharvest stresses associated with senescence and susceptibility to rind breakdown in mandarin fruit.

- *Within a row and within an orchard or field*: Orchard and field management practices (nutrition, irrigation) and growing conditions (row orientation, elevation) can affect the final produce growth, quality, and physiological maturity. For example, nitrogen nutrition has been shown to affect the level of storage disorders in numerous crops, such as apples (Little and Holmes, 2000). Although there may be similar grower management practices (such as same fertilizer and irrigation management) in an entire row/orchard, there may be different soils or elevations, which may affect plant growth and fruit maturity and quality at harvest and subsequent postharvest storage behavior.
- *Within growing regions and between different growing regions/countries*: The variation in produce quality and postharvest behavior can be further exaggerated by different grower practices (irrigation, fertilizer, thinning, etc.), rootstocks and growing conditions between geographical regions. All

these differences in produce quality may affect postharvest storage behavior in different ways and underline the need to appreciate and manage produce variability. Experiments claiming to show regional effects of quality based on a single orchard should be interpreted with caution. Similar care should be taken when comparing production methods such as organic versus conventional on postharvest quality.

- *Variation in the postharvest environment and unknown sources of variation*: Examples of this type of variability can occur within the postharvest cool room or shipping container, where the delivery air from the refrigeration unit can be substantially cooler than the return air. While this may be acknowledged, the accumulated differences in storage temperature over the whole life of the experiment can be significant. If the experiment is not designed correctly, treatments may be confounded with sources of variability. For example, if an entire treatment was stored in the return-air-stream, this treatment would be subject to higher storage temperature for the entire storage time, compared to those stored in the delivery air. The effect of treatment temperatures within a pallet stack should also be managed. In addition to temperature, it is also important to manage and record relative humidity (RH). RH is directly responsible for a wide range of postharvest issues such as weight loss, firmness, and incidence of decay, and can regularly vary by up to 35% between storage environments. Therefore, it is important to control and accurately monitor RH in the storage environment.

It is also important to appreciate and manage other sources of potential contamination within a cool room such as ethylene, where there may also be other commodities or experiments in the same cool room. Some commodities, such as green beans, are sensitive to low ethylene levels (0.01 $\mu L/L$), and these levels of ethylene have been shown to accumulate in cool rooms where other high-ethylene-producing commodities (such as apples) are also stored.

1.2.4 FAILING TO ACKNOWLEDGE SCOPE OF INFERENCE

It is a common practice to exclude variability from a sample population when selecting fruit for an experiment, which results in a uniform sample population at the start of the experiment. Whilst this does eliminate "outliers," the use of limited sample populations also limits the inference and conclusions. Authors and readers often unintentionally extend inference of these results to the entire population instead of just to the uniform part chosen for the study. To ensure that the results of postharvest experiments are widely applicable, it is important to sample from broad populations. The scope of inference provides the context in which the results of an experiment apply to a broader population of fruit. The scope of inference should be clearly defined and applied in each study. For example, a fruit-softening experiment that used only uniformly large-sized peaches, produced results that are applicable only to similarly sized fruit of that variety. Further extension to other peach types and sizes need to be conducted to confirm these findings.

1.2.5 FAILING TO USE A GOOD EXPERIMENTAL DESIGN

As stated by Rubin (2008), *"Design Trumps Analysis."* A good experimental design minimizes bias and incorporates known sources of variation, which may affect the response of interest. When sources of variation have been identified, a randomized complete block (RCB) design is often used. A "block" is a group of experimental units that are as similar as possible and/or experience similar conditions. The word "complete" refers to the condition that the whole (or complete) set of treatments being evaluated is represented within each block. It is important to recognize and correctly use an RCB design. With the correct use of an RCB, we can incorporate potential variability between farms or orchards by forming the blocks based on fruit source (where a block represents a farm or orchard), and allocating a complete set of treatments to the experimental units within each block. In the subsequent statistical analysis, comparing treatment means differences between farms will cancel out, and unexplained (residual) variation is reduced.

In the laboratory setting, an example of efficient use of an RCB design is when blocks are based on a time order. For example, when assessing produce quality after a storage experiment instead of assessing all of "Treatment 1" at once, then assessing "Treatment 2," it is better to use blocks. This is where a complete set of treatments (Block 1) is assessed in the morning and another (Block 2) at mid-day and a third (Block 3) in the afternoon. Any bias due to operator (e.g., fatigue) or time of day will be accounted for in the subsequent statistical analysis by the blocking structure.

Many authors often state that they use an RCB when they have actually used a completely randomized design (CRD). A CRD is recommended when no known sources of variation—that may affect the experiment—can be identified, and the experimental material is homogeneous.

1.2.6 POOR DESCRIPTION OF EXPERIMENTAL DESIGN

A clear description of the type of experimental design, the number of replicates, a definition of the experimental unit, and type of statistical analysis method employed, is essential. This should provide enough detail so that someone else could repeat the experiment at another time and place, without the need to make any assumptions about how the experiment was setup or conducted.

In any postharvest experiment, it is important to clearly define the experimental unit for each treatment. The experimental unit is the physical unit to which a treatment is independently applied, usually through a randomization process. An experimental unit may be an individual (a single apple) or a group (five apples), depending on the experimental protocol. The identification of experimental units, the independence of the experimental unit and the consequent scope of inference are all important concepts to understand and apply to all postharvest experiments.

1.2.7 UNDERSTANDING REPLICATION

True replication occurs when the same treatment is independently applied to a number of experimental units. Replicating (or repeating the application of) treatments leads to better estimates of the treatment means and allows the variability of the population to be estimated. With several replicates of a treatment, variability about the mean can be estimated and the credibility of the mean is established. It is the variability about the treatment mean that is used to compare treatment performance. Treatments should be replicated for a minimum of three times. Increasing the number of replicates improves the credibility of the results but also increases the cost and time involved in running an experiment. These factors can lead to researchers unknowingly allowing their experiments to be pseudoreplicated.

1.2.8 PSEUDOREPLICATION

The term "pseudoreplication" is used to describe the situation where there is repeated application of treatments but not "true" replication of treatment for reasons of lack of independence or fallible randomization (Hurlbert, 1984). This can lead to potentially incorrect statistical analysis and misinterpretation of the results.

In our survey of published papers, only 22% of papers studied could confidently be classified as not incorporating pseudoreplication. Pseudoreplication was clearly identified in 6% of the studied papers. The remaining 72% of papers failed to provide sufficient detail to determine if pseudoreplication had or had not occurred. This lack of clarity in descriptions of experimental designs made the validity of conclusions of each experiment difficult to assess.

The following example demonstrates a common pseudoreplication scenario in postharvest experiments. The experiment aimed to compare the effect of two storage temperatures on banana ripening (Figure 1.2). In this case, there are four replicates of the two storage temperatures. Each treatment was applied to an experimental unit comprised of a tray of 10 bananas. Measurements were made on individual bananas and an average of 10 measurements used to represent each replicate. In the statistical analysis, it is important to correctly identify the experimental unit as 10 bananas, not an individual fruit.

Ideally, we would have enough cool rooms to supply true replicates of the temperature treatments. In this case, four replicates of the two temperatures require eight different cool rooms (Figure 1.2a). This is often impractical and researchers may resort to a pseudoreplicated experimental design, where the different replicates are located within the same treatment cool room (Figure 1.2b). Whilst we recognize this is the day-to-day reality for many researchers, the temptation to draw inferences as if the treatments were properly replicated should be avoided or at the very least acknowledged.

Arguments against conducting pseudoreplication usually address three main areas of statistical concern: (1) variation, (2) independence of experimental units, and (3) undetected events.

1. *Variation*: Treatments are usually assessed by comparing their observed impact on the variability inherent in the experimental material. The purpose of replication is to allow a measure of that variability to be made.

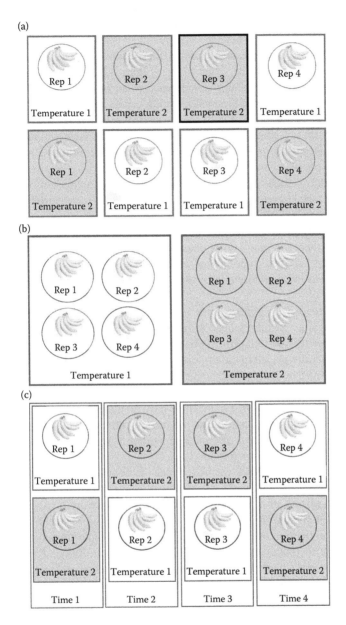

FIGURE 1.2 (a) Representation of bananas inside eight different cool rooms set at two different temperatures. (b) Representation of bananas (four hands/"replicates") inside two cool rooms set at two different temperatures. (c) Representation of banana storage experiment replicated at four different times in two different cool rooms at each time.

A pseudoreplicated experiment will tend to underestimate the random variation that may occur when the same treatment is applied at another time and place. In the banana storage example, this occurs when the method of treatment application (in the single room—Figure 1.2b) produces pseudoreplicates—the four "replicates" inside each room have actually received the same treatment. When pseudoreplication occurs, the variability of a single treatment is confounded with the variability inherent for the single room in which it is contained. There is no estimate of the background variability that relates to the process of supplying the temperature that is independent of the specific chamber in which it is located. Replicate rooms are needed for each treatment (Figure 1.2a) to distinguish the between-room variability from the between-"bananas within trays"-within-room variability (Riley and Edwards, 1998).

Apart from the difference in temperature between the different cool rooms, there are many other characteristics such as lighting, temperature gradients, and the presence of ethylene and volatile organics that may vary from room to room, even though attempts are made to control these features.

2. *Independence of experimental units or true replicates*: Another argument against pseudoreplication involves the concept of independence of the experimental units. In the banana temperature experiment, when the bananas were placed in the two different cool rooms, there may have been a lack of independence present, and consideration must be given to the extent to which one fruit's response influences the responses of the other fruit. For example, one banana in a bunch may have an unnoticed bruise, which may initiate wound-induced ethylene production. This may induce general ripening of the whole hand (depending on maturity) and could extend to other replicates within the cool room.

Lack of independence can also be seen in controlled-atmosphere storage experiments where a gas for a treatment is supplied from a single source. Each replicate of a treatment should ideally have its own independent management system. In that way a mechanical failure, technical failure, or contamination event will only affect a single experimental unit and be unlikely to produce a "treatment effect" (Hurlbert, 1984). Although this may be impractical, these limitations should be recognized and the conclusions interpreted in the light of possible pseudoreplication.

3. *Undetected events*: When a treatment is prepared only once, and this singular source is then supplied to all the different "replicates," an undetected problem with the treatment application is more likely to affect all the repeats. An observed treatment effect may be biased, either positively or negatively, without the experimenter's knowledge. For example, if a postharvest dip solution is inadvertently made up incorrectly, then all the different replicates are treated from this one incorrectly prepared solution.

Arguments in favor of accepting pseudoreplication in postharvest experiments usually include logistic and cost considerations. For example, it may be impractical to have multiple cool rooms set at the same temperature, since resources are limited. It can be argued that experimental methods used are so well monitored with data loggers that any unusual events will be recorded and not attributed to particular treatments. The technical expertise possessed by postharvest researchers is also put forward as a justification for the practice of pseudoreplication. Although mistakes in treatment application are rare, they can happen.

In many areas of science, similar treatments are often included in experiments conducted by other researchers in different places and times. The pseudoreplicated experiment can then become a "true" replicate in a broader body of knowledge using meta-analysis (Cottenie and De Meester, 2003; Hannah, 1999).

Pseudoreplication should be avoided in postharvest experiments, although, when it is present in a study, there are different ways to deal with it: (1) acknowledgment, (2) limit the inference, (3) replicate in time, and/or (4) use of regression models in the analysis.

1. *Acknowledgment*: Some people acknowledge that they have pseudorep-licated their experiment but still analyze their data as if these were true replicates. Osunkoya and Ash (1991) reported: "*As the shade house com-partments were not replicated, differences among treatments may be con-founded with other sources of error. In the statistical analysis of the data, it is assumed that the differences in seedling responses among compart-ments are due solely to the differences in light regime.*" This acknowledg-ment allows the reader to assess these assumptions and consider the paper accordingly.

2. *Limit the inference*: In some pseudoreplicated experiments, other authors clearly limit the inference drawn from the results. Shah et al. (2000) included the following statement: "*ANOVA could not be performed to directly com-pare the effects of either different drying humidities or storage conditions. Such analyses would have been statistically inappropriate and examples of pseudoreplication due to isolative segregation. The present work should, however, be seen as indicative of drying and storage conditions which should be further investigated to obtain more robust guidelines.*"

   If the scope of inference of an experiment is limited to the particular time and place in which the experiment was conducted, then the experiment is more accurately described as a pilot study. Pilot studies are often under-taken to assess possible promising treatments for further research, so true replication is not as important. The scope of inference is confined to the time, place, and individuals used in the pilot study (Phillips, 2002). "*It is an acceptable analysis if it is recognised that the results are only relevant to the individual or group of individuals used for the study*" (Phillips, 2002).

3. *Replicate in time*: If resource availability does not permit true replica-tion of treatments, repeating the experiments in time can resolve this issue. Treatment allocations are rerandomized at each time (Figure 1.2c). Conclusions will be more robust with, say, a different batch of fruit, or fruit from a different season, and the scope of inference will be wider. However, a constant issue in postharvest science is seasonality of horticultural produce.

Generally, there is only a narrow window during which harvesting occurs and experiments can be conducted. Even if the experiment is repeated at different times in the same growing season, the same variety may not be available and the conditions under which the produce has been grown are likely to be different. Each of these factors may affect storage behavior and are often given as the reason for pseudoreplication. On the other hand, if a treatment effect persists across a season and/or between varieties, the treatment effect would be more robust.

4. *Regression models*: Statistical analysis methods, such as linear regression, are suitable to use when treatments are quantitative levels (e.g., storage responses at 0°C, 2°C, 5°C, 10°C, and 15°C) and there has been no replication. Linear mixed effects regression models, which allow for correlation among residuals, may also be used to analyze pseudoreplicated data (Pinheiro and Bates, 2000).

1.2.9 SUB-SAMPLING

It is common that journal articles report experiments that were conducted "*in triplicate.*" A reader is left to wonder whether there were three true replicates of each treatment, or whether a single batch of experimental material was prepared for each treatment and then divided into three sub-samples for analysis (triplicates). The distinction between the two scenarios is critical for the correct data analysis and interpretation of results. An example is the extraction and measurement of anthocyanins, where an experimental unit comprised of five apples that were blended into a liquid, from which three sub-samples were taken and measured. This is a triplicate sub-sample from one replicate, and measures the variability of the analytical method. Data from each sub-sample triplicate is not independent and needs to be averaged to obtain one value to represent the experimental unit (five apples). Unfortunately this is an extremely common error. To validly compare the effect of treatments, independent replications are necessary. For example, for each treatment, extractions from three different replicates of five apples (where each replicate has been treated independently) are required. Pooling of variation is further discussed in Section 16.2.4.

1.2.10 INCORRECT STATISTICAL ANALYSIS

The most appropriate type of statistical analysis to use is determined by the experimental design and the type of data collected in the study. Some recommended texts to assist experimental design and statistical analysis include Steel and Torrie (1980), Ramsey and Schafer (2002), Moore et al. (2012), and Welham et al. (2014).

Data arising from a single experiment is just one realization of the many, many experiments that could have occurred under similar conditions. Statistical analysis allows statements to be made about the population of experiments of which the observed data is only a sample (Snedecor and Cochran, 1967). The data from one single experiment provides information about the particular system that we are studying and it is the system that is the focus of modeling, not the data. A discussion and summary of current techniques in postharvest modeling are presented in Chapter 16.

In postharvest experiments, there are generally four types of data collected: (1) binomial (e.g., yes/no), (2) ordinal (e.g., quality scores), (3) discrete (e.g., count of rotten fruit), and (4) continuous (e.g., weight in grams). It is important to identify the data type as each requires a specialized statistical analysis method and analysis of variance (ANOVA) is not always the most appropriate data analysis method to use. Methods for statistical analysis of each data type are described below.

1. *Binomial data*: Binomial data has traditionally been transformed using the arcsine (or angular) transformation of proportions prior to ANOVA. However, better methods such as generalized linear regression models (GLMs) can be used to analyze this type of data. Residuals follow the binomial distribution and a logistic (logit) link function is commonly used. Instead of analyzing the actual proportion, a GLM with a logistic link function analyzes the logarithm of the odds of the event occuring. All the statistical analysis and treatment comparisons are made on the logistic scale and can be back transformed to proportions for presentation. The probit link function is also available for binomial data.

2. *Ordinal data*: Subjective assessment of fruit and vegetables using scoring techniques is common in many postharvest experiments. This data type has a defined scale, often on a 1–5 subjective scale. This data is not continuous and ANOVA assumptions are unlikely to be met. However, when an experimental unit comprises multiple single fruit and, having several different people independently assess each fruit, the average of all the scores for each unit can be assumed continuous and analyzed using ANOVA after carefully checking that the residuals assumptions are met. Ordinal linear regression methods are also available to analyze score data.

3. *Discrete (count) data*: While ANOVA of log-transformed count data is common, a GLM with Poisson error distribution and log link function is recommended. When counts are converted to percentage data, it is preferable to analyze the raw count data using a GLM with binomial errors.

4. *Continuous data*: Linear models ANOVA has been a standard method for comparing treatments from designed experiments with continuous data. ANOVA is a type of linear model widely used to test treatment effects when data originates from a continuous scale of measurement (e.g., fruit weight). ANOVA is a method for comparing means and the term "variance" should not be misleading. The word "variance" relates to the approach to assess differences in treatment means by comparing amounts of variability explained by different sources. The overall variability is calculated by squaring differences between each data value and the overall mean, and then adding each squared difference together.

After using ANOVA to determine differences between treatment means, the unexplained variability (residuals) requires consideration. There are three assumptions, related to residuals from the model, need to be met for a valid ANOVA. These are

1. Independence (residuals from one experimental unit are not influenced by residuals from any other unit).

2. Identical (residuals have a constant variance regardless of the size of the fitted value).
3. Normally distributed (a histogram of residuals resembles the bell shape of the normal distribution).

All of these assumptions need to be tested and met to conduct a valid ANOVA. A correct analysis of the data ensures that sound conclusions can be drawn from the data.

Permutation test: The permutation test is an alternate method for testing treatment effects when either the normality assumption is not met, there are outliers, or the sample size is too small. Although this test is not commonly used in postharvest statistical analysis, it could be used with data that does not meet the assumptions of ANOVA (Welham et al., 2014). This approach has been applied in postharvest science, where Bobelyn et al. (2010) examined apple storage behavior, which was predicted by near infra red (NIR) analyses.

Mixed effects models: Linear modeling is a subset of a larger group of models known as mixed-effects models (Pinheiro and Bates, 2000). Mixed-effects models include both fixed and random effects and provide a way to investigate complex patterns of variability that are representative of real world variability (Zuur et al., 2009). A fixed effect (or factor) is one that the research question identifies (e.g., 1-MCP rate), while random effects are usually ones that are assumed to represent a broader population (e.g., physical blocks). Random factors can also take into account the lack of independence in residuals, for example, from a repeated-measures experiment where the data comprises responses from exactly the same apple taken on several occasions over a period of time. Mixed-effect models also allow the analysis of data from unbalanced experimental designs where there are unequal numbers of replicates of each treatment.

1.2.11 Reporting Statistical Analysis Methods and Results

In our survey of published postharvest papers, we found that 62% of the papers had either no or poor statistical methods descriptions, while 56% had poor presentation of the results. Common failings in the reporting of statistical analysis methods and results include:

1. Not detailing the actual treatment factors being tested and experimental design structure used; the reader needs to have confidence that the factors were correctly structured and tested in the analysis. An incorrect design structure in the model will lead to incorrect conclusions.
2. Not identifying measures of variability—for example, stating whether the variability is a standard deviation, standard error, or least-significant difference.
3. Not using appropriate wording to describe the statistical analysis—for example, saying that "*data were analyzed using Statistica*" is not sufficient. In addition, terms encountered in published papers, such as "*complete randomized block*," are incorrect.

4. Exaggerating the precision of results by using too many decimal places—for example, the average SSC has been reported as 13.256 Brix. This measurement should be reported as 13.3, since the precision of most refractometers is in increments of 0.1.

Statistical significance was historically indicated using as asterisk system where * represented $p < 0.05$, etc. However when reporting on the results, it is more useful to supply the actual test statistic (e.g., F value), appropriate degrees of freedom and actual probability value for all treatment effects where appropriate. For example, in a trial of avocado storage, where the results were critical to industry, the treatment effect and treatment means could be reported, even though the calculated F value was not significant ($F_{3,6} = 0.69$, $P = 0.593$). When actual probability values are supplied, a single experiment is more easily incorporated into any future meta-analysis, thus increasing the value of the research (Fidler, 2010). Indeed, some online journals provide the opportunity to include supplementary data that support the manuscript. The supply of this additional data not only allows the reader to scrutinize the data to confirm results and conclusions, it also allows the potential for future meta-analysis and more open research.

Another approach to data-reporting is presenting effect sizes, by supplying treatment means within the text and providing confidence intervals (usually 95%) for means, rather than just stating that an effect was significant (Cummings, 2012). The confidence interval for the mean provides a range that is highly likely (often 95% or 99%) to contain the true population mean.

1.2.12 STATISTICAL SIGNIFICANCE VERSUS BIOLOGICAL SIGNIFICANCE OF RESULTS

On some occasions, the differences between postharvest treatments that have been declared significant according to a statistical test do not always relate to biological or commercial differences. For example, there may be a statistically significant treatment effect on peel color (as measured by the Minolta color meter), but the actual difference in color between the treatments may not be detectable to the eye or have any commercial significance. It is important to relate the statistically analyzed results to biological and commercial significance.

After conducting a statistical analysis, researchers commonly investigate differences between treatments and want to know if an observed effect is "statistically significant." The statistician R.A. Fisher recommended a probability level (P value) of 1 in 20 (or 5%) in his classic book *Statistics Methods for Research Workers* as being "convenient to take this point as a limit in judging whether a deviation is considered significant or not. Deviations exceeding twice the standard deviation are thus formally regarded as significant" (Fisher, 1938). This convention has since been widely adopted in science.

P is the probability (assuming there is no difference between treatment means) of obtaining a test statistic (e.g., F value) as large, or even larger, than the one that was observed. According to the Fisher convention, a treatment effect with a P value greater than 0.05 would be reported as being "not significant." However, it may be appropriate to report that there was evidence of a treatment effect and, by including the actual P value, a reader can interpret the treatment effect in the context of the replicated and designed experiment, for example, when P = 0.051.

When a test statistic is not significant (usually P > 0.05), instead of writing "there was no difference between the treatments," it is more appropriate to write "no significant difference was detected between the treatments." In a small experiment, say, with only three replicates of four treatments, there will sometimes be a "nonsignificant" treatment effect that is real and important. However, if the experiment had been larger, say, six replicates of the same four treatments, this effect may have been detected as being statistically significant.

Although using P values provides a convenient guide to help delineate statistical significance, other knowledge about the system being studied and the context of the experiment is also important. Cummings (2012) claims that P values are unreliable and, instead, suggests that confidence intervals provide a better estimate of the true treatment effect, and also indicate the extent of uncertainty in the result.

People also sometimes misinterpret the P value as the probability that the treatment being studied actually works.

1.3   CONCLUSIONS

A sound postharvest experiment manages variability and clearly articulates the population to which the conclusions will apply. It provides sufficient details of the experimental design, protocols, statistical analysis, and results. When all of these factors are all appropriately addressed, conclusions based on the results from the experiment will be robust and valid. Subsequent postharvest treatment recommendations can then be assessed with confidence by industry, thus potentially improving the availability and quality of horticultural produce.

ACKNOWLEDGMENT

The authors gratefully acknowledge the contributions of Beverley Orchard and Sharon Nielsen as codevelopers of the NSW Department of Primary Industries Biometrics Basic Statistics course where many of the ideas expressed in this chapter originated. They also acknowledge the critical reading of this chapter by Daniel Valero and Chris Watkins.

REFERENCES

Bobelyn E, Serban AS, Nicu M, Lammertyn J, Nicolai BM, Saeys W. 2010. Postharvest quality of apple predicted by NIR-spectroscopy: Study of the effect of biological variability on spectra and model performance. *Postharv Biol Technol* 55: 133–143.

Cottenie K, De Meester L. 2003. Comments to Oksanen (2001): Reconciling Oksanen (2001) and Hurlbert (1984). *Oikos* 100: 394–396.

Cummings G. 2012. *Understanding the New Statistics*. Taylor & Francis Group, LLC. New York.

Fidler F. 2010. The American Psychological Association *Publication Manual* sixth edition: Implications for Statistics Education. ICOTS8. International Association of Statistical Education (IASE) www.stat.auckland.ac.nz/~iase/(accessed July 14, 2014).

Fisher RA. 1938. *Statistics Methods for Research Workers*. Seventh edition. Oliver and Boyd. London. p 46.

Golding JB. 2008. Scoping study to identify and quantify factors to improve summer-fruit quality and consistency. Horticulture Australia Project SF 06013. ISBN 0 7241 1909 7. p 161.

Golding JB, Satyan S, Liebenberg C, Walsh K, McGlasson WB. 2006. Application of portable NIR for measuring sugar concentrations in peaches. *Acta Hort* 713: 461–464.

Hannah MC. 1999. Usefully combining a series of unreplicated cheesemaking experiments. *J Dairy Res* 66: 365–374.

Hurlbert SH. 1984. Pseudoreplication and the design of ecological field experiments. *Ecol Mono* 54: 187–211.

Jajo A, Ziliotto F, Rasori A, Bonghi C, Holford P, Jones M, Golding J, Tonutti P, McGlasson B. 2012. Transcriptome analysis of the differential effect of 1-MCP on the development of chilling injury in peaches harvested at early and late maturities. *Acta Hort* 934: 1003–1009.

Little CR, Holmes RJ. 2000. *Storage Technology for Apples and Pears. A guide to production, postharvest treatment and storage of pome fruit in Australia*. DNRE, Victoria, Australia. p 528.

Lopresti J, Goodwin I, McGlasson B, Holford P, Golding J. 2014. Variability in size and soluble solids concentration in peaches and nectarines. *Hort Reviews* 42: 253–311.

Magwaza LS, Opara UL, Cronje PRJ, Landahl S, Terry LA. 2013. Canopy position affects rind biochemical profile of "Nules Clementine" Mandarin fruit during postharvest storage. *Postharv Biol Technol* 86: 300–308.

Moore DS, McCabe GP, Craig BA. 2012. *Introduction to the Practice of Statistics*. Seventh edition. W.H. Freeman and Company. New York.

Osunkoya OO, Ash JE. 1991. Acclimation to a change in light regime in seedlings of six Australian rainforest tree species. *Aust J Bot* 39: 591–605.

Phillips CJC. 2002. Further aspects of the use of individual animals as replicates in statistical analysis. *App Animal Behav Sci* 75: 265–268.

Pinheiro JC, Bates DM. 2000. *Mixed Effects Models in S and S-Plus*. Springer-Verlag, New York.

Ramsey FL, Schafer DW. 2002. *The Statistical Sleuth. A Course in Methods of Data Analysis*. Second edition. Duxbury, Thomson Learning. Pacific Grove, CA.

Riley J, Edwards P. 1998. Statistical aspects of aquaculture research: Pond variability and pseudoreplication. *Aquacult Res* 29: 281–288.

Rubin DB. 2008. For objective causal inference, design trumps analysis. *Ann App Stat* 2: 808–840.

Shah PA, Aebi M, Tuor U. 2000. Drying and storage procedures for formulated and unformulated mycelia of the aphid-pathogenic fungus *Erynia neoaphidis*. *Mycol Res* 104: 440–446.

Snedecor GW and Cochran WG. 1967. *Statistical Methods*. Iowa University Press. Ames, IA.

Steel, GD, Torrie JH. 1980. *Principles and Procedures of Statistics. A Biometrical Approach*. Second edition. McGraw-Hill Inc. New York.

Stenzel NMCS, Neves CSV, Marur CJ, Scholz MBS, Gomes JC. 2006. Maturation curves and degree-days accumulation for fruits of "Folha Murcha" orange trees. *Sci Agric (Brazil)* 63: 219–225.

Welham SJ, Gezan SA, Clark SJ, Mead A. 2014. *Statistical Methods in Biology: Design and Analysis of Experiments and Regression*. Taylor & Francis. Boca Raton, FL.

Woolf AB, Bowen JH, Ferguson IB. 1999. Preharvest exposure to the sun influences postharvest responses of 'Hass' avocado fruit. *Postharv Biol Technol* 15: 143–153.

Zuur AF, Ieno EN, Walker NJ, Saveliev AA, Smith GM. 2009. *Mixed Effects Models and Extensions in Ecology*. R. Springer. New York.

2 Recent Research on Calcium and Postharvest Behavior

Babak Madani and Charles F. Forney

CONTENTS

2.1 INTRODUCTION

Calcium is essential for the growth and development of plants. Calcium moves in soil mostly by mass-flow, and its uptake by plants is passive and restricted to the tips of young roots with nonsuberized endodermal cell walls. Calcium is translocated in the xylem mostly through the transpiration stream. Calcium moves upward in the xylem by adsorption onto exchange sites and by chelation with organic acids in the xylem sap. High concentration of calcium in the xylem sap facilitates its movement to the shoot apex (Biddulph et al., 1961; Bell and Biddulph, 1963). Preharvest application of calcium can be applied to the soil or foliage. Application of calcium in the soil will not always increase the concentration of calcium in fruits or vegetables, because of the immobility of calcium inside the plant and competition for calcium among different plant parts. Also, calcium can take up to four years to move from the roots to the leaves or fruits (Jones et al., 1983; Yuen, 1994). Calcium can also be applied as a foliar spray or by postharvest application to the produce. When calcium is applied to fruit, it enters through trichomes, stomata, and lenticels (Scott and Wills, 1975; Glenn et al., 1985). Calcium uptake into leaves also depends on stomata channels. Transportation of calcium to nonvascular flesh tissue is by diffusion through the apoplast. Stomata density also can influence calcium uptake by fruits or leaves (Harker et al., 1989). In the European Union, there are many commercial foliar fertilizers containing various forms of calcium, but their efficiency of uptake as a preharvest spray is seldom more efficient than calcium chloride or nitrate (Wójcik and Szwonek, 2002).

Postharvest application of calcium to fruits and vegetables is generally accomplished by dipping in solutions that primarily utilize calcium chloride, but lactate, propionate, gluconate, and nitrate salts have also been reported to be effective. Vacuum infiltration enhances the penetration of calcium solutions into cells through the pressure generated after release of the partial vacuum and has been shown to be very effective in modifying postharvest changes in a range of produce (Scott and Wills, 1975, 1977a,b, 1979; Wills and Scott, 1980; Tirmazi and Wills, 1981; Wills and Tirmazi, 1982; Wills et al., 1982, 1988a,b; Wills and Sirivatanapa, 1988; Martin-Diana et al., 2007).

2.2 POSTHARVEST DISEASES

Economic losses caused by postharvest pathogens have resulted in the development of a range of cultural, physical, and chemical treatments to control postharvest diseases. The application of synthetic fungicides has been a major treatment for controlling many diseases, but use of any compound for an extended time invariably leads to an increase in fungicide-resistant fungal strains. In addition, potential fungicide residues on fruit can be an issue for consumers. These concerns have encouraged the development of safer approaches to disease management.

The application of calcium is an alternative to the use of fungicides. Calcium has been shown to reduce disease in crops by inhibiting spore germination and stimulating host resistance. Calcium also acts by stabilizing cell walls to make them more resistant to harmful enzymes released by fungi (Sams and Conway, 1984; Wisniewski et al., 1995; Biggs, 1999). In addition, the calcium ion inhibits the action of ethylene on cell membranes and thereby inhibits the onset of senescence (Torre et al., 1999). High concentrations of cytosolic calcium have been shown to enhance the synthesis of phytoalexins and phenolic compounds that decrease the activity of pathogenic pectolytic enzymes (Miceli et al., 1999).

In this section, we focus on the effect of calcium on anthracnose (*Colletotrichum* spp.), brown rot (*Monilinia fructicola* (G. Wint.) Honey), gray mold (*Botrytis cinerea*), green mold (*Penicillium digitatum*), and sour rot (*Geotrichum citriauranti*).

Anthracnose is a postharvest disease that affects a range of fruits, including papaya, banana, dragonfruit, and strawberry (Chau and Alvarez, 1983; Kim et al., 1992; Ghani et al., 2011). In papaya, the effect of pre- and postharvest calcium has been studied by Madani et al. (2014a) who reported that 1.5% and 2% calcium chloride significantly reduced anthracnose incidence and severity during five weeks of storage while Mahmud et al. (2008) found that the lowest incidence of anthracnose was obtained in fruit infiltrated with 2.5% calcium chloride. Awang et al. (2011) further reported that postharvest dipping of dragonfruit in 1–4 g/L calcium chloride did not have significant effect on the incidence of anthracnose but the size of lesions was reduced with increasing calcium concentrations. On the other hand, Ghani et al. (2011) indicated that preharvest 1%–4% calcium chloride dips reduced the severity of anthracnose disease in dragonfruit. Chillet et al. (2000) showed that the susceptibility of the banana fruit to anthracnose was lower in fruit with higher calcium levels.

However, Nam et al. (2006) did not find any effect of 1–6 mM calcium in the nutrient solution used for growing strawberry plants on the severity of anthracnose.

Brown rot affects most commercially grown *Prunus* spp. and can result in extensive crop losses (Adaskaveg et al., 2008). Early studies showed that the application of calcium chloride did not reduce the incidence of postharvest brown rot in peaches (Conway, 1987). However the use of calcium chloride as a foliar spray applied up to six times was shown to increase fruit Ca concentration and reduce the occurrence of brown rot (Manganaris et al., 2005; Elmer et al., 2007). In New Zealand, foliar sprays of calcium have been widely adopted by stonefruit growers as a practical tool to reduce brown rot. Biggs et al. (1997) tested the effects of several calcium salts on the *in vitro* growth of *Monilinia fructicola* and found that calcium propionate strongly inhibited fungal growth.

Gray mold rots can cause severe postharvest losses in grapes and strawberry fruit (Bulger et al., 1987; Wilcox and Seem, 1994; de Kock and Holz, 1994; Hernandez-Munoz et al., 2006). Spraying calcium chloride 2–3 times before veraison, increased the Ca concentration of grape berries and reduced rot caused by gray mold (Amiri et al., 2009; Ciccarese et al., 2013). Singh et al. (2007) also showed that spraying calcium on strawberry could decrease gray mold. The lower incidence of rot might be due to calcium slowing the senescence processes in many fruit, including strawberry (Ferguson, 1984; Poovaiah, 1986; Sharma et al., 2006). Wójcik and Lewandowski (2003) and Naradisorn et al. (2006) have also reported that the fruit receiving calcium is much firmer and less affected by gray mold. Similarly, Lara et al. (2004) reported a lower incidence of gray mold in strawberry after postharvest calcium dipping.

Green mold (*Penicillium digitatum*) and sour rot (*Geotrichum citriauranti*) are the most economically important postharvest diseases of citrus in the arid fruit-growing regions of the world (Powell, 1908; Eckert and Brown, 1986). Smilanick and Sorenson (2001) indicated that incidence of green mold was greatly reduced by dipping lemons or oranges in a 40°C liquid lime–sulfur solution that contained 0.75% (w/v) calcium polysulfide. The incidence of sour rot was also reduced by this treatment. Efficiency was higher on lemons than oranges, and on green- compared to yellow-lemons. Youssef et al. (2012) reported that the efficiency of preharvest sprays of calcium chloride and calcium chelate against postharvest green and blue mold on "Comune" clementine and "Valencia late" orange fruit were greater than using a postharvest dip. They concluded that field application of calcium should be included in an integrated approach for controlling postharvest diseases of citrus fruit.

2.3 PHYSIOLOGICAL DISORDERS

Physiological disorders involve the breakdown of plant tissue that is not directly caused by pests and diseases or by mechanical damage. They may develop in response to various pre- and postharvest conditions, including nutrient accumulation during organ development or low temperature stresses during storage (Wills et al., 2007). Several disorders have been associated with calcium deficiency, which include blossom end rot (BER) in tomato, glassiness in melon, bitter pit in apple, cracking in cherry, and chilling injury of various commodities in storage (Wills et al., 2007).

Bitter pit (BP) is a major problem for an apple industry, especially where growing conditions are dry. Application of calcium either pre- or postharvest has been known for many years to reduce the susceptibility of apples to develop BP (Shear, 1975; Wills et al., 1976; Scott and Wills, 1977b, 1979; Ferguson and Watkins, 1989; Conway et al., 1991, 1994; Fallahi et al., 1997; Ferguson et al., 1999; Blanco et al., 2010). In some recent studies, Benavides et al. (2001) showed that the greatest increase in calcium concentration in apples was achieved when calcium was sprayed six times at 15-day intervals, from 60 days after full bloom, while Lötze and Theron (2007) showed that application of foliar calcium during midseason (40 days after full bloom) was shown to be more effective in increasing fruit calcium and reducing BP than a later application. Neilsen et al. (2005) showed that sprays of calcium chloride in the early growing season were as effective as in the late season for reducing BP, in spite of low calcium concentration in the harvested fruit. The reason for this effect is unclear. Blanco et al. (2010) showed that spraying the apple trees with calcium chloride or calcium propionate in combination with carboxymethyl ether of cellulose can favor the distribution of calcium into the apple fruit and help to reduce the incidence of calcium-related disorders during cold storage.

Internal browning (IB) is a low-temperature disorder affecting a number of fruits with the browning of the flesh tissue occurring after removal from cool storage to room temperature (Teisson et al., 1979; Smith, 1983; Paull and Rohrbach, 1985). Youryon et al. (2013) mentioned that calcium infiltration of pineapple via the peduncle for three days increased calcium concentrations in the core and adjacent flesh tissue and reduced IB in "Trad-Srithong" pineapples stored at 13°C. Wójcik (2012) showed that six sprays of calcium chloride on pear fruit—from six weeks after full bloom to two weeks before harvest—resulted in firmer and greener fruits that were less sensitive to IB after storage. Manganaris et al. (2007) found that dipping peach in calcium chloride for 5 min decreased flesh browning after storage at 5°C.

Cracking in sweet cherries and litchi, caused by rainfall shortly before harvest, limits the production of sweet cherry in many parts of the world (Sekse, 1998; Huang, 2005). Many studies have examined the potential to reduce cracking in cherry fruit by spray-application of calcium compounds such as calcium chloride, hydroxide, and nitrate (Callan, 1986; Meheriuk et al., 1991; Yamamoto et al., 1992; Rupert et al., 1997; Marshall and Weaver, 1999). Erogul (2014) sprayed sweet cherry trees with calcium nitrate, chloride, caseinate and hydroxide 30, 20, and 10 days before harvest, and showed the most effective applications to decrease cracking were calcium hydroxide and calcium chloride. However, Huang et al. (2008) found that spraying with calcium chloride three times did not decrease cracking in litchi.

It is widely accepted that BER of tomato is caused by a combination or sequence of environmental and physiological factors related to calcium uptake and transport within the plant, which affect the fruit's growth rate and/or size (Ho and White, 2005). The majority of studies identify a local calcium deficiency in distal fruit tissue during the period of rapid cell expansion (7–21 days after anthesis), when calcium demand exceeds supply, as the primary cause of BER (Bradfield and Guttridge, 1984; Adams and Ho, 1993; Ho and White, 2005). Additional application of calcium is commonly considered as a preventive measure to overcome local calcium deficiency in tomato (Wada et al., 1996; Ho, 1999; Schmitz-Eiberger et al., 2002).

Liebisch et al. (2009) showed that spraying tomato plants with calcium chloride and boric acid decreased BER, but increased cracking in the fruit. They concluded that since sprays are costly, labor-intensive, and did not reduce the proportion of nonmarketable fruit, the selection of cultivars, not susceptible to BER and cracking is more effective than sprays, when conditions favor these disorders. This is particularly true for protected cultivation in central Thailand. However, Saure (2014) suggested that the actual causes of BER are the effects of abiotic stress (such as salinity, drought, high-light intensity, heat, and ammonia nutrition) resulting in an increase of reactive oxygen species, high oxidative stress, and, finally, cell death. Cell death results in a disintegration of the plasma membrane and tonoplast and a breakdown of the endoplasmic reticulum, thus not following but preceding ion leakage, including Ca^{2+} leakage, and loss of turgor. With this approach, a better understanding and a more efficient control of BER in tomato and pepper fruit is envisaged.

Glassiness is a disorder in which the melon flesh appears water-soaked; this has been related to low calcium concentration in fruit tissue (Lester and Grusak, 1999, 2001; Madrid et al., 2004). Serrano et al. (2002) confirmed that this disorder is related to calcium deficiency by decreasing calcium in the nutrient solution, which caused a more rapid onset of softening of the fruit and increasing glassiness.

2.4   QUALITY AND RIPENING OF CLIMACTERIC FRUIT

In climacteric fruit, ripening is accompanied by a rise in respiration rate and ethylene production (Wills et al., 2007). It has been proposed that calcium delays ethylene production by preventing solubilization of calcium binding sites in cell walls, which activates the ethylene generation system located in the cell-wall plasma membrane complex (Mattoo and Lieberman, 1997). Wang et al. (2006) found that the expression of LeACO1 gene (related to changing ACC to ethylene) was inhibited with calcium application in never-ripe mutant and wild-type tomatoes at the green stage. Calcium also decreases the respiration rate in climacteric fruit by regulation of membrane fluxes of initiator molecules like phosphate or respiratory substrates like malate (Ferguson, 1984). Phosphate flux changes across the tonoplast and plasmalemma have been associated with an onset of the climacteric respiratory rise. It is suggested that the increased tonoplast and plasmalemma fluxes of phosphate and increased cytoplasmic phosphate may be critical for climacteric respiration (Woodow and Rowan, 1979). The tonoplast and plasmalemma could be the possible site for calcium to regulate phosphate. Other substrates such as malate, which, through decarboxylation, may be a source for CO_2 in the climacteric phase, may be regulated by calcium in terms of compartmentation between vacuole and cytoplasm and, thus, decrease the respiration rate (Ferguson, 1984).

This section will discuss recent findings on the role of calcium in the postharvest quality of selected pome fruits (apple and pear), stone fruits (plum and peach), tropical fruit (papaya), small fruit (blueberry), and tomato and melon.

In apples, the effects of pre- and postharvest calcium in postharvest quality were evaluated in many early studies (Wills, 1972; Mason, 1976; Scott and Wills, 1977a; Tirmazi and Wills, 1981; Sams and Conway, 1984). More recently, Kadir (2005) showed that spraying apple trees six times with calcium chloride improved

"Jonathan" apple quality in terms of fruit weight, size, appearance, and color, but had no effect on soluble solids content. Similar findings were noted by Wójcik and Borowik (2013), who found that apples sprayed with a range of calcium compounds had higher titratable acidity and were firmer, but did not differ in soluble solids. Val et al. (2008) found that 1% calcium chloride sprays had no effect on the concentration of calcium in the flesh of "Fuji" apples and no effect on fruit quality traits; Wu et al. (2012) showed that postharvest calcium dipping of "Fuji" with 0.5% calcium chloride was effective in maintaining the firmness of cut-apple products and intact fruit. Similarly, Zheng et al. (2014) reported a 2% calcium chloride dip on intact "Fuji" maintained firmness and improved surface color and titratable acidity of cut apple fruit. Hussain et al. (2012) showed that "Red Delicious" apple fruit dipped with 2% calcium chloride and then irradiated with 0.4 kGy gamma radiation had higher ascorbic acid levels and extended shelf-life. Ortiz et al. (2010) showed that most of the compounds that contribute to flavor in ripe apple were improved in response to a postharvest calcium chloride dip. For example, they showed that the release of acetate esters was favored, and acetaldehyde content was enhanced, in apple fruit treated with calcium (Ortiz et al., 2010). They further suggested that this technique may be an appropriate method to improve the fruit's aroma where these effects were from an increase in pyruvate decarboxylase and alcohol dehydrogenase activities in the presence of calcium. Ortiz et al. (2011) indicated that preharvest calcium sprays enhanced most of the compounds contributing to overall flavor in ripe fruit, suggesting that this procedure may be suitable for improving the fruit's aroma at harvest.

The positive effects of calcium in enhancing postharvest quality of pear have confirmed earlier studies by Wills et al. (1982), Raese and Drake (1993, 1995) and Raese et al. (1999). Wójcik et al. (2014) examined the impacts of autumn sprays of calcium as a supplement to summer-time calcium sprays on "Conference" pear quality and storability. They found that the greatest increase in calcium status was in fruit treated with calcium in the summer and in the fall. After storage, pears sprayed with calcium in the summer and in the fall produced less ethylene, had lower respiration, contained more organic acids and were firmer (Wójcik et al., 2014). They concluded that calcium chloride at 20 or 25 kg/ha should be used in "Conference" pear orchards as a supplement to summer-time calcium sprays to improve fruit storability. Mahajan and Dhatt (2004) studied the effects of different dip concentrations of calcium chloride and showed that application of 4% calcium chloride was the most effective treatment in reducing the weight-loss of pear and in maintaining the firmness and quality of fruits up to 75 days in storage at 0–1°C.

In plum, the beneficial effects of calcium treatment—in terms of prolonging storability, delaying the ripening process and maintaining fruit quality—has been reported by Valero et al. (2002). Serrano et al. (2004) studied the role of postharvest treatments with calcium or heat on reducing mechanical damage during storage. They showed that 1 mM calcium chloride dips or hot-water dips at 45°C for 10 min led to a reduction of mechanical damage, and in turn alleviated the physiological responses that occurred in mechanically damaged plums. Alcaraz-Lopez et al. (2003) studied the combined effects of foliar applications of Ca, Mg, and Ti on the calcium nutrition and fruit quality of plum and found that titanium application increased the calcium concentrations in the fruit's peel and flesh.

Recent studies on calcium in peach have confirmed earlier reports by Abdalla and Childers (1973) and Wills and Mahendra (1990). Research on peach in the eastern USA showed that calcium sprays were effective, while, in western USA, these treatments did not increase the fruit's calcium content. Thus, growing conditions and cultivar determine calcium absorption into fruit (Johnson et al., 1998; Crisosto et al., 2000). Manganaris et al. (2005) found that preharvest calcium sprays increased calcium in peach, but did not change acidity, soluble solids, and ethylene production after harvest. Manganaris et al. (2007) found that peach dipped in calcium lactate, propionate, and chloride solutions, maintained firmness during storage. Gupta et al. (2011) showed that peach dipped in calcium chloride had reduced weight-loss and maintained firmness, acidity, and vitamin A content during storage.

Mahmud et al. (2008) studied the effect of dipping or infiltration with calcium chloride on the postharvest quality of papaya and found that vacuum infiltration was more effective than dipping at ambient pressure for maintaining quality of papaya "cv. Eksotika II." However, Qiu et al. (1995) demonstrated that spraying or dipping papaya ("cv. Kapoho Solo") in calcium chloride did not increase calcium concentration in the fruit's mesocarp. In contrast, Madani et al. (2014b) concluded that the foliar sprays of papaya ("cv. Eksotika II") with calcium chloride resulted in an increased calcium concentration in its peel and pulp tissues, firmness, and titratable acidity and overall fruit quality, but reduced respiration rate, ethylene production, and soluble solids concentrations.

A factor limiting the postharvest life of blueberries is excessive softening (Lambert, 1990) and this has been shown to be positively affected by dipping in calcium chloride (Hanson et al., 1993). Stuckrath et al. (2008) studied the effect of foliar application of calcium under three different growing conditions (tunnel, mesh, and ambient) and found that the fruit's firmness was higher at the beginning of the cell-expansion period in the tunnel-grown fruit and a linear correlation between calcium concentration and firmness was established. Angeletti et al. (2010) evaluated the effect of calcium fertilization of "O'Neal" and "Bluecrop" blueberry on quality during storage and showed that calcium-treated fruit for both varieties had less softening and weight-loss compared with fruit without calcium fertilization. Angeletti et al. (2010) also showed that although the respiration rate increased during storage, this was lower in calcium-treated blueberries. They also showed that the calcium treatments did not alter hemicellulose content, but, in some cases, reduced solubilization of pectic polymers (Angeletti et al., 2010).

Cut melon is prone to softening during storage, even under modified atmosphere packaging; calcium has been shown to decrease softening (Madrid et al., 2004). Moreover, Saftner et al. (2003) showed that honeydew fruit dipped in calcium propionate, calcium amino acid chelate formulation, or calcium chloride, decreased respiration and ethylene production during storage. However, Aguayo et al. (2008) demonstrated that cut melon dipped in calcium chloride at 60°C maintained firmness during eight days of storage and Luna-Guzmán and Barret (2000) found that cut cantaloupe dipped in calcium lactate had firmer tissue without bitterness. Johnstone et al. (2008) found fertigation with calcium chloride had no effect on quality or calcium concentration of Californian honeydew or muskmelon, probably due to the physiological limitation of calcium movement into the fruit tissue. However,

Lester and Grusak (2004) reported that foliar application of calcium metalosate—or "Folical"—increased the honeydew fruit's calcium and storage life, but did not affect netted cantaloupe.

Early research on calcium and postharvest quality of tomato was reported by Wills et al. (1977), Wills and Rigney (1979), Wills and Tirmazi (1979), Rigney and Wills (1981), Minamide and Ho (1993), and Paiva et al. (1998). A recent study by Senevirathna and Daundasekera (2010) on mature-turning tomato fruit dipped in calcium chloride at ambient and partial pressure, found vacuum infiltration to be the most effective treatment with respect to shelf-life extension and reduced ethylene production. Coolong et al. (2014) further found that foliar application of calcium chloride increased fruit soluble solids content and dry weight, but did not affect texture while weight loss during storage increased. Dong et al. (2004) found that calcium chloride sprayed during anthesis and on one and three-week old fruit improved vitamin C content and reduced titratable acidity of the fruit.

2.5 QUALITY OF NONCLIMACTERIC FRUIT

The effects of calcium on the postharvest quality of the nonclimateric fruits such as cherry, pomegranate, grape, and strawberry with emphasis on recent findings will be discussed in this section.

Studies on the effects of preharvest calcium on quality parameters in cherries during storage are relatively limited (Lidster et al., 1979). Tsantili et al. (2007) investigated effects of preharvest sprays of calcium chloride on the quality of "Vogue" cherries during storage and found calcium-treated fruit had greater firmness, lower soluble pectin content, and greater resistance to stem removal, but did not affect respiration and showed only slight effects on reducing ethylene production. Wang et al. (2014) futher showed that a postharvest calcium chloride dip treatment reduced the fruit's respiration and ethylene production rates.

Ramezanian et al. (2009) examined the effect of calcium on pomegranate quality and showed that preharvest calcium chloride sprays at full bloom and one month later increased fruit weight, soluble solids content, and ascorbic acid in the aril. Kazemi et al. (2013) showed that postharvest dipping in calcium chloride and found better retention of firmness, vitamin C, and titratable acidity. Ramezanian et al. (2010) further showed that pomegranate dipped in calcium chloride or infiltrated with calcium chloride and spermidine exhibited reduced weight loss and increased ascorbic acid.

As a consequence of early research on calcium in grapes (Schaller et al., 1992), calcium chloride is now commonly applied to grapes in Italy and is reported to decrease decay and delay senescence (Romanazzi et al., 2012). Ciccarese et al. (2013) studied the effect of spraying calcium EDTA before and after veraison and showed that the fruit maintained berry firmness. The applications were particularly efficacious if carried out between fruit set and veraison when stomata are functional and calcium was better adsorbed. Marzouk and Kassem (2011) showed that vines sprayed with calcium chloride during fruit development until harvest had higher firmness and improved quality after harvest. Amiri et al. (2009) examined calcium chloride sprays from fruit set until veraison and showed that grape quality parameters such as juice pH, soluble solids, and titratable acidity were not affected, whereas berry firmness, berry color, and appearance improved at harvest. However, contradictory results were shown by Bonomelli and Rafael (2010) who found foliar and soil application of calcium chloride before veraison had no influence on calcium and sugar concentration in fruit.

Strawberry shelf-life is limited mostly due to susceptibility to fungus disease. Singh et al. (2007) studied effects of preharvest foliar application of calcium and boron on fruit yield and quality of "Chandler" strawberry. They found no effect on individual berry weight, but marketable fruit yield increased and was highest in plants sprayed with calcium plus boron. Calcium-treated fruits were firmer, had lower soluble solids, and higher acidity and ascorbic acid at harvest (Singh et al., 2007). Figueroa et al. (2012) further evaluated the effect of calcium chloride dipping and/or naphthalene acetic acid (NAA) at 45°C and found lower soluble solids in calcium and NAA treated fruits during cold storage, suggesting that these compounds could alter the normal breakdown of cell-wall polysaccharides during postharvest. After storage, calcium chloride in combination with NAA produced a reduction in the transcriptional level of polygalacturonase, pectate lyase, and EG1 genes. Figueroa et al. (2012) concluded that calcium and auxin could individually alter fruit ripening, preventing the normal degradation of cell walls during cold storage, while the combination of calcium and NAA reduced the transcript level of cell-wall degrading genes after cold storage, albeit no differences in firmness were recorded. Hernandez-Munoz et al. (2006) studied the effect of postharvest calcium gluconate and chitosan coatings on quality of strawberry fruit and showed that the calcium dips were effective in maintaining firmness, while chitosan coatings markedly slowed ripening as shown by their retention of firmness and delayed changes in their external color. Whilst addition of calcium gluconate to the chitosan coating formulation did not further extend the shelf-life of the fruit, the amount of calcium retained by strawberries was greater than that obtained with calcium dips alone (Hernandez-Munoz et al., 2006).

2.6 FRESH CUTS

In recent years, the demand for fresh-cut products has increased as a consequence of changes in consumer attitudes (Martin-Diana et al., 2007). However consumers have the expectation that processing and storage of fresh-cut products will not alter the sensory properties. While freshness is the main consumer concern on product quality, subsequent purchases depend upon the satisfaction with texture (Kays, 1999; Beaulieu and Baldwin, 2002; Kader, 2002; Ngamchuachit et al., 2014). Toivonen and Brummell (2008) reported that flesh firmness is a major textural property of mouth-feel and calcium treatments have been shown to be effective at maintaining firmness during storage in many fresh-cut fruits. Endogenous and added calcium can make plant tissue firmer by binding to the pectin carboxyl groups that are generated through the action of pectin methylesterase (PME) (Conway and Sams, 1983; Stanley et al., 1995).

Calcium ions can passively diffuse within the cell-wall structure because plant cell wall porosity is approximately 3.5–9.2 nm (Read and Bacic, 1996), while calcium ions are about 0.1 nm (Gillard, 1969). When fruit parenchyma cells are dipped in a calcium salt solution, calcium ions are primarily transported through the apoplast, or intercellular spaces, where they are attracted by negatively charged carboxyl groups in the homogalacturonan that constitutes pectin in the middle lamella and cell wall. The negatively charged chloride or lactate ions remain unbound in solution (Harker and Ferguson, 1988; Hasegawa, 2006). There are several ways that calcium could be applied to fresh-cut fruits such as in conjunction with packaging in low oxygen, osmotic dehydration, adding to hot water or with irradiation. The combination of a calcium chloride treatment and packaging with a low O_2 concentration was shown to be more effective than the use of calcium chloride alone to maintain firmness of fresh-cut "Piel de Sapo" melons (Oms-Oliu et al., 2010) and "Golden Delicious" apples (Soliva-Fortuny and Martin-Belloso, 2003). Quiles et al. (2004) observed that calcium chloride maintained the structure of "Granny Smith" apples during the process of osmotic dehydration. (Rico et al., 2007) further showed that warm treatment temperatures (40–60°C) increase the beneficial effects of calcium treatment due to higher washing solution retention inside the product. Melon pieces dipped at 60°C in calcium chloride, lactate, or ascorbate showed good firmness retention (Silveira et al., 2011) as did kiwi fruit dipped in calcium chloride at 45°C (Beirao-da-Costa et al., 2008). The primary benefit of irradiation was found to be reducing microbial growth and delaying ripening of fresh-cut fruits and vegetables although irradiation can cause undesirable textural changes. Prakash et al. (2002) showed that irradiation substantially decreased the firmness of fresh-cut tomatoes. Magee et al. (2003) reported that dipping tomato discs in either calcium chloride or calcium lactate solution and exposed to irradiation at 1.25 kGy increased the levels of calcium and firmness. While Fan et al. (2005) found apple slices treated with calcium ascorbate followed by irradiation at 0.5 and 1.0 kGy decreased softening and increased firmness during storage. In fresh-cut pear cubes, Alandes et al. (2009) showed that dipping in calcium lactate strengthened the structure of the fruit at microscopic level by maintaining the fibrillar packing in the cell walls and the cell-to-cell contacts. Ngamchuachit et al. (2014) also showed that mango cubes that had been dipped in calcium chloride and lactate were firmer during storage.

2.7   CONCLUSIONS

The application of calcium is a simple and safe technology which has been known for many years to improve the postharvest performance of a range of fruit and vegetables. However, due to increasing consumer trends to minimize the use of synthetic compounds in foods, calcium treatments are receiving more attention. An important area of potential application is in reducing microbial growth. However, further studies are needed to investigate the effects of calcium in reducing bacteria or other fungi. When calcium is used at an appropriate dosage and timing, it has been shown to improve product quality and enhance the nutritional value of a range of intact and fresh-cut produce. Further studies are needed to optimize the uptake of calcium by managing the form of added calcium and application method. In particular, more research is required on the effects of calcium in vegetables.

REFERENCES

Abdalla DA, Childers NF. 1973. Calcium nutrition of peach and prune relative to growth, fruiting, and fruit quality. *J Am Soc Hort Sci* 98: 517–522.

Adams P, Ho LC. 1993. Effects of environment on the uptake and distribution of calcium in tomato and on the incidence of blossom-end rot. *Plant Soil* 154: 127–132.

Adaskaveg JE, Schnabel G, Foerster H. 2008. Diseases of peach caused by fungi and fungal-like organisms: Biology, epidemiology and management. *The Peach: Botany, Production and Uses*. (eds D Layne, D Bassi), CAB International, Wallingford, UK, 352–406.

Aguayo E, Escalona VH, Artés F. 2008. Effect of hot water treatment and various calcium salts on quality of fresh-cut "Amarillo" melon. *Postharv Biol Technol* 3: 397–406.

Alandes L, Pérez-Munuera I, Llorca E, Quiles A, Hernando I. 2009. Use of calcium lactate to improve structure of Flor de Invierno fresh-cut pears. *Postharv Biol Technol* 53: 145–151.

Alcaraz-Lopez CA, Botia M, Carlos F, Alcaraz CF, Riquelme F. 2003. Effects of foliar sprays containing calcium, magnesium and titanium on plum (*Prunus domestica* L.) fruit quality. *J Plant Physiol* 160: 1441–1446.

Amiri EM, Fallahi E, Safari G. 2009. Effects of preharvest calcium sprays on yield, quality and mineral nutrient concentrations of "Asgari' table grape". *Int J Fruit Sci* 3: 294–304.

Angeletti P, Castagnasso H, Miceli E, Terminiello L, Concellón A, Chaves A, Vicente AR. 2010. Effect of preharvest calcium applications on postharvest quality, softening and cell-wall degradation of two blueberry (*Vaccinium corymbosum*) varieties. *Postharv Biol Technol* 58: 98–103.

Awang Y, Ghani MAA, Sijam K, Rosli B, Mohamad RB. 2011. Effect of calcium chloride on anthracnose disease and postharvest quality of red-flesh dragon fruit (*Hylocereus polyrhizus*). *Afr J Microbiol Res* 29: 5250–5259.

Beaulieu JC, Baldwin EA. 2002. Flavor and aroma of fresh-cut fruits and vegetables. *Fresh-Cut Fruits and Vegetables: Science, Technology, and Market* (ed. O Lamikanra), Boca Raton, FL, CRC Press, 387–421.

Beirao-da-Costa S, Cardoso A, Martins LL, Empis J, Moldao-Martins M. 2008. The effect of calcium dips combined with mild heating of whole kiwifruit for fruit slices quality maintenance. *Food Chem* 108: 191– 197.

Bell CW, Biddulph O. 1963. Translocation of calcium. Exchange versus mass flow. *Plant Physiol* 38: 610–614.

Benavides A, Recasens I, Casero T, Puy J. 2001. Chemometric analyses of "Golden Smoothee" apples treated with two preharvest calcium-spray strategies in the growing season. *J Sci Food Agric* 81: 943–952.

Biddulph O, Nakayama FS, Cory R. 1961. Transpiration stream and ascension of calcium. *Plant Physiol* 36: 429–436.

Biggs AR. 1999. Effects of calcium salts on apple bitter rot caused by two *Colletotrichum* spp. *Plant Dis* 83: 1001–1005.

Biggs AR, El-Kholi MM, El-Neshawy S, Nickerson R. 1997. Effects of calcium salts on growth, polygalacturonase activity, and infection of peach fruit by *Monilinia fructicola*. *Plan Dis* 81: 399–403.

Blanco A, Fernández V, Val J. 2010. Improving the performance of calcium-containing spray formulations to limit the incidence of bitter pit in apple (*Malus x domestica* Borkh.). *Scientia Hortic* 127: 23–28.

Bonomelli C, Rafael R. 2010. Effects of foliar and soil calcium application on yield and quality of table grape "cv. Thompson" seedless. *J Plant Nutr* 3: 299–314.

Bradfield EG, Guttridge CG. 1984. Effects of night-time humidity and nutrient solution concentration on the calcium content of tomato fruit. *Scientia Hortic* 22: 207–217.

Bulger MA, Ellis MA, Madden LV. 1987. Influence of temperature and wetness duration on infection of strawberry flowers by *Botrytis cineria* and disease incidence of fruit originating from infected flowers. *Phytopathology* 77: 1225–1230.

Callan NC. 1986. Calcium hydroxide reduces splitting of "Lambert" sweet cherry. *J Am Soc Hortic Sci* 111: 173–175.

Chau KF, Alvarez AM. 1983. A histological study of anthracnose on *Carica papaya*. *Phytopathology* 73: 1113–1116.

Chillet M, Delaperyre de Bellaire L, Dorel M, Joas J, Dubois C, Marchal J, Perrier X. 2000. Evidence for the variation in susceptibility of bananas to wound anthracnose due to *Colletotrichum musae* and the influence of edaphic conditions. *Scientia Hortic* 86: 33–47.

Ciccarese A, Stellacci M, Gentilesco G, Rubino P. 2013. Effectiveness of pre- and post-veraison calcium applications to control decay and maintain table grapefruit quality during storage. *Postharv Biol Technol* 75: 135–141.

Conway W. 1987. Effects of preharvest and postharvest calcium treatments of peaches on decay caused by *Monilinia fructicola*. *HortSci* 32: 820–823.

Conway WS, Sams CE, Abbott JA, Bruton BD. 1991. Postharvest calcium treatment of apple fruit provides broad spectrum protection against postharvest pathogens. *Plant Dis* 75: 620–622.

Conway WS, Sams CE. 1983. Calcium infiltration of "Golden Delicious" apples and its effect on decay. *Phytopathology* 73: 1068–1071.

Conway WS, Sams CE, Kelman A. 1994. Enhancing natural mechanisms of resistance to postharvest diseases through calcium applications. *HortSci* 29: 751–754.

Coolong T, Mishra S, Barickman C, Sams C. 2014. Impact of supplemental calcium chloride on yield, quality, nutrient status, and postharvest attributes of tomato. *J Plant Nutr* 14: 2316–2330.

Crisosto CH, Day KR, Johnson RS, Garner D. 2000. Influence of in-season foliar calcium sprays on fruit quality and surface discoloration incidence in peaches and nectarines. *J Am Pomol Soc* 54: 118–122.

de Kock PJ, Holz G. 1994. Application of fungicides against postharvest Botrytis cinerea bunch rot of table grapes in the Western Cape. *S Afr J Enol Vitic* 15: 33–40.

Dong CX, Zhou JM, Fan XH, Wang HY, Duan ZQ, Tang C. 2004. Application methods of calcium supplements affect nutrient levels and calcium forms in mature tomato fruits. *J Plant Nutr* 27: 1443–1455.

Eckert JW, Brown GE. 1986. Evaluation of postharvest treatments for citrus fruits. *Methods for Evaluating Pesticides for Control of Plant Pathogens*. (eds K D Hickey), American Phytopathological Society, St Paul, MN, 93–97.

Elmer PAG, Spiers TM, Wood PN. 2007. Effects of preharvest foliar calcium sprays on fruit calcium levels and brown rot of peaches. *Crop Prot* 26: 11–18.

Erogul D. 2014. Effect of preharvest calcium treatments on sweet cherry fruit quality. *Not Bot Horti Agrobo* 1: 150–153.

Fallahi E, Conway WS, Hickey KD, Sams CE. 1997. The role of calcium and nitrogen in postharvest quality and disease resistance of apples. *HortSci* 32: 831–835.

Fan X, Niemera BA, Mattheis JP, Zhuang H, Olson DW. 2005. Quality of fresh-cut apple slices as affected by low-dose ionizing radiation and calcium ascorbate treatment. *J Food Sci* 2: 143–148.

Ferguson IB. 1984. Calcium in plant senescence and fruit ripening. *Plant Cell Environ* 7: 477–489.

Ferguson I, Volz R, Woolf A. 1999. Preharvest factors affecting physiological disorders of fruit. *Postharv Biol Technol* 15: 255–262.

Ferguson IB, Watkins CB. 1989. Bitter pit in apple fruit. *Hortic Rev* 11: 289–355.

Figueroa CR, Opazo MC, Vera P, Arriagada O, Díaz M, Moya-León MA. 2012. Effect of postharvest treatment of calcium and auxin on cell-wall composition and expression of cell-wall-modifying genes in the Chilean strawberry (*Fragaria chiloensis*) fruit. *Food Chem* 132: 2014–2022.

Ghani MAA, Awang Y, Sijam K. 2011. Disease occurrence and fruit quality of preharvest calcium treated red-flesh dragon fruit (*Hylocereus polyrhizus*). *Afr J Biotechnol* 10: 1550–1558.

Gillard RD. 1969. The simple chemistry of calcium and its relevance to biological systems. *A Symposium on Calcium and Cellular Function.* (eds A W Cuthbert), St Martin's Press, New York, 3–9.

Glenn GM, Poovaiah BW, Rasmussen HP. 1985. Pathway of calcium penetration through isolated cuticles of "Golden Delicious" apple fruit. *J Am Soc Hort Sci* 110: 166–171.

Gupta N, Jawandha SK, Singh Gill P. 2011. Effect of calcium on cold storage and post-storage quality of peach. *J Food Sci Technol* 2: 225–229.

Hanson EJ, Beggs JL, Beaudry RM. 1993. Applying calcium chloride postharvest to improve blueberry firmness. *HortSci* 28: 1033–1034.

Harker FR, Ferguson IB. 1988. Calcium ion transport across discs of the cortical flesh of apple fruit in relation to fruit development. *Physiol Plant* 74: 695–700.

Harker FR, Ferguson IB, Dromgoole FI. 1989. Calcium ion transport through tissue disks of the cortical flesh of apple fruit. *Physiol Plant* 74: 688–694.

Hasegawa PM. 2006. Stress physiology. *Plant Physiology* (eds L Taiz, E Zeiger) 4th ed. Sinauer Associates, Sunderland, MA, 671–705.

Hernandez-Munoz P, Almenar E, Ocio MJ, Gavara R. 2006. Effect of calcium dips and chitosan coating on postharvest life of strawberries (*Fragaria ananassa*). *Postharv Biol Technol* 39: 247–253.

Ho LC. 1999. The physiological basis for improving tomato fruit quality. *Acta Hortic* 487: 33–40.

Ho LC, White PJ. 2005. A cellular hypothesis for the induction of blossom-end rot. *Ann Bot* 95: 571–581.

Huang XM. 2005. Fruit disorders. *Litchi and Longan-Botany, Production and Uses.* (eds C Menzel, G K Waite) CABI Publishing, London, 141–152.

Huang XM, Wang C, Zhong WL, Yuan WQ, Lu JM, Li JG. 2008. Spraying calcium is not an effective way to increase structural calcium in litchi pericarp. *Scientia Hortic* 117: 39–44.

Hussain PR, Meena RS, Dar MA, Wani AM. 2012. Effect of postharvest calcium chloride dip treatment and gamma irradiation on storage quality and shelf-life extension of Red delicious apple. *J Food Sci Technol* 4: 415–426.

Johnson RS, Crisosto CH, Beede R, Andris H, Day K. 1998. Foliar calcium sprays to improve peach quality. California Tree Fruit Agreement, Reedley, CA, Report No 98-12.

Johnstone PR, Hartz TK, May DM. 2008. Calcium fertigation ineffective at increasing fruit yield and quality of muskmelon and honeydew melons in California. *HortTech* 18: 685–689.

Jones HG, Higgs KH, Samuelson TJ. 1983. Calcium uptake by developing apple fruits. I. Seasonal changes in calcium content of fruits. *J Hortic Sci* 58: 173–182.

Kader AA. 2002. Quality parameters of fresh-cut fruit and vegetable products. *Fresh-Cut Fruits and Vegetables: Science, Technology, and Market.* (eds O Lamikanra), CRC Press, Boca Raton, FL, 11–20.

Kadir SA. 2005. Fruit quality at harvest of "Jonathan" apple treated with foliarly-applied calcium chloride. *J Plant Nutr* 27: 1991–2006.

Kays SJ. 1999. Preharvest factors affecting appearance. *Postharv Biol Technol* 15: 233–247.

Kazemi F, Jafararpoor M, Golparvar A. 2013. Effects of sodium and calcium treatments on the shelf-life and quality of pomegranate. *Int J Farm Alli Sci* 22: 1375–1378.

Kim WG, Cho WD, Lee YH. 1992. Anthracnose of strawberry caused by Colletotrichum gloeosporioides Penz. *Korean J Plant Pathol* 8: 213–215.

Lambert DH. 1990. Postharvest fungi of lowbush blueberry fruit. *Plant Dis* 74: 285–287.

Lara I, Garcia P, Vendrell M. 2004. Modifications in cell-wall composition after cold storage of calcium-treated strawberry (*Fragaria ananassa* Duch.) fruit. *Postharv Biol Technol* 34: 331–339.

Lester GE, Grusak MA. 1999. Postharvest application of calcium and magnesium to honeydew and netted muskmelons: Effects on tissue ion concentrations, quality, and senescence. *J Am Soc Hortic Sci* 124: 545–552.

Lester GE, Grusak MA. 2001. Postharvest application of chelated and non-chelated calcium dip treatments to commercially grown honeydew melons: Effects on peel attributes, tissue calcium concentration, quality and consumer preference following storage. *HortTech* 11: 561–566.

Lester GE, Grusak MA. 2004. Field application of chelated calcium: Postharvest effects on cantaloupe and honeydew fruit quality. *HortTech* 14: 29–38.

Lidster PD, Tung MA, Yada RG. 1979. Effects of preharvest and postharvest calcium treatments on fruit calcium content and susceptibility of "Van" cherry to impact damage. *J Am Soc Hortic Sci* 104: 790–793.

Liebisch F, Max JFJ, Heine G, Horst WJI. 2009. Blossom end rot and fruit cracking of tomato grown in net-covered greenhouses in central Thailand can partly be corrected by calcium and boron sprays. *J Plant Nutr Soil Sci* 172: 140–150.

Lötze E, Theron KI. 2007. Evaluating the effectiveness of preharvest calcium applications for bitter pit control in Golden Delicious apples under South African conditions. *J Plant Nutr* 30: 471–485.

Luna-Guzmán I, Barret DM. 2000. Comparison of Ca chloride and Ca lactate effectiveness in maintaining shelf stability and quality of fresh-cut cantaloupes. *Postharv Biol Technol* 19: 61–72.

Madani B, Mohamed MTM, Biggs AR, Kadir J, Awang Y, Tayebimeigooni A, Roodbar Shojaei T. 2014a. Effect of preharvest calcium chloride applications on fruit calcium level and post-harvest anthracnose disease of papaya. *Crop Prot* 55: 55–60.

Madani B, Mohamed MTM, Watkins,CB, Kadir J, Awang Y, Roodbar Shojaei T. 2014b. Preharvest calcium chloride sprays affect ripening of "Eksotika II" papaya fruits during cold storage. *Scientia Hortic* 171: 6–13.

Madrid R, Valverde M, Alcolea V, Romojaro F. 2004. Influence of calcium nutrition on water-soaking disorder of ripening muskmelon melon. *Scientia Hortic* 101: 69–79.

Magee RL, Caporaso F, Prakash A. 2003. Effects of exogenous calcium salt treatments on inhibiting irradiation-induced softening in diced Roma Tomatoes. *J Food Sci* 8: 2430–2435.

Mahajan BVC, Dhatt AS. 2004. Studies on postharvest calcium chloride application on storage behavior and quality of Asian pear during cold storage. *J Food Agric Environ* 3&4: 157–159.

Mahmud TMM, Eryani-Raqeeb AA, Sayed Omar SR, Zaki ARM, Al Eryani AR. 2008. Effects of different concentrations and applications of calcium on storage life and physiochemical characteristics of papaya (*Carica papaya* L.). *Am J Agric Biol Sci* 3: 526–533.

Manganaris GA, Vasilakakis M, Diamantidis G, Mignani I. 2007. The effect of postharvest calcium application on tissue calcium concentration, quality attributes incidence of flesh browning and cell-wall physicochemical aspects of peach fruits. *Food Chem* 100: 1385–1392.

Manganaris GA, Vasilakakis M, Mignani I, Diamantidis G, Tzavella-klonari K. 2005. The effect of preharvest calcium sprays on quality attributes physic–chemical aspects of cell-wall components and susceptibility to brown rot of peach fruit. *Scientia Hortic* 107: 43–50.

Marshall R, Weaver E. 1999. An update on the use of calcium chloride to reduce rain cracking of cherries—Year 2 results. *Orchardist* 72: 34–36.

Martin-Diana AB, Rico D, Frias JM, Barat JM, Henehana GTM, Barry-Ryan C. 2007. Calcium for extending the shelf-life of fresh whole and minimally processed fruits and vegetables: A review. *Trends Food Sci Technol* 18: 210–218.

Marzouk HA, Kassem HA. 2011. Improving yield, quality, and shelf-life of "Thompson" seedless grapevine by preharvest foliar applications. *Scientia Hortic* 130: 425–430.

Mason JL. 1976. Calcium concentration and firmness of stored "McIntosh" apples increased by calcium chloride solution plus thickener. *Hort Sci* 11: 504–505.

Mattoo AK, Lieberman M. 1997. Localization of the ethylene-synthesizing system in apple tissue. *Plant Physiol* 60: 794–799.

Meheriuk M, Neilsen GH, Mckenzie DL. 1991. Incidence of rain-splitting in sweet cherries treated with calcium or coating materials. *Can J Plant Sci* 71: 231–234.

Miceli A, Ippolito A, Linsalata V, Nigro F. 1999. Effect of preharvest calcium treatment on decay and biochemical changes in table grape during storage. *Phytopathol Mediterr* 38: 47–53.

Minamide RT, Ho LC. 1993. Deposition of calcium compounds in tomato fruit in relation to calcium transport. *J Hortic Sci* 68: 755–762.

Nam MH, Jeong S, Lee YS, Choi JM, Kim HG. 2006. Effects of nitrogen, phosphorus, potassium and calcium nutrition on strawberry anthracnose. *Plant Pathol* 55: 246–249.

Naradisorn M, Klieber A, Sedgley M, Scott E, Able AJ. 2006. Effect of preharvest calcium application on grey mould development and postharvest quality in strawberries. *Acta Hortic* 708: 147–150.

Neilsen G, Neilsen D, Dong S, Toivonen P, Peryea, F. 2005. Application of $CaCl_2$ sprays earlier in the season may reduce bitter pit incidence in "Braeburn" apple. *Hort Sci* 40: 1850–1853.

Ngamchuachit P, Sivertsen HK, Mitcham EJ, Barrett DM. 2014. Effectiveness of calcium chloride and calcium lactate on maintenance of textural and sensory qualities of fresh-cut mangos. *J Food Sci* 79: 786–794. doi: 10.1111/1750-3841.12446.

Oms-Oliu G, Rojas-Graü MA, González LA, Varela P, Soliva-Fortuny R, Hernando MIH, Munuera IP, Fiszman S, Martín-Belloso O. 2010. Recent approaches using chemical treatments to preserve quality of fresh-cut fruit: A review. *Postharv Biol Technol* 57: 139–148.

Ortiz A, Graell J, Lara I. 2011. Preharvest calcium sprays improve volatile emission at commercial harvest of "Fuji Kiku-8" apples. *J Agric Food Chem* 59: 335–341.

Ortiz A; Echeverría G, Graell J, Lara I. 2010. The emission of flavour-contributing volatile esters by "Golden Reindeers" apples is improved after mid-term storage by postharvest calcium treatment. *Postharv Biol Technol* 57: 114–123.

Paiva EAS, Aampaio RA, Martinez HEP. 1998. Composition and quality of tomato fruit cultivated in nutrient solutions containing different calcium concentrations. *J Plant Nutr* 12: 2653–2661.

Paull RE, Rohrbach KG. 1985. Symptom development of chilling injury in pineapple fruit (*Ananas comosus*). *J Am Soc Hortic Sci* 1: 100–105.

Poovaiah IB. 1986. Role of calcium in prolonging storage life of fruits and vegetables. *Food Technol* 40: 86–89.

Powell GH. 1908. The decay of oranges while in transit from California. *Bulletin No. 123, Bureau of Plant Industry.* United States Department of Agriculture, Washington, DC.

Prakash A, Manley J, DeCosta S, Caporaso F, Foley DM. 2002. The effects of gamma irradiation on the microbiological, physical, and sensory qualities of diced tomatoes. *Radiat Phys Chem* 63: 387–90.

Qiu Y, Nishina MS, Paull RE. 1995. Papaya fruit growth, calcium uptake, and fruit ripening. *J Am Soc Hortic Sci* 120: 246–253.

Quiles A, Hernando I, Perez-Munuera I, Llorca E, Larrea V, Lluch MA. 2004. The effect of calcium and cellular permeabilization on the structure of the parenchyma of osmotic dehydrated "Granny Smith" apple. *J Sci Food Agric* 84: 1765–1770.

Raese JT, Drake SR. 1993. Effects of preharvest calcium sprays on apple and pear quality. *J Plant Nutr* 16: 1807–1819.

Raese JT, Drake SR, Staiff DC. 1999. Calcium sprays, time of harvest, and duration in cold storage affects fruit quality of "d'Anjou" pears in a critical year. *J Plant Nutr* 12: 1921–1929.

Raese JT, Drake SR. 1995. Calcium sprays and timing affect fruit calcium concentrations, yield, fruit weight, and cork spot of "d'Anjou" pears. *Hort Sci* 30: 1037–1039.

Ramezanian A, Rahemi M, Maftoun M, Kholdebarin B, Eshghi S, Safizadeh MR, Tavallali V. 2010. The ameliorative effects of spermidine and calcium chloride on chilling injury in pomegranate fruits after long-term storage. *Fruits* 3: 169–178.

Ramezanian A, Rahemi M, Vazifehshenas MR. 2009. Effects of foliar application of calcium chloride and urea on quantitative and qualitative characteristics of pomegranate fruits. *Sci Hortic* 121: 171–175.

Read S, Bacic A. 1996. Cell-wall porosity and its determination. *Plant Cell-Wall Analysis* (eds H Linskens, J Jackson), Springer, Berlin, Heidelberg, 63–80.

Rico D, Martín-Diana AB, Frías JM, Barat JM, Henehan GTM, Barry-Ryan C. 2007. Improvement in texture using calcium lactate and heat shock treatments for stored ready-to-eat carrots. *J. Food Eng* 79: 1196–1206.

Rigney CJ, Wills RBH. 1981. Calcium movement, a regulating factor in the initiation of tomato fruit ripening. *Hort Sci* 16: 550–551.

Romanazzi G, Lichter A, Gabler FM, Smilanick JL. 2012. Recent advances on the use of natural and safe alternatives to conventional methods to control postharvest gray mold of table grapes. *Postharv Biol Technol* 63: 141–147.

Rupert M, Southwick S, Weis K, Vikupitz J, Flore J, Zhou H. 1997. Calcium chloride reduces rain-cracking in sweet cherries. *Calif Agric* 51: 35–40.

Saftner RA, Bai J, Abbott JA, Lee YS. 2003. Sanitary dips with calcium propionate, valcium chloride, or a calcium amino acid chelate maintain quality and shelf stability of honeydew chunks. *Postharv Biol Technol* 29: 257–269.

Sams CE, Conway WS. 1984. Effect of calcium infiltration on ethylene production, respiration rate, soluble polyuronide content and quality of "Golden Delicious" apple fruit. *J Am Soc Hortic Sci* 109: 53–77.

Saure MC. 2014. Why calcium deficiency is not the cause of blossom end rot in tomato and pepper fruit—A reappraisal. *Scientia Hortic* 174: 151–154.

Schaller K, Lohnebtz O, Chikkasubbanna V. 1992. Calcium absorption by the grape berries of different cultivars during growth and development. *Wein-Wiss* 47: 62–65.

Schmitz-Eiberger M, Haefs R, Noga G. 2002. Reduction of calcium-deficiency symptoms by exogenous application of calcium chloride solutions. *Acta Hortic* 594: 535–540.

Scott KJ, Wills RBH. 1975. Postharvest application of calcium as a control for storage breakdown of apples. *Hort Sci* 10: 75–76.

Scott KJ, Wills RBH. 1977a. Deep dipping for more calcium in stored apples. *Agric Gaz NSW* 6: 47.

Scott KJ, Wills RBH. 1977b. Vacuum infiltration of calcium chloride: A method for reducing bitter pit and senescence of apples during storage at ambient temperatures. *Hort Sci* 12: 71–72.

Scott KJ, Wills RBH. 1979. Effects of vacuum and pressure infiltration of calcium chloride and storage temperature on the incidence of bitter pit and low temperature breakdown of apples. *Aust J Agric Res* 30: 917–28.

Sekse L. 1998. Fruit cracking mechanisms in sweet cherries (*Prunus avium* L.)—A review. *Acta Hortic* 468: 637–648.

Senevirathna P, Daundasekera W. 2010. Effect of postharvest calcium chloride vacuum infiltration on the shelf-life and quality of tomato ("cv. Thilina"). *Ceylon J Sci Biol Sci* 1: 35–44.

Serrano M, Amoros A, Pretel MT, Martinez-Madrid MC, Madrid R, Romojaro F. 2002. Effect of calcium deficiency on melon (*Cucumis melo* L.) texture and glassiness incidence during ripening. *Food Sci Technol* 8: 147–154.

Serrano M, Martinez-Romero D, Castillo S, Guillen F, Valero D. 2004. Role of calcium and heat treatments in alleviating physiological changes induced by mechanical damage in plum. *Postharv Biol Technol* 34: 155–157.

Sharma RR, Krishna H, Patel VB, Dahuja A, Singh R. 2006. Fruit calcium content and lipoxygenase activity in relation to albinism disorder in strawberry. *Scientia Hortic* 107: 150–154.

Shear CB. 1975. Calcium-related disorders of fruit and vegetables. *Hort Sci* 4: 361–365.

Silveira AC, Aguayo E, Chisari M, Artes F. 2011. Calcium salts and heat treatment for quality retention of fresh-cut "Galia" melon. *Postharv Biol Technol* 62: 77–84.

Singh R, Sharma RR, Tyagi SK. 2007. Preharvest foliar application of calcium and boron influences physiological disorders, fruit yield and quality of strawberry (*Fragaria ananassa* Duch.) *Scientia Hortic* 112: 215–220.

Smilanick JL, Sorenson D. 2001. Control of postharvest decay of citrus fruit with calcium polysulfide. *Postharv Biol Technol* 21: 157–168.

Smith LG. 1983. Cause and development of blackheart in pineapples. *Trop Agric (Trinidad)* 1: 31–35.

Soliva-Fortuny RC, Martin-Belloso O. 2003. New advances in extending the shelf-life of fresh-cut fruits: A review. *Trends Food Sci Technol* 14: 341–353.

Stanley DW, Bourne MC, Stone AP, Wismer WV. 1995. Low temperature blanching effects on chemistry, firmness and structure of canned green beans and carrots. *J Food Sci* 60: 327–333.

Stuckrath R, Quevedo R, de la Fuente L, Hernandez A, Sep ulveda V. 2008. Effect of foliar application of calcium on the quality of blueberry fruits. *J Plant Nutr* 31: 1299–1312.

Teisson C, Combres JC, Martin PP, Marchal J. 1979. Internal browning of pineapple fruits. *Fruits* 4: 245–261.

Tirmazi SIH, Wills RBH. 1981. Retardation of ripening of mangoes by postharvest application of calcium. *Tropic Agric* 58: 137–41.

Toivonen PMA, Brummell DA. 2008. Biochemical bases of appearance and texture changes in fresh-cut fruit and vegetables. *Postharv Biol Technol* 48: 1–14.

Torre S, Borochov A, Halevy AH. 1999. Calcium regulation of senescence in roses. *Physiol Plantarum* 107: 214–219.

Tsantili E, Rouskas D, Christopoulos, MV, Stanidis V, Akrivos J, Papanikolaou D. 2007 Effects of two preharvest calcium treatments on physiological and quality parameters in "Vogue" cherries during storage. *J Hortic Sci Biotechnol* 4: 657–663.

Val J, Monge E, Risco D, Blanco A. 2008. Effect of preharvest calcium sprays on calcium concentrations in the skin and flesh of apples. *J Plant Nutr* 31: 1889–1905.

Valero D, Perez-Vicente A, Martinez-Romero D, Castillo S, Guillen F, Serrano M. 2002. Plum storability improved after calcium and heat treatments: role of polyamines. *J Food Sci* 67: 2571–2575.

Wada T, Ikeda H, Ikeda M, Furukawa H. 1996. Effects of foliar application of calcium solutions on the incidence of blossom end rot of tomato fruit. *J Jpn Soc Hort Sci* 65: 553–558.

Wang WY, Zhu BZ, Lu J, Luo YB. 2006. No difference in the regulation pattern of calcium on ethylene biosynthesis between wild-type and never-ripe tomato fruit at immature green stage. *Russ J Plant Physiol* 53: 60–67.

Wang Y, Xingbin X, Long LE. 2014. The effect of postharvest calcium application in hydro-cooling water on tissue calcium content, biochemical changes, and quality attributes of sweet cherry fruit. *Food Chem* 160: 22–30.

Wilcox WF, Seem RC. 1994. Relationship between strawberry grey mould incidence, environmental variables and fungicide applications during different periods of the fruiting season. *Phytopathology* 84: 264–270.

Wills RBH, MacGlasson WB, Graham D, Joyce DC. 2007. *Postharvest: An Introduction to the Physiology and Handling of Fruit, Vegetables and Ornamentals*, 5th ed. CABI, Australia.

Wills RBH, Mahendra MS. 1990. Effect of postharvest application of calcium on ripening of peach. *Aust J Expt Agric* 29: 751–753.

Wills RBH, Rigney CJ. 1979. Effect of calcium on activity of mitochondria and pectic enzymes isolated from tomato fruits. *J Food Biochem* 3: 103–110.

Wills RBH, Scott KJ. 1980. The use of calcium chloride dipping to reduce wastage in stored apples. *Food Technol Aust* 32: 412–413.

Wills RBH, Sirivatanapa S. 1988. Evaluation of postharvest infiltration of calcium to delay the ripening of avocados. *Aust J Expt Agric* 28: 801–804.

Wills RBH, Tirmazi SIH. 1979. Effect of calcium and other minerals on ripening of tomatoes. *Aust J Plant Physiol* 6: 221–627.

Wills RBH, Tirmazi SIH. 1982. Inhibition of ripening of avocados with calcium. *Scientia Hortic* 16: 323–326.

Wills RBH. 1972. Effect of calcium on production of volatiles by apples. *J Sci Food Agric* 23: 1131–1134.

Wills RBH, Scott KJ, Lyford PB, Smale PE. 1976. Prediction of bitter pit with calcium content of apple fruit. *NZ J Agric Res* 19: 513–519

Wills RBH, Tirmazi SIH, Scott KJ. 1977. Use of calcium to delay ripening of tomatoes. *Hort Sci* 12: 551–552.

Wills RBH, Tirmazi SIH, Scott KJ. 1982. Effect of postharvest application of calcium on ripening rates of pears and bananas. *J Hortic Sci* 57: 431–435.

Wills RBH, Yuen MCC, Murtiningsih UD. 1988a. Effect of calcium infiltration on delayed ripening of "Minyak" avocado. *ASEAN Food J* 4: 43–44.

Wills RBH, Yuen MCC, Sabari Laksmi LDS, Suyanti. 1988b. Effect of calcium infiltration on delayed ripening of three mango cultivars in Indonesia. *ASEAN Food J* 4: 67–68.

Wisniewski M, Droby S, Chalutz E, Eilam Y. 1995. Effects of Ca and Mg on Botrytis cinerea and Penicillium expansum *in vitro* and on the biocontrol activity of *Candida oleophila*. *Plant Pathol* 44: 1016–1024.

Wójcik P. 2012. Quality and "Conference" pear storability as influenced by preharvest sprays of calcium chloride. *J Plant Nutr* 35: 1970–1983.

Wójcik P, Borowik M. 2013. Influence of preharvest sprays of a mixture of calcium formate, calcium acetate, calcium chloride and calcium nitrate on quality and "Jonagold" apple storability. *J Plant Nutr* 36: 2023–2034.

Wójcik P, Lewandowski M. 2003. Effect of calcium and boron sprays on yield and quality of Elsanta strawberry. *J Plant Nutr* 3: 671–682.

Wójcik P, Skorupinska A, Filipczak J. 2014. Impacts of preharvest fall sprays of calcium chloride at highrates on quality and "Conference" pear storability. *Scientia Hortic* 168: 51–57.

Wójcik P, Szwonek E. 2002. The efficiency of different foliar-applied calcium materials inimproving apple quality. *Acta Hortic* 594: 563–567.

Woodow IE, Rowan KS. 1979. Change of flux of orthophosphate between cellular comartments in ripening tomato fruit in relation to the climateric rise in respiration. *Aust J Plant Physiol* 6: 39–46.

Wu ZS, Zhang M, Wang S. 2012. Effects of high pressure argon treatments on thequality of fresh-cut apples at cold storage. *Food Control* 23: 120–127.

Yamamoto T, Satoh H, Watanabe S. 1992. The effects of calcium and naphtalene acetic acid sprays on cracking index and natural rain-cracking in sweet cherry fruits. *Jap Soc Hortic Sci* 3: 507–511.

Youryon P, Wongs-Aree C, McGlasson WB, Glahan S, Kanlayanarat S. 2013. Alleviation of internal browning in pineapple fruit by peduncle infiltration with solutions of calcium chloride or strontium chloride under mild chilling storage. *Int Food Res J* 1: 239–246.

Youssef K, Ligorio A, Sanzani SM, Nigro F, Ippolito A. 2012. Control of storage diseases of citrus by pre-and post-harvest application of salts. *Postharv Biol Technol* 72: 57–63.

Yuen CMC. 1994. Calcium and fruit storage potential: Postharvest handling of tropical fruits. *ACIAP Proc* 50: 218–227.

Zheng WW, Chun IJ, Hong SB, Yun-Xiang Z. 2014. Quality characteristics of fresh-cut "Fuji" apple slices from 1-methylcyclopropene-, calcium chloride-, and rare earth-treated intact fruits. *Scientia Hortic* 173: 100–105.

3 Nondestructive Assessment of Fruit Quality

Kerry Walsh

CONTENTS

3.1   INTRODUCTION

Noninvasive tools have become increasingly available for medical assessment of human "internal quality" over the last century, starting with the tool that symbolizes a medical doctor, the stethoscope. Today, a health assessment may involve noninvasive assessment using acoustic (e.g., stethoscope or ultrasound imaging), radiography (x-ray imaging), magnetic resonance imaging, positron emission tomography, etc. In this chapter, we explore the relevance and adoption of such technologies for grading of fruit on quality attributes.

Quality is a concept that has different meanings to each actor in the value chain, from grower to consumer, depending on the attributes that create value. Attributes of relevance include those associated with

1. Aesthetic appearance (such as size, weight, external color, external blemishes)
2. Internal defects (such as tissue browning, including rots, granulation, translucency)
3. Maturity and ripeness, influencing eating quality (specific gravity, flesh color [FC], dry matter content [DM], starch content, total soluble solids [TSS], titratable acidity, flesh firmness, and presence of volatiles)
4. Safety issues (such as the presence of foreign bodies and chemical contamination)

Ideally, a value chain would assess every item for a selection of attributes relevant to the produce type (Walsh, 2014).

Large retail chains set specifications for a number of above attributes on a product basis (e.g., Woolworths, 2014), which generally mirror those set by Codex Alimentarus (Walsh, 2014). These specifications are set around the needs of a value chain that requires long shelf-life. Other parts of the value chain, or other value chains, may give weight to other features. For example, harvest can be delayed to maximize sugar import, at the expense of firmness and shelf-life. Such a product targets a market that values eating quality (Figure 3.1). The use of fruit or vegetables as nutraceutics (the fusion of "nutrition" with "pharmaceutical," also often referred to as "nutraceuticals") represents a different value chain. Flavonoids, carotenoids, anthocyanins, and other bioactive phenolics have a range of claimed health benefits. For example, acai (palm) fruit of Brazil contains a high level of anthocyanin and is marketed for associated health benefits. Logically, such produce should be graded for content of active ingredient before undertaking value adding activity (e.g., freeze drying).

FIGURE 3.1 Two ends of the quality spectrum: subsistence mandarin production in Nepal and Japanese gift fruit, with "light" sorted label (circle at top right, indicating TSS assessed using near-infrared spectroscopy), with fruit selling at approx. and US$0.70/kg and $40/tray, respectively.

Thus, the list of attributes associated with "fruit quality" is long, requiring a range of instrumentation for assessment (quality control) of the varied attributes.

3.2 AVAILABLE TECHNOLOGIES

A seminal review of the range of noninvasive technologies appropriate to the sorting of fruit on internal quality attributes was produced by Abbot in 1999, and other reviews have updated this assessment (e.g., Ruiz-Altisent et al., 2010). The principle of these technologies is briefly described in this section, while following sections highlight current uses of these technologies in orchards, in pack-houses, and during transport and ripening.

3.2.1 FLOTATION

The specific gravity of fruit generally increases during maturation as carbohydrate (dry matter) accumulates. However, the change is slight (0.97–1.04, Abbot, 1999), such that measurement uncertainties and changes in density due to other reasons, for example, various fruit disorders, generally render this measure an unreliable maturity index. However, density can be a useful index of disorders such as spongy tissue, and voids in watermelons. Fruit can be separated based on specific gravity by flotation in a fluid with an appropriate specific gravity. A given application will require choice of the separation medium, control of medium density and washing/drying of the fruit. However, for routine grading, calculation of density from fruit weight and estimated volume is preferred (see below).

3.2.2 LOAD CELL

The humble load cell is now central to electronic fruit-grading equipment, although mechanical level/counterweight-balance mechanical grading systems are still produced. The load-cell systems offer increased speed and accuracy of measurement (typically to 1–2 g). Various load-cell technologies exist, for example, hydraulic, pneumatic and strain, with the "button and washer" type of strain gauge offering the highest accuracy and precision. Applied force changes the electrical resistance of a wire, with a typical load cell containing four strain gauges in a Wheatstone bridge configuration. Cell output is normally in the order of a few millivolts, with an amplifier incorporated into the load-cell design.

3.2.3   RGB CAMERA

The technology of digital imaging has become ubiquitous (e.g., embedded in phones) and ever cheaper. These systems are based on silicon charge-coupled device (CCD) arrays. The CCD elements are sensitive to radiation through the visible range (approximately 400–700 nm) and into the Herschel (or short wave near-infrared) region (to 1100 nm), producing a grey-scale image. In typical RGB cameras, adjacent detector pixels are coated with a red, green or blue filter, respectively, at the sacrifice of some image resolution. The red filter commonly employed in cameras allows Herschel-region radiation to reach the detector. Many cameras employ a short-wave pass filter (that blocks wavelengths above 700 nm) to produce images based on "visible" light only. A long-wave pass filter can be used to produce images only in the Herschel region. A quick test of whether a camera possesses such a filter is to use the camera to image the operating end of an infrared remote controller—the camera will image a spot if it does not have a 700 nm long pass filter fitted.

Not only has camera technology surged, but so has machine vision, in terms of the development of algorithms to identify objects in an image (e.g., review by Payne and Walsh, 2014). Such systems offer the promise of replacing human labor wherever routine identification tasks are required, for example, in object counting or surface color characterization. The image analysis task involves "segmentation" of the image, that is, classification of those pixels that are associated with fruit, and "blobbing," that is, association of fruit pixels into objects (fruit in this case). Typical features include color, texture (variance between adjacent pixels), shape (perimeter features) and size (blob pixel count).

3.2.4   X-RAY RADIOGRAPHY AND TOMOGRAPHY

X-ray transmission through a material is decreased by absorption and elastic or inelastic scattering. Transmission is thus determined by the photon energy, path length and average atomic number of the material. The medical fields commonly employ "hard" x-ray imaging (generator energies >100 kV), but also use "soft" energy x-rays (e.g., 20–50 kV) for imaging of lower-density material, for example, bones of infants. Dual energy systems, typified by airport security systems, can detect objects of a range of densities. "Hard" x-ray radiography is used in fresh-cut salad packaging operations to detect dense foreign objects, for example, stones. The technology has potential for the characterization of fruit quality, although fruit density is low (similar to water), and therefore "soft" x-ray imaging solutions are required. However, tissue density varies surprisingly little within a fruit (e.g., between skin, flesh and seed), except in relation to air spaces (voids). Thus, this technique is best suited to the detection of voids within fruit (Figure 3.2). The first report of the detection of a disorder using x-ray technology was that of potatoes with hollow "hearts" (Finney and Norris, 1978); commercial adoption followed, based on a simple absorption index rather than imagery. Other application reports have followed, for example, split-pit peaches (Han et al., 1992), watercore, rot, insect damage in apple (Schatzki et al., 1997), and mango spongy tissue and seed weevil (Thomas et al., 1995). These reports involve imaging, but the recognition rate of (apple) defects by a human operator is poor at commercial speeds, demonstrating the need for recognition by machines (Schatzki et al., 1997; Shahin et al., 2001; Mathanker et al., 2011).

X-ray imaging involving computed tomography (CT) based on multiple views of an object from different directions represents the "high end" of this technology (e.g., Figure 3.2). Virtual slices and 3D representation of an object are created, although scan times of minutes are required (i.e., too slow for in-line sorting) (Zwiggelaar et al., 1997; Barcelon et al., 1999). In the last two decades, the scientific literature has emphasized use of x-ray CT, for example, in assessment of tomato maturity (Brecht et al., 1991),

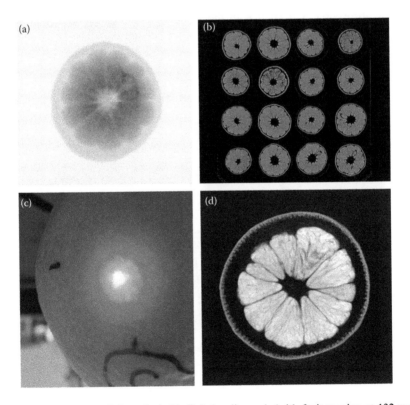

FIGURE 3.2 Images of citrus fruit (a) digital radiograph (with fruit moving at 132 mm/s using a XR3000 operating at 50 kV, Applied Sorting, Bulleen, Australia), (b) 2 mm 2D CT-xray of a tray of mandarin fruit, imaged using an accelerating voltage of 35 kV and a current of 2 mA on a Philips CT system; (c) light scatter from point source (640 nm laser) on fruit surface; (d) 2D MRI slice. (Adapted from https://www.flickr.com/photos/56604318@N00/4528898158.)

water content in apple (Tollner et al., 1992; Shahin et al., 1999, 2001), and woolly breakdown (chilling disorder) in peach and nectarine (Sonego et al., 1995) (Figure 3.2). Shahin et al. (1999) achieved detection (90% correct, with <10% false positives) of the internal apple defect, water-core. Microfocus systems have enabled visualization of intercellular spaces in small samples (e.g., Verboven et al., 2008). In a quest for practical application, rather than attempting image analysis of the CT tomograph, some authors report use of a "CT number." This number is an x-ray absorption coefficient for the object. For example, Suzuki et al. (1994) calculated CT numbers for papaya subject to vapor heat treatment which causes nonripening of internal flesh, while Barcelon et al. (1999) related CT numbers of mango fruit to moisture content.

3.2.5 Visible and Near Infrared Spectroscopy

Nicolai et al. (2007) and Herold et al. (2009) provide useful reviews of this topic. The visible (400–700 nm) spectrum of fruit is dominated by pigments such as chlorophyll and carotenoids, while the near-infrared (700–2500 nm) spectra are dominated by absorption associated with the vibration and rotation of O–H and C–H (dipolar) bonds of organic compounds in water (Figure 3.3). The O–H and C–H

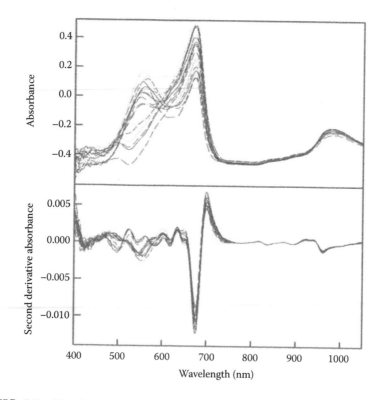

FIGURE 3.3 Absorbance (top panel) and second derivative of absorbance (bottom panel) spectra of mango fruit collected using interactance optics (Nirvana unit, Integrated Spectronics, Australia).

bonds produce fundamental frequencies in the infrared (>2500 nm), and a series of overtone and combination bands in the near-infrared and Herschel (700–1100 nm) regions. These bands are of progressively decreasing intensity and increasing width at decreasing wavelengths. The net effect is that infrared wavelengths are absorbed very strongly and have little effective penetration in biological materials, including fruit, which are 80%–90% water. Given the greater effective depth of penetration of the Herschel region (700–1100 nm), this wavelength range is of particular interest for the assessment of intact fruit, although the absorption features are broader and overlapped. Further, the Si diode and Si CCD detectors that service this region are lower in cost than the detectors required for other ranges.

NIRS instrumentation can come in many formats—from full transmission systems operating at two wavelengths to reflectance-based line-scan hyperspectral imaging (e.g., Sankaran et al., 2010). Unfortunately, many scientific publications appear to involve use of a particular instrument simply because that was available. Certainly, any scientific publication of the application of near-infrared spectroscopy to assessment of fruit should document characteristics of the system in use, and justify its match to the application (as discussed in Walsh, 2005).

Instrumentation that exploits the near-infrared spectrum to assess fruit quality must address the following design parameters for the assessment of fruit dry matter or TSS (other criteria can be set for other applications):

1. Wavelength range (typically 700–1000 nm, but up to 2500 nm)
2. Wavelength resolution (typically FWHM 10 nm)
3. Detector signal-to-noise ratio (at least 10,000:1)
4. Light source (typically halogen lamp, light emitting diode or Xe lamp) stability
5. Repeatability (standard deviation of less than 10 mA on repeated measures of a white tile)
6. Source-sample-detector geometry (reflectance, partial transmittance, or complete transmittance)
7. Matching the reference sampling volume to the optically sampled volume
8. Chemometrics (pre-pretreatment of spectra using techniques such as derivatives, analysis using discriminant analysis or multivariate regression approaches)

Consider optical geometry: Reflectance geometry, where the detector views an illuminated region of the sample, is appropriate for an assessment of skin attributes, such as color, but information will be collected only to a depth of some millimeters into the fruit. A partial or full transmittance geometry (i.e., a detector-sample-detector geometry between 0 and 180°, but in which the detector does not directly view an illuminated area of the sample) is appropriate for assessment of an internal attribute such as flesh dry matter content (DM). In this geometry, the detector views a non-illuminated region of the sample, such that light reaching the detector has passed through some volume of the fruit.

Consider wavelength range and resolution: Some spectrometer systems (e.g., CCD arrays) offer wavelength pixel spacing in the order of 1 nm. This apparent wealth of information must be interpreted in the context of the system's optical resolution (how broad a line wavelength is when viewed by the spectrometer). Also, the data used with care as absorbance data of adjacent pixels will be highly intercorrelated and, as pixel number is high, there is great risk of the over-fitting of regression models. Use of restricted wavelength ranges can improve the robustness of a model (ability to predict new sets of data), and allows for the development of lower cost "multispectral" measurement systems.

The risk of over-fitting multivariate regression models cannot be over-emphasized. The scientific literature is full of reports in which groups of fruit were randomly divided into a calibration set and a validation set for model development. Such a model will predict the attribute of interest within that population, but is likely to fail spectacularly on a new, independent set. Practical application relies on demonstration of the robustness of the model in prediction of independent populations—following the adage of "no prediction without interpretation, and no interpretation without prediction."

Near infrared spectroscopy (NIRS) is, in general, suited to use with thin skinned fruit (such as apples, stonefruit, and mango), and to the measure of constituents that present at macro levels in the fruit (Golic and Walsh, 2005). For example, TSS and DM levels in fruit typically range from 10% to 20%, with a typical error of measurement when using NIRS and intact fruit of around 1% (although the accuracy of TSS estimation is compromised for samples in which starch levels vary, e.g., in ripening fruit that convert starch to sugar).

A number of authors have claimed measurement of fruit acidity and firmness. However, organic acids are present in most fruit at around 1% w/v, and so their near-infrared signature (of C–H and O–H bonds) will be confounded with that of sugar, and any correlation with NIRS is likely to be indirect (e.g., higher Brix, lower acid, with NIRS measurement of Brix) (Subedi et al., 2012). Of course, fruit such as lemon have acids at high levels and sugars at low levels, so, in this case, direct measure of acidity by NIRS is possible. Firmness is a physical characteristic, related to cell-wall composition and structure. NIRS is unlikely to be able to differentiate cell-wall composition in intact (high-moisture) fruit, so a reported NIRS correlation between absorption and firmness is likely to represent an indirect relationship (Subedi and Walsh, 2008). Cell-wall changes may impact light-scattering rather than absorption coefficients of a fruit.

For example, the detection of "water core" defect and the internal breakdown defect within apple fruit was first reported by Francis et al. (1965) ($R = -0.81$ and 0.91, respectively, between defect level and absorbance difference between 740 and 805 nm), based on work with a complete transmission geometry and a single beam Biospec spectrometer. However, detection of internal breakdown was masked by internal browning. Upchurch et al. (1997) reported discrimination of apples with internal browning ($R^2 = 0.71$, with bruises causing misclassification) while McGlone et al. (2005) reported a R^2 between 0.7 and 0.9 for detection of brown heart disorder on fruit moving at 500 mm/s. However, such reports are often based on a very limited population—practical adoption relies on the robustness of the method, for example, in correct identification of fruit with the disorder in the presence of other, acceptable, fruit features.

3.2.6 Time or Spatial Resolved Spectroscopy

Beers Law, which related concentration of solute to absorbance, assumes no scattering of light (or at least constant scatter). In practise, the measure of light passage through fruit involves the measure of an "apparent absorbance," comprising both absorbance and scattering features. The amount of scatter may change as cell and intercellular air-space sizes change.

Two methods have been proposed to allow estimation of the scattering coefficient (μ_s')—in time-resolved reflectance spectroscopy (TRS), this is estimated from the time (picoseconds) for a photon to travel through the fruit and to the detector (more scattering, greater time), while in spatially-resolved spectroscopy it is estimated from the distance light is scattered from a point source of illumination (e.g., Figure 3.2; Seifert et al., 2015). Given the estimation of a scattering coefficient for a given fruit, a true absorption coefficient (μ_a) can also be calculated. Unfortunately, the initial hope that μ_s' might be related to firmness has not been demonstrated.

3.2.7 Raman Spectroscopy

Raman spectroscopy involves inelastic scattering of monochromatic light, with excitation and relaxation of electrons of target analytes. The sample is illuminated with a (near infrared, visible or ultraviolet) laser beam, and light from the illuminated spot is dispersed on a monochromator. Wavelengths close to the laser line are due to elastic Rayleigh scattering, and are filtered out, while the rest (Raman scattered light) is assessed, with the vibrational information specific to the chemical bonds present. Raman scattering is very weak compared to the intense Rayleigh scattered laser light, and application has come only with the relatively recent development of sensitive detectors.

Reported applications include assessment of carotenoid levels, and detection of pesticides on fruit surfaces (for a review, see Yang and Ying, 2011). For example, Bicanic et al. (2010) reported use for assessment of carotenoid levels in mango fruit homogenates. Indeed, Raman spectroscopy has become a method for measuring carotenoid levels in humans using a through skin measurement, as a biomarker of fruit/vegetable intake (Scarmo et al., 2012). However, there are no commercial applications in postharvest operations at this time.

3.2.8 Fluorescence

Fluorescence involves the excitation of a fluorophore with a specific wavelength, causing excitation of electrons, and emission of a longer wavelength associated with a drop in electron energy level. Chlorophyll, anthocyanins, carotenoids and flavonols in intact fruit will fluoresce in the visible range on exposure to UV/violet light. For example, fruit lycopene content has been related to autofluoresence under excitation wavelengths of 275 and 400 nm, while chlorophyll concentration is related to emission at 735 nm given excitation typically around 450 nm. Fluorescence levels are low, usually below 1% of total absorbed light.

Chlorophyll fluorescence is widely used to assess chlorophyll content and plant physiological status (stress level) (see review by Maxwell and Johnson, 2000).

Chlorophyll fluorescence occurs when there is no call for reductive power of the light reactions, for example, stomata are closed and the dark reactions of photosynthesis do not occur. The ratio of fluorescence at 735 nm and at 700 nm is linearly proportional to chlorophyll content. Other parameters are calculated from the induction kinetics for chlorophyll florescence (i.e., the time course of fluorescence from illumination), such as minimal (F_o), maximal (Fm), and variable ($F_v = F_m - F_o$) chlorophyll fluorescence. From these values, the potential quantum yield of photosystem (PS) II, given by the ratio of F_v/F_m is assessed. F_o represents the fluorescence yield when PSII is able to pass on almost all the electrons excited by light. F_v will vary depending on the state of photosystem II. Quantum yield (F_v/F_m) is a measure of the efficiency of the energy transfer process and chloroplast activity. For most species, the optimal value of quantum yield is 0.83. Stresses like chilling or high-temperature injuries can reduce PSII function, thereby lowering photochemical efficiency. In stored fruit, some physiological disorders will result from interruption to PSII such that chlorophyll fluorescence is high (DeEll and Toivonen, 2003). Measurement of these parameters can be made using a pulse amplitude-modulated fluorometer, which involves dark adaption of the sample, followed by a short-duration pulse of excitation wavelengths, with subsequent measure of fluorescence.

3.2.9 THERMAL IMAGING

Thermal imaging is commercially used in audits of cool-store insulation efficiency, or temperature distribution within a cool-room. The wavelengths and intensity emitted by a "blackbody" alter with the temperature of the object, as explained by Planck relationship. Less than perfect blackbodies have an emissivity coefficient of $\gg 1$; however biological material with water content above 80% w/w effectively has a high and constant emissivity factor. For blackbody objects at 30°C, the peak wavelength of emission is around 10 μm (i.e., infrared). A thermal image is very much a surface temperature measurement as wavelengths in the order of micrometers are absorbed very strongly by water-containing objects (such as fruit).

The most accurate thermal imagers use cooled detectors—making for expensive and power hungry instruments. Typical detector materials include lead sulfide and mercury cadmium telluride. However, major advances have occurred in the development of uncooled detectors, for example, silicon microbolometer arrays, which can measure a "noise equivalent temperature difference" of 20 mK. Specifications on such units include: (1) the spectral band sensed by the detector; (2) the detector life; (3) noise equivalent temperature difference of the detector; (4) minimum resolvable temperature difference; (5) the number of pixels in the array; (6) the frame refresh rate; (7) field of view; and (8) power requirements.

3.2.10 MAGNETIC RESONANCE AND RESONANCE IMAGING

A useful recent review of this topic is provided by McCarthy and Zhang (2012). Nuclear magnetic resonance (NMR) spectroscopy and magnetic resonance imaging (MRI) involve use of a radio frequency pulse to excite spin of the nucleus, with a radio signal emitted as the system returns to equilibrium over a finite period of time. Signal intensity is recorded over time, providing information about the environment of the nucleus. Relaxation is generally a function of T_1 (spin-lattice or longitudinal relaxation) and T_2 (spin-spin or transverse relaxation). Information of the bulk sample can be collected (NMR) or 2D or 3D images can be produced (MRI), and simultaneous measure of a number of fruit quality parameters is possible.

Various nuclei (e.g., C or P) can be excited to obtain information on chemical composition ("metabolic profiling"), but most use in fruit imaging has been in terms of proton mobility (e.g., Figure 3.2d). For example, when H^+ nuclei are excited, information related to the water content and mobility of the sample can be obtained. T_1 and T_2 decrease as bulk water becomes viscous, until a point at which T_1 increases again while T_2 continues to decrease. Air–water interfaces increase dephasing (the rate at which the NMR signal decreases), such that T_2 can become very short while T_1 is unchanged. Thus images based on T_2 relaxation-time can provide information on the distribution of air spaces within a fruit. 1H NMR sequences have been also used to study distribution of mobile water with respect to the ripening and defects of fruit. For some intriguing "fly through" images, see http://i.imgur.com/jTHnQ8W.gif and http://insideinsides.blogspot.com.au/p/3d-interactive-fruits-and-veggies.html.

However, cost and speed of assessment have limited the adoption of magnetic resonance imaging. Recent advances in the technology, including improvement in magnet strength and development of benchtop machines, provide a foundation for future adoption by the horticultural industry.

3.2.11 ELECTRONIC AND CHEMICAL NOSES

A characteristic aroma is produced by many fruit during ripening, a character that can be used as an index of ripeness. However, measurement of component volatiles using chromatographic separation and mass spectroscopy (GC-MS) is technically demanding and so not suited to application in the horticultural industry. Electronic noses use an array of electrochemical sensors to create a fingerprint for a given headspace of atmosphere (e.g., Cyranose 320, Sensigent, California, USA; www.sensigent.com/products/cyranose.html). However, the sensors used by such systems are prone to drift, and such systems have not found practical application in the fruit industry. A simpler technology based on a colorimetric reaction for detecting the presence of specific volatiles is discussed under the section on "Ripening Technologies."

3.2.12 FORCE–DEFORMATION

The firmness of fruit is traditionally destructively assessed with a penetrometer, involving measurement of the force to push a probe of known diameter at a known velocity a set distance into the fruit flesh. A range of noninvasive techniques have been proposed to assess aspects of the firmness, although, as the principles of measurement are different to the penetrometer, different attributes are being assessed (see reviews by García-Ramos et al., 2005; Khalifa et al., 2011). The search for a robust noninvasive technique remains something of a holy grail for the horticultural industry.

One noninvasive estimate of the fruit's firmness involves measurement of the micro-deformation of the fruit's surface under an applied force, or the force required to deform the surface a set distance. As a variant of this approach, Chen and Ruiz-Altisent (1996) described an arrangement in which the impact head carried a piezo-electric accelerometer (which outputs a voltage signal in proportion to applied force), with the rate of deacceleration of the impact head related to the fruit's firmness.

3.2.13 ACOUSTIC MEASURES

A fruit will vibrate when subject to a small force. The response is related to the modulus of elasticity, mass and shape of the object. The elastic modulus (E; Pa) of a fruit can be calculated from the highest amplitude resonant frequency (f; Hz) as:

$$E = C m^{2/3} f^2 \rho^{1/3}$$

Here, C is a constant related to shape; m, mass of fruit, kg; and ρ, density of fruit, kg m^{-3} (Studman, 2001).

The vibration of the fruit will impact on the surrounding air to create an acoustic signal. This signal can be detected, Fourier transformed, and the dominant (resonant) frequency identified. The movement of the fruit surface can also be assessed with a single-point laser. Alternately, the resonant frequency(ies) of fruit can be obtained by vibrating the fruit with a sine wave sweep of a range of frequencies (e.g., 100–3000 Hz) (e.g., the Vibsoft 4.8, a laser doppler vibrometer from Polytec Ltd., UK). Another approach was proposed by Sugiyama et al. (1994), in which two directional microphones were placed at fixed distances away from a point at which a small impact occurred, and the speed of transmission through the fruit of the dominant pressure wave was assessed. Other technologies have also been proposed, such as the use of an "air puff" to cause vibration.

Ultrasound systems operate in an inaudible frequency range (around 20 Hz) and offer potential for the detection and imaging of disorders which result in a change in fruit density (e.g., Jha et al., 2010). However, the method relies on the presence of a liquid medium between the transducer and the sample surface, and the signal suffers high attenuation due to the numerous air–water interfaces in fruit. A useful review is presented by Mizrach (2008).

3.3 TECHNOLOGIES: IN PACK-HOUSE

This section documents the adoption of noninvasive technologies into the commercial horticultural packing line (Figure 3.4).

3.3.1 EVOLUTION OF TECHNOLOGY

The story of the grading of fruit (reviewed by Walsh, 2005) begins with the sorting of fruit by color, size and shape by humans assessing by sight, and feel. The driver for mechanization of this sorting process in developed countries has been the increasing

FIGURE 3.4 Example in-line sorting equipment: (a) typical multi-lane electronic grading platform, with assessment of fruit for color, shape, weight and internal quality parameters; (b) demonstration unit from MAF Roda, incorporating transmission optics and a visible-short wave near infra red spectrometer ("InSight 2") and a four LED multispectral system for defect sorting ("IDD"); (c) images of orange from the Greefa iPIX, which uses UV illumination to image skin defects.

cost of labor and the demand for consistency in product quality. Indeed, the development of pack-house technology over the last 50 years has been nothing short of spectacular. These developments are poorly reported in the scientific literature, presumably as the advances have been largely translations of technologies developed in other disciplines and because they have been driven by private, inhouse development teams.

The first step in the automation of fruit-grading involves human sorters working alongside a conveyor belt that is carrying fruit, to enhance the efficient flow of the product. By the early twentieth century, simple mechanical designs were in use to sort fruit, for example, diverging belts that carry fruit to different positions on the conveyor depending on fruit size, or screens with slots, graded in size from small to large, to separate fruit on the basis of diameter. Such systems work well for spherical fruit. Later came the counterweighted mechanical tipping bucket graders used to categorize fruit weight.

A large advance occurred in the early 1970s, with the adoption of electronic load cells to gauge fruit weight more accurately (i.e., to 1–2 g) and more quickly than the counterweighted bucket. The load-cell systems involved a shift from a mechanical to an electronic platform, in which conveyor drop points were electrically actuated. Other sensor systems can easily be added to an electronic grading platform. By the late 1970s, RGB camera systems were used to grade fruit on the basis of color. Conditions within the camera box present a structured environment optimized for machine vision. For example, uniform lighting can be provided, mirrors can be used to image the sides of fruit, rollers can be used to rotate fruit passing through the field of view of the camera, and the background color of the conveyor can be chosen to maximize contrast with fruit. Such systems can also be used to assess fruit size. In some systems a second camera operating at Herschel wavelengths (to 1100 nm) was employed to improve the detection of fruit edges and some types of blemishes.

By the 1990s, more sophisticated algorithms were in use to process images and categorize features, allowing for detection of blemishes and recognition of a non-acceptable skin blemish from an acceptable skin mark of stem scar. During the first decade of the twenty-first century, a range of other sensor technologies were adapted for application on pack-lines, as discussed below.

As with other industries, with increasing sophistication of the product comes rationalization of production, with fewer but larger manufacturers of grading equipment with each passing decade. Initially, effectively every production area supported a local manufacturer with graders of simple design. However, as production volumes increased, and as market specifications tightened, packing house preference shifted to graders of more sophisticated design, that offered more ancillary equipment (e.g., bin unloading, hydrocoolers, washing or hot dipping, singulators, drying, bagging, and bar-code labeling for inventory control). Today a multilane grader is in use, a sophisticated instrument involving investment of hundreds of thousands of dollars. The internationally dominant manufacturers include MAF, Greefa, and Aweta P/L.

More recently, grading technologies have begun to move into the field and into use during fruit transport or ripening. These advances are elaborated in later sections.

3.3.2   Weight and Density

Grading by weight, using load cells (e.g., bending beam load cell), is now a standard feature of a modern grader. Innovation is required in the application of this technology as the fruit cup is moving during measurement, with noise superimposed on the weight signal from the natural resonant frequency of the load-cell mechanics. Various solutions have been found to this, generally involving rapid measurement (e.g., every millisecond), with filtering of higher frequency components.

Density-grading on the basis of specific gravity can be achieved using a solution of known specific gravity. However, such systems are infrequently used, due to issues associated with changing solution specific gravity and the need to wash and dry the fruit. For more spherical fruit, for example, oranges, machine-vision-based estimates of fruit volume and load-cell measurement of fruit weight can be used to estimate the fruit's specific gravity. The extension of such technologies to nonspherical fruit may become possible as image analysis routines improve.

3.3.3 Machine Vision Using RGB Cameras

As noted above, by the late 1970s, CCD camera systems had been added to fruit pack-lines, allowing grading of the fruit by size and color. By the early 1990s, more sophisticated algorithms and procedures were in use, for example, to calculate fruit volume and, thus, density, given fruit weight, and neural network analyses to determine whether a given image feature was acceptable (e.g., a fruit stalk) or unacceptable (e.g., skin blemishes over more than 2% of fruit surface area). Common features include the use of conveyor rollers to rotate the fruit while it passes through the camera's field of vision, coupled with the use of side mirrors, to image all sides of a fruit, the use of diffuse lighting and a color for the conveyor that contrasts with the fruit. Such a structured environment simplifies the machine's vision task.

A long-wave pass filter on a silicon CCD camera can be used to produce images only in the Herschel region. Such images are useful for the detection of blemishes, bruises, or diseased areas on fruit.

3.3.4 Near-Infrared Spectroscopy

NIR spectroscopy based on "point" assessment has achieved a high level of adoption in the fruit packing industry of Japan and a small level of adoption worldwide. With a conveyor belt moving at 1 m s^{-1}, a spectrometer integration time of 30 ms incurs a fruit movement of 30 mm, so a "point" measurement is effectively an average of at least part of a fruit. These systems typically involve assessment of >200 wavelength points between 400 and 1000 nm. Typical applications include assessment of fruit DM and TSS, and certain internal defects. Power is not a limiting factor to the pack-line application, compared with hand-held assessment, such that temperature stabilization features for the detector and lamp are often present. Furthermore, as pack-in-line installations do not experience the range of light level variations encountered by portable field instrumentation, less frequent referencing is required during application, which increases the throughput to several pieces of fruit per second.

NIRS technology was first applied to commercial fruit-grading in Japan, from as early as 1990. The Japanese value chain reward of high-quality fruit (Figure 3.1) provided the driver for adoption by pack-houses, supported by a range of manufacturers such as Fantec, Mitsui, Eminet, and Saika, all of whom based products on a halogen lamp/photodiode array detector system (Table 3.1). Sumitomo P/L made an interesting entry to this field in the late 1990s, marketing a system based on the use of Herschel region diode lasers. The product was subsequently withdrawn from the market without explanation, and it is a point of conjecture whether this decision was linked to technical issues. For example, output stability is difficult to achieve with diode lasers. Further, the much cheaper halogen lamp is a broad emitter, remarkably effective at producing Herschel region wavelengths, and thus a good alternative to a laser. For example, a 200 W lamp emitting equally across the range 400–2400 nm will produce 100 mW over a 1 nm range.

Adaption to the faster grading rates required in Western world agriculture followed from 2000, with release of equipment by Colour Vision Systems P/L (Australia; now MAF-Oceania, www.maf.com) and Taste Technology (New Zealand) (Table 3.1).

TABLE 3.1
Manufacturers of SWNIRS-Based Fruit Grading Units, Claiming Capability for Assessment of TSS, DM, Flesh Color, Internal Defects Such as Apple Flesh Browning, Maturity, or Acidity

Company (Location)	Website	Product/Comment
In-line Application		
Aweta (Holland)	http://www.inscan-iqa.com/	"IQA"
Eminet (Japan)	http://www.eminet.co.jp/web/ hard/index.html	
Greefa (Holland)	www.greefa.nl	"iFA," "iPIX," "iQS"
Brettech (Australia)	http://www.bret-tech.com.au/	"Hypervision"
MAF (France)	www.maf.com	"InSight," "IDD"
Mitsui (Japan)	http://www.mitsui-kinzoku. co.jp/en/seihin/s_sozai/other/	
Saika (Japan)	http://www.saika.or.jp/aguri/ seihin.html (in Japanese)	
Sacmi (Italy)	www.sacmi.it Sumitomo (Japan)	Ceased production diode laser-based system for melon TSS
Taste Tech—Compac (NZ)	www.taste-technologies.com	T1 R2 M2
Hand-Held (In-field) Application		
Astem (Japan)	http://www.astem-jp.com/ english/products.html	
Fantec (Japan)	No longer trading	"NIR Gun"; early innovator in this field
Felix instruments (USA)	www.felixinstruments.com	"F750," operating principles of Nirvana
Force-A (France)	www.force-a.eu	"Multiplex" absorbance and fluorescence spectroscopy, also in tractor-mounted format
Integrated spectronics (Australia)	No longer trading	"Nirvana"
Sunforest (Korea)	http://sunforest.en.ec21.com/	"H-100C," newest market entrant

Note: Location is of company headquarters.

Other grader manufacturers followed, for example, Greefa (Netherlands), Aweta (Netherlands) and Unitec (Italy). Interestingly, while the Japanese manufacturers of NIRS equipment for fruit-grading were independent of the conveyor manufacturers, in the Western world it is the grader-manufacturers themselves that have developed NIRS products. The list of manufacturers in Table 3.1 is not complete, with new entrants to this field occurring continually. However, rationalization of the field is also expected, as seen in all manufacturing fields in a global economy.

Equipment design differs in terms of light sources (halogen, LED, laser), optical geometry (complete or partial transmission, or reflectance), detector figures of merit (such as wavelength resolution, wavelength range, pixel resolution, signal to noise, stray light), and chemometrics (such as multiple linear regression, partial least squares, discrimination analysis).

Recently, focus has shifted from the capability to sort by "positive" quality attributes such as DM and TSS, to sorting by "negative" attributes (internal defects). For example, a number of commercial providers of sorting equipment (Table 3.1) claim detection of water content and browning in apple, internal browning in kiwifruit, and granulation in citrus and clementine. All such systems are based on transmission spectroscopy, but there is no characterization of the systems or validation of their efficacy in the scientific literature.

Multispectral and hyperspectral imaging are further levels of sophistication that have yet to find commercial application in fruit grading (Sankaran et al., 2010), although a range of hyperspectral systems are commercially available (e.g., www.zolix.com.cn/en/prodcon_370_375_359.html). Exceptions include the HyperVision in-line grading system (Table 3.1), which acquires reflectance hyperspectra using a line-scan camera (charge couple device or CCD) operating over the spectral range of 450–1150 nm and 6×300 W tungsten halogen lamps, with a distance of 800 mm from the CCD to the conveyor belt (e.g., Wedding et al., 2011), and the Helios (Helios, http://www.hyperspectral-imaging.com/products.html) "chemical imaging" system used in detection of "sugar end" in potato.

A converse trend is to produce equipment capable of assessing only a few wavelengths, reducing the cost and complexity of instrumentation. This type of example often relies on the use of LEDs, now available at a range of wavelengths through the 400–1100 nm spectrum. For example, the "DA-head" (Turoni, Italy) provides a measurement of fruit chlorophyll content, based on absorption at two wavelengths around the chlorophyll peak, with the claim that segregation of harvested fruit (e.g., peach, nectarine) into homogeneous classes of ripening can be achieved, at a speed of 4–5 fruits/s. In another example, the "IDD2" unit (MAF, France) utilizes four wavelengths in assessment of internal defects such as diffused browning of the apple.

3.3.5 X-Ray Imaging

Line-scan systems allow the acquisition of transmission images of fruit passing on a conveyor. The best and perhaps only commercial application of this technology in the context of intact produce is the assessment of potato tubers for the presence of internal voids, based on the seminal work of Finney and Norris (1978). Several manufacturers service this market (e.g., BEST, 2014). The potential health hazards and cost of x-ray imaging are limits to the adoption of this technology.

3.3.6 Magnetic Resonance

Cost and speed of assessment have limited the adoption of magnetic resonance imaging. However, recent advances in magnet technology have allowed increased field strength at reduced cost (Zhang and McCarthy, 2013). For example, Aspect Imaging (Israel) offers a prototype magnetic resonance imaging scanner coupled with a conveyor belt for measuring fruit quality attributes in-line, sorting fruit in groups of 10–12 fruits/s.

3.3.7 FORCE–DEFORMATION

The Sinclair (2014) "iQ" technology is an impact system with an accelerometer, assessing the deceleration rate of a light object striking the surface of the fruit. An air-actuated bellows (as used in the Sinclair fruit labeling system) carries the accelerometer into contact with the fruit. Four measurements of the one piece of fruit are made as it travels down the conveyor.

The Greefa "iFD" ("intelligent Firmness Detector") on-line firmness-testing unit operates on a similar principle, with a set of sensor heads that impact the top of the rotating fruit, taking up to 20 measurements around each fruit. Effective use with apples, avocados, mangos, peaches, and kiwis has been claimed (Greefa, 2014).

3.3.8 ACOUSTIC MEASURES

Given the level of vibrations present in an operating grading machine it is perhaps not surprising that no acoustic-based technology has achieved commercial success in on-line grading. For example, Aweta (www.aweta.com) offers equipment to measure the resonant frequency of vibrations within a fruit produced by a light tap of the fruit. This equipment is available in a benchtop format and is not available in on-line applications, in which environmental vibrations must complicate use.

3.4 TECHNOLOGIES: IN FIELD

3.4.1 AUTOMATED HARVEST AND CROP-LOAD ASSESSMENT

The last half century has seen great adoption of technology within the pack-house, driving labor costs down. The logical next areas for attention are in reduction of harvest costs (which account for roughly 40% of the orchard's operating costs) and in moving aspects of the sorting function of the pack-house into the field. The first steps in this direction have appeared as harvest aids manned by crews that perform sorting functions in the field. Automated harvesters are not yet commercially viable for fleshy fruits, but their development can be expected to continue, spurred by development of ancillary devices, for example, the vacuum-assisted harvest device (US patent 4,501,113; Brown, 2014). Another target for automation is the assessment of fruit load per tree. Ideally, such a system would allow identification and counting of fruit on a per tree basis, estimation of fruit size, recording of fruit location on a tree to allow estimation of increase in fruit size between repeat measurement events, and data warehousing of information to allow comparison of yields per tree across years.

Automation of these orchard tasks is dependent on machine vision systems that allow recognition of fruit within the "unstructured" and variably illuminated environment of the tree canopy (e.g., see review by Payne and Walsh, 2014). The default sensor for a machine-vision system is a RGB camera, given comparative cost and availability, but other technologies also have merits (e.g., LIDAR, time of flight imaging). Any action to improve the "structure" of the environment will improve the machine-vision result, for example, use of production systems that present fruit in 2D arrays, or night imaging with use of artificial illumination, in which the background is effectively removed.

Of course, some applications will be more difficult than others. The detection of green fruit set amidst a green canopy will be difficult, compared to colored fruit arranged on a plane surface. For example, in kiwifruit orchards, the kiwifruit hang suspended under vine canopies, and the fruit contrast in color with the foliage. Wijethunga et al. (2008) reported automated counting with accuracy levels >90% for a gold kiwifruit image data, with CIE Lab color space input.

3.4.2 Estimation of Crop Maturity: DM and Flesh Color by Spectroscopy

Indices of fruit maturity should be easy to assess, objective and preferably nondestructive, informing decisions on timing of harvest. In some fruit, external color and shape are adequate measures, but in others this is not the case. Other indices include temperature integrals (degree days), flesh color, skin and flesh pigmentation (e.g., decrease in skin chlorophyll and increase in skin carotenoid and anthocyanin concentration).

Several technologies are available for in-field use. Various absorption indices involving assessment of fruit on tree at several wavelengths have been proposed. For example, the Pigment Analyzer (PA1101, CP, Germany) has been used to assess shift in the chlorophyll red edge position (the wavelength at which the first derivative of the chlorophyll peak is maximal) as fruit mature (Zude, 2003; Seifert et al., 2014, 2015). The "DA meter" (Turoni, Forli, Italy) calculates the difference of absorption at 720 and 680 nm (i.e., indexes chlorophyll content). The DA index has been related to stonefruit maturation (Ziosi et al., 2008), although the index can be varied and site-specific (e.g., sun exposure can cause loss of chlorophyll).

Two hand-held Herschel region spectroscopy instruments have been applied to the assessment of mango fruit quality attributes: the FT20 (Fantec, Japan) (Saranwong et al., 2003) and the Nirvana (Integrated Spectronics, Sydney) (Subedi et al., 2007) (Table 3.1, Figure 3.5). For example, Subedi et al. (2007) reported on the use of a hand-held spectrometry system to assess mango fruit DM and internal flesh color, in the context of informing harvest scheduling, by

1. Assessing variation within a given tree canopy.
2. Defining areas within an orchard containing fruit of similar maturity.
3. Repeated measures over time to provide information on the rate of fruit maturation (e.g., Loefflen and Jordan, 2013, report use of such data in predictive modeling in kiwifruit and mango supply chains).

The value of such technology is achieved when used to reduce variability in fruit maturity or ripeness.

3.4.3 Estimation of Crop Maturity: Fluorescence, Volatiles

FORCE-A (Orsay, France) markets a fluorescence-based optical sensor (MULTIPLEX®) suited to in-field fruit measurement of anthocyanin, chlorophyll

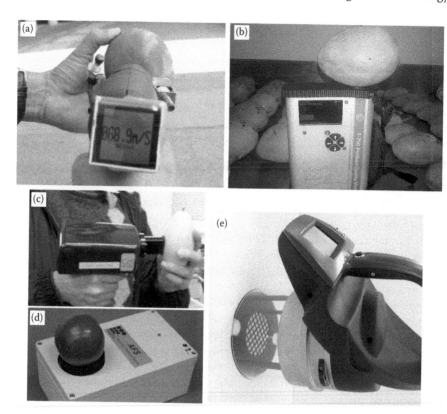

FIGURE 3.5   Example hand-held devices: (a) acoustic device measuring time delay between impact to fruit and detection of resonant vibration by microphone at a known distance from impact; (b, c) hand-held near-infrared spectrometers for fruit TSS or DM from Felix Instruments (USA) and Fantec (Japan), respectively; (d) benchtop device for assessment of acoustic resonant frequency from Aweta (Holland), and (e) fluorometer for anthocyanin and chlorophyll assessment, from Force-A (France).

and flavonol contents as maturity and quality indices (Figure 3.5). For example, use of this technology to monitor grape and oil palm fruit maturation by anthocyanin accumulation was reported by Ghozlen et al. (2010) and Hazir et al. (2012), respectively, while assessment of kiwifruit quality and maturity of apples, based on flavonol content, was reported by Pinelli et al. (2012) and Betemps et al. (2011), respectively. This technology is also available as a mounted sensor for agricultural machines.

Cyranose (USA) offers an electronic nose for volatile assessment as an alternate method for assessment of fruit maturity (Figure 3.6).

3.5   TECHNOLOGIES: IN TRANSPORT

Wireless networks allow transfer of information—allowing data access to instrumentation placed in orchards and in transport systems. A useful review of the current state of this developing technology is provided by Ruiz-Garcia et al. (2009).

(a)   (b)

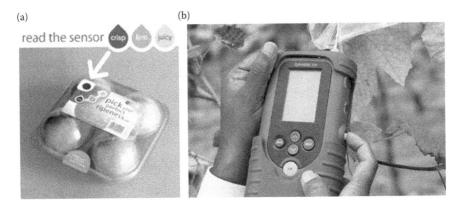

FIGURE 3.6 (a) Pears in clamshell with chemical sensing labels (red spot changes from red to yellow as fruit ripens (RipeSense, www.ripesense.com); (b) Cyranose320 electronic nose (Sensigent, USA http://www.sensigent.com/products/cyranose.html) in trials for assessment of grape maturity (http://www.vt.edu/spotlight/innovation/2013-12-09-nose/M_device.jpg).

Fruit temperature control during transport and storage is critical to the postharvest shelf-life. Unfortunately, information on breakdowns in the cool chain has in the past only been available "after the fact." The use of temperature-logging devices within loads and wireless networking offers potential for a new level of management, including manipulation of ripening during transport, on the basis that quality can be predicted from temperature history. For example, Stepac P/L offers "Xsense"—a wireless network of temperature and relative humidity sensors placed within pallets of a consignment coupled to a GPS device, with data relayed to a centralized, web-based database that can model remaining shelf-life (http://www.po.stepac.com/; https://www.xsensesystem.com/doa 14/2/14). Managers at different locations can access the system, which can be set to issue alerts, for example, once threshold temperatures are exceeded.

3.6   TECHNOLOGIES: IN RIPENING

Controlled ripening of climacteric fruit requires control of temperature and a system to introduce ethylene. Fruit are typically monitored for their stage of ripening as assessed by skin or flesh color, or firmness, with the aim of moving the fruit while a reasonable shelf-life still remains. Several measurement technologies are available for installation into sealed storage or ripening chambers, allowing for continuous measurements, while other technologies are available in an "at-line" form, for measurement of representative sample numbers of fruit. Continuous monitoring systems rely on assessment of representative fruit within the consignment. For example, HarvestWatch P/L offers a system for continuous monitoring of chlorophyll fluorescence of stored product for dynamic controlled atmosphere storage of fruits and vegetables (http://www.harvestwatch.net/doa. 14/2/2014). Installation in >1000 controlled atmosphere rooms in >15 countries, storing >200,000 t of apple, is claimed. The DAFL (Difference Absorbance Fruit Logger, Turoni P/L, Italy) is another device for continuous monitoring, in this case, based on the difference of absorption at 720 and 680 nm, with daily transmission of results to a central server outside the cold-storage room.

Electronic noses also have potential for assessment of ripening, but commercial adoption has not occurred. In contrast to the chlorophyll and volatile assessment technologies, RipeSense P/L (New Zealand; www.ripesense.com) offers a "simple" volatile detection technology based on a colorimetric reaction for specific volatiles. This technology is incorporated into a label and included within packaging material (Figure 3.6). A "clamshell" packaging is used that allows gaseous compounds to accumulate, as well as providing protection to the fruit. The first commercial application of this system was for pears.

Several "at-line" noninvasive measurement technologies for firmness have seen limited commercial use at-line in ripening rooms. Each of these techniques measure parameters linked to mechanical properties of fruit, although each measures somewhat different properties, and all are subtly different from that measured by a penetrometer.

Agro Technologie (Les Eaux, France) offers a micro-deformation device, the "Durofel," which uses a flat-ended probe (three sizes, matched to fruit type). Its use with apricots, tomatoes, cherries and other soft fruits has been reported (Agro Technologie, 2014). Technology marketed by Aweta (Acoustic Firmness Sensor, or AFS; Aweta, 2104) is based on measurement of the resonant frequency of sound emitted by a fruit following a light tap. It is recommended for measurement of the firmness of apples and tomatoes. Technology developed by Sugiyama et al. (1994) and Subedi and Walsh (2008) evaluates the velocity of a pressure wave travelling through the fruit from the point of impact of a low force (Figure 3.5). The Sinclair accelerometer system (as used on-line) has also been available in an "at-line" format.

3.7 TECHNOLOGY ADOPTION

In this chapter we have reviewed the applicability and availability of a range of instrumentation for nondestructive assessment of fruit quality. Adoption of this technology will be driven by need—when a fruit quality issue becomes a severe problem for marketability, there is a driver for uptake of a relevant technology. Of course, as with any commercial decision, a sound cost-benefit analysis is required for use of a given sorting technology. For example, consider a packing house operating for a harvest period of 100 days per annum, packing 10 t per lane per day or 1000 t of fruit over the season and selling fruit at $1/kg. If the cost of a piece of sorting equipment is US$30,000 pa (asset value US$80,000 leased or depreciated at US$20,000 per annum, plus operator and maintenance costs), then a cost of US$0.03/kg is incurred. Thus the cost of such technology is not trivial but neither is it prohibitive. The extra cost/complexity of such a system can only be justified where a clear supply chain advantage exists. For example, based on the figures above, a farm-gate value of produce packed in a single day by a single pack line is US$10,000. The rejection by a supply chain of eight days of production because of presence of a defect incurs a financial loss equivalent to the cost of the technology.

Noninvasive sorting capabilities can be used to reinforce an existing marketing system, for example, to meet specifications, or to realize "disruptive" applications, opening new markets. The introduction of new technologies also creates different opportunities—for example the range of technologies employed in the value chain may reduce the need for physical labor but increase the requirement for a range of professionally trained staff (e.g., in information technology and mechatronics).

Let us finish this chapter with a case study: the adoption of an NIRS technology into a supply chain. Harvest Fresh P/L attempted implementation of in-line Brix and DM sorting systems in the early 2000s, marketing lines of fruit that more than met retailer specifications on these attributes. However, premiums were not consistently offered by retailers, such that the extra costs incurred in growing to and sorting for higher specifications was not economic. Instead, adoption of the technology occurred in support of quality control for a Plant Variety Rights protected variety. The fruit of this variety presented well in terms of skin color but the normal visual clues on fruit maturity were not present—leading to regular harvesting of immature fruit with less than ideal eating quality. A quality control system was implemented using hand-held near-infrared spectrometers (Nirvana units from Integrated Spectronics, Sydney) with monitoring of fruit for dry matter content in all orchard blocks on a weekly basis in the lead up to harvest. The quantile-functions approach developed by Loeffen and Jordan (2013) was implemented in a decision support system, managed across all plantations, in which the percentage of fruit under specification for a DM criterion is monitored and modeled, allowing estimate of the optimal harvest date. The system allows an estimate of appropriate harvest date, supporting decisions on contracts for harvest labor, fruit transport, and marketing. Further, as DM at maturity is linked to Brix at fully ripe in this climacteric fruit, fruit of high eating quality are also produced.

ACKNOWLEDGMENT

Support of Horticulture Australia Ltd., Hortical P/L, MAF Oceania P/L, Integrated Spectronics P/L, and OneHarvest is acknowledged, and Bed Khatiwada for input in the preparation of tables.

REFERENCES

Abbot JA. 1999. Quality measurements of fruits and vegetables. *Postharv Biol Technol* 15: 207–225.

Agro Technologie. 2014. http://www.agro-technologies.com/ang/produits/societe-english.htm

Barcelon E, Tojo S and Watanabe K. 1999. X-ray CT imaging and quality detection of peach at different physiological maturity. *Trans ASAE* 42: 435–441.

BEST. 2014. http://www.bestsorting.com/sorting-food/sorters/ixus-bulk-x-ray-sorter/doa 12/7/2014.

Betemps DL, Fachinello JC, Galarça SP, Portela NM, Remorini D, Massai R and Agati G. 2011. Non-destructive evaluation of ripening and quality traits in apples using a multi-parametric fluorescence sensor. *J Sci Food Agric* 92: 1855–1864.

Bicanic D, Dimitrovski D, Luterotti S, Twisk C, Buijnsters G and Doha O. 2010. Estimating rapidly and precisely the concentration of beta carotene in mango homogenates by measuring the amplitude of optothermal signals, chromaticity indices and the intensities of Raman peaks. *Food Chem* 121: 832–838.

Brecht JK, Shewfelt RL, Garner JC and Tollner EW. 1991. Using x-ray computed tomography to non-destructively determine maturity of green tomatoes. *J Am Soc Hort Sci* 26: 45–47.

Brown J. 2014. Prototype harvester http://www.youtube.com/watch?v=B2JYRmDpBpE.

Chen P and Ruiz-Altisent M. 1996. A low-mass impact sensor for high-speed firmness sensing of fruits. *Proc International Conference Agricultural Engineering*. Madrid, Spain. September 23–26, Paper 96F-003.

DeEll JR and Toivonen PMA. 2003. Use of chlorophyll fluorescence in postharvest quality assessment of fruits and vegetables, Ch. 6, in *Practical Applications of Chlorophyll Fluorescence in Plant Biology*. pp. 203–242. Springer Publications, ISBN 978-1-4615-0415-3.

Finney E and Norris K. 1978. X-ray scans for detecting hollow heart in potatoes. *Am J Potato Res* 55: 95–105.

Francis FJ, Bramlage WJ and Lord W. 1965. Detection of watercore and internal breakdown in delicious apples by light transmittance. *Proc Am Soc Hortic Sci* 87: 78–84.

García-Ramos FJ, Valero C, Homer I, Ortiz-Canavate J and Ruiz-Altisent M. 2005. Non-destructive fruit firmness sensors: A review. *Spanish J Agric Res* 3: 61–73.

Ghozlen NB, Cerovic ZG, Germain C, Toutain S, Latouche G. 2010. Non-destructive optical sensing of grape maturation by proximal sensing. *Sensors* 10: 40–68.

Golic M and Walsh KB. 2005. Robustness of calibration models based on near infrared spectroscopy to the in-line grading of stonefruit for total soluble solids. *Anal Chim Acta* 555: 286–291.

Greefa. 2014. http://www.greefa.nl/UK/products-measuring-systems.htm, http://www.greefa.nl/UK/products-grading-machines-combisort.htm, http://www.greefa.nl/UK/products-grading-machines-geosort.htm.

Han Y, Bowers S and Dodd R. 1992. Nondestructive detection of split-pit peaches. *Trans ASAE* 35: 2063–2067.

Hazir MHM, Shariff ARM and Amiruddin MD. 2012. Determination of oil palm fresh fruit bunch ripeness based on flavonoids and anthocyanin content. *Ind Crops Prod* 36: 466–475.

Herold B, Kawano S, Sumpf B, Tillmann P and Walsh KB. 2009. Chapter 3. VIS/NIR spectroscopy, in *Optical Monitoring of Fresh and Processed Agricultural Crops*, ed. M. Zude. pp. 141–249. CRC Press, Boca Raton, FL.

Jha SN, Narsaiah K, Sharma AD, Singh M, Bansal S and Kumar R. 2010. Quality parameters of mango and potential of non-destructive techniques for their measurement—A review. *J Food Sci Technol* 47: 1–14.

Khalifa S, Komarizadeh MH and Touisi B. 2011. Usage of fruit response to both force and forced vibration applied to fruit firmness—A review. *Aust J Crop Sci* 5: 516–522.

Loefflen MPF and Jordan RB. 2013. A new method for modelling biological variation using quantile functions. *Postharv Biol Technol* 86: 387–401.

Mathanker SK, Weckler PR, Bowser T, Wang N and Maness NO. 2011. AdaBoost classifiers for pecan nut classification. *Trans ASABE* 53: 961–969.

Maxwell K and Johnson GN. 2000. Chlorophyll fluorescence—A practical guide. *J Exp Bot* 51: 659–668.

McCarthy MJ and Zhang L. 2012. *Food Quality Assurance and Control*. eMagRes. DOI:10.1002/9780470034590.emrstm1295.

McGlone VA, Martinsen P, Clark C and Jordan R. 2005. On-line detection of brownheart in Braeburn apples using near infrared transmission measurements. *Postharv Biol Technol* 37: 142–151.

Mizrach A. 2008. Ultrasonic technology for quality evaluation of fresh fruit and vegetables in pre- and post-harvest processes. *Postharv Biol Technol* 48: 315–330.

Nicolai BM, Beullens K, Bobelyn E, Peirs A and Saeys W. 2007. Non-destructive measurement of fruit and vegetable quality by means of NIR spectroscopy. *Postharv Biol Technol* 46: 99–118.

Payne A and Walsh K. 2014. Machine vision in estimation of crop yield, in *Plant Image Analysis: Fundamentals and Applications*, eds. S. Dutta-Gupta and Yasuomi Ibaraki. CRC Press, Boca Raton, FL. ISBN13:978-1-4665-8302-3.

Pinelli P, Romani A, Fierini E, Remorini D and Agati G. 2012. Characterisation of the polyphenol content in the kiwifruit exocarp for the calibration of a fruit-sorting optical sensor. *Phytochem Anal* 24: 460–466.

Ruiz-Garcia L, Lunadei L, Barreir P and Robla JI. 2009. A review of wireless sensor technologies and applications in agriculture and food industry: State of the art and current trends. *Sensors* 9: 4728–4750.

Ruiz-Altisent M, Ruiz-García L, Moreda GP, Lu R, Hernández-Sanchez N, Correa EC, Diezma B, Nicolai B and García-Ramos J. 2010. Sensors for product characterisation and quality of speciality crops—A review. *Comp Elect Agric* 74: 176–194.

Sankaran S, Mishra A, Ehsani R and Davis C. 2010. A review of advanced techniques for detecting plant diseases. *Comp Elect Agric* 72: 1–13.

Saranwong S, Sornsrivichai J and Kawano S. 2003. Performance of a portable near infrared instrument for Brix value determination of intact mango fruit. *J Near Infrared Spect* 11: 175–182.

Scarmo S, Henebery K, Peracchio H, Cartmel B, Lin H, Ermakov IV, Gellermann W, Bernstein PS, Duffy VB and Mayne ST. 2012. Skin carotenoid status measured by resonance Raman spectroscopy as a biomarker of fruit and vegetable intake in preschool children. *Eur J Clinic Nut* 66: 555–560.

Schatzki T, Haff R, Young R, Can I, Le L and Toyofuku N. 1997. Defect detection in apples by means of x-ray imaging. *Trans ASAE* 40: 1407–1415.

Seifert B, Pflanz M and Zude M. 2014. Spectral shift as advanced index for fruit chlorophyll breakdown. *Food Bioprocess Technol* 7: 2050–2059.

Seifert B, Zude M, Spinelli L and Torricelli A. 2015. Optical properties of developing pip and stone fruit reveal under lying structural changes. *Physiol Plant* 153: 327–336.

Shahin M, Tollner E, Evans M and Arabnia H. 1999. Water core features for sorting red delicious apples: A statistical approach. *Trans ASAE* 42: 1889–1896.

Shahin M, Tollner E and McClendon R. 2001. AE-automation and emerging technologies: Artificial intelligence classifiers for sorting apples based on Watercore. *J Agric Eng Res* 79: 265–274.

Sinclair. 2014. IQ system http://www.sinclair-intl.com/pages/iq_main.html doa 10/7/2014.

Sonego L, Ben-Arie R, Raynal J and Pech J. 1995. Biochemical and physical evaluation of textural characteristics of nectarines exhibiting woolly breakdown: NMR imaging, X-ray computed tomography and pectin composition. *Postharv Biol Technol* 5: 187–198.

Studman CJ. 2001. Computers and electronics in postharvest technology—A review. *Comput Electron Agr* 30: 109–124.

Subedi PP and Walsh KB. 2008. Non-invasive measurement of fresh fruit firmness. *Postharv Biol Technol* 51: 297–304.

Subedi PP, Walsh KB and Hopkins DW. 2012. Assessment of titratable acidity in fruit using short wave near infrared spectroscopy. Part B: Intact fruit studies. *J Near Infrared Spectrosc* 20: 459–463.

Subedi P, Walsh K and Owens G. 2007. Prediction of mango eating quality at harvest using short-wave near infrared spectrometry. *Postharv Biol Technol* 43: 326–334.

Sugiyama J, Otobe K, Hayashi S and Usui S. 1994. Firmness measurement of musk melons by acoustic impulse transmission. *Trans ASAE* 37: 1235–1241.

Suzuki K, Tajima T, Takano S, Asano T, Hasegawa T. 1994. Nondestructive methods for identifying injury to vapor heat treated papaya. *J Food Sci* 59: 855–857.

Thomas P, Kannan A, Degwekar VH and Ramamurthy MS. 1995. Non-destructive detection of seed weevil-infested mango fruits by X-ray imaging. *Postharv Biol Technol* 5: 161–165.

Tollner E, Hung Y, Upchurch B and Prussia S. 1992. Relating X-ray absorption to density and water content in apples. *Trans ASAE* 35: 1921–1928.

Upchurch BL, Throop JA and Aneshansley DJ. 1997. Detecting internal breakdown in apples using interactance measurements. *Postharv Biol Technol* 10: 15–19.

Verboven P, Kerckhofs G, Mebatsion H, Ho QT and Temst K. 2008. Three dimensional gas exchange pathways in pome fruit characterized by synchrotron X-ray computed tomography. *Plant Physiol* 147:518–527.

Walsh KB. 2005. Commercial Adoption of Technologies for Fruit Grading, with Emphasis on NIRS. FRUTIC 05, September 12–16, 2005, Montpellier France. http://www.symposcience.net/exl-doc/colloque/ART-00001679.pdf.

Walsh KB, Golic M and Greensill CV. 2004. Sorting of fruit and vegetables using near infrared spectroscopy: Application to soluble solids and dry matter content. *J Near Infrared Spectrosc* 12: 141–148.

Walsh KB. 2014. Chapter 9. Postharvest regulation and quality standards on fresh produce, in *Postharvest Handling—A Systems Approach*. Third edition, eds. W.J. Florkowski, R.L. Shewfelt, B. Brueckner and S.E. Prussia. Academic Press Inc, San Diego, ISBN 13: 9780124081376, pp. 205–245.

Wedding B, Wright C, Grauf S, White R and Gadek P. 2011. Non-invasive assessment of avocado quality attributes. *VII World Avocado Congress* www.worldavocadocongress2011.com.

Wijethunga P, Samarasinghe S, Kulasiri D and Woodhead I. 2008. Digital image analysis-based automated kiwifruit counting technique. Image and Vision Computing New Zealand IVCNZ, 2008. *23rd International Conference*, Lincoln University, Christchurch, pp. 1–6. http://ieeexplore.ieee.org/stamp/stamp.jsp?tp=&arnumber=4762061.

Woolworths. 2014. http://www.wowlink.com.au/wps/portal.

Yang D and Ying Y. 2011. Applications of Raman spectroscopy in agricultural products and food analysis: A review. *Appl Spectrosc Rev* 46: 539–560.

Zhang L and McCarthy MJ. 2013. Assessment of pomegranate postharvest quality using nuclear magnetic resonance. *Postharv Biol Technol* 77: 59–66.

Ziosi V, Noferini M, Fiori G, Tadiello A, Trainotti L, Casadoro G and Costa G. 2008. A new index based on vis spectroscopy to characterize the progression of ripening in peach fruit. *Postharv Biol Technol* 49: 319–329.

Zude M. 2003. Comparison of indices and multivariate models to non-destructively predict the fruit chlorophyll by means of visible spectrometry in apple fruit. *Anal Chim Acta* 481: 119–126.

Zwiggelaar R, Bull CR, Mooney MJ and Czarnez S. 1997. The detection of "soft" materials by selective energy X-ray transmission imaging and computer tomography. *J Agric Eng Res* 66: 203–212.

4 Biological Control of Postharvest Diseases

Giuseppe Lima, Simona Marianna Sanzani,
Filippo De Curtis, and Antonio Ippolito

CONTENTS

4.1 INTRODUCTION

Postharvest diseases are responsible for consistent losses of fresh fruits and vegetables, with up to 50%–60% of fresh produce discarded because of postharvest spoilage. Pathogenic microorganisms, such as fungi, mainly belonging to the genera *Alternaria, Aspergillus, Botrytis, Fusarium, Monilinia, Penicillium, Rhizopus*, are the main causal agents of postharvest deterioration (Snowdon 1990). Some of these pathogenic fungi such as *Alternaria alternata, Aspergillus* spp., *Fusarium* spp. and *Penicillium expansum* also produce toxic metabolites known as mycotoxins.

Despite the use of modern storage facilities and techniques, synthetic fungicides are still the main control means for reducing rots over extended periods of storage and/or transportation. However, increasing global concerns about environmental and human health risks due to pesticide residues has consistently reduced their use in the field near harvest and/or in postharvest situations (Droby et al. 2009). The use of synthetic fungicides is also discouraged by the increase of fungicide-resistant pathogen strains in packing houses as a consequence of their prolonged and intensive use

(FRAC 2013), the very low or even "zero" chemical residues required by fruit retailers and consumers (Cross and Berrie 2008), as well as more restrictive international regulations (e.g., EC Directive 2009/128).

Such issues, jointly with the latest worldwide guidelines on integrated and sustainable diseases management, is increasing research efforts to find more environmentally acceptable and safer measures for controlling postharvest disease. Use of natural bioactive compounds, microbial antagonists, and physical means are the main approaches for finding new solutions to ensure fruit quality and safety (Palou et al. 2008; Mari et al. 2009). Research with microbial biocontrol agents (BCAs) in pre and/or postharvest situations of fruit and vegetables started around 30 years ago (Wilson and Pusey 1985) against brown rot (*Monilinia fructicola*) of peach with a strain of *Bacillus subtilis*. This line of research is still of intense worldwide interest (Janisiewicz and Korsten 2002; Ippolito et al. 2004; Lima and De Cicco 2006; Wisniewski et al. 2007; Droby et al. 2009; Liu et al. 2013). The high amount of research carried out in this period (with more than one thousand such scientific articles published!) has shown the effectiveness of a large number of BCAs against numerous and harmful postharvest pathogens (Lima et al. 1999; Ippolito and Nigro 2000; Janisiewicz and Korsten 2002; Liu et al. 2013). However, BCAs, when applied as stand-alone treatment under commercial conditions, sometimes fail to control postharvest pathogens at a satisfactory level and this is a major obstacle for their large-scale implementation (Droby et al. 2009; Janisiewicz 2013).

In the present chapter, we supply fundamental information on (1) the basis of biological control of postharvest diseases; (2) the nature and characteristics of the most studied BCAs; (3) key strategies for improvement and optimized use of BCAs formulation; (4) constraints and obstacles to be overcome for the effective and large-scale implementation of BCAs; and (5) future perspectives in the biological control of postharvest diseases.

4.2 ANTAGONIST MICROORGANISMS AS BIOCONTROL AGENTS

In the biological control of postharvest diseases, the natural epiphytic microflora of the surfaces of fruit and vegetable is the main source for selection of BCAs (Droby et al. 2009; Liu et al. 2013). The main antagonists selected include bacteria, yeasts, and yeast-like fungi.

The most investigated biocontrol bacteria are *Bacillus subtilis* (Wilson and Pusey 1985), *Pseudomonas cepacia* (Janisiewicz and Roitman 1988), *P. syringae* (Bull et al. 1997), *Erwinia herbicola* (Bryk et al. 1998), *Pantoea agglomerans* (Usall et al. 2008), *B. amyloliquefaciens* (Arrebola et al. 2010), *B. megaterium* (Kong et al. 2010), *P. aeruginosa* (Shi et al. 2011), *Citrobacter freundii* (Janisiewicz et al. 2013). The main mechanisms of action exerted by selected biocontrol bacteria are (1) secretion of bioactive molecules such as antibiotics and cell-wall degrading enzymes (Ongena and Jacques 2008); (2) stimulation of the plant's defensive capacity (Jones and Dangl 2006); and (3) competition for space and nutrients (Poppe et al. 2003). In several studies, selected biocontrol bacteria, in addition to competition for space and nutrients, exerted a strong inhibitory effect on fungal pathogens by affecting conidial germination and/or germ tube elongation through production of bacterial lipopeptides such as fengycins, iturins, and surfactins (Ongena and Jacques 2008).

Currently bacterial-based products available for commercial use in pre- and postharvest operations are (1) BioSave® (JET Harvest Solutions, Longwood, FL, USA), which is based on *P. syringae* as an active ingredient and is used for the control of sweet potato and potato diseases; (2) Serenade® (Bayer, Leverkusen, Germany), based on *B. subtilis* and used against pre- and postharvest diseases of stone and pome fruits, strawberry, and tomato; (3) Pantovital® (Domca, Granada, Spain), based on *P. agglomerans* and used against pre- and postharvest diseases of different fruit; and (4) Amylo-X® (Biogard CBC, Grassobbio, Italy), based on *B. amyloliquefaciens* and used against fungal and bacterial diseases of different vegetables.

Concerning yeasts, numerous species have been isolated from a variety of sources, including fruit and leaf surfaces, soil, and seawater, and their potential as BCAs has been extensively investigated (Droby et al. 2009; Liu et al. 2013). The most studied yeasts are *Pichia guilliermondii* (Wilson et al. 1993), *M. pulcherrima* (De Curtis et al. 1996), *Candida oleophila* (Lima et al. 1997), *Saccharomyces cerevisiae* (Mari and Carati 1997), *C. laurentii* and *Rhodotorula glutinis* (Lima et al. 1998), *C. saitoana* (El Ghaouth et al. 2000), *P. membranaefaciens* (Fan and Tian 2000), *C. sake* (Nunes et al. 2001), *Metschnikowia fructicola* (Kurtzman and Droby 2001), *Debaryomyces hansenii* (Hernández-Montiel et al. 2010), *Rhodosporidium paludigenum* (Wang et al. 2010), *R. kratochvilovae* (Castoria et al. 2011), *Cryptococcus humicola* (Bonaterra et al. 2012), *Kloeckera apiculata* (Bonaterra et al. 2012), *Cystofilobasidium infirmominiatum* (Vero et al. 2013).

The main mechanisms of action of antagonist yeasts are (1) competition for space and nutrients; (2) activation of host defences; (3) secretion of bioactive molecules such as antibiotics and cell-wall degrading enzymes; (4) direct physical interaction with fungal hyphae; and (5) tolerance of the biocontrol yeast to reactive oxygen species (ROS) produced in the fruit in response to wounding (Castoria et al. 2003; Droby et al. 2009).

Inspite of the high number of selected strains, very few yeast-based commercial biofungicides are available for pre- and/or postharvest use. These include (1) YieldPlus® (Lallemand, Montreal, Canada), based on *Cryptococcus albidus*; (2) Candifruit® (IRTA, Lleida, Spain), based on *C. sake*; 3) Nexy® (BioNext, Paris, France), based on *C. oleophila*; and 4) Shemer® (Koppert, The Netherlands) based on *M. fructicola*.

Research on yeast-like fungi includes studies on *Aureobasidium pullulans* (de Bary) Arnaud, which is effective in preventing postharvest fungal diseases of several crops in the field and in postharvest conditions (Leibinger et al. 1997; Lima et al. 1997, 1999, 2003). Of great interest is the use of *A. pullulans* isolates during flowering, a phase in which some necrotrophic pathogens can attack the host tissues, as evidenced by Lima et al. (1997) using strain L47 of *A. pullulans* for preventing postharvest grey mold (*B. cinerea*) rots on strawberries. *A. pullulans* also proved to be a very effective antagonist when used in orchard trials in combination with reduced rates of fungicides to control rots and extend the shelf-life of apples (Leibinger et al. 1997; Lima et al. 2003).

The main modes of action reported to play an important role in the biocontrol activity of *A. pullulans* isolates are (1) competition for space and nutrients; (2) stimulation of host defences (Castoria et al. 1997); and (3) production of extracellular depolymerase enzymes such as chitinases and glucanases, which act on pathogen cell-walls (Castoria et al. 2001). Currently, there are three commercial products based on *A. pullulans* for use from the flowering to the postharvest stages: (1) Boni Protect®; (2) Blossom protect®; and (3) Botector® (Bioferm, Tulln, Austria).

4.3   ROLE OF MICROBIAL ECOLOGY IN POSTHARVEST DISEASES

While the aerial plant surface is not an ideal environment, it is populated by a variety of epiphytic nonpathogenic microorganisms playing a central role in different ecological interactions. The epiphytic microbial communities include different genera of bacteria, yeasts, yeast-like fungi, and filamentous fungi. Bacteria, the most abundant inhabitants of phylloplane and carpoplane, differ among plant species, as per the growing season, and the physical and nutritional conditions of plant surface (Lindow and Brand 2003). Yeasts (white and pink yeasts) and yeast-like fungi are widespread and active phylloplane and carpoplane colonizers, effective natural buffers against phyllosphere and/or carposphere plant pathogens (Andrews and Harris 2000). Antagonists of postharvest pathogens have also been isolated from other sources, as roots, soil, and seawater (Liu et al. 2013). Yeasts and yeast-like fungi seem ideal candidates for an efficient biocontrol of postharvest pathogens since they colonize efficiently and steadily wounded and nonwounded plant surfaces, even under unfavorable conditions (Lima et al. 1998; Ippolito et al. 2005; De Curtis et al. 2012).

The size and composition of epiphytic microflora populations depend on the host species, nutrients, rainfall, humidity, and temperature, but are also influenced by human activities such as fertilizer and pesticide application (Nix-Stohr et al. 2008; Droby et al. 2009). Moreover, the growth of the epiphytic microflora is supported by a variety of organic compounds, as vegetal exudates and pollen and honeydew deposits. Epiphytic microorganisms are able to deplete from plant surfaces nutrients essential for both growth of fungal pathogens and beginning of infection process into vegetal tissues. Yeasts and yeast-like-fungi, in general, are tolerant to low nutrients concentration and are able to utilize and metabolize various carbon sources, inorganic and organic nitrogen, lipids and other exogenous nutrients like pollen and aphid honeydew (Nix-Stohr et al. 2008).

The type and availability of nutrients affect the microbial population present on the phylloplane (Andrews and Harris 2000; Nix-Stohr et al. 2008). In general, all yeasts require sources of carbon, nitrogen, minerals, and vitamins for maintenance and growth, and this need varies considerably among yeast species. Yeasts preferentially utilize simple sugars versus polysaccharides, proteins, and lipids. Nevertheless, some yeasts isolated from phylloplane are tolerant to very low nutrient concentration (Nix-Stohr et al. 2008) and, in particular conditions of starvation, they can show high and specific extracellular enzymatic activities (e.g., chitinases and glucanases), which depolymerize the cell-wall structure of fungal pathogen to metabolize the derived simple carbon compounds (Castoria et al. 1997, 2001). Several yeasts used as BCAs were isolated from samples of a variety of fruit and vegetables or even from other environmental matrices as seawater and soil (Liu et al. 2013). As regards yeast-like fungi, *A. pullulans* is the predominant and widespread organism in various environments (Blakeman and Fokkema 1982).

4.4 INFLUENCE OF AGROCHEMICALS ON USEFUL MICROBIAL POPULATIONS OF PHYLLOSPHERE AND CARPOSPHERE

The term agrochemical in this context mainly includes biocide products as fungicides, bactericides, acaricides, herbicides, insecticides, and nematocides. Knowledge of the effects of commonly used agrochemicals on nontarget saprophytic microflora of plant surfaces, as well as on selected antagonist microorganisms, is essential for efficient management of postharvest diseases from an integrated point of view. This information can identify agrochemicals with or without a negative impact on useful nontarget epiphytic microorganisms as well as on select BCAs.

Since the target pathogens responsible for postharvest diseases are mainly fungi, particular attention is given to interactions with fungicides. The compatibility of BCAs with fungicides is a key prerequisite for their successful use in integrated pest management schedules, since the survival and colonization of a BCA after its introduction into the hard niche of the phylloplane may be influenced by their interaction with chemicals. The effect of fungicides on naturally occurring saprophytic microorganisms was assessed in various studies (Dik and van Pelt 1992; Southwell et al. 1999; Buck and Burpee 2002; Lima et al. 2003; Cadez et al. 2010; De Curtis et al. 2012). For instance, Dik and van Pelt (1992) observed that the fungicides prochloraz and triadimenol had no effect on pink and white yeasts, while maneb drastically inhibited the yeasts growth; they also observed that a combination of insecticides with fungicides caused a reduction in the presence of honeydew on the phyllosphere and, consequently, drastically reduced the presence of yeasts.

Recently, Debode et al. (2013) observed that the population dynamic of the main epiphytic yeast species on strawberry leaves and fruit was not affected by the application of the fungicides cyprodinil, fludioxonil, boscalid, and pyraclostrobin. These results are in agreement with those of Cadez et al. (2010) who observed a higher presence of yeasts on the surface of the berries of grapes treated with the commonly used fungicides iprodione, pyrimethanil, and cyprodinil plus fludioxonil. Conversely, Buck and Burpee (2002) showed that the combination cyprodinil plus fludioxonil on epiphytic yeasts of grapes and grasses resulted in a dramatic reduction in yeast density. Our recent studies evidenced a decrease in the natural yeast population of wheat phyllosphere caused by tebuconazole and tetraconazole; such decrease was significantly higher than that caused by azoxystrobin or sulphur (De Curtis et al. 2012).

Concerning the epiphytic bacteria, different researchers reported that the fungicides prochloraz, triadimenol, maneb, mancozeb, and triadimefon had no negative effect on the bacterial population of wheat phyllosphere (Dik and van Pelt 1992; Southwell et al. 1999). By contrast, Walter et al. (2007) showed that the majority of tested fungicides negatively affected the natural bacterial population on apple surfaces.

4.5 INTEGRATION OF ANTAGONIST MICROORGANISMS WITH OTHER CONTROL SYSTEMS

BCAs, when applied as stand-alone treatments under commercial conditions, rarely exhibited high efficacy and consistency against postharvest diseases. The low persistence, a narrow spectrum of activity and a failure to control previously established infections (e.g., latent, quiescent, and incipient infections) are considered the main limiting factors (Droby et al. 2009). Thus, it is generally accepted that a combination of BCAs with other methods is necessary to improve their efficacy (Ippolito and Nigro 2000; Lima et al. 2006; Droby et al. 2009). Such an integrated approach, due to additive or synergistic effects of combined or sequenced treatments, can even provide rates of disease control comparable to or better than synthetic fungicides applied alone at label dosages. Therefore, integrating BCAs with other control means can increase their acceptance by packing houses, retailers, and consumers. Integrative strategies include their combination with physical means, natural-derived compounds, or even low dosage of synthetic fungicides (Lima and De Cicco 2006; Feliziani and Romanazzi 2013). One of the most effective strategies to optimize the efficacy of BCAs and control latent infections is the application of antagonists in preharvest situations (Ippolito and Nigro 2000).

4.5.1 BCAs and Physical Means

The integration of BCAs with physical means such as high/low temperature, UV-C light, ozone, and modified or controlled atmosphere has received increasing attention, since, presumably, the control effect is due not only to direct inhibition of the pathogen but also to the induction of resistance in fruit (Droby et al. 2002). Although cold storage retards spoilage of many produce, it is often not sufficient to fully inhibit the postharvest pathogens throughout the required storage period. Morales et al. (2008) reported that two BCAs (*C. sake* CPA-2 and *P. agglomerans* CPA-1) were able to control *P. expansum* growth and patulin accumulation in cold-stored (1°C) apples. In particular, *C. sake* was more effective on fungal growth, whereas *P. agglomerans* reduced toxin accumulation. However, both BCAs could not control blue rot and patulin accumulation during storage at 20°C; in some cases, they even increased *P. expansum* aggressiveness.

Use of high temperature can contribute to biocontrol success. Heat (38–60°C) may be applied to fruit and vegetable by hot-water dips (HW), steam, hot dry air, hot-water rinsing and brushing (HWRB). For example, D'Hallewin et al. (1998) reported that the combination of a heat treatment (37°C, 95% relative humidity (RH) for 72 h) with *C. famata* produced a synergistic effect against *P. digitatum* on grapefruit comparable to that of the fungicide imazalil, when the heat treatment was performed within 36 h after inoculation with the antagonist. Whereas, Casals et al. (2010) tested HW (60°C for 40 s) with *B. subtilis* CPA-8, separately or in combination, against *Monilinia* spp. infections during the postharvest storage of stone fruit. When HW treatment was followed by CPA-8 application, a significant additive control effect on *Monilinia laxa* was detected. Similarly, Massignan et al. (2005) observed a significant reduction of green/blue mold only on oranges treated by HW (57°C, 1 min) in combination with the antagonist *A. pullulans*, L47, as compared to the control or HW and antagonist alone, but also found the highest reduction on fruit held at 95% RH as compared to fruit stored at the same temperature (7°C), but at 85% RH during storage and shelf-life. Also, treatment by *B. amyloliquefaciens* combined with 2% sodium bicarbonate and HW (45°C for 2 min) was as effective as the fungicide treatment and reduced postharvest decay of mandarin fruit by more than 80% as compared to the control. This combination significantly reduced postharvest decay without impairing fruit quality after storage at 25°C for four weeks or at 6°C for eight weeks (Hong et al. 2014). Promising results were obtained even with HWRB where the fresh product is rinsed with pressurized hot water for no longer than 30 s at 48–60°C by nozzles while rolling on brushes. Treating stonefruit by HWRB at 60°C for 20 s and then dipping into a cell suspension of the yeast *Candida* sp. 24 h after inoculation with *P. expansum* reduced decay development by 60% compared with the control (Karabulut et al. 2002).

Controlled atmosphere (CA) or modified atmosphere (MA) storage can delay the onset of ripening and senescence and hence postpone the time at which produce become more susceptible to decay, thus enhancing the biocontrol ability of antagonists. For example, yeasts *Trichosporon* sp. and *C. albidus* were more effective against *B. cinerea* and *P. expansum* on apples and pears under CA (3% O_2 and 3% CO_2 or 3% O_2 and 8% CO_2) conditions than in air (Tian et al. 2002). *C. oleophila* strain O in combination with 2% calcium chloride and modified atmosphere packaging (MAP) in nonperforated polyethylene bags, reduced by 53% crown rot of banana caused by *Colletotrichum musae*, exerting a synergistic effect as compared to the single treatment (Bastiaanse et al. 2010).

Since ozone treatment, in many instances, significantly reduces populations of filamentous fungi, yeasts, and bacteria after 10 min exposure, it could negatively affect BCAs applied to the fruit surface. However, the application of *Muscodor albus*, which produces inhibitory volatiles, reduced decay incidence on grape berries artificially infected with *B. cinerea* from 92% (control) to 21%. The ozone treatment alone or in combination with *M. albus* reduced incidence of decay to 19% or 10%, respectively (Mlikota Gabler et al. 2010). On organically grown grapes, where the natural incidence of grey mold was 31%, treatment with a combination of *M. albus* and ozone reduced decay incidence after one month of storage at 0.5°C to 3%, a greater reduction than either treatment alone. Ozone treatment also reduced decay on berries inoculated with *R. stolonifer* before or after treatment, indicating a possible resistance induction.

Among eradicative treatments, 2450 MHz microwave (used in kitchen-type microwave ovens) was successfully used to control postharvest decay, although it had no residual protection (Karabulut and Baykal 2002). Zhang et al. (2006) reported the effect of microwave and *C. laurentii*, singly and in combination, on pears. The incidence of blue mold by *P. expansum* was reduced from 100% in the control to 73% after microwave treatment, to 66% using the antagonist only, and to 20% after application of both treatments. Microwaves proved not to impair major fruit-quality indices.

Recently, UV-C irradiation and the yeast antagonist *Candida guilliermondii* were used against artificial and natural infection by *P. expansum* and *B. cinerea* in pear fruit stored at 20°C (Xu and Du 2012). Applied separately, both *C. guilliermondii* and UV-C (5 kJ/m²) effectively inhibited rots; however, their combination showed better control efficacy. Application of UV-C proved to not affect the BCA growth in pear fruit wounds, while it induced a significant increase in the activities of chitinase, β-1,3-glucanase, catalase, and peroxidase in fruit. The elicitation of the defense responses in pear fruit might be accountable for the enhanced *C. guilliermondii* biocontrol efficacy.

4.5.2 BCAs and Natural Compounds

Natural compounds include a wide range of products of animal, microbial, plant, or mineral origin, many of which are included in the GRAS (Generally Recognized as Safe) list. Several of these compounds, characterized by low toxicity toward mammalian and environment, display antimicrobial activity. Moreover, many are classified as food-grade additives and thus particularly suitable for combination with BCAs (Lima et al. 2005; Sanzani and Ippolito 2011).

A successful example of a combination of an antagonist with a compound of microbial origin is that of xanthan gum (a polysaccharide secreted by the bacterium *Xanthomonas campestris*) with the yeast-like fungus *A. pullulans* to control postharvest table grape and strawberry rots. On both commodities, the activity of the antagonist was significantly improved when applied in combination with the polysaccharide at 0.5% (w/v) (Ippolito et al. 1997). The higher activity of *A. pullulans* combined with xanthan gum was related to its greater survival on the fruit surface. Single applications of ethanol (8%–20%) and *S. cerevisiae* were not effective in reducing grey mold on apples (Mari and Carati 1997), but their combination reduced the incidence of disease by over 90%. Similar results were obtained against lemon green mold by combining ethanol with *C. oleophila*. Infections were reduced from 82% (control) to 17% (ethanol alone), 40% (yeast alone), and 3.3% (ethanol-yeast) with no appreciable differences compared with the fungicide imazalil (Lanza et al. 1997).

Within the plant-origin category, essential oils can be included. The combination of *B. amyloliquefaciens*, strain PPCB004, with thyme (TO) and lemongrass (LO) oils was evaluated against *B. cinerea*, *P. expansum*, and *R. stolonifer* on peach fruit (Arrebola et al. 2010). The biofilm formation of PPCB004 was significantly higher in LO than TO. LO and the BCA completely inhibited pathogen mycelial growth. Fruit inoculation trials with PPCB004+LO in NatureFlex™ Modified Atmosphere Packaging (MAP) showed lower disease incidence and severity at 25°C for five days than other treatment combinations or stand-alone MAP. In particular, the combination of PPCB004+LO in NatureFlex™ MAP showed the absence of disease and off-flavor development, retained the overall appearance and increased the acceptance at market-shelf conditions after cold storage at 4°C for 14 days.

The effects of the plant product methyl jasmonate (MeJA) and the yeast *C. laurentii,* alone or in combination against postharvest diseases in peach fruit, and the possible mechanisms involved, were investigated by Yao and Tian (2005). MeJA enhanced the population of *C. laurentii* and inhibited mycelial growth of *P. expansum*. The MeJA and *C. laurentii* combination induced higher activities of chitinase, β-1,3-glucanase, phenylalanine ammonia-lyase, and peroxidase than applying the yeast or MAJA alone and reduced the lesion-diameter of brown rot and blue mold caused by *M. fructicola* and *P. expansum*.

Good results were obtained combining compounds of animal origin such as chitosan, a polysaccharide obtained from crustacean shells, with several antagonists. A combination of chitosan and the antagonist *C. utilis* proved to be effective in controlling postharvest pathogens on tomato (Sharma et al. 2006). Table grape bunches sprayed 10 days before harvest with the combination of 0.1% chitosan and *C. laurentii* and stored for 42 days at 0°C, followed by a three day-shelf life, had a decay index of grey mold of 0.15 (based on a 0–1 scale) compared to 0.30 recorded in the control. Bunches treated preharvest with the same antagonist, dipped in 1% chitosan solution after harvest and stored in the above conditions, had a decay index of 0.35 in the control and 0.15 in treated bunches (Meng et al. 2010). Among grapes stored at 2°C for 15 days, the decayed berries were reduced from 22% in the control to 10% by using the combined treatment of two yeasts (*Pichia anomala* and *Cryptococcus humicolus*) with potassium caseinate and calcium chloride (Ligorio et al. 2007).

Mineral-derived compounds are more extensively used in combined treatment studies. For instance, *A. pullulans*, in combination with calcium chloride or sodium bicarbonate (SBC), was found to be effective against postharvest pathogens on sweet cherries (Ippolito et al. 2005). Moreover, calcium chloride infiltrations combined with antagonist application on apples increased control of *P. expansum* upto six months of storage at 1°C, compared to biological treatment alone (Janisiewicz et al. 1998).

The activity of 2 min dips in 3% sodium carbonate or sodium bicarbonate aqueous solutions heated to 40°C, alone or followed by the application of *P. agglomerans* CPA-2 (BA) were simultaneously evaluated against citrus green mold. The combination of the salts and BCA gave the best activity in preexisting wounds, but not in new wounds (Usall et al. 2008). In addition, Spadaro et al. (2002) found that heat treatment (HT) and SBC significantly improved the efficacy of the BCA *M. pulcherrima* against blue mold. Similarly, enhancement of biocontrol activity was achieved in pome fruits combining L-serine, L-aspartic acid, or ammonium molybdate with *C. sake* (Nunes et al. 2001), in papaya SBC with *C. oleophila* (Gamagae et al. 2004). Twelve compounds (organic and inorganic calcium salts, natural gums, and antioxidants) currently used as food additives, in combination with three BCAs (*R. glutinis*, *C. laurentii*, and *A. pullulans*), dramatically improved the antagonistic activity of one or more of the tested BCA against *P. expansum* on apples, with additive or synergistic effects (Lima et al. 2005).

The efficacy of antagonistic yeast *C. oleophila* strain O, 2% calcium chloride, and MAP in nonperforated polyethylene bags, applied alone or in various combinations, was evaluated by Bastiaanse et al. (2010) under conditions highly conducive to the development of crown rot on banana fruit artificially inoculated with *C. musae*. Both antagonistic yeast and storage under MAP, applied separately, reduced crown rot significantly whereas calcium chloride had no effect on *C. musae*. The yeast showed a 16% higher biocontrol activity when applied with calcium chloride, achieving the same protective effect with a lower yeast concentration. However, the highest synergistic efficacy (53%) was achieved combining the three alternative control means. A synergistic effect of the BCAs *C. laurentii* and *R. glutinis* applied in combination with silicon (Si, 2%) against *A. alternata* and *P. expansum* was observed in jujube fruit stored at 20°C but not at 0°C (Tian et al. 2005). Similarly, in a study by Droby et al. (2003), 2% SBC enhanced the performance of the yeast-based bioproduct Aspire (curative and protective effect) against Botrytis and Penicillium rot in apple and Monilinia and Rhizopus rot in peach. The result of this integrated approach is the development of a second generation of products such as "Biocoat," whose main components are *C. saitoana* and chitosan, or "Biocure" with *C. saitoana* and lysozyme (Micro Flo, Memphis, TN, USA). Both products contain other additives such as sodium bicarbonate (SBC) (Wisniewski et al. 2007). The bioactive coating activity was found to be superior to each component in controlling decay of several varieties of sweet orange, lemons, and apples, with a control level comparable to that of the fungicides imazalil and thiabendazole (El Ghaouth et al. 2000; Schena et al. 2005).

4.5.3 BCAs and Low Dosage of Synthetic Fungicides

Most BCAs show good resistance *in vitro* to fungicides (e.g., anilides, anilinopyrimidines, benzimidazoles, chlorothalonil, copper salts, dicarboxymides, dithiocarbamates, strobilurines, sulphur, and triazoles) commonly applied on fruit and vegetables in the field or postharvest (Omar et al. 2006; Lima et al. 2008). Fungicides based on copper or sulphur (which are also allowed in organic agriculture), in several cases showed a slight effect on yeasts and yeast-like fungi, whereas products based on synthetic chemical fungicides have an activity that depends on the species of the BCA and the chemical group to which the fungicide belongs.

In specific studies, BCAs compatible with low doses of chemical fungicides were evaluated. Lima et al. (2003) observed that some strains of *C. laurentii* and *A. pullulans* were more resistant to high concentrations of benzimidazoles and dicarboximides than strains of *R. glutinis*, while the sensitivity to triazole fungicides was high and similar for all three BCAs.

Compatibility with chemicals is an important trait for a more appropriate utilization of a BCA in an integrated control schedule. An antagonist can interact with the pathogen and also with fungicides applied before or after harvesting (Lima et al. 1997; Buck and Burpee 2002; Ippolito et al. 2004). Therefore, the optimization of antagonist-efficacy depends on its survival and colonization of unwounded and wounded plant surfaces in the presence of fungicides. Numerous studies performed in commercial packing houses demonstrated that integrating BCAs or their based commercial formulates with small quantities of synthetic fungicides showed higher efficacy and persistence against postharvest decay of several important fruit, sometimes displaying an efficacy comparable to the fungicide applied alone at the full label rate (Lima et al. 2003). Furthermore, studies have also shown that a combination of biocontrol yeasts with small quantities of synthetic fungicides exerted a more efficient control of both fungicide-sensitive and fungicide-resistant strains of fungal pathogens (Lima et al. 2006) and reduced fungicide residues and accumulation of mycotoxins in fruit (Lima et al. 2011).

4.5.4 Preharvest Application of BCAs

Postharvest quality of fresh fruit and vegetables depends on the quality of produce at harvest. Important components of the preharvest environment include plant pathogens; indeed, produce that appears healthy at harvest may harbor latent, quiescent, or incipient infections capable of causing significant losses during storage. Their impact has generally been underestimated due to postharvest application of fungicides with a strong curative activity. In addition, many fungicides applied just before harvest can control postharvest diseases (Feliziani and Romanazzi 2013). However, as stated above, nowadays their use has been strongly restricted, the time interval between application and harvest has been increased, and residues are no longer accepted. Available alternative control measures, including BCAs, fail to control infections initiated before treatment (Sanzani et al. 2009). Latent and quiescent infections in fruit and vegetables (Snowdon 1990; Sanzani et al. 2012) in some instances can prevail over infections occurring during and after harvest (Ippolito and Nigro 2000). Based on this consideration, the best system to control preharvest infection is prevention, including application of BCAs before harvest. There are few published examples of preharvest applications of BCAs but there is an increasing interest towards preharvest treatments to reduce field latent/quiescent infections or induce fruit resistance as part of an integrated diseases management program (Feliziani et al. 2013). It was shown that antagonists preemptively colonized flower parts to an extent that their activity prevented the colonization of senescent stamens by *B. cinerea* (Lima et al. 1997). The application of *A. pullulans* and *Epicoccum purpurascens* to sweet cherry blossoms reduced the number of latent infections by *M. laxa* in green fruits (Wittig et al. 1997). Also, in case of stem-end rot of avocado, the application of *B. subtilis* strain B246 at flowering was effective in reducing the incidence of disease (Korsten et al. 1997); the antagonist extensively and steadily colonized flower parts, multiplying in specific niches and preventing pathogen establishment and germination (Demoz and Korsten 2006). These findings suggest that the application of BCAs at flowering stage is a good control strategy when the pathogen gains entrance through flower parts. However, depending on the epidemiology of the disease, other moments of preharvest application can be effective. In apples, late application (August) of a mixture of antagonists was found to effectively suppress postharvest rots, mainly those due to *Pezicula* spp., *Penicillium* spp., and *M. fructigena* (Leibinger et al. 1997). Four strain of *Bacillus* spp. applied in May, or May and June, significantly reduced fruit and foliar apple-scab severity (Poleatewich et al. 2012), while the May + June + postharvest application of *Bacillus megaterium,* and isolate A3-6, resulted in the greatest suppression of bitter rot by *C. acutatum*, with an average of 45% and 95% reduction in the lesion size compared to nontreated apples (Poleatewich et al. 2012). On sweet cherries, *A. pullulans* and salts applied alone and in combination, one week before harvest, were as effective as the fungicide tebuconazole in controlling postharvest rots (Ippolito et al. 2005).

Other benefits in applying antifungal preharvest treatments are to reduce field populations of the pathogen, suppress the pathogen at the source, and induce resistance in the fruit (Palou et al. 2008). Regarding the first aspect, Lima et al. (1997) observed a significant decrease of filamentous fungi population including *B. cinerea*, only on *A. pullulans* L47 treated strawberries but not in those fruit treated with a weak antagonist (*C. oleophila*, strain L66). Regarding the second aspect, an example of the induction of resistance is the combination of preharvest application with *C. laurentii* and postharvest chitosan coating, which was found to significantly decrease table grape decay index; the results were ascribed to the strongest increase of activity of polyphenol oxidase (PPO) and phenylalanine ammonia-lyase (PAL) in combined treatments (Meng et al. 2010).

The preharvest application of BCAs seems an appropriate strategy also, where postharvest handling is unacceptable because the produce may appear less appealing. This relates to fruit that are easily damaged or cannot be exposed to water-based treatment such as strawberries, and fruit with a waxy bloom on the surface such as table grapes. Postharvest treatment with *P. guilliermondii* or *Hanseniaspora uvarum* was found to markedly suppress the postharvest decay of table grapes; but it also tended to remove the surface bloom. This problem was avoided by applying *P. guilliermondii* three days before harvest (Ippolito and Nigro 2000). In various parts of the world, to avoid handling, table grape bunches are directly packed in the field and stored at low temperature. In this situation, only fumigation with gaseous substances such as sulfur dioxide or ozone is feasible. For similar reasons, it is preferable to harvest soft fruit and pack them directly into plastic baskets for marketing through to consumer. In these situations, spraying suspensions of antagonist propagules one day before harvest, using the same equipment as for pesticide applications, is feasible.

Knowledge of the epidemiology of the target disease is crucial for choosing the right time for the application of BCAs. Under commercial conditions, the antagonist may encounter wounds colonized by resident microbial flora. In such situations, a preharvest or near-harvest application of the antagonist could permit a preemptive colonization of the wound immediately after it has been inflicted, saturating the fresh wound before the arrival of the pathogen (Janisiewicz and Korsten 2002). Competitive or preemptive exclusion is considered to be the primary mechanism by which pear and apple blossoms are protected against fire blight by applying *E. herbicola* or *P. fluorescens* (Stockwell et al. 1998). On strawberries treated with *A. pullulans* immediately before harvest, Rhizopus rot was reduced by 72% although only slight activity was observed against Botrytis rot (Lima et al. 1997). This difference can be explained on the basis of a different disease cycle of the pathogens: *B. cinerea* was present inside the fruit as a latent infection, while *R. stolonifer* was an external contaminating pathogen infecting ripe strawberry through wounds on the surface.

Finally, application of BCAs before harvest implies that the antagonist should have the capability to survive at a high population rate, despite difficulties encountered in the field environment. Some BCAs, such as *A. pullulans*, have these characteristics (Ippolito and Nigro 2000). In other cases, environmental stress tolerance has been enhanced, gaining good rot control performance, as for *C. sake* strain CPA-1 (Teixidó et al. 2010).

4.5.5 Multifaceted Approaches

A multifaceted approach can be defined as the combined or sequenced integration of the most common and available field and postharvest control methods (e.g., agronomic, biological, chemical, genetic, and physical). According to the integrated pest management system, this multistrategy approach, if mainly addressed to a preventive control of postharvest diseases, can maximize the efficacy of BCAs by reducing, to the lowest possible level, the use of synthetic fungicides. As also evidenced by various examples of combined treatment reported in other sections of this chapter, multifaceted integrated approaches can provide additive or synergistic activity to biocontrol treatment, so that the BCA can completely control the development of postharvest decay (Lima and De Cicco 2006; Palou et al. 2008; Janisiewicz 2013).

4.6 BCA FORMULATION

Formulation of microorganisms for biocontrol of plant pathogens is undeveloped, if compared to other applications of microorganisms (Janisiewicz and Korsten 2002). However, considering that most of the positive results on biological control of postharvest diseases of fruit and vegetables were obtained by applying microbial antagonists as acqueous cell suspension, the potential to set, up efficient formulates of microbial antagonist is very high.

Data on the formulation of the most promising BCAs is often not accessible to the public, since it is obtained by a private company and seldom published. However, as evidenced in other sections of this chapter, research activity aimed at enhancing antagonist viability, efficacy and shelf-life is increasing and the presence of specific compounds seems to be an essential prerequisite for the commercial success of antagonist-based biofungicides. Adjuvants are compounds of various origins which, generally, if used alone at a concentration compatible with fruit, do not exert any direct activity against pathogens. However, they can exert additive or synergistic effects on biocontrols, if combined with BCAs. Several natural molecules such as food-grade additives, edible coatings, plant or animal extracts, enzymes or antioxidants were found to improve biocontrol efficacy against a range of postharvest pathogens (see also Section 4.5.2).

Among these compounds, antioxidants are of particular interest. In fact, the demonstration of the role of resistance to oxidative stress, as a key mechanism in the biocontrol exerted by antagonistic yeasts against postharvest wound pathogens, paves the way to the improvement of the activity of BCAs in wounds by combining with antioxidants (Castoria et al. 2003).

Manipulation and storage of microbial antagonist formulates are more problematic than formulates of chemical pesticides. Performance and stability of BCA-based products can be greatly affected by a variety of factors, such as water, food, and environment. Another crucial factor for some antagonists is the high cost of their biomass production. However, when applied in combination with effective adjuvants and/or low doses of fungicides, a lower BCA cell concentration can be required to obtain high antagonistic efficacy. This could, in turn, reduce the cost of fermentation for biomass production, making more realistic the potential for developing new and highly effective BCA formulations for large-scale applications against postharvest diseases.

4.7 GENOMIC APPROACHES TO STUDY BCA–HOST–PATHOGEN INTERACTIONS

In recent years, advanced molecular techniques greatly contributed to improving knowledge on the mode of action of BCAs and providing insights into their detection and population dynamics. Indeed, conventional detection methods often do not enable the identification of specific strains and the population of BCAs can be influenced by several factors, including time and mode of application, capability for colonization, survival in unfavorable conditions, and tolerance to chemical treatments (Sanzani et al. 2014). Furthermore, a prerequisite for the registration of effective BCAs is the assessment of environmental risks related to their distribution, since they should not have nontarget effects on the environment and/or nontarget organisms (Gullino et al. 1995). A decisive boost was given by the invention in 1984 of polymerase chain reaction (PCR) by Kary Mullis. For example, the repetitive sequence-based PCR (rep-PCR) uses primers targeting several of these repetitive elements to generate DNA profiles or "fingerprints" of individual microbial strains (Ishii and Sadowsky 2009). In screening strategies, these fingerprints can be applied to differentiate strains at population level and to select unique isolates; in *in vivo* assays, they can be used for identity and quality control purposes (Berg et al. 2006). Random amplified polymorphic DNA (RAPD) markers proved also to be useful in estimating genetic variations; for instance, they were applied to obtain fingerprinting patterns among 16 strains of *Trichoderma asperellum*, *T. atroviride*, *T. harzianum*, *T. inhamatum,* and *T. longibrachiatum* previously selected as BCAs (Hermosa et al. 2001). The obtained SCAR (sequence-characterized amplified region) marker clearly distinguished *T. atroviride* strain 11 from other closely related *Trichoderma* strains. Finally, the usefulness of the amplified fragment-length polymorphism (AFLP) technique for the genetic analysis of 26 *M. pulcherrima* strains, isolated from different sources in different geographical regions, was confirmed by Spadaro et al. (2008). Genetic relationships between strains were also estimated using AFLP: all the isolates, previously tested as BCAs, were grouped in a single cluster with a high bootstrap value indicating robustness and reproducibility. Fluorescent AFLP (fAFLP) analysis was used to investigate the intraspecific variability of the yeast-like fungus *A. pullulans* strain LS30, in order to identify specific molecular markers for tracking this agent in the environment (De Curtis et al. 2004). Forty-eight isolates of *A. pullulans* from phyllosphere and carposphere of several crops from different sites of Greece and southern Italy were analyzed. Most of the isolates grouped into three main clusters, but only two (AU73 and AU91) were very similar in all fAFLP patterns. Three DNA fragments that appeared to be specific for strain LS30 were found.

However, the potential of conventional PCR was greatly increased by the development of real-time quantitative amplification technologies (qPCR) (Schena et al. 2013). For example, qPCR methods based on the use of SCAR regions have been utilized to differentiate field-applied biocontrol strains from autochthonous wild populations of the same species or genus (Sanzani et al. 2014). A strain of *A. pullulans* (L47), was monitored and quantified on the carposphere of table grapes and sweet cherries, demonstrating that its population increased soon after application and remained high over the growing season (Schena et al. 2002). Vallance et al. (2009) studied the influence of the BCA *Pythium oligandrum* on fungal and oomycete population dynamics of the rhizosphere and found that, with few exceptions, there were no significant differences between the microbial ecosystems inoculated with *P. oligandrum* and untreated systems.

A recent evolution of qPCR, High Resolution Melting (HRM), based on the analysis of the melt peak, has significant potential to enhance the currently available detection protocols. Indeed, different regions within a PCR amplicon sometimes denature from double-stranded to single-stranded at different temperatures because of varying thermal stabilities between regions. This results in a unique melt profile for the amplicon. HRM application for a simple and efficient detection of BCAs has been recently reviewed by Monk et al. (2011).

Genome sequencing also offers a tool to study BCAs in great detail. Strains of *P. fluorescens* were the first to be sequenced (Paulsen et al. 2005). Indeed, genomic information allows the analysis of the mode of action and interactions, as well as the optimization of formulation processes (Gross and Loper 2009). Moreover, as a consequence of genomic information availability, gene inactivation and over-expression studies can be performed, providing information on the transcription and regulation of these genes. De Bruijn et al. (2007) used genome mining to discover unknown gene clusters and traits highly relevant to *P. fluorescens* SBW25. Catalano et al. (2011) used gene deletion for studying the involvement of laccases in the degradation of sclerotia of plant pathogenic fungi by the mycoparasitic fungus *T. virens*. The laccase gene lcc1, expressed after the interaction with sclerotia of the plant pathogenic fungi *B. cinerea* and *Sclerotinia sclerotiorum*, was deleted, obtaining a mutant altered in its ability to degrade the sclerotia. Interestingly, while the decaying ability for *B. cinerea* sclerotia was significantly decreased, that to degrade *S. sclerotiorum* sclerotia was enhanced, suggesting different mechanisms in the mycoparasitism of these two sclerotia types by lcc1.

Biocontrol activity could even involve reduction of mycotoxin levels. Mycotoxins are not only a safety issue but may be involved in the pathogen virulence/pathogenicity (Sanzani et al. 2012). Molecular tools can also be useful in evaluating the detoxifying activity of BCAs. For instance, *Sporobolomyces* sp. IAM 13481 and *R. kratochvilovae* strains were found to be able to degrade patulin to less toxic breakdown products, desoxypatulinic acid and (Z)-ascladiol (Castoria et al. 2011; Ianiri et al. 2013). To gain insight into the genetic basis of tolerance and degradation of patulin, mutants were generated and screened. The patulin-sensitive mutants also exhibited hypersensitivity to ROS, and genotoxic and cell-wall-destabilizing agents, suggesting that inactivated genes are essential for overcoming the toxicity of patulin.

Transcriptomic studies are particularly useful to study the function of BCAs. For example, Garbeva et al. (2011), studying the transcriptional and antagonistic responses of *P. fluorescens* Pf0-1 to phylogenetically different bacterial competitors, demonstrated the existence of a species-specific response. Perazzolli et al. (2011)

showed that *T. harzianum* T39 reduced downy mildew severity on susceptible grape-vines under controlled greenhouse conditions by a direct modulation of defense-related genes and the activation of priming for enhanced expression of these genes after pathogen inoculation. Transcriptomic studies can also lead to new insights into plant responses on BCAs: Pseudomonas-primed barley genes indicated that jasmonic acid plays a role in host responses (Petti et al. 2010).

Microarray-based experiments have focused on model organisms whose genomes have been completely sequenced. Hassan et al. (2010) developed a whole genome oligonucleotide microarray for *P. fluorescens* Pf-5 to assess the consequences of a gacA mutation; gadA significantly influenced transcript levels of genes involved in the production of hydrogen cyanide, pyoluteorin and the extracellular protease. Similarly, microarrays were used to gain insight into the mechanism by which *Trichoderma hamatum* 382 induced resistance in tomato, high-density oligonucle-otide (Alfano et al. 2007). It proved to have the ability to consistently modulate the expression in tomato leaves of genes associated to biotic or abiotic stress, as well as RNA, DNA, and protein metabolism.

Methods for the *in situ* analysis of antifungal gene expression using green fluores-cent protein (GFP)-based reporter fusions have been established. Marker genes were introduced in the biocontrol strain *Clonostachys rosea* IK726 as a tool for monitor-ing the strain in ecological studies (Lübeck et al. 2002). The β-glucuronidase (GUS) reporter gene and a green fluorescent protein (GFP) encoding gene were, in separate experiments, integrated into the genome of IK726. Compared to the wild type, the two selected GUS and GFP transformants maintained the ability to colonize barley roots, and to efficiently reduce the severity of *Fusarium culmorum* without affect-ing plant emergence. Nigro et al. (1999) transformed a strain of *M. pulcherrima* with the yeast-enhanced GFP (yEGFP). The activity of two obtained transformants was indistinguishable from that of the parental strain, significantly reducing Botrytis storage rot.

The innovative Next Generation Sequencing (NGS) approaches could greatly affect the efficacy and registration of BCAs, providing a plethora of information in a single experiment. Hershkovitz et al. (2013) performed a transcriptome analysis using RNA-Seq technology to examine the response of the BCA *M. fructicola* to citrus fruit and to the postharvest pathogen *Penicillium digitatum*. An analysis of differential expression, when the yeast was interacting with the fruit vs. the patho-gen, revealed more than 250 genes with specific expression responses. In the antago-nist–pathogen interaction, genes related to transmembrane, multidrug transport, and amino acid metabolism were induced. In the antagonist–fruit interaction, expression of genes involved in oxidative stress, iron and zinc homeostasis, and lipid metabo-lism were induced.

4.8 CONCLUDING REMARKS

The use of microbial antagonists has been intensively explored and considered over the last three decades as a promising method to control postharvest diseases as an alternative to the use of synthetic fungicides. However, despite the extensive research, very few BCA-based products are commercially available for use against postharvest pathogens. This contrasts with the high number of biopesticides registered for field application against rhizosphere or phyllosphere pathogens.

Various obstacles and constraints limiting the commercial implementation of BCAs in the control of postharvest diseases have been identified by researchers and fruit-industry operators. However, the most recent research on biological control of postharvest diseases has given various suggestions to overcome most of the limiting factors, raising the potential of BCAs as a valuable alternative to synthetic pesticides.

For a satisfactory biocontrol of a postharvest plant pathogen, it is of paramount importance to determine the most appropriate strategy of application according to the disease cycle. Further studies on microbial ecology of plant surfaces in the field can also contribute to improve survival, colonization, and preventive activity of microbial antagonists. Another crucial aspect to enlarge the market potential of biocontrol products is the integration of antagonists with various control means, including natural compounds as well as genetic, agronomic, chemical and/or physical tools. Moreover, to optimize fruit protection and quality, the global chain of fruit and vegetables production and manipulation must be considered by combining preharvest and postharvest interventions.

On the improvement of effectiveness and shelf-life of BCA formulations, considerable progress is expected in the coming years by adding biocontrol formulations with suitable new adjuvants that can markedly reduce the amount of microbial biomass in BCA formulates without reducing biocontrol efficacy.

In integrated agriculture systems, combining low dosages of fungicide with compatible antagonists seems one of the most immediate possibilities for large-scale use of BCAs by growers and packing houses. Such integration not only consistently reduces the amount of fungicides needed and the related risks, but can also control both sensitive and resistant isolates of fungal pathogens. In organic agriculture systems of horticultural commodities to be exported to markets that require zero pesticide residues and organically labeled fruit, the integration of BCAs with other nonchemical alternatives can allow a satisfactory control of postharvest rots.

In conclusion, a great deal of research has shown that using antagonistic microorganisms in integrated approaches can lead to results comparable to or even better than chemicals, thus increasing interest for implementation and large-scale applications of BCA-based products and stimulating the registration of new antagonist-based formulates.

REFERENCES

Alfano G, Ivey ML, Cakir C. 2007. Systemic modulation of gene expression in tomato by *Trichoderma hamatum* 382. *Phytopathology* 97: 429–437.

Andrews JH, Harris RF. 2000. The ecology and biogeography of microorganisms on plant surfaces. *Annu Rev Phytopathol* 38: 145–180.

Arrebola E, Sivakumar D, Bacigalupo R, Korsten L. 2010. Combined application of antagonist *Bacillus amyloliquefaciens* and essential oils for the control of peach postharvest diseases. *Crop Protect* 29: 369–377.

Bastiaanse H, de Lapeyre de Bellaire L, Lassois L, Misson C, Jijakli MH. 2010. Integrated control of crown rot of banana with *Candida oleophila* strain O, calcium chloride and modified atmosphere packaging. *Biol Control* 53: 100–107.

Berg G, Opelt K, Zachow C. 2006. The rhizosphere effect on bacteria antagonistic towards the pathogenic fungus *Verticillium* differs depending on plant species and site. *FEMS Microbiol Ecol* 56: 250–261.

Blakeman JP, Fokkema NJ. 1982. Potential for biological control of plant diseases on the phylloplane. *Ann Rev Phytopathol* 20: 167–192.

Bonaterra A, Badosa E, Cabrefiga J, Frances J, Montesinos E. 2012. Prospects and limitations of microbial pesticides for control of bacterial and fungal pome fruit tree diseases. *Trees* 26: 215–226.

Bryk H, Dyki B, Sobiczewski P. 1998. Antagonistic effect of *Erwinia herbicola* on *in vitro* spore germination and germ tube elongation of *Botrytis cinerea* and *Penicillium expansum*. *Biocontrol* 43: 97–106.

Buck JW, Burpee LL. 2002. The effects of fungicides on the phylloplane yeast populations of creeping bentgrass. *Can J Microbiol* 48: 522–529.

Bull CT, Stack JP, Smilanick JL. 1997. *Pseudomonas syringae* strains ESC-10 and ESC-11 survive in wounds on citrus and control green and blue molds of citrus. *Biol Control* 8: 81–88.

Cadez N, Zupan J, Raspor P. 2010. The effect of fungicides on yeast communities associated with grape berries. *FEMS Yeast Res* 10: 619–630.

Casals C, Teixidó N, Viñas I, Silvera E, Lamarca N, Usall J. 2010. Combination of hot water, *Bacillus subtilis* CPA-8 and sodium bicarbonate treatments to control postharvest brown rot on peaches and nectarines. *Eur J Plant Pathol* 128: 51–63.

Castoria R, Caputo L, De Curtis F, De Cicco V. 2003. Resistance of postharvest biocontrol yeasts to oxidative stress: A possible new mechanism of action. *Phytopathology* 93: 564–572.

Castoria R, De Curtis F, Lima G, Caputo L, Pacifico S, De Cicco V. 2001. *Aureobasidium pullulans* (LS30) an antagonist of postharvest pathogens of fruits: Study on its modes of action. *Postharv Biol Technol* 22: 7–17.

Castoria R, De Curtis F, Lima G, De Cicco V. 1997. ß-1,3-Glucanase activity of two saprophytic yeasts and possible mode of action as biocontrol agents against postharvest diseases. *Postharv Biol Technol* 12: 293–300.

Castoria R, Mannina L, Duràn-Patròn R, Maffei F, Sobolev AP, De Felice DV, Pinedo-Rivilla C, Ritieni A, Ferracane R, Wright SAI. 2011. Conversion of the mycotoxin patulin to the less toxic desoxypatulinic acid by the biocontrol yeast *Rhodosporidium kratochvilovae* strain LS11. *J Agric Food Chem* 59: 11571–11578.

Catalano V, Vergara M, Hauzenberger JR. 2011. Use of a nonhomologous end joining deficient strain (delta-ku70) of the biocontrol fungus *Trichoderma virens* to investigate the function of the laccase gene lcc1 in sclerotia degradation. *Curr Genet* 57: 13–23.

Cross JV, Berrie AM. 2008. Eliminating the occurrence of reportable pesticide residues in apples. CIGR e-journal, http://cigrjournal.org/index.php/Ejounral/article/viewFile/1242/1099.

De Bruijn I, De Kock MJ, Yang M, De Waard P, van Beek TA, Raaijmakers JM. 2007. Genome-based discovery, structure prediction and functional analysis of cyclic lipopeptide antibiotics in *Pseudomonas* species. *Mol Microbiol* 63: 417–428.

De Curtis F, Caputo L, Castoria R, Lima G, Stea G, De Cicco V. 2004. Use of fluorescent amplified fragment length polymorphism (fAFLP) for molecular characterization of the biocontrol agent *Aureobasidium pullulans* strain LS30. *Postharv Biol Technol* 34: 179–186.

De Curtis F, Lima G, De Cicco V. 2012. Efficacy of biocontrol yeasts combined with calcium silicate or sulphur for controlling durum wheat powdery mildew and increasing grain yield components. *Field Crops Res* 134: 36–46.

De Curtis F, Torriani S, Rossi F, De Cicco V. 1996. Selection and use of *Metschnikowia pulcherrima* as a biological control agent for postharvest rots of peaches and table grapes. *Ann Microbiol Enzymol* 46: 45–55.

Debode J, Van Hemelrijck W, Creemers P, Maes M. 2013. Effect of fungicides on epiphytic yeasts associated with strawberry. *Microbiol Open Access* 2: 482–491.

Demoz BT, Korsten L. 2006. *Bacillus subtilis* attachment, colonization, and survival on avocado flowers and its mode of action on stem-end rot pathogens. *Biol Control* 37: 68–74.

D'Hallewin G, Arras G, Dessì R, Dettori A, Schirra M. 1998. Citrus green mould control in stored 'Star Ruby' grapefruit by the use of a bio-control yeast under curing conditions. *Acta Hortic* 495: 111–115.

Dik AJ, van Pelt JA. 1992. Interaction between phyllosphere yeasts, aphid honeydew and fungicide effectiveness in wheat under field conditions. *Plant Pathol* 41: 661–675.

Droby S, Porat S, Wisniewski M, El-Gaouth A, Wilson C. 2002. Integrated control of postharvest decay using yeast antagonists, hot water and natural materials. *IOBC Bull* 25: 25–28.

Droby S, Wisniewski M, El Ghaouth A, Wilson C. 2003. Influence of food additives on the control of postharvest rots of apple and peach and efficacy of the yeast-based biocontrol product Aspire. *Postharv Biol Technol* 27: 127–135.

Droby S, Wisniewski M, Macarisin D, Wilson C. 2009. Twenty years of postharvest biocontrol research: Is it time for a new paradigm? *Postharv Biol Technol* 52: 137–145.

EC Directive 2009/128. 2009. Establishing of a framework for Community action to achieve the sustainable use of pesticide. Europ Parl Council, Oct. 2, 2009.

El Ghaouth A, Smilanick JL, Wilson CL. 2000. Enhancement of the performance of *Candida saitoana* by the addition of glycolchitosan for the control of postharvest decay of apple and citrus fruit. *Postharv Biol Technol* 19: 103–110.

Fan Q, Tian S. 2000. Postharvest biological control of Rhizopus rot of nectarine fruits by *Pichia membranaefaciens*. *Plant Disease* 84: 1212–1216.

Feliziani E, Romanazzi G. 2013. Preharvest application of synthetic fungicides and alternative treatments to control postharvest decay of fruit. *Stewart Postharv Rev* 3: 4.

Feliziani E, Santini M, Landi L, Romanazzi G. 2013. Pre- and postharvest treatment with alternatives to synthetic fungicides to control postharvest decay of sweet cherry. *Postharv Biol Technol* 78: 133–138.

FRAC. 2013. List of plant pathogenic organisms resistant to disease control agents. FRAC Monographs. http://frac.info (accessed March 22, 2014).

Gamagae S, Sivakumar D, Wijesundera R. 2004. Evaluation of post-harvest application of sodium bicarbonate-incorporated wax formulation and *Candida oleophila* for the control of anthracnose of papaya. *Crop Protect* 23: 575–579.

Garbeva P, Silby MW, Raaijmakers JM, Levy SB, Boer WD. 2011. Transcriptional and antagonistic responses of *Pseudomonas fluorescens* Pf0-1 to phylogenetically different bacterial competitors. *ISME J* 5: 973–985.

Gross H, Loper JE. 2009. Genomics of secondary metabolite production by *Pseudomonas* spp. *Nat Product Rep* 26: 1408–1446.

Gullino ML, Migheli Q, Mezzalama M. 1995. Risk analysis in the release of biological control agents. *Plant Disease* 79: 1193–1201.

Hassan KA, Johnson A, Shaffer BT. 2010. Inactivation of the GacA response regulator in *Pseudomonas fluorescens* Pf-5 has far reaching transcriptomic consequences. *Environ Microbiol* 12: 899–915.

Hermosa MR, Grondona I, Diaz-Minguez JM, Iturriaga EA, Monte E. 2001. Development of a strain-specific SCAR marker for the detection of *Trichoderma atroviride* 11, a biological control agent against soil-borne fungal plant pathogens. *Curr Genet* 38: 343–350.

Hernández-Montiel LG, Ochoa JL, Troyo-Diéguez E, Larralde-Corona CP. 2010. Biocontrol of postharvest blue mold (*Penicillium italicum* Wehmer) on Mexican lime by marine and citrus *Debaryomyces hansenii* isolates. *Postharv Biol Technol* 56: 181–187.

Hershkovitz V, Sela N, Taha-Salaime L, Liu J, Rafael G, Kessler C, Aly R, Levy M, Wisniewski M, Droby S. 2013. De-novo assembly and characterization of the transcriptome of

Metschnikowia fructicola reveals differences in gene expression following interaction with *Penicillium digitatum* and grapefruit peel. *BMC Genomics* 14: 168.

Hong P, Hao W, Luo J, Chen S, Hu M, Zhong G. 2014. Combination of hot water, *Bacillus amyloliquefaciens* HF-01 and sodium bicarbonate treatments to control postharvest decay of mandarin fruit. *Postharv Biol Technol* 88: 96–102.

Ianiri G, Idnurm A, Wright SA. 2013. Searching for genes responsible for patulin degradation in a biocontrol yeast provides insight into the basis for resistance to this mycotoxin. *Appl Environ Microbiol* 79: 3101–3115.

Ippolito A, Nigro F. 2000. Impact of preharvest application of biological control agents on postharvest diseases of fresh fruits and vegetables. *Crop Protect* 19: 715–723.

Ippolito A, Nigro F, Romanazzi G, Campanella V. 1997. Field application of *Aureobasidium pullulans* against Botrytis storage rot of strawberry. In: *Proc COST 915, Non Conventional Methods for the Control of Postharvest Disease and Microbiological Spoilage*, Bologna, Italy, pp. 127–134.

Ippolito A, Nigro F, Schena L. 2004. Control of postharvest diseases of fresh fruit and vegetables by preharvest application of antagonistic microorganisms. In: *Crop Management and Postharvest Handling of Horticultural Products. Vol. 4 Disease and Disorders of Fruits and Vegetables*. eds R Dris, R Niskanen, SM Jain. Science Publisher Inc, Enfield, NH, pp. 1–30.

Ippolito A, Schena L, Pentimone I, Nigro F. 2005. Control of postharvest rots of sweet cherries by pre- and postharvest applications of *Aureobasidium pullulans* in combination with calcium chloride or sodium bicarbonate. *Postharv Biol Technol* 36: 245–252.

Ishii S, Sadowsky MJ. 2009. Applications of the rep-PCR DNA fingerprinting technique to study microbial diversity, ecology and evolution. *Environ Microbiol* 11: 733–740.

Janisiewicz WJ. 2013. Biological control of postharvest diseases: Hurdles, successes and prospects. *Acta Hortic* 1001: 273–284.

Janisiewicz WJ, Conway WS, Glenn DM, Sams CE. 1998. Integrating biological control and calcium treatment for controlling postharvest decay of apples. *HortSci* 33: 105–109.

Janisiewicz WJ, Jurick WM, Vico I, Peter KA, Buyer JS. 2013. Culturable bacteria from plum fruit surfaces and their potential for controlling brown rot after harvest *Postharv Biol Technol* 76: 145–151.

Janisiewicz WJ, Korsten L. 2002. Biological control of postharvest diseases of fruits. *Ann Rev Phytopathol* 40: 411–441.

Janisiewicz WJ, Roitman J. 1988. Biological control of blue mold and gray mold on apple and pear with *Pseudomonas cepacia*. *Phytopathology* 78: 1697–1700.

Jones JDG, Dangl JL. 2006. The plant immune system. *Nature* 444: 323–329.

Karabulut OA, Baykal N. 2002. Evaluation of the use of microwave power for the control of postharvest diseases of peaches. *Postharv Biol Technol* 26: 237–240.

Karabulut OA, Cohen L, Wiess B, Daus A, Lurie S, Droby S. 2002. Control of brown rot and blue mold of peach and nectarine by short hot water brushing and yeast antagonists. *Postharv Biol Technol* 24: 103–111.

Kong Q, Shan S, Liu Q, Wang X, Yu F. 2010. Biocontrol of *Aspergillus flavus* on peanut kernels by use of a strain of marine *Bacillus megaterium*. *Int J Food Microbiol* 139: 31–35.

Korsten L, De Villiers EE, Wehner FC, Kotze JM. 1997. Field sprays of *Bacillus subtilis* and fungicides for control of preharvest fruit diseases of avocado in South Africa. *Plant Disease* 81: 455–459.

Kurtzman CP, Droby S. 2001. *Metschnikowia fructicola*, a new ascosporic yeast with potential for biocontrol of postharvest fruit rots. *System Appl Microbiol* 24: 393–399.

Lanza G, Di Martino Aleppo E, Strano MC. 1997. Evaluation of integrated approach to control postharvest green mould of lemons. In: *Proc Cost 914-915, Non-Conventional Methods for the Control of Postharvest Disease and Microbiological Spoilage*, Bologna, Italy, pp. 111–114.

Leibinger W, Breuker B, Hahn M, Mendgen K. 1997. Control of postharvest pathogens and colonization of the apple surface by antagonistic microorganisms in the field. *Phytopathology* 87: 1103–1110.

Ligorio A, Platania G, Schena L, Castiglione V, Pentimone I, Nigro F, Di Silvestro I, Ippolito A. 2007. Control of table grape storage rots by combined applications of antagonistic yeasts, salts and natural substances. In: *Proc COST 924, Novel Approaches for the Control of Postharvest Diseases and Disorders*, Bologna, Italy, pp. 124–128.

Lima G, Arru S, De Curtis F, Arras G. 1999. Influence of antagonist, host fruit, and pathogen on the control of postharvest fungal diseases by yeasts. *J Indust Microbiol Biotechnol* 23: 223–229.

Lima G, Castoria R, De Curtis F, Raiola A, Ritieni A, De Cicco V. 2011. Integrated control of blue mould using new fungicides and biocontrol yeasts lowers levels of fungicide residues and patulin contamination in apples. *Postharv Biol Technol* 60: 164–172.

Lima G, De Cicco V. 2006. Integrated strategies to enhance biological control of postharvest diseases. In: *Advances in Postharvest Technologies for Horticultural Crops*. eds N Berkebliaand, N Shiomi. Research Signpost, Kerala, India, pp. 173–194.

Lima G, De Curtis F, Castoria R, De Cicco V. 1998. Activity of the yeasts *Cryptococcus laurentii* and *Rhodotorula glutinis* against postharvest rots on different fruits. *Biocontrol Sci Technol* 8: 257–267.

Lima G, De Curtis F, Castoria R, De Cicco V. 2003. Integrated control of apple postharvest pathogens and survival of biocontrol yeasts in semi-commercial conditions. *Eur J Plant Pathol* 109: 341–349.

Lima G, De Curtis F, De Cicco V. 2008. Interaction of microbial biocontrol agents and fungicides in the control of postharvest diseases. *Stewart Postharv Rev* 1: 1–7.

Lima G, De Curtis F, Piedimonte D, Spina AM, De Cicco V. 2006. Integration of biocontrol yeast and thiabendazole protects stored apples from fungicide sensitive and resistant isolates of *Botrytis cinerea*. *Postharv Biol Technol* 40: 301–307.

Lima G, Ippolito A, Nigro F, Salerno M. 1997. Effectiveness of *Aureobasidium pullulans* and *Candida oleophila* against postharvest strawberry rots. *Postharv Biol Technol* 10: 169–178.

Lima G, Spina AM, Castoria R, De Curtis F, De Cicco V. 2005. Integration of biocontrol agents and food-grade additives for enhancing protection of apples from *Penicillium expansum* during storage. *J Food Protect* 68: 2100–2106.

Lindow SE, Brand MT. 2003. Microbiology of the phyllosphere. *Appl Environ Microbiol* 69: 1875–1883.

Liu J, Sui Y, Wisniewski M, Droby S, Liu Y. 2013. Utilization of antagonistic yeasts to manage postharvest fungal diseases of fruit. *Int J Food Microbiol* 167: 153–160.

Lübeck M, Knudsen I, Jensen B, Thrane U, Janvier C, Jensen DF. 2002. GUS and GFP transformation of the biocontrol strain *Clonostachys rosea* IK726 and the use of these marker genes in ecological studies. *Mycol Res* 106: 815–826.

Mari M, Carati A. 1997. Use of *Saccharomyces cerevisiae* with ethanol in the biological control of grey mould on pome fruits. In: *Proc Cost 914-915, Non-Conventional Methods for the Control of Postharvest Disease and Microbiological Spoilage*, Bologna, Italy, pp. 85–91.

Mari M, Neri F, Bertolini P. 2009. Management of important diseases in Mediterranean high value crops. *Stewart Postharv Rev* 5: 1–10.

Massignan L, Lovino RF, De Cillis M, Ligorio A, Nigro F, Ippolito A. 2005. Termoterapia, micro-antagonisti ed elevata umidità relativa per una migliore conservazione delle arance Tarocco. *Frutticoltura* 5: 52–55.

Meng XH, Qin GZ, Tian SP. 2010. Influences of preharvest spraying *Cryptococcus laurentii* combined with postharvest chitosan coating on postharvest diseases and quality of table grapes in storage. *LWT Food Sci Technol* 43: 596–601.

Mlikota Gabler F, Mercier J, Jimenez JI, Smilanick JL. 2010. Integration of continuous bio-fumigation with *Muscodor albus* with pre-cooling fumigation with ozone or sulphur dioxide to control postharvest gray mold of table grapes. *Postharv Biol Technol* 55: 78–84.

Monk J, Young SD, Vink CJ, Winder LM, Hurst MRH. 2011. q-PCR and high-resolution DNA melting analysis for simple and efficient detection of biocontrol agents. In: *Paddock to PCR: Demystifying Molecular Technologies for Practical Plant Protection*. eds HJ Ridgway, TR Glare, SA Wakelin. New Zealand Plant Protection Society, New Plymouth, pp. 117–124.

Morales H, Sanchis V, Usall J, Ramos AJ, Marín S. 2008. Effect of biocontrol agents *Candida sake* and *Pantoea agglomerans* on *Penicillium expansum* growth and patulin accumula-tion in apples. *Int J Food Microbiol* 122: 61–67.

Nigro F, Finetti Sialer MM, Gallitelli D. 1999. Transformation of *Metschnikowia pulcherrima* 320, biocontrol agent of storage rot, with the green fluorescent protein gene. *J Plant Pathol* 81: 205–208.

Nix-Stohr S, Burpee LL, Buck JW. 2008. The influence of exogenous nutrients on the abun-dance of yeasts on the phylloplane of turfgrass. *Microbial Ecol* 55: 15–20.

Nunes C, Usall J, Teixidó N, Miró M, Viñas I. 2001. Nutritional enhancement of biocontrol activity of *Candida sake* CPA-1 against *Penicillium expansum* on apples and pears. *Eur J Plant Pathol* 107: 543–551.

Omar I, O'Neill TM, Rossall S. 2006. Biological control of Fusarium crown and root rot of tomato with antagonistic bacteria and integrated control when combined with the fungi-cide carbendazim. *Plant Pathol* 55: 92–99.

Ongena M, Jacques P. 2008. Bacillus lipopeptides: Versatile weapons for plant disease biocon-trol. *Trends Microbiol* 16: 115–125.

Palou L, Smilanick AL, Droby S. 2008. Alternatives to conventional fungicides for the control of citrus postharvest green and blue moulds. *Stewart Postharv Rev* 2: 1–16.

Paulsen IT, Press CM, Ravel J. 2005. Complete genome sequence of the plant commensal *Pseudomonas fluorescens* Pf-5. *Nat Biotechnol* 23: 873–878.

Perazzolli M, Roatti B, Bozza E, Pertot I. 2011. *Trichoderma harzianum* T39 induces resis-tance against downy mildew by priming for defense without costs for grapevine. *Biol Control* 58: 74–82.

Petti C, Khan M, Doohan F. 2010. Lipid transfer proteins and protease inhibitors as key factors in the priming of barley responses to Fusarium head blight disease by a biocontrol strain of *Pseudomonas fluorescens*. *Funct Integrat Genomics* 10: 619–627.

Poleatewich AM, Ngugi HK, Backman PA. 2012. Assessment of application timing of *Bacillus* spp. suppress pre- and postharvest diseases of apple. *Plant Disease* 96: 211–220.

Poppe L, Vanhoutte S, Höfte M. 2003. Modes of action of *Pantoea agglomerans* CPA-2, an antagonist of postharvest pathogens on fruits. *Eur J Plant Pathol* 109: 963–973.

Sanzani SM, Ippolito A. 2011. State of the art and future prospects of alternative control means against postharvest blue mould of apple: Exploiting the induction of resistance. In: *Fungicides—Beneficial and Harmful Aspects*. ed N Thajuddin. InTech, Croatia, pp. 117–132.

Sanzani SM, Li Destri Nicosia MG, Faedda R, Cacciola SO, Schena L. 2014. Use of quantita-tive PCR detection methods to study biocontrol agents and phytopathogenic fungi and oomycetes in environmental samples. *J Phytopathol* 162: 1–13.

Sanzani SM, Nigro F, Mari M, Ippolito A. 2009. Innovation in the management of postharvest diseases. *Arab J Plant Protect* 27: 240–244.

Sanzani SM, Schena L, De Cicco V, Ippolito A. 2012. Early detection of *Botrytis cinerea* latent infections as a tool to improve postharvest quality of table grapes. *Postharv Biol Technol* 68: 64–71.

Schena L, Finetti Sialer M, Gallitelli D. 2002. Molecular detection of the strain L47 of *Aureobasidium pullulans*, a biocontrol agent of postharvest diseases. *Plant Disease* 86: 54–60.

Schena L, Li Destri Nicosia MG, Sanzani SM, Faedda R, Ippolito A, Cacciola SO. 2013. Development of quantitative PCR detection methods for phytopathogenic fungi and oomycetes. *J Plant Pathol* 95: 7–24.

Schena L, Nigro F, Soleti Ligorio V, Yaseen T, Ippolito A, El Ghaouth A. 2005. Biocontrol activity of Bio-coat and Biocure against postharvest rots of table grapes and sweet cherries. *Acta Hortic* 682: 2115–2120.

Sharma N, Verma U, Awasthi P. 2006. A combination of the yeast *Candida utilis* and chitosan controls fruit rot in tomato caused by *Alternaria alternata* (Fr. Keissler) and *Geotrichum candidum* Link ex Pers. *J Hortic Sci Biotechnol* 81: 1052–1056.

Shi J, Liu A, Li X, Chen W. 2011. Identification and antagonistic activities of an endophytic bacterium MGP3 isolated from papaya fruit. *Acta Microbiol Sinica* 51: 1240–1247.

Snowdon AL. 1990. *A Color Atlas of Postharvest Diseases and Disorders of Fruit and Vegetables*. Vols. 1 and 2. CRC Press, Boca Raton, FL.

Southwell RJ, Brown JF, Welsby SM. 1999. Microbial interactions on the phylloplane of wheat and barley after applications of mancozeb and triadimefon. *Aust Plant Pathol* 28: 139–148.

Spadaro D, Piano S, Duverney C, Gullino ML. 2002. Use of microorganisms, heat treatment, and natural compounds against Botrytis rot on apple. In: *Proc 2nd Internat Conf Alternative Control Methods against Plant Pests and Diseases*, AFPP, Paris, pp. 446–453.

Spadaro D, Sabetta W, Acquadro A, Portis E, Garibaldi A, Gullino ML. 2008. Use of AFLP for differentiation of *Metschnikowia pulcherrima* strains for postharvest disease biological control. *Microbiol Res* 163: 523–530.

Stockwell VO, Johnson KB, Loper JE. 1998. Establishment of bacterial antagonists of *Erwinia amylovora* on pear and apple blossoms as enhanced by inoculum preparation. *Phytopathology* 88: 506–516.

Teixidó N, Usall J, Nunes C, Torres R, Abadias M, Viñas I. 2010. Preharvest strategies to control postharvest diseases in fruits. In: *Postharvest Pathology*. eds D Prusky, ML Gullino. Springer, London, pp. 89–106.

Tian S, Fan Q, Yong X, Liu HB. 2002. Biocontrol efficacy of antagonist yeasts to gray mold and blue mold on apples and pears in controlled atmospheres. *Plant Disease* 86: 848–853.

Tian SP, Qin GZ, Xu Y. 2005. Synergistic effects of combining biocontrol agents with silicon against postharvest diseases of jujube fruit. *J Food Protect* 68: 544–550.

Usall J, Smilanick J, Palou L. 2008. Preventive and curative activity of combined treatments of sodium carbonates and *Pantoea agglomerans* CPA-2 to control postharvest green mold of citrus fruit. *Postharv Biol Technol* 50: 1–7.

Vallance J, Le Floch G, Déniel F, Barbier G, Lévesque CA Rey P. 2009. Influence of *Pythium oligandrum* biocontrol on fungal and Oomycete population dynamics in the rhizosphere. *Appl Environ Microbiol* 75: 4790–4800.

Vero S, Garmendia G, González MB, Bentancur O, Wisniewski M. 2013. Evaluation of yeasts obtained from Antarctic soil samples as biocontrol agents for the management of postharvest diseases of apple (*Malus × domestica*). *FEMS Yeast Res* 13: 189–199.

Walter M, Frampton CM, Body-Wilson KSH, Harris-Virgin P, Walpara NW. 2007. Agrichemical impact on growth and survival of non-target apple phyllosphere microorganisms. *Can J Microbiol* 53: 45–55.

Wang Y, Wang P, Xia J, Yu T, Luo B, Wang J, Zheng X. 2010. Effect of water activity on stress tolerance and biocontrol activity in antagonistic yeast *Rhodosporidium paludigenum*. *Int J Food Microbiol* 143: 103–108.

Wilson CL, Pusey PL. 1985. Potential for biological control of postharvest plant diseases. *Plant Disease* 69: 375–378.

Wilson CL, Wisniewski M, Droby S, Chalutz E. 1993. A selection strategy for microbial antagonist to control postharvest diseases of fruits and vegetables. *Sci Hortic* 53: 183–189.

Wisniewski M, Wilson C, Droby S, Chalutz E, El Ghaouth A, Stevens C. 2007. Postharvest biocontrol: New concepts and applications. In: *Biological Control: A Global Perspective.* eds C Vincent, MS Goettal, G Lazarovits. CABI, Cambridge, pp. 262–273.

Wittig HPP, Johnson KB, Pscheidt JW. 1997. Effect of epiphytic fungi on brown rot blossom blight and latent infections in sweet cherry. *Plant Disease* 81: 383–387.

Xu L, Du Y. 2012. Effects of yeast antagonist in combination with UV-C treatment on postharvest diseases of pear fruit. *BioControl* 57: 451–461.

Yao HJ, Tian SP. 2005. Effects of a biocontrol agent and methyl jasmonate on postharvest diseases of peach fruit and the possible mechanisms involved. *J Appl Microbiol* 98: 941–950.

Zhang H, Fu C, Zheng X, Su D. 2006. Postharvest control of blue mold rot of pear by microwave treatment and *Cryptococcus laurentii. J Food Eng* 77: 539–544.

5 Physical and Chemical Control of Postharvest Diseases

Alice Spadoni, Fiorella Neri, and Marta Mari

CONTENTS

5.1 INTRODUCTION

Fruit and vegetables are highly perishable produce and, after harvest, are subjected to considerable quantitative losses. Since they are rich in water and nutrients, such produce are ideal substrates during storage for the development of pathogenic microorganisms such as *Penicillium* spp., *Monilinia* spp. and *Botrytis cinerea*. Waste, produced during the postharvest phase, also represents an important economic loss, considering the added value of fruit after harvest, storage, transport, and commercialization. The extent of postharvest losses varies, depending on the commodities and countries in question, and although little uptodate data is available, losses can be estimated to range from a minimum of 10%–15% in countries with advanced technologies to over 50% in developing countries (Wilson and Wisniewski, 1989). In addition, qualitative losses such as loss in edibility, nutritional quality, caloric value, and consumer acceptability of produce are much more difficult to assess than quantitative losses (Kader, 2005). Moreover, some postharvest pathogenic fungi such as *Penicillium* spp. and *Aspergillus* spp. represent a serious concern to human health since they are producers of several toxic compounds (e.g., patulin, citrinin, chaetoglobosins, ochratoxin A) that can affect the safety of both fresh fruit and fruit-based products (Andersen et al., 2004; Guzev et al., 2008).

The natural resistance of fruit to disease declines with ripeness and storage duration, although the use of appropriate postharvest technologies has greatly reduced losses from production to retail in the more developed countries (Kader, 2005). Fungicide treatments still remain one of the most effective methods to reduce postharvest decay, since they protect the fruit from infections occurring before treatment, including quiescent infections, as well as from infections during storage, handling, and marketing (Adaskaveg and Forster, 2010). However, the repeated and continuous use of fungicides has led to a strong selection pressure in pathogen populations, resulting in the development of resistance to some common fungicides like benzimidazole, demethylation inhibitors, and dicarbossimide. Moreover, in the last 25 years, concern about the effect of chemicals on human health and the environment has intensified consumer demand for fruit and vegetables with low fungicide residues and grown with sustainable agriculture, integrated crop management and organic production (Directive2009/128/EC) practices. Multiple retailers have identified the occurrence of pesticide residues as one of the prime concerns of consumers regarding fresh produce, leading them to pursue a policy of residue reduction to 3–4 active ingredients or to their complete elimination (Cross and Berrie, 2008). In this context, considerable efforts are devoted by researchers to finding safer methods for disease control. This review deals with the substantial progress achieved in the use of physical and chemical control measures, and also considers constraints and obstacles that make their widespread diffusion and practical application difficult.

5.2 PHYSICAL CONTROL OF POSTHARVEST DISEASES

5.2.1 HEAT TREATMENTS

Heat treatments to control postharvest disease are usually applied at temperatures above 40°C for a relative short time (usually 55–60°C for 20–60 s, or 45–50°C for 5–20 minutes). The inactivation of fungal propagules usually increase with temperature and the duration of the treatment; however, excessive temperatures should be avoided due to detrimental effects on produce quality (i.e., phytotoxic symptoms, discoloration, water loss) and short treatment durations are preferable in packing houses. The choice of the most suitable heat treatment to control postharvest pathogens is related to the thermal sensitivity of the commodity and the target pathogen, and is also influenced by the location and form of the pathogen in or on the host (Barkai-Golan and Phillips, 1991; Lurie, 1998; Janisiewicz and Conway,

2010; Vigneault et al., 2012). Hot-water treatments were first reported in 1922 to control decay on citrus fruit (Fawcett, 1922), but extensive reviews have subsequently dealt with the possibility of using heat to control quarantine insects, prolong the shelf-life of fruit and vegetables, and to prevent the development of disorders triggered by cold-storage, like chilling injury (Shellie and Mangan, 2000; Fallik, 2004). Heat treatments can be applied to fruit and vegetables in the form of hot water, vapor heat, hot air or hot water rinse-brushing. Hot-water treatment appears to be one of the most effective and promising methods and is especially useful for organic crops to control relatively high rates of postharvest decay in environmentally-friendly ways (Mari et al., 2007; Karabulut et al., 2010; Liu et al., 2012, 2013). The activity of hot-water dipping (HWD) depends on at least two components: the first is a direct and lethal action of heat on fungal inoculum as spores and mycelium present on the fruit surface or in the first layers of peel; the second component could be an indirect action of heat on the host, mediated by a stress-induced response of fruit (Maxin et al., 2012), including resistance to disease. The efficacy of fungal eradication depends on genetic differences among fungi. For example, *M. fructicola*, is more heat sensitive than *P. expansum* (Barkai-Golan and Phillips, 1991) but more resistant than *M. laxa* and *M. fructigena* (Spadoni et al., 2013). In terms of host response, several studies have shown that heat treatments can induce a stress response in fruit (Liu et al., 2012); however, more investigations are required to better elucidate the molecular mechanisms involved.

It has been reported that appropriate heat treatments can delay ripening in several fruit species (Paull, 1990; Lurie, 1998; Martinez and Civello, 2008), resulting in lower softening rates in fruit and was found to have extended the postharvest life of nectarine (Fruk et al., 2012), peach (Bustamante et al., 2012), papaya (Li et al., 2013), banana (Promyou et al., 2008), and citrus (Yun et al., 2013). Heat treatments are also effective for better preparing peach fruit for subsequent storage (Lauxmann et al., 2012), avoiding or reducing the effect of some abiotic or biotic stress on fruit quality. At the molecular level, the main factor involved seems to be the expression of heat shock genes encoding different heat shock proteins (HSPs). It is known that HSPs perform a critical function in refolding partially denatured protein, completing degradation of denatured proteins, assisting the *ex novo* protein synthesis and averting protein aggregation (Aghdam et al., 2013). HWD might be a good solution to improve fruit postharvest life but some negative aspects have to be considered. As reported by Schirra and D'hallewin (1997), prestorage dipping of "Fortune" mandarins in water at 50–54°C significantly inhibited chilling injury but increased the fruit's weight loss. Moreover, while this method has the potential to reduce the natural inoculum presents on the fruit surface, it is not always compatible with fresh-cut plant foods since it may reduce consumer acceptability due to undesirable changes of flavor, texture, color and nutritional quality (Maghoumi et al., 2013). Although HWD requires specialized equipment, it might result in substantial economic saving since, in the industrial packing house, the heating of the water can be obtained by refrigerant gas coming from the cooling plant of the stores; indeed, the heat of this gas is currently usually dispersed in the environment through the condensers.

A more recent technology is the hot-water rinsing and brushing (HWRB). HWRB treatments are extensively studied because they require a higher temperature and a shorter exposure time than traditional hot-water dips. HWRB treatments could not only remove the heavy dirt, pesticides and fungal spores on freshly harvested produce but they could also improve general produce appearance and maintain produce quality (Fallik, 2004). The method involves the fruit rolling over brushes directly into the pressurized recycled hot-water rinse at temperatures between 48°C and 63°C for 10–20 s (Fallik, 2004). HWRB treatments possess the ability to remove fungal pathogens from the fruit surface as a result of the brushing effect, while the natural wax platelets could be melted and smoothed to seal stomata openings on the fruit surface (Fallik, 2004). It has been demonstrated that the use of HWRB at 55°C for 12 s for sweet peppers is a practical strategy for reducing weight loss, chilling injury and softening and preserving nutritional quality, especially antioxidant activity (Ilić et al., 2012). However, this methodology did not have such good results as HWD when applied to apple. In fact, the main advantage of hot-water immersion is a uniformly consistent temperature profile throughout the treatment tank at or slightly above the set point temperature, while in the case of HWRB a possible nonuniform coverage of fruit by hot water could result in limited pathogen control. However, HWRB has the potential to become a sustainable alternative for disease control in fruit because it is less costly than HWD and because its shorter treatment time enables it to be integrated into existing fruit-grading (Maxin et al., 2012).

Heat can also be applied for an extended period of time at lower temperatures. Hot-air treatment, called "curing," consists of holding the fruit at 30 to 37°C and at high relative humidity (90%–98%) for 65–72 h. Forced hot air is preferred in many countries for the development of quarantine procedures (Vigneault et al., 2012). However, the method based on curing treatments has also been evaluated to control green and blue molds on citrus fruit (Zhang and Swingle, 2005; Montesinos-Herrero et al., 2011), proving an effective alternative to fungicides. In apple fruit, hot-air treatment at 38°C for four days has been considered the optimum to preserve postharvest storage quality. It can delay ripening and maintain firmness of the apple fruit to improve consumer acceptability, also controlling the development of common postharvest diseases caused by *P. expansum*, *B. cinerea*, and *C. acutatum* (Klein et al., 1997; Conway et al., 2005; Shao et al., 2007). As recently reported by Zhang et al. (2013), the application of curing to tomato fruit (38°C for 12 h) activated arginine catabolism and consequently reduced the susceptibility to chilling injury. In addition, forced-hot-air treatment for prolonging postharvest shelf-life of mango did not cause injury to the fruit but the moisture content of the heating air differentially modulated the postharvest ripening; in particular, moist air temporarily slowed down the ripening process of mangoes (Ornelas-Paz and Yahia, 2013). Curing can also induce increased residual activity (Shao et al., 2009) and enhance the wound-healing process (Shao et al., 2010) in controlling fungal diseases in "Fuji" apple fruit. However, some negative impacts, including enhanced yellowing of peel, reduced titratable acidity and weight loss were observed after heat treatment in apples. These three factors are regarded as the signals of senescence for fruit. In a previous study, Lurie and Klein (1992) suggested that these adverse outcomes of heat treatment could be reduced when combined with a calcium dip. A specific curing application to kiwifruits against stem-end rot caused by *B. cinerea* is considered an important tool in view of a reduction in fungicide treatment. In this case, curing involves keeping fruits at ambient temperature for at least two days before cold storage (Mari et al., 2009). During the early phases of wound-healing, active phenol metabolism was observed in the fruits and it probably provided the monomers required for the suberization of picking wounds, the main infection site of *B. cinerea* in kiwifruits. The notable increase in phenylalanine ammonia-lyase and polyphenolxidase activity shown by cured fruits, respectively indicated the activation of a resistance mechanism and an oxidative phenolic polymerization, which further increased tissue resistance to the pathogen (Ippolito et al., 1995). Currently, for grey mold management in kiwifruit, curing is carried out directly in cold-storage rooms, decreasing the temperature from 10 to 1°C in eight days and delaying establishment of the controlled atmosphere regimes to 30–40 days after harvesting. This method reduced the incidence of stem-end rot but no negative effects were observed in fruit quality (Brigati et al., 2003).

Heat-based treatment has advantages over other nonconventional control methods of fruit disease such as natural and GRAS (generally recognized as safe) compounds (described and discussed in Section 5.3.1 below), since it does not require any registration from the European Community and ready to use. In addition, HWD or curing appear particularly attractive to the fruit industry since it could be immediately utilized and incorporated into handling practices before storage and without extensive technical modifications.

5.2.2 Irradiation

Irradiation involves exposing the produce to controlled levels of radiation for a specified period. The effects of irradiation on food have been extensively studied for many years (Burton et al., 1959) and are well reviewed by Arvanitoyannis et al. (2009). Only in the last few years, due to the strong demand for reducing the application of chemical fungicides to fruits and vegetables, investigations on the use of different types of irradiation have been increasing and radiation technology has now been shown to be effective in reducing postharvest losses and controlling the stored produce against insects and microorganisms. Internationally, the World Health Organization (WHO), the Food and Agriculture Organization (FAO), and the International Atomic Energy Agency in Vienna (El-Samahy et al., 2000) have considered food irradiation a safe and effective technology. Among the ionizing radiation technologies, gamma ray irradiation has been widely investigated on different fruits and vegetables against postharvest fungal pathogens as *B. cinerea* (Jung et al., 2014), *P. expansum* (Mostafavi et al., 2011) and *Colletotrichum gloeosporioides* (Cia et al., 2007); these generally show a dose rate limitation to control postharvest fruit diseases. However, depending on the irradiation dose, changes in product quality have been observed. Doses above 2 kGy of gamma irradiation are necessary to control some postharvest pathogens, but "Fuji" apple and "Niitaka" pear irradiated at the doses above 0.8 kGy showed an alteration in the physiological progress of firmness (Jung et al., 2014). In addition, sensory qualities of irradiated grapefruit were comparable to the untreated control fruit after 35 days' storage only when the dose of the gamma irradiation was 0.4 kGy (Patil et al., 2004), while the appearance of ten citrus cultivars were negatively affected by the loss of glossiness after treatment with 0.45 kGy (Miller et al., 2000). In order to overcome this issue, the integration of gamma irradiation at low doses with other treatments such as BCAs was explored, showing a suitable method to sustain fruit quality and reduce fruit losses during storage (Mostafavi et al., 2013).

The use of nonionizing radiation has also been of interest. The application of UV-C radiation has been shown to control *P. expansum* of apples (de Capdeville et al., 2002), postharvest rot of strawberries (Marquenie et al., 2002) and Rhizopus soft rot of tomatoes (Stevens et al., 1998). The activity of UV-C, assayed in *in vitro* trials, showed an inhibition of mycelial growth of *Monilinia* spp. after long-wave (320–380 nm) UV exposure (De Cal and Melgarejo, 1999) but had no effect against brown rot after UV-C light treatment was observed in cherries (Marquenie et al., 2002) and in peaches (Bassetto et al., 2007). In addition, radio frequency energy was proposed not only to heat foods but also to disinfect commodities and to control stonefruit postharvest diseases such as brown rot (Casals et al., 2010). The effect of UV radiation has been associated with both germicidal properties as well as physiological response such as activation of defense mechanisms. Charles et al. (2009) found changes in the protein content and profile of tomato fruit treated with UV-C (3.7 kJ/m^2) at the mature green stage. In particular, the results showed the synthesis repression of two proteins detected in senescing control fruit and the induction of stress- and/or pathogenesis-related proteins—both aspects could be involved in the reduction of *B. cinerea* incidence. Moreover, UV-C can affect quality attributes in relation to species and doses: treated papaya maintained a higher firmness than the control (Zhao et al., 1996) while strawberry and apple firmness decreased as the irradiation dose increased (Yu et al., 1996; Drake et al., 1999).

5.3 CHEMICAL CONTROL OF POSTHARVEST DISEASES

5.3.1 GENERALLY REGARDED AS SAFE

Chemical compounds with low toxicity and generally recognized as safe (GRAS) compounds have received increasing interest for the control of postharvest fruit disease. GRAS is a US Food and Drug Administration designation and denotes a substance added to food that is considered safe by experts and for this reason is exempt from the usual Federal Food, Drug, and Cosmetic Act food additive tolerance requirements (Senti, 1981). GRAS compounds are allowed with very few restrictions for many industrial and agricultural applications by regulations worldwide and offer a considerable promise in postharvest technology, showing antimicrobial, antifungal, and insecticidal properties (Gregori et al., 2008; Fagundes et al., 2013). Peracetic acid, K-sorb, sodium bicarbonate, and calcium salts are some examples of GRAS widely used in the food industry for leavening, pH control, taste and texture development. They demonstrate a broad spectrum of antimicrobial activity and have been shown to inhibit the development of postharvest diseases such as blue mold apple decay (Janisiewicz et al., 2008), *M. laxa* and *Rhizopus stolonifer* of stonefruits (Mari et al., 2004) and green mold and sour rot of citrus (Smilanick et al., 2008). In spite of interesting results obtained in laboratory and small-scale experiments, commercial usage limitations arise from inconsistent activity and limited persistence, lack of preventive effect, risk of fruit injury (Larrigaudiere et al., 2002; Palou et al., 2002).

However, preharvest salt treatments (a few days before harvest) seem to overcome these issues, completely inhibiting the incidence of decay in citrus fruits (Youssef et al., 2012) and have been shown a higher efficacy or similar efficacy to that of conventional chemical treatments against storage rots of table grapes (Nigro et al., 2006). The antifungal activity of salts against several pathogens is correlated to a direct inhibition of spore germination, germ-tube elongation and production of pectinolytic enzymes (Hervieux et al., 2002). However, indirect effects must also be taken into consideration, such as a possible increase of tissue resistance with structural changes in cell-wall or phytoalexin production induced, for example, by sodium carbonate (Venditti et al., 2005). In addition, the osmotic stress caused by the presence of salts in field applications could contribute to a decrease in epiphytic fungal populations, including *Pencillium* spp. species (Youssef et al., 2012).

Other chemicals, such as chlorine dioxide, hydrogen peroxide, citric acid, and ethanol, are listed as GRAS substances, and their application in the postharvest phase, by fogging, could be very useful, since some fruits like strawberries require that handling and wetting are minimized. Vardar et al. (2012) obtained a significant reduction of postharvest decay in strawberries treated by fogging with hydrogen peroxide (2000 µL/L), which resulted in reduced microbial populations on the surface of the treated fruit and in the storage atmosphere. Moreover, fruit and vegetables are considered major vehicles for transmission of food-borne enteric viruses since they are easily contaminated at pre- and postharvest stages and the sanitizers commonly used are relatively ineffective for removing human norovirus surrogates from fresh produce. Some GRAS compounds, such as polysorbates, were able to achieve a 3-log reduction in virus titre in strawberries and an approximately 2-log reduction in virus titre in lettuce, cabbage, and raspberries (Predmore and Li, 2011).

In 1997, ozone was declared to be GRAS for food contact applications (EPRI Expert Panel, 1997) and since then ozone-based treatment of fresh fruit and vegetables has been used in the postharvest handling industry with satisfactory results. In a preliminary report, the inhibition of conidia of *B. cinerea*, *M. fructicola*, *P. digitatum*, and *R. stolonifer* required an ozone concentration of more than 200 µL/L under humid conditions, and 4000 µL/L under dry conditions (Margosan and Smilanick, 1998). Unfortunately, it was found that the concentrations of ozone that inactivated conidia were relatively high and could not be used without complete containment of the gas and protection of workers. Under conditions where ozone is present during an 8 h workday, gas concentrations cannot exceed 0.075 µL/L (USEPA, 2008). Ozone was extensively tested for the control of table-grape decay. Although it is fungistatic, dose-dependent, and can be phytotoxic at high concentrations (above 5000 µL/L), a treatment with 5000 µL/L ozone in a commercial chamber reduced grey mold incidence from natural inoculum by about 50% after six weeks' storage at 0°C (Gabler et al., 2010). Similarly, kiwifruit continuously exposed for four months to gaseous ozone (0.3 µL/L) showed a delayed development of stem-end rot and a 56% reduction of grey mold incidence. The observed disease suppression strongly suggests that ozone treatments induce resistance of kiwifruit to the pathogen. Measurements of antioxidant substances and antioxidant activity on fruit exposed to ozone for the same time intervals showed a strong negative correlation between disease incidence or severity and phenol content (Minas et al., 2010).

5.3.2 ELECTROLYZED WATER

Electrolyzed water (EW) is generated by the electrolysis of salt solution through an electrolytic cell where the anode and cathode are separated by nonselective membranes. EW was initially developed in Japan as a medical product and was successfully applied as a sanitizer to inactivate microorganisms on food and food processing equipment surfaces (Al-Haq et al., 2002). Several studies have also documented the strong antifungal activity of EW against postharvest diseases such as brown rot in peaches (Guentzel et al., 2010), green mold in tangerine (Whangchal et al., 2010) and blue mold in apple (Okull and Laborde, 2004). In all of these investigations, electrolysis was conducted with the addition of sodium chloride as the electrolyte, with a consequent formation of free-chlorine and chlorinated organic compounds like chloramines, dichloramines and trichloromethanes, creating drawbacks for handlers and consumers. Moreover, free chlorine is quickly inactivated by the heavy inorganic load present in the wash-water of commercial packing houses; therefore, the use in the electrolysis reaction of salts not containing chlorine might be particularly interesting (Fallanaj et al., 2013). Compared to other conventional methods of disinfection, EW reduces the treatment time, is easy to obtain, has very few side effects, and is relatively cheap (Tanaka et al., 1999). In addition, it does not negatively affect the organoleptic properties, color, scent, flavor, or texture of the various food commodities (Al-Haq et al., 2005). However, the main advantage of EW is its safety for humans and the environment; when it comes into contact with organic matter, or is diluted with tap-water or water produced by reverse osmosis, it reverts to normal tap water (Huang et al., 2008). In 2002, Japan officially approved EW as a food additive (Al-Haq et al., 2005).

The main mechanism of action of EW relates to oxidation that could damage cell membranes, create disruption in cell metabolic processes and, essentially, kill the cell. Spore treatment with EW involves cell structural changes as well as reactive oxygen species (ROS) accumulation, mitochondrial membrane integrity and ATP production. A significant increase of ROS accumulation was observed in *P. digitatum* spores exposed to the electrolyzed salt solution for 15 min. However, other factors can be involved in the production of ROS, such as pH increase; in fact, the pH gradient was in the range 8.5–9.0 after the electrolysis process (Fallanaj, 2012).

5.4 NATURAL COMPOUNDS

5.4.1 VOLATILE ORGANIC COMPOUNDS FROM PLANTS

Plants produce an amazing diversity of secondary metabolites (more than 100,000 have been identified and at least 1700 are volatile) having a wide range of biological activities (Boulogne et al., 2012; Bitas et al., 2013). Some secondary metabolites occurring in plants in their active forms or as inactive precursors, are associated with the defense system and have shown potential for the control of postharvest diseases. Many have been shown to directly inhibit pathogens by disrupting or damaging fungal membranes (Fallik et al., 1998; Arroyo et al., 2007), but some, such as flavonoids, enhanced the defense responses of the host (Sanzani et al., 2010).

Plant volatile organic compounds (VOCs) are substances with low molecular weight and high vapor pressure that are naturally emitted by different organs (i.e., leaves, buds, flowers, fruits, bark, wood, roots). Their high volatility at ambient temperature makes them suitable for postharvest biofumigation, a technique for disease-control that offers the advantage of minimal handling and absence of fruit-wetting. Studies carried out in the last 20 years have produced significant progress in our knowledge of the antifungal activity of plant metabolites, and more than 20 volatile compounds from edible plants (spices, herbs, fruits, vegetables) were found to be particularly interesting as novel means for decay-control because of their safety at low concentrations. Most of these compounds are also widely used as food additives, and the Joint FAO/WHO Expert Committee on Food Additives expressed no safety concerns at current levels of intake for allyl-isothiocyanate (AITC), *p*-anisaldehyde, carvacrol, (–) carvone, *trans*-cinnamaldehyde, hexanal, *trans*-2-hexenal, 2-nonanone, terpineol and thymol, when used as flavoring agents. Different forms of application (liquid or vapor phase) and measurements of pathogen inhibition (mycelial growth and/or conidial germination) applied in the studies often make it difficult to compare the minimal inhibitory concentrations (MICs) obtained. The most consistent fungicidal activity by plant bioactive compounds was found with some isothiocyanates (ITCs), followed by *trans*-2-hexenal, *trans*-2-nonenal, carvacrol, thymol, citral and *trans*-cinnamaldehyde and they have been reviewed by Mari et al. (2011). The main factors involved in the antimicrobial activity of the compounds proved to be functional groups, hydrophobicity and vapor pressure (Andersen et al., 1994; Caccioni et al., 1997; Arfa et al., 2006). The *in vitro* inhibition by plant compounds has not always been confirmed in *in vivo* assays. Besides chemical characteristics, other factors proved to influence the effectiveness of antifungal compounds in disease-control, including treatment conditions (form of application, concentration, temperature, exposure time, time of treatment, formulation) and characteristics of the pathogen (age and form of infection structures, location of pathogen in host tissue). Different levels of sensitivity to treatments were found among fruit species or cultivars, and detrimental effects on sensory traits (odor, texture, and flavor) or phytotoxic symptoms on fruit have been observed in some studies with treatments effective in disease control (Vaughn et al., 1993; Neri et al., 2006c, 2007; Mehra et al., 2013).

5.4.2 ISOTHIOCYANATES

Isothiocyanates (ITCs) derive mainly from hydrolysis of *Brassica carinata* defatted meal containing glucosinolates. The ITCs have shown strong activity against a wide range of food pathogens in specific biological tests (Delaquis and Mazza, 1995). Some ITCs are volatile substances and could potentially be successfully employed in treatments such as biofumigation. The postharvest phase, characterized by restricted environment parameters such as temperature, relative, humidity, and composition of atmospheric gas, represents an advantage for biofumigation of fresh fruit before storage. In *in vitro* tests, allyl-isothiocyanate (AITC), a volatile ITC, showed significant inhibition of conidia germination and/or mycelial growth of *M. laxa*, *P. expansum* (Mari et al., 1993), *Fusarium oxysporum* (Ramos-Garcia et al., 2012), *B. cinerea* (Ugolini et al., 2014) while benzyl ITC inhibited *Alternaria alternata* mycelial growth (Troncoso-Rojas et al., 2005). In *in vivo* trials, ITCs were found to be active against numerous postharvest pathogens and on different hosts. However, their activity did not always confirm the results obtained in preliminary *in vitro* tests, showing that the treatment conditions should often be established not only in relation to the active substance and fungal pathogen but also to the fruit and vegetables response to treatment (Mari et al., 2008). While many ITCs have been synthesized, little data has reported on the activity of ITCs produced *in situ*, although their effectiveness was similar; in fact, synthetic and glucosinolate-derived AITC vapors were evaluated against *B. cinerea* on strawberries and no significant differences were found between two origins (Ugolini et al., 2014). This is an important aspect, showing that biofumigation could be used for industrial applications; the use of bio-based chemicals obtained from renewable natural resources also fits well with the goals of a sustainable agriculture system. The mechanism by which ITCs inhibit fungal growth is not yet completely known. Probably, a nonspecific and irreversible interaction of the ITC with the sulphydryl groups, disulphide bonds and amino groups of proteins and amino acids residues can take place (Banks et al., 1986). Despite much data on antifungal, antibacterial, anti-nematode and anti-insect activities of ITCs, only a few investigations reported their effects on treated fruit quality and residue content. The postharvest quality of bell peppers represented by general appearance (absence of phytotoxic symptoms), weight loss, firmness, titratable acidity, and total soluble solids, was not affected by the mixture of ITC treatment (Troncoso-Rojas et al., 2005). Similar results were obtained in strawberries, where total phenolic content and antioxidant capacity estimated in treated and untreated fruits showed no significant difference. In addition, residue analysis performed on fruit at the end of storage (7 days after treatment) showed very low values (<1 mg/kg) (Ugolini et al., 2014).

5.4.3 *TRANS*-2-HEXENAL

trans-2-Hexenal is C_6 an α,β-unsaturated aldehyde naturally occurring in olive oil, tea and most fruits and vegetables. The compound is also used as a flavoring agent and its estimated daily intake in Europe and the United States at 791 and 409 µg/person per day, respectively, is below the threshold of concern (1800 µg/person per day), showing no safety concerns at current levels of intake. The compound is known for its antimicrobial properties and is thought to be involved in the defense mechanism of plants. Its production, together with other C_6 aldehydes and alcohols named "green leafy volatiles", increases rapidly in damaged plant tissues as a result of the activation of the lipoxygenase hydroperoxide lyase enzymatic pathway in response to wounding and herbivore or pathogen attack (Matsui, 2006).

trans-2-Hexenal was shown to have strong antimicrobial activity against many postharvest pathogens (*B. cinerea, C. acutatum, Helminthosporium solani, M. laxa, Pectobacterium atrosepticum, P. expansum*) *in vitro* and *in vivo*, and its inhibition has been found particularly marked against the fungal conidial form (Vaughn et al., 1993; Fallik et al., 1998; Neri et al., 2006a, 2007; Arroyo et al., 2007; Wood et al., 2013). The high electrophilic properties of the carbonyl group adjacent to the double-bond, makes *trans*-2-hexenal particularly reactive with nucleophiles such as protein sulphydryl and amino groups of the pathogens. In a study where *B. cinerea* was exposed to a radio-labeled mixture of cis-2-hexenal and trans-2-hexenal, it was demonstrated that fungal proteins, and particularly proteins of the surface, were targets of the C_6 aldehydes (Myung et al., 2007). C_6 aldehydes were preferentially incorporated into conidia rather than mycelia, a result correlated to the greater sensitivity of spore germination than mycelial growth to trans-2-hexenal. Postharvest exposure of pome fruits to trans-2-hexenal vapor, even for short times (2–4 h at 20°C), was found to significantly reduce the infections of P. expansum, while 8 h treatment was needed to significantly reduce patulin content in "Conference" pears (Neri et al., 2006a,b). Treatment with trans-2-hexenal showed a curative activity up to 72 h in control of blue mold, and cold-storage temperature, after exposure of the fruit to trans-2-hexenal, tested in "Conference" pears, did not affect the activity of the compound (Neri et al., 2006b). The timing of treatment was particularly important. Fumigation with trans-2-hexenal applied immediately after inoculation (2 h) was effective against M. laxa and B. cinerea, but was generally not useful to control P. expansum. In "Golden Delicious" apples, P. expansum and patulin content were greatly reduced by trans-2-hexenal treatment (12.5 µL/L) applied 24 h after inoculation without negative effects on quality traits (Neri et al., 2006c). In contrast, concentrations effective in decay-control caused phytotoxic effects in apricots, peaches, nectarines, "Abate Fetel" pears and strawberries, and they affected fruit flavor in plums, strawberries, "Conference" and "Bartlett" pears and "Royal Gala" apples (Neri et al., 2006b,c, 2007).

The corresponding saturated aldehyde, hexanal was shown to inhibit the growth of several postharvest pathogens in vitro but at concentrations higher than trans-2-hexenal (Caccioni et al., 1995; Neri et al., 2006a, 2007, 2009; Arroyo et al., 2007; Song et al., 2007). Fumigation with hexanal (900 µL/L for 24 h at 20°C) significantly reduced the incidence of decayed fruit in raspberry and blueberry and lesion development in peaches artificially inoculated with M. fructicola (Song et al., 2007, 2008). Although 40 µmol/L of hexanal was effective in killing the majority of P. expansum spores, "Golden Delicious" apples exposed to this concentration of hexanal for 48 h showed symptoms of phytotoxicity (Fan et al., 2006). Continued exposure to hexanal (40–70 µL/L for 7 days) effectively suppressed grey mold in tomato although tomato respiration increased about 50% and fruit reddening was slowed (Utto et al., 2008).

5.4.4 CARVACROL AND THYMOL

Carvacrol, a monoterpenoid phenol with a warm and pungent odor, is the character-impact constituent of oregano essential oils (Origanum vulgare, O. onites, Thymus capitatus) in which it occurs at concentrations of 60%–80%. It exhibits fungicidal activity against a wide range of postharvest pathogens (B. cinerea, G. candidum, M. laxa, N. alba, P. expansum, R. stolonifer) and in particular shows consistent inhibition of mycelial growth (Plotto et al., 2003; Neri et al., 2006a, 2007, 2009). Comparable fungicidal activity was exhibited by the carvacrol isomer thymol (Plotto et al., 2003). This compound has a strongly aromatic, burnt, medicinal odor and occurs mainly in thyme. The antimicrobial activity of carvacrol and thymol has been ascribed to the hydrophobicity of the compounds, the presence of a phenolic hydroxyl group in these molecules and an adequate system of delocalized electrons (double bonds) that allow the OH group to release its proton (Arfa et al., 2006). The chemical structure of these molecules would allow these compounds to act as proton exchangers, reducing the gradient across the cytoplasmic membrane. The resulting collapse of the proton motive force and depletion of the ATP pool eventually lead to cell death. Postharvest fumigation with carvacrol or thymol was effective in controlling *B. cinerea* and *M. fructicola* on cherries (Tsao and Zhou, 2000) and *M. fructicola* on apricots and plums (Liu et al., 2002). However, they caused phytotoxic symptoms and off-flavors in cherries and a firmer texture and phytotoxicity in apricots. Phytotoxicity after carvacrol or thymol exposure was also observed in oranges (Arras and Usai, 2001) and tomatoes (Plotto et al., 2003). The addition of a mixture of carvacrol, thymol and eugenol inside active packaging was effective in reducing decay in table grapes (Guillén et al., 2007) but, as confirmed also in our unpublished trials on cv Italia, fumigation with these compounds caused off-flavors in grapes. Exposure to carvacrol vapours failed to control blue mold in pears and only slightly controlled brown rot in peaches and lenticel rot in apples (Neri et al., 2006a, 2007, 2009).

5.4.5   CITRAL

Citral, which occurs naturally as the isomers neral and geranial, is an acyclic α,β-unsaturated aldehyde mainly contained inside oil glands of lemon and lime peel and in the essential oils of many plants. The acceptable daily intake established for citral by the Joint FAO/WHO Expert Committee on Food Additives (1980) is ≤0.5 mg/kg body weight and the compound is considered a GRAS compound in the United States. Nevertheless, the antifungal activity of citral against several postharvest pathogens has been well documented in *in vitro* trials (Wuryatmo et al., 2003; Palhano et al., 2004; Neri et al., 2006a, 2007, 2009; Zhou et al., 2014). Postharvest fumigation of fruit with citral showed a low degree of efficacy in the control of blue mold and brown rot (Neri et al., 2006a, 2007) and failed to control lenticel rot (Neri, 2009). Treatment with citral caused severe phytotoxicity in tomato fruits, either when evaluated as a pure compound or as the main constituent of lemongrass essential oil (Plotto et al., 2003). As for *trans*-2-hexanal, the fungicidal activity of citral has been ascribed to the high electrophilic properties of the carbonyl group adjacent to the double-bond.

5.4.6   *TRANS*-CINNAMALDEHYDE

trans-Cinnamaldehyde, an aromatic aldehyde with the typical odor of cinnamon, is the main constituent of essential oils of cinnamon bark and cassia bark and leaves. Its fungicidal activity against postharvest pathogens (*Botryodiplodia theobromae*, *C. gloeosporioides*, *C. musae*, *Glicephalotrichum microchlamydosporum*, *M. laxa*, *N. alba*, *P. expansum*, *P. digitatum*, *R. stolonifer*) has been demonstrated in *in vitro* studies (Sivakumar et al., 2002; Utama et al., 2002; Neri et al., 2006a, 2007, 2009). Exposure of rambutan to *trans*-cinnamaldehyde vapor significantly reduced fungal infection without any negative effect on fruit, while treatment in an aqueous solution of *trans*-cinnamaldehyde caused phytotoxic symptoms (Sivakumar et al., 2002).

Vapor treatment with *trans*-cinnamaldehyde failed to control blue mold in pears, brown rot in peaches and lenticel rot in apples (Neri et al., 2006a, 2007, 2009).

5.4.7 ESSENTIAL OILS

Essential oils are extracts, generally obtained by steam distillation or cold-pressing various organs of aromatic plants, and contain a complex mixture of compounds (up to 100) with diverse chemical structures, many of which are volatile. Some essential oils have shown inhibitory activity on postharvest pathogens although at concentrations usually much higher than single-plant bioactive compounds. Most essential oils are characterized by one to three main components, which impart the characteristic odor or flavor to the oil and are generally the bioactive ingredients. The achievement of essential oils with constant composition could be a critical aspect for their practical use since quantity and quality of components of essential oils can vary depending on climate, soil composition, harvest period, chemical polymorphism in populations and method of extraction, and this may influence the antimicrobial properties of essential oils. Instead, the possible synergism in antimicrobial activity among different components, also occurring as minor molecules, could be an advantage in using essential oils. In addition, the mixture of a variety of functional groups could reduce the risk of resistance developed by pathogens.

In studies comparing the antifungal activity of several essential oils, those containing mainly carvacrol (*T. capitatus*, *O. compactum*) or thymol (*T. glandolosus*, *T. vulgaris*, *O. vulgare*, and *Syzygium aromaticum*) have been shown to exhibit the highest inhibitory influences among many postharvest pathogens (*A. citri*, *B. cinerea*, *Geotrichum candidum*, *P. digitatum*, *P. italicum*, *R. stolonifer*) (Daferera et al., 2000; Arras and Usai, 2001; Bouchra et al., 2003; Plotto et al., 2003; Barrera-Necha et al., 2008). Essential oils rich in citral also showed a broad spectrum of antifungal activity (Shahi et al., 2003; Palhano et al., 2004; Lazar-Baker et al., 2011). Different results with the application of essential oils have been found in *in vivo* assays. Exposure to vapors of thyme oil caused severe phytotoxic symptoms on peaches (Arrebola et al., 2010). Among essential oils rich in citral, lemongrass (*Cymbopogon citratus*) completely controlled the development of *P. expansum* and *B. cinerea* infections on apples (Shahi et al., 2003) while it showed a low reduction of *B. cinerea*, *P. expansum* and *R. stolonifer* infections in peaches (Arrebola et al., 2010). In addition, the application of lemon myrtle (*Backhousia citriodora*) essential oil (250 µL/L) reduced the incidence of *M. fructicola* rot only on noninoculated nectarines (Lazar-Baker et al., 2011). Among other essential oils, spray application of laurel oil (3 mg/mL), containing several components (mainly 1,8-cineole, linalool, terpineol acetate and methyl eugenol), showed good antifungal activity against *M. laxa* in peaches while it exhibited less control of *B. cinerea* in kiwifruit and *P. digitatum* in oranges and lemons (De Corato et al., 2010). Spray emulsion of cinnamon (*Cinnamomum zeylanicum*) essential oil on bananas showed better control of crown rot than clove (*S. aromaticum*) essential oil (0.2%) while it failed to control anthracnose disease (Ranasinghe et al., 2005). Vice versa, dip treatment with clove essential oil (50 µg/L) showed a higher efficacy in reducing natural infections than cinnamon oil in papayas (Barrera-Necha et al., 2008). Two plant oil-based fungicides have been recently labeled: Sporatec (Brandt Consolidated, Springfield, IL, containing rosemary, clove and thyme oils) and Sporan (EcoSmart Technologies, Franklin, TN, containing rosemary and wintergreen oils) are now commercially available. However, a study on blueberry to control *A. alternata*, *B. cinerea* and *C. acutatum* diseases showed that only biofumigation with Sporatec resulted in significant reduction of *C. acutatum* disease; in addition, both biofumigants negatively affected the sensory quality of treated blueberry (Mehra et al., 2013). Some promising results in the disease-control of citrus fruit were found by incorporation of essential oils in wax coatings. Incorporation of *Lippia scaberrima* essential oil (2500 μL/L) in Carnauba Tropical coating (du Plooy et al., 2009) or *Cinnamomun zeylanicum* (0.5%) in shellac and/or carnauba (Kouassi et al., 2012) led to a significant reduction of Penicillium disease in orange fruit, without detrimental effects on the fruit. An advantage of using coatings with essential oils could be the close contact between the essential oils and the fruit's surface over a long period.

5.4.8 VOLATILE ORGANIC COMPOUNDS FROM MICROORGANISMS

Recently, increasing interest has also been devoted to biofumigation with VOCs produced by some microorganisms, such as *Muscodor albus* (Mercier and Jiménez, 2004; Mercier and Smilanick, 2005; Gabler et al., 2006), *Candida intermedia* (Huang et al., 2011), *Streptomyces* spp. (Wan et al., 2008; Li et al., 2010, 2012) and *P. expansum* (Rouissi et al., 2013), showing fungicidal or fungistatic activity against a variety of postharvest pathogens. *M. albus* is an endophytic, nonspore producing fungus originally isolated from a cinnamon tree. Biofumigation with VOCs produced by *M. albus*, applied within 24 h from inoculation-controlled brown rot on peach, grey mold on apple and table grape (Mercier and Jiménez, 2004; Gabler et al., 2006), blue mold on apple (Mercier and Jiménez, 2004) and green mold on lemon (Mercier and Smilanick, 2005), while only if the treatment was applied immediately after inoculation did it provide some control of sour rot on lemon (Mercier and Smilanick, 2005). Species of *Streptomyces* are gram-positive, filamentous and spore-forming bacteria and volatiles from *S. globisporus* JK-1 (120 or 240 g/L) significantly reduced the incidence and severity of blue mold in Shatang mandarin (Li et al., 2010) and grey mold in tomato (Li et al., 2012). Although *P. expansum* is a common postharvest pathogen, VOCs emitted by *P. expansum* strain R82 completely inhibited the development of *B. cinerea*, *C. acutatum*, and *M. laxa* (Rouissi et al., 2013). Many species of *Candida* yeasts are effective in the control of fruit disease and VOCs of *C. intermedia* C410 isolated from healthy strawberry leaf were effective in reducing grey mold in strawberry (Huang et al., 2011). Some components of VOCs emitted from these microorganisms are common to different species of fungi, bacteria, yeasts or plants, whereas others seem to be unique for one species or isolate. The antifungal activity of these VOCs seemed to be not related to the relative abundance of the single components of the mixture. Geosmin, for example, is the primary component associated with the musty-earthy odor produced by many fungi and actinomycetes and it has been found to be the most abundant component also of *S. globisporus* JK-1; however, minor components (dimethyl trisulphide, dimethyl disulphide and acetophenone) led to complete inhibition of blue mold in "Shatang" mandarin (Li et al., 2010).

An advantage of the use of volatiles produced by microorganisms is that these VOCs occur as a mixture of compounds belonging to different chemical classes and these mixtures could have synergistic or additive properties that cannot be achieved by any single component alone. Phenylethyl alcohol is probably one of most effective antifungal components, and is common to several microorganisms, including *M. albus*, *P. expansum* R82, *C. intermedia* and *S. globisporus* JK-1. However, when tested as a single component, it completely inhibited *B. cinerea*, *C. acutatum* and *M. laxa* mycelium growth and conidial germination at concentrations 2000 times higher than that naturally released from *P. expansum* R82 in culture (Rouissi et al., 2013). Moreover, the quantity and quality of VOC production by some microorganisms depend on the growth medium, as reported by Huang et al. (2012) for *Sporidiobolus pararoseus* strain YCXT3. The results suggested that *S. pararoseus* is incapable of accumulating highly toxic substances in nutrient yeast dextrose agar and potato dextrose agar media inhibitory to *B. cinerea*, while, when it is cultured on yeast extract peptone agar, it produced VOCs highly effective against both the conidial germination and the mycelial growth of the pathogen. Among 39 VOCs emitted by yeast and identified using gas chromatography-mass, authentic 2-ethyl-hexanol was found to have a strong antifungal activity against *B. cinerea*, suggesting that the strain YCXT3 of *S. pararoseus* is a promising agent for the control of grey mold under air-tight conditions. Disease-control with VOCs produced by microorganisms is still at experimental/levels and focused on antifungal activity, while the evaluation of effects on fruit quality or residues of treatment in fruit have not yet been carried out. These aspects should be taken into consideration, since the volatiles emitted from these microorganisms are characterized by strong odors (for example, the musty-earthy odor conferred by geosmin). The use of VOCs for postharvest disease control of fruit on a commercial scale have also to be evaluated in relation to human safety, and more in-depth investigations on their toxicity for humans have to be performed. In fact, the occurrence of human asthma symptoms, for example, has been related to the emission of 2-ethyl-hexanol from dampness-related alkaline degradation of plasticizer floor material (Norbäck et al., 2000).

5.4.9 CHITOSAN

Chitosan is an edible coating derived from natural sources by deacetylation of chitin and is considered harmless to humans and the environment. It has been studied for efficacy in inhibiting decay and extending the shelf-life of various fruits (Aider, 2010). A large amount of data is available on the effectiveness of chitosan in pre- and postharvest treatments on fresh produce and this was recently reviewed by Shiekh et al. (2013). On temperate fruit such as strawberries, a chitosan coating controlled *Rhizopus* rot and also reduced total aerobic count, coliforms, and weight-loss of fruit during storage (Park et al., 2005). Similar results were obtained for small bunches of table grape dipped in 0.5% and 1% chitosan solutions. The treatment decreased the spread of grey mold infection from a berry to close neighbors (nesting) (Romanazzi et al., 2002). The control of brown rot caused by *Monilinia* spp. was achieved in peach and cherry using the application of 0.1% and 1%, respectively, of chitosan (Li and Yu, 2000; Feliziani et al., 2013). The infections caused by *R. stolonifer* on tomato fruits were inhibited after the application of chitosan, although the severity of soft rot symptoms was not related to the molecular weight of chitosan (Hernández-Lauzardo et al., 2012). *In vitro* trials showed the ability of chitosan to reduce mycelial growth of decay-causing fungi, comparable to the reduction obtained with synthetic fungicides (Feliziani et al., 2013). However, other results showed high levels of anti-oxidants, antioxidant activity, ascorbic acid, glutathione and high activity of β-1,3-glucanase in chitosan-treated strawberries, proving a reinforced microbial defense mechanism of the fruit and an accentuated resistance against fungal invasion (Wang and Gao, 2013).

5.5   COMPOUNDS INDUCING RESISTANCE

Several chemical compounds are known for their ability to induce disease-resistance in treated plants, especially in weedy species, such as cotton (Colson-Hanks et al., 2012), and sunflower (Tosi et al., 1998), but also in woody plants, like grapevines (Reuveni et al., 2001). Recently, inducers of resistance have been tested directly on fruit in order to induce resistance against postharvest fruit pathogens.

β-aminobutyric acid (BABA), a nonprotein amino acid, applied to specific wound sites on the peel surface of grapefruit is able to induce resistance against *P. digitatum* (Porat et al., 2003). Its activity is concentration-dependent, being most effective at a concentration of 20 mM. BABA acts directly against pathogens; in fact, at increasing concentrations of 1–100 mM, it showed a marked reduction in percentage of *P. digitatum* conidia germination and germ-tube elongation, in agreement with other data on *B. cinerea* and *P. expansum* on *in vitro* growth (Quaglia et al., 2011). However, when BABA was used as a postharvest treatment and fruits were dipped in the inducer solution, it had no effect or reduced the development of *P. italicum* and *P. digitatum* of orange only slightly (Moscoso-Ramirez and Palau, 2013). Probably, when BABA is used with the aim of inducing resistance, it does not reduce the percentage of infections or the lesion diameters; however, it can induce the activation of chitinase gene expression and protein accumulation (Porat et al., 2003). Similarly, acibenzolar-S-methyl (ASM) induced a significant increase in the levels of PR-1a (antifungal), PR-2 (b-1,3-glucanase), PR-5 (Thaumatin-like protein), and PR-8 (chitinase) gene transcripts in apple treated with increasing ASM concentrations (Quaglia et al., 2011).

Salicylic acid (SA) is an endogenous hormone having a key role in various species of plant growth. Positive SA effects have been reported for control of *P. expansum* in sweet cherry (Xu and Tian, 2008), grey mold in peach (Zhang et al., 2008) and fungal decay in persimmon fruit (Khademi et al., 2012). A stimulation of the antioxidant enzyme activities reported in sweet cherry treated with SA (2 mM) suggested that the activation of antioxidant defense plays the main role in resistance against *P. expansum* (Xu and Tian, 2008). However, the disease reduction obtained in treated fruit was very low (<15%) and SA cannot be considered satisfactory and recommended for inclusion in postharvest decay management programs for fruit packing houses.

5.6 INTEGRATED METHODS

All the alternative methods (physical and chemical) cited above did not always achieve an acceptable level of postharvest disease-reduction when used individually, and some had a poor effect against future infection, which can occur after treatment, or *vice versa* against established infection before treatment (Droby et al., 2009). To overcome this drawback, integrated strategies have been proposed and widely investigated. Treatments with GRAS such as ammonium molibdate or sodium bicarbonate improved the efficacy of some microbial antagonists (Qin et al., 2006; Torres et al., 2007), although the effectiveness of combined treatments depends upon the mutual compatibility, duration, and time at which they are applied. It is postulated that the enhancement of disease-control is directly caused by the inhibitory effects of the salts on pathogen growth, and indirectly because of the relatively small influence of the GRAS compounds on the growth of the antagonist (Qin et al., 2006). Physical methods like heat could enhance the bioefficacy of microbial antagonists such as *B. subtilis* (Obagwu and Korsten, 2003), *Pseudomonas syringae* (Conway et al., 2005), *Cryptococcus laurentii* (Zhang et al., 2007), providing, in some cases, an equivalent control to synthetic fungicides (Hong et al., 2014).

The combination of HWD and ethanol improved the control of *M. fructicola* in peaches and nectarines compared to HWD and ethanol alone (Margosan et al., 1997). Similarly, a combination of peracetic acid and hot water showed a greater reduction of brown rot in "Mountain Gold" and "Rome Star" peaches inoculated with *M. fructicola* and treated with 40°C-heated peracetic acid (200 mg/L[1]) than that observed in fruit treated only with hot water or peracetic acid (Sisquella et al., 2013). Recently, Dessalegn et al. (2013) demonstrated a high effectiveness in integrating plant defense inducing chemicals (PDIC), inorganic salts and hot-water treatments for the management of postharvest mango anthracnose. Additive and synergistic increases in effectiveness were also observed by integrating heat therapy with various fungicides, thus leading to significant reductions in the application of active ingredients (Schirra et al., 2011). Imazalil (IMZ), a synthetic fungicide employed to control *P. italicum* and *P. digitatum* on citrus, applied at 490 mg/L in aqueous solution at 37.8°C, was found more effective in decay-control than IMZ in a wax mixture at 4200 mg/L sprayed on fruit at ambient temperature (Smilanick et al., 1997). On the other hand, the use of a fungicide in combination with hot water is not always preferable, since pyraclostrobin, a fungicide belonging to the anilinopyrimidine class, showed residues in orange and lemons that were dependent on treatment temperature. They approximately doubled for each 5°C increase in solution temperature above 30°C (Smilanick et al., 2006).

A possible synergistic effect between *Debaromyces hansenii*, an antagonist yeast, and UV-C irradiation in controlling brown rot incidence on both artificially inoculated and naturally infected peaches was observed by Stevens et al. (1997). The superiority of the combined treatment was probably due to the ability of UV-C to control deep-seated infections such as latent infections, whereas the yeast controlled only superficial infections originating in recent wounds. The integrated treatment of gamma irradiation and a biocontrol agent (*Pseudomonas fluorescens*) on "Golden Delicious" apple against blue mold allows the use of irradiation at lower doses (200 and 400 Gy) with a dual benefit of no detrimental effects on fruit quality but significant control of disease (Ahari Mostafavi et al., 2013). Combined heat and UV treatment reduced postharvest decays, and maintained kumquat and orange quality. Heat treatment followed by UV-C radiation was the most effective combination, while, where UV treatment preceded heat application, the elicitation of phytoalexins was inhibited (Ben-Yehoshua et al., 2005).

5.7 CONCLUSIONS

Microbial decay is a major factor responsible for postharvest losses and compromises to the quality of fresh produce. In the past, the use of new fungicides has extended the shelf-life of fresh fruits by reducing losses, but in the last two or three decades, concerns about public health and the environment has considerably limited their use after harvest. Future scenarios are tending increasingly more towards integrated crop management and organic fruit production, with a consequent reduced use of fungicides, for a sustainable agriculture system. This goal requires new technologies to control postharvest disease. Intense research in the last 30 years has produced numerous studies that show significant progress in the reduction of pesticide use for disease control, although some critical points have still to be considered. It is unrealistic to assume that the physical and chemical methods described above have the same fungicidal activity as pesticides, and an integrated approach appears the best method to obtain acceptable results. However, further research is needed to investigate the activity of GRAS compounds, natural compounds and VOCs, in large-scale experiments, their mode of action and their degradation in organisms that are still not fully understood. Physical methods probably have a better chance of prompt application on a commercial scale since some of these, like heat, do not require any registration. Nevertheless, also in this case, more investigations have to provide additional information on appearance, texture, flavors, and storability of treated fruit. Finally, research should lead to the development of appropriate tools to tailor a complete integrated disease management strategy specific for each situation that takes into account factors such as species, climate and seasonal conditions, and the market destination.

REFERENCES

Adaskaveg JE, Forster H. 2010. New developments in postharvest fungicide registrations for edible horticultural crops and use strategies in the United States. In: *Postharvest Pathology Series: Plant Pathology in the 21st Century* (eds D Prusky, ML Gullino). Springer, New York, Vol.2, pp. 107–111.

Aghdam MS, Sevillano L, Flores FB, Bodbodak S. 2013. Heat shock proteins as biochemical markers for postharvest chilling stress in fruits and vegetables. *Sci Hortic* 160: 54–64.

Ahari Mostafavi H, Mirmajlessi SM, Fathollahi H, Shahbazi S, Mirjalili SM. 2013. Integrated effect of gamma radiation and biocontrol agent on quality parameters of apple fruit: An innovative commercial preservation method. *Radiation Phys Chem* 91: 193–199.

Aider M. 2010. Chitosan application for active bio-based films production and potential in the food industry: Review. *LTW-Food Sci Technol* 43: 837–842.

Al-Haq MI, Sea Y, Oshita S, Kawagoe Y. 2002. Disinfection effect of electrolyzed oxidizing water on suppressing fruit rot of pear caused by *Botryosphaeria berengeriana*. *Food Res Int* 35: 657–664.

Al-Haq MI, Sugiyama J, Isobe S. 2005. Applications of electrolyzed water in agriculture and food industries. *Food Sci Technol Res* 11: 135–150.

Andersen RA, Hamilton-Kemp TR, Hildebrand DF, McCracken CT, Collins RW, Fleming PD. 1994. Structure-antifungal activity relationships among volatile C_6 and C_9 aliphatic aldehydes, ketones, and alcohols. *J Agric Food Chem* 42: 1563–1568.

Andersen B, Smedsgaard J, Frisvad JC. 2004. *Penicillium expansum*: Consistent production of patulin, chaetoglobosins, and other secondary metabolites in culture and their natural occurrence in fruit product. *J Agric Food Chem* 52: 2421–2428.

Arfa AB, Combes S, Preziosi-Belloy L, Gontard N, Chalier P. 2006. Antimicrobial activity of carvacrol related to its chemical structure. *Lett Appl Microbiol* 43: 149–154.

Arras G, Usai M. 2001. Fungitoxic activity of 12 essential oils against four postharvest citrus pathogens: Chemical analysis of *Thymus capitatus* oil and its effect in subatmospheric pressure conditions. *J Food Prot* 64: 1025–1029.

Arrebola E, Sivakumar D, Bacigalupo R, Kosten L. 2010. Combined application of antagonist *Bacillus amyloliquefaciens* and essential oils for the control of peach postharvest diseases. *Crop Prot* 29: 369–77.

Arroyo FT, Moreno J, Daza P, Boianova L, Romero F. 2007. Antifungal activity of strawberry fruit volatile compounds against *Colletotrichum acutatum*. *J Agric Food Chem* 55: 5701–5707.

Arvanitoyannis IS, Stratakos AC, Panagiotis T. 2009. Irradiation applications in vegetables and fruits: A review. *Crit Rev Food Sci Nutrit* 49: 427–62.

Banks JG, Board RG, Sparks NHC. 1986. Natural antimicrobial systems and their potential in food preservation of the future. *Biotechnol Appl Biochem* 8: 103–107.

Barkai-Golan R, Phillips DJ. 1991. Postharvest heat treatment of fresh fruits and vegetables for decay control. *Plant Dis* 75: 1085–1089.

Barrera-Necha L, Bautista Banos S, Flores-Moctezuma HE, Estudillo AR. 2008. Efficacy of essential oils on the conidial germination, growth of *Colletotrichum gloeosporioides* (Penz.) Penz. and Sacc and control of postharvest diseases in papaya (*Carica papaya* L.). *Plant Pathol J* 7: 174–178.

Bassetto E, Amorim L, Benato EA, Gonçalves FP, Lourenço SA. 2007. Effect of UV-C irradiation on postharvest control of brown rot (*Monilinia fructicola*) and soft rot (*Rhizopus stolonifer*) of peaches. *Fitopatol Brasil* 32: 393–399.

Ben-Yehoshua S, Rodov V, D'Hallewin G, Dore A. 2005. Elicitation of resistance against pathogens in citrus fruits by combined UV illumination and heat treatments. *Acta Hort* 682: 2013–2020.

Bitas V, Kim HS, Bennett JW, Kang S. 2013. Sniffing on microbes: Diverse roles of microbial volatile organic compounds in plant health. *Mol Plant-Microbe Interact* 26: 835–843.

Bouchra C, Achouri M, Hassani LMI, Hmamouchi M. 2003. Chemical composition and antifungal activity of essential oils of seven Moroccan Labiatae against *Botrytis cinerea* Per: Fr. *J Ethopharm* 89: 165–169.

Boulogne I, Petit P, Ozier-Lafontaine H, Desfontaines L, Loranger-Merciris G. 2012. Insecticidal and antifungal chemicals produced by plants: A review. *Environ Chem Lett* 10: 325–347.

Brigati S, Gualanduzzi S, Bertolini P, Spada G. 2003. Influence of growing techniques on the incidence of *Botrytis cinerea* in cold stored kiwifruit. *Acta Hort* 610: 275–281.

Burton WG, Horne T, Powell DB. 1959. The effect of γ-radiation upon the sugar content of potatoes. *Eur Potato J* 2: 105–111.

Bustamante CA, Budde CO, Borsani J. 2012. Heat treatment of peach fruit: Modifications in the extracellular compartment and identification of novel extracellular proteins. *Plant Physiol Biochem* 60: 35–45.

Caccioni DRL, Gardini F, Lanciotti R, Guerzoni E. 1997. Antifungal activity of natural compounds in relation to their vapour pressure. *Sci Aliments* 17: 21–34.

Caccioni DRL, Deans SG, Ruberto G. 1995. Inhibitory effect of citrus fruit essential oil components on *Penicillium italicum* and *Penicillium digitatum*. *Petria* 5: 177–182.

Casals C, Vinas I, Landl A, Picouet P, Torres R, Usall J. 2010. Application of radio frequency heating to control brown rot on peaches and nectarines. *Postharvest Biol Technol* 58: 218–224.

Charles MT, Tano K, Asselin A, Arul J. 2009. Physiological basis of UV-C induced resistance to *Botrytis cinerea* in tomato fruit. V. Constitutive defense enzymes and inducible pathogenesis-related proteins. *Postharvest Biol Technol* 51: 414–424.

Cia P, Pascholati SF, Benato EA, Camili EC, Santos CA. 2007. Effects of gamma and UV-C irradiation on the postharvest control of papaya anthracnose. *Postharvest Biol Technol* 43: 366–373.

Colson-Hanks ES, Allen SJ, Devrall BJ. 2012. Effect of 2,6-dichloroisonicotinic acid or benzothiadiazole on Alternaria leaf spot, bacterial blight and Verticillium wilt in cotton under field conditions. *Australas Plant Pathol* 29: 170–177.

Conway WS, Leverentz B, Janisiewicz WJ, Saftner RA, Camp MJ. 2005. Improving biocontrol using antagonist mixtures with heat and/or sodium bicarbonate to control postharvest decay of apple fruit. *Postharvest Biol Technol* 36: 235–244.

Cross JV, Berrie AM. 2008. Eliminating the occurrence of reportable pesticide residues in apples. *Agricultural Engineering International: CIGR eJournal*, Manuscript ALNARP 08 004. Vol. X (May). http://cigrjournal.org/index.php/Ejounral/article/view/1240/1097.

Daferera D, Ziogas BN, Polissiou MG. 2000. GC-MS analysis of essential oils from Greek aromatic plants and their fungi toxicity on *Penicillium digitatum*. *J Agric Food Chem* 48: 2576–2581.

de Capdeville G, Wilson CL, Beer SV, Aist JR. 2002. Alternative disease control agents induce resistance to blue mold in harvested "Red Delicious" apple fruit. *Phytopathology* 92: 900–908.

De Cal A, Melgarejo P. 1999. Effects of long-wave UV light on *Monilinia* growth and identification of species. *Plant Dis* 83: 62–65.

De Corato U, Maccioni O, Trupo M, di Sanzo G. 2010. Use of essential oil of *Laurus nobilis* obtained by means of a supercritical carbon dioxide technique against postharvest spoilage fungi. *Crop Prot* 29: 142–147.

Delaquis PJ, Mazza G. 1995. Antimicrobial properties of isothiocyanates in food preservation. *Food Technol* 49: 73–84.

Dessalegn Y, Ayalew A, Woldetsadik K. 2013. Integrating plant defense inducing chemical, inorganic salt and hot water treatments for the management of postharvest mango anthracnose. *Postharvest Biol Technol* 85: 83–88.

Directive2009/128/EC 2009 http://eurlex.europa.eu/LexUriServ/LexUriServ.do?uri = CONS LEG:2009L0128:20091125:EN:PDF

Drake SR, Sanderson PG, Nexen LG. 1999. Response of apple and winter pear quality to irradiation as a quarantine treatment. *J Food Process Preserv* 23: 203–216.

Droby S, Wisniewski M, Macarisin D, Wilson C. 2009. Twenty years of postharvest biocontrol research: Is it time for new paradigm? *Postharvest Biol Technol* 52: 137–145.

du Plooy W, Regnier T, Combrinck S. 2009. Essential oil amended coatings as alternatives to synthetic fungicides in citrus postharvest management. *Postharvest Biol Technol* 53: 117–122.

El-Samahy SK, Youssef BM, Aaskarand AA, Swailam HMM. 2000. Microbiological and chemical properties of irradiated mango. *J Food Saf* 20: 139–156.

EPRI Expert Panel. 1997. Expert Panel Report: Evaluation of the history and safety of ozone in processing food for human consumption. Electric Power Research Institute Pub TR-108026-V1-4827, R&D Enterprises, Walnut Creek, CA.

Fagundes C, Pérez-Gago MB, Monteiro AR, Palou L. 2013. Antifungal activity of food additives *in vitro* and as ingredients of hydroxypropyl methylcellulose-lipid edible coatings against *Botrytis cinerea* and *Alternaria alternata* on cherry tomato fruit. *Int J Food Micro Biol* 166: 391–398.

Fallanaj F. 2012. Activity and mode of action of electrolyzed water for controlling postharvest diseases of citrus fruit. PhD diss., Bari University "Aldo Moro"

Fallanaj F, Sanzani SM, Zavanella C, Ippolito A. 2013. Salts addition to improve the control of citrus postharvest diseases using electrolysis with conductive diamond electrodes. *J Plant Pathol* 95: 373–383.

Fallik E. 2004. Prestorage hot water treatments (immersion, rinsing and brushing). *Postharvest Biol Technol* 32: 125–134.

Fallik E, Archbold DD, Hamilton-Kemp TR, Clements AM, Collins RW. 1998. (E)-2-hexenal can stimulate *Botrytis cinerea* growth *in vitro* and on strawberries *in vivo* during storage. *J Am Soc Hortic Sci* 123: 875–881.

Fan L, Song J, Beaudry RM, Hildebrand PD. 2006. Hexanal vapour on spore viability of *Penicillium expansum*, lesion development on whole apples and fruit volatile biosynthesis. *J Food Sci* 71: M105–109.

FAO/WHO. 1980. Technical Report Series N. 648.

Fawcett HS. 1922. Packinghouse control of brown rot. *Citrograph* 7: 232–234.

Feliziani E, Santini M, Landi L, Romanazzi G. 2013. Pre and postharvest treatments with alternative to synthetic fungicides to control postharvest decay of sweet cherry. *Postharvest Biol Technol* 78: 133–138.

Fruk G, Niševič L, Sever Z, Milićević T, Jemrić T. 2012. Effects of different postharvest heat treatments on decreasing decay, reducing chilling injury and maintaining quality of nectarine fruit. *Agric Conspectus Scientif* 77: 27–30.

Gabler FM, Fassel R, Mercier J, Smilanick JL. 2006. Influence of temperature, inoculation interval, and dosage on biofumigation with *Muscodor albus* to control postharvest gray mold on grapes. *Plant Dis* 90: 1019–1025.

Gabler FM, Smilanick JL, Mansour MF, Karaca H. 2010. Influence of fumigation with high concentrations of ozone gas on postharvest gray mold and fungicide residues on table grapes. *Postharvest Biol Technol* 55: 85–90.

Gregori R, Borsetti F, Neri F, Mari M, Bertolini P. 2008. Effects of potassium sorbate on postharvest brown rot of stone fruit. *J Food Prot* 71: 1626–1631.

Guentzel JL, Lam KL, Callan MA, Emmons SA, Dunham VL. 2010. Postharvest management of gray mold and brown rot on surfaces of peaches and grapes using electrolyzed oxidizing water. *Int J Food Microbiol* 143: 54–60.

Guillén F, Zapata PJ, Martinez-Romero D, Castillo S, Serrano M, Valero D. 2007. Improvement of the overall quality of table grapes stored under modified atmosphere packaging in combination with natural antimicrobial compounds. *J Food Sci* 72: S185–S190.

Guzev L, Danshin A, Zahavi T, Ovadia A, Lichter A. 2008. The effects of cold storage of table grapes, sulphur dioxide and ethanol on species of black *Aspergillus* producing ochratoxin A. *Internat J Food Sci Technol* 143: 1187–1194.

Hervieux V, Yaganza ES, Arul J, Tweddell RJ. 2002. Effect of organic and inorganic salts on the development of *Helminthosporium solani*, the causal agent of potato silver scurf. *Plant Dis* 86: 1014–1018.

Hernández-Lauzardo AN, Guerra-Sánchez MG, Hernández-Rodríguez A, Heydrich-Pérez M, Vega-Pérez J, Velázquez-del Valle MG. 2012. Assessment of the effect of chitosan of different molecular weights in controlling Rhizopus rots in tomato fruits. *Arch Phytopathology Plant Protect* 45: 33–41.

Hong P, Hao W, Luo J, Chen S, Hu M, Zhong G. 2014. Combination of hot water, *Bacillus amyloliquefaciens* HF-01 and sodium bicarbonate treatments to control postharvest decay of mandarin fruit. *Postharvest Biol Technol* 88: 96–102.

Huang R, Che HJ, Zhang J, Yang L, Jiang DH, Li GQ. 2012. Evaluation of *Sporidiobolus pararoseus* strain YCXT3 as biocontrol agent of *Botrytis cinerea* on post-harvest strawberry fruits. *Biol Control* 62: 53–63.

Huang R, Li GQ, Zhang J. 2011. Control of postharvest Botrytis fruit rot of strawberry by volatile organic compounds of *Candida intermedia*. *Phytopathology* 101: 859–869.

Huang YR, Hung YC, Hsu SY, Huang YW, Hwang DF. 2008. Application of electrolyzed water in the food industry. *Food Control* 19: 329–345.

Ilić ZS, Trajković R, Pavlović R, Alkalai-Tuvia S, Perzelan Y, Fallik E. 2012. Effect of the heat treatment and individual value of bell pepper stored at suboptimal temperature. *Int J Food Sci Technol* 47: 83–90.

Ippolito A, Lattanzio V, Nigro F. 1995. Mechanisms of resistance to *Botrytis cinerea* in wounds of cured kiwifruits. *Acta Hort* 444: 719–22.

Janisiewicz WJ, Conway WS. 2010. Combining biological control with physical and chemical treatments to control fruit decay after harvest. *Stewart Postharv Rev* 1: 3.

Janisiewicz WJ, Saftner RA, Conway WS, Yoder KS. 2008. Control of blue mold decay of apple during commercial controlled atmosphere storage with yeast antagonists and sodium bicarbonate. *Postharvest Biol Technol* 49: 374–378.

Jung K, Yoon M, Park HJ, Lee KY, Jeong RD, Song BS, Lee JW. 2014. Application of combined treatment for control of *Botrytis cinerea* in phytosanitary irradiation processing. *Radiat Phys Chem* 99: 12–17.

Kader AA. 2005. Increasing good availability by reducing postharvest losses of fresh produce. *Acta Hortic* 682: 2169–2176.

Karabulut OA, Smilanick JL, Crisosto CH, Palou L. 2010. Control of brown rot of stone fruits by brief heated water immersion treatments. *Crop Prot* 29: 903–906.

Khademi O, Zamani Z, Mostofi Y, Kalantari S, Ahmadi A. 2012. Extending storability of persimmon fruit cv. Karaj by postharvest application of salicylic acid. *J Agric Sci Technol* 14: 1067–1074.

Klein JD, Conway WS, Whitaker BD, Sams CE. 1997. Botrytis cinerea decay in apples is inhibited by postharvest heat and calcium treatments. *J Am Soc Hortic Sci* 122: 91–94.

Kouassi KHS, Bajji M, Jijakli H. 2012. The control of postharvest blue and green molds of citrus in relation with essential oil-wax formulations, adherence and viscosity. *Postharvest Biol Technol* 73: 122–128.

Larrigaudiere C, Pons J, Torres R, Usall J. 2002. Storage performance of Clementines treated with hot water, sodium carbonate and sodium bicarbonate dips. *J Hortic Sci Biotech* 77: 314–319.

Lauxmann MA, Brun B, Borsani J. 2012. Transcriptomic profiling during the post-harvest of heat-treated Dixiland *Prunus persica* fruits: Common and distinct response to heat and cold. *Plos One* 7: e51052.

Lazar-Baker EE, Hetherington SD, Ku VVV, Newman SM. 2011. Evaluation of commercial essential oil samples on the growth of postharvest pathogen *Monilinia fructicola* (G. Winter) Honey. *Lett Appl Microbiol* 52: 227–232.

Li Q, Ning P, Zheng L, Huang J, Li G, Hsiang T. 2010. Fumigant activity of volatiles of *Streptomyces globisporus* JK-1 against *Penicillium italicum* on *Citrus microcarpa*. *Postharvest Biol Technol* 58: 157–165.

Li Q, Ning P, Zheng L, Huang J, Li G, Hsiang T. 2012. Effects of volatile substances of *Streptomyces globisporus* JK-1 on control of *Botrytis cinerea* on tomato fruit. *Biol Control* 61: 113–120.

Li H, Yu T. 2000. Effect of chitosan on incidence of brown rot, quality and physiological attributes of postharvest peach fruit. *J Sci Food Agric* 81: 269–274.

Li X, Zhu X, Zhao N, Fu D. 2013. Effect of hot water treatment on anthracnose disease in papaya fruit and its possible mechanism. *Postharvest Biol Technol* 86: 437–446.

Liu J, Sui Y, Wisniewski M. 2012. Effect of heat treatment on inhibition of *Monilinia fructicola* and induction of disease resistance in peach fruit. *Postharvest Biol Technol* 65: 61–68.

Liu WT, Chu CL, Zhou T. 2002. Thymol and acetic acid vapors reduce postharvest brown rot of apricots and plums. *HortScience* 37: 151–156.

Lurie S. 1998. Postharvest heat treatments. *Postharvest Biol Technol* 14: 257–269.

Lurie S, Klein JD. 1992. Calcium and heat treatments to improve storability of 'Anna' apples. *HortScience* 27: 36–39.

Maghoumi M, Gòmez PA, Mostofi Y, Zamani Z, Artès-Hernàndez F, Artès F. 2013. Combined effect of heat treatment, UV-C and superatmospheric oxygen packing on phenolics and browning related enzymes of fresh-cut pomegranate arils. *LWT Food Sci Technol* 54: 389–396.

Margosan DA, Smilanick JL. 1998. Mortality of spores of *Botrytis cinerea, Monilinia fructicola, Penicillium digitatum,* and *Rhizopus stolonifer* after exposure to ozone under humid conditions. *Phytopathology* 88: S58.

Margosan DA, Smilanick JL, Simmons GF, Henson DJ. 1997. Combination of hot water and ethanol to control postharvest decay of peaches and nectarines. *Plant Dis* 81: 1405–1409.

Mari M, Gregori R, Donati I. 2004. Postharvest control of *Monilinia laxa* and *Rhizopus stolonifer* in stone fruit by peracetic acid. *Postharvest Biol Technol* 33: 319–325.

Mari M, Iori R, Leoni O, Marchi A. 1993. *In vitro* activity of glucosinolaes-derived isothyocianates against postharvest fruit pathogens. *Ann Appl Biol* 123: 155–164.

Mari M, Leoni O, Bernardi R, Neri F, Palmieri S. 2008. Control of brown rot on stone fruit by synthetic and glucosinolate-derived isothiocyanates. *Postharvest Biol Technol* 47: 61–67.

Mari M, Neri F, Bertolini P. 2007. Novel approaches to prevent and control postharvest diseases of fruits. *Stewart Postharv Rev* 6: 4.

Mari M, Neri F, Bertolini P. 2009. Management of important diseases in Mediterranean high value crops. *Stewart Postharv Rev* 2: 2.

Mari M, Neri F, Bertolini P. 2011. Fruit postharvest disease control by plant bioactive compounds. In: *Natural Antimicrobials in Food Safety and Quality* (eds M Rai, M Chikindas) Wallingford, CABI, 2011, pp. 242–260.

Marquenie D, Michiels CW, Geeraerd, AH. 2002. Using survival analysis to investigate the effect of UV-C and heat treatment on storage rot of strawberry and sweet cherry. *Postharvest Biol Technol* 73: 187–196.

Martinez G, Civello PM. 2008. Effect of heat treatments on gene expression and enzyme activities associated to cell wall degradation in strawberry fruit. *Postharvest Biol Technol* 49: 38–45.

Matsui K. 2006. Green leaf volatiles: Hydroperoxide lyase pathway of oxylipin metabolism. *Current Opinion Plant Biol* 9: 274–280.

Maxin P, Weber RWS, Pederson HL, Williams M. 2012. Control of a wide range of storage rots in naturally infected apple by hot-water dipping and rinsing. *Postharvest Biol Technol* 70: 25–31.

Mehra LK, MacLean DD, Shewfelt RL, Smith KC, Scherm H. 2013. Effect of postharvest biofumigation on fungal decay, sensory quality, and antioxidant levels of blueberry fruit. *Postharvest Biol Technol* 85: 109–115.

Mercier J, Jiménez JI. 2004. Control of fungal decay of apples and peaches by fumigant fungus *Muscodor albus. Postharvest Biol Technol* 31: 1–8.

Mercier J, Smilanick JL. 2005. Control of green mold and sour rot of stored lemon by biofumigation with *Muscodor albus. Biol Control* 32: 401–407.

Miller WR, McDonald RE, Chaparro J. 2000. Tolerance of selected orange and mandarin hybrid fruit to low-dose irradiation for quarantine purposes. *HortScience* 35: 1288–1291.

Minas IS, Karaoglanidis GS, Manganaris GA, Vasilakakis M. 2010. Effect of ozone application during cold storage of kiwifruit on the development of stem-end rot caused by *Botrytis cinerea. Postharvest Biol Technol* 58: 203–210.

Montesinos-Herrero C, Smilanick JL, Tebbets JS, Walse S, Palou L. 2011. Control of citrus postharvest decay by ammonia gas fumigation and its influence on the efficacy of the fungicide imazalil. *Postharvest Biol Technol* 59: 85–93.

Moscoso-Ramirez P, Palau L. 2013. Evaluation of postharvest treatments with chemical resistance inducers to control green and blue molds on orange fruit. *Postharvest Biol Technol* 85: 132–135.

Mostafavi HA, Mirmajlessi SM, Fathollahi H, Minassyan V, Mirjalli SM. 2011. Evaluation of gamma irradiation effect and Pseudomonas fluorescens against *Penicillium expansum*. *Afr J Biotechnol* 10: 11290–11293.

Mostafavi HA, Mirmajlessi SM, Fathollahi H, Shahbazi S, Mirjalli SM. 2013. Integrated effect of gamma radiation and bioconrol agent on quality parameters of apple fruit: An innovative commercial preservation method. *Radiat Phys Chem* 91: 193–199.

Myung K, Hamilton-Kemp T, Archbold D. 2007. Interaction with and effects on the profile of proteins of *Botrytis cinerea* by C_6 aldehydes. *J Agric Food Chem* 55: 2182–2188.

Neri F, Mari M, Brigati S. 2006a. Control of *Penicillium expansum* by plant volatile compounds. *Plant Pathol* 55:100–05.

Neri F, Mari M, Brigati S, Bertolini P. 2009. Control of *Neofabraea alba* by plant volatile compounds and hot water. *Postharvest Biol Technol* 51: 425–430.

Neri F, Mari M, Menniti AM, Brigati S. 2006b. Activity of *trans*-2-hexenal against *Penicillium expansum* in "Conference" pears. *J Appl Microbiol* 100: 1186–1193.

Neri F, Mari M, Menniti AM, Brigati S, Bertolini P. 2006c. Control of *Penicillium expansum* disease in pears and apples by *trans*-2-hexenal vapours. *Postharvest Biol Technol* 41: 101–108.

Neri F, Mari M, Menniti AM, Brigati S, Bertolini P. 2007. Fungicidal activity of plant volatile compounds for controlling *Monilinia laxa*. *Plant Dis* 91: 30–35.

Nigro F, Schena L, Ligorio Pentimone AI, Ippolito A, Salerno MG. 2006. Control of table grape storage rots by preharvest applications of salts. *Postharvest Biol Technol* 42: 142–149.

Norbäck D, Wieslander G, Nordstrom K, Walinder R. 2000. Asthma symptoms in relation to measured building dampness in upper concrete floor construction, and 2-ethyl-hexanol in indoor air. *Int J Tuberculosis Lung Dis* 4: 1016–1025.

Obagwu J, Korsten L. 2003. Integrated control of citrus green and blue molds using *Bacillus subtilis* in combination with sodium bicarbonate or hot water. *Postharvest Biol Technol* 28: 187–194.

Okull DO, LaBorde LF. 2004. Activity of electrolyzed oxidizing water against *Penicillium expansum* in suspension and on wounded apples. *J Food Sci* 69: 23–27.

Ornelas-Paz JdeJ, Yahia EM. 2013. Effect of the moisture content of forced hot air on the postharvest quality and bioactive compounds of mango fruit (*Mangifera indica* L. cv. Manila). *J Sci Food Agric* 94: 1078–83.

Palou L, Usall J, Smilanick JL, Aguilar MJ, Vinas I. 2002. Evaluation of food additives and low-toxicity compounds as alternative chemicals for the control of *Penicillium digitatum* and *Penicillium italicum* on citrus fruit. *Pest Management Sci* 58: 459–466.

Palhano FL, Vilches TTB, Santos RB, Orlando MTD, Ventura JA, Fernandes PMB. 2004. Inactivation of *Colletotrichum gloeosporoides* spores by high hydrostatic pressure combined with citral or lemongrass essential oil. *Int J Food Microbiol* 95: 61–66.

Park SI, Stan SD, Daescheler MA, Zhao Y. 2005. Antifungal coatings on fresh strawberries (*Fragaria* x *ananassa*) to control mold growth during cold storage. *J Food Sci* 70: 202–207.

Patil BS, Vanamala J, Hallman G. 2004. Irradiation and storage influence on bioactive components and quality of early and late season "Rio Red" grapefruit (*Citrus paradisi* Macf.). *Postharvest Biol Technol* 34: 53–64.

Paull RE. 1990. Postharvest heat treatments and fruit ripening. *Postharv News Information* 1: 355–361.

Plotto A, Roberts RG, Roberts DD. 2003. Evaluation of plant essential oils as natural postharvest control of tomato (*Lycopersicon esculentum*). *Acta Hortic* 628: 737–745.

Porat R, Vinokur V, Weiss B. 2003. Induction of resistance to *Penicillium digitatum* in grapefruit by β-aminobutyric acid. *Eur J Plant Pathol* 109: 901–917.

Predmore A, Li J. 2011. Enhanced removal of a human norovirus surrogate from fresh vegetables and fruits by a combination of surfactants and sanitizers. *Appl Environ Microbiol* 77: 4829–4838.

Promyou S, Ketsa S, Doorn WG. 2008. Hot water treatment delay cold-induced banana peel blackening. *Postharvest Biol Technol* 48: 132–138.

Quaglia M, Ederli L, Pasqualini S, Zazzerini A. 2011. Biological control agents and chemicals inducers of resistance for postharvest control of *Penicillium expansum* on apple fruit. *Postharvest Biol Technol* 59: 307–315.

Qin GZ, Tian SP, Xu Y. 2006. Integrated control of brown rot of sweet cherry fruit with food additives in combination with biological control agents. *J Appl Microbiol* 100: 508–515.

Ramos-Garcia M, Hernández LM, Barrera NLL, Bautista BS, Troncoso RR, Bosquez ME. 2012. *In vitro* response of *Fusarium oxysporum* isolates to isothiocyanates application. *Revista Mexic Fitopatol* 30: 1–10.

Ranasinghe L, Jayawardena B, Abeywickrama K. 2005. An integrated strategy to control postharvest decay of Emul banana by combing essential oils with modified atmosphere packaging. *Int J Food Sci Technol* 40: 97–103.

Reuveni M, Zahavi T, Cohen Y. 2001. Controlling downy mildew (*Plasmopara viticola*) in field grown grapevine with β-aminobutiric acid. *Phytoparasitica* 29: 125–133.

Romanazzi G, Nigro F, Ippolito A, di Venere D, Salerno M. 2002. Effect of pre- and post-harvest chitosan treatments to control storage gray mold of table grapes. *J Food Sci* 67: 1862–1867.

Rouissi W, Ugolini L, Martini C, Lazzeri L, Mari M. 2013. Control of postharvest fungal pathogens by antifungal compounds from *Penicillium expansum*. *J Food Prot* 76: 1879–1886.

Sanzani SM, Schena L, de Girolamo A, Ippolito A, González-Candelas L. 2010. Characterization of genes associated to induced resistance against *Penicillium expansum* in apple fruit treated with quercetin. *Postharvest Biol Technol* 56: 1–11.

Schirra M, D'hallewin G. 1997. Storage performance of Fortune mandarins following hot water dips. *Postharvest Biol Technol* 10: 229–238.

Schirra M, D'Aquinio S, Cabras P, Angioni A. 2011. Control of postharvest diseases of fruit by heat and fungicides: Efficacy, residual/levels and residues persistence. A review. *J Agric Food Chem* 59: 8531–8547.

Senti FR. 1981. Food additive and food safety. *Ind Eng Chem Res* 20: 267–246.

Shahi SK, Patra M, Shukla AC, Sikshit A. 2003. Use of essential oil as botanical-pesticide against postharvest spoilage in *Malus pumilo* fruits. *Bio Control* 48: 223–232.

Shao X, Tu K, Zhao Y, Chen L, Chen YY, Wang H. 2007. Effects of pre-storage heat treatment on fruit ripening and decay development in different apple cultivars. *J Hortic Sci Biotech* 82: 297–303.

Shao X, Tu K, Zhao Y. 2009. Increased residual activity controls disease in "Red Fuji" apples (Malus domestica) following postharvest heat treatment. *NZ J Crop Hortic Sci* 37: 375–381.

Shao X, Tu K, Tu S, Su J, Zhao Y. 2010. Effects of heat treatment on wound healing in Gala and Red Fuji apple fruits. *Agric Food Chem* 58: 4303–4309.

Shellie KC, Mangan RL. 2000. Postharvest disinfestation heat treatments: response of fruit and fruit larvae to different heating media. *Postharvest Biol Technol* 21: 51–60.

Shiekh RA, Malik MA, Al-Thabaiti SA, Shiekh MA. 2013. Chitosan as a novel edible coating for fresh fruits. *Food Sci Technol Res* 19: 139–155.

Sisquella M, Casals C, Vinas I, Teixido N, Usall J. 2013. Combination of peracetic acid and hot water treatment to control postharvest brown rot on peaches and nectarines. *Postharvest Biol Technol* 83: 1–8.

Sivakumar D, Wijeratnam RSW, Wijesundera RLL, Abeyesekere M. 2002. Control of postharvest diseases of rambutan using cinnamaldehyde. *Crop Prot* 21: 847–852.

Smilanick JL, Mansour MF, Gabler FM, Sorenson D. 2008. Control of citrus postharvest green mold and sour rot by potassium sorbate combined with heat and fungicides. *Postharvest Biol Technol* 47: 226–238.

Smilanick JL, Mansour MF, Gabler FM, Goodwin WR. 2006. The effectiveness of pyrimethanil to inhibit germination of *Penicillium digitatum* and to control citrus green mould after harvest. *Postharvest Biol Technol* 42: 75–85.

Smilanick JL, Michael IF, Mansour MF. 1997. Improved control of green mould of citrus with imazalil with warm water compared its use in wax. *Plant Dis* 84: 1299–1304.

Song J, Fan L, Forney CF, Campbell-Palmer L, Fillmore S. 2008. Effect of hexanal vapour to control postharvest decay and extend shelf-life of highbush blueberry fruit during controlled atmosphere storage. *Can J Plant Sci* 90: 359–365.

Song J, Hildebrand PD, Fan L. 2007. Effect of hexanal vapour on the growth of postharvest pathogens and fruit decay. *J Food Sci* 72: M108–112.

Spadoni A, Neri F, Bertolini P, Mari M. 2013. Control of Monilinia rots on fruit naturally infected by hot water treatment in commercial trials. *Postharvest Biol Technol* 86: 280–284.

Stevens C, Khan VA, Lu JY. 1997. Integration of ultraviolet (UV-C) light with yeast treatment for control of postharvest storage rots of fruits and vegetables. *Biol Control* 10: 98–103.

Stevens C, Liu J, Khan VA. 1998. Application of hormetic UVC for delayed ripening and reduction of Rhizopus soft rot in tomatoes: The effect of tomatine on storage rot development. *J Pathol* 146: 211–221.

Tanaka N, Fujisawa T, Daimon T, Fujiwara K, Tanaka N, Yamamoto M. 1999. The effect of electrolyzed strong acid aqueous solution on hemodialysis equipment. *Arti Organs* 23: 1055–1062.

Torres R, Nunes C, Garcia JM. 2007. Application of *Pantoea agglomerans* CPA-2 in combination with heated sodium bicarbonate solutions to control the major postharvest diseases affecting citrus fruit at several Mediterranean locations. *Eur J Plant Pathol* 118: 73–83.

Tosi L, Luigetti R, Zazzerini A. 1998. Induced resistance against *Plasmopara helianthi* in sunflower plants by DL-b-amino-n-butyric acid. *J Pathol* 146: 295–299.

Troncoso-Rojas R, Espinoza C, Sanchez-Estrada A, Tiznado-Hernandez M, Garcia HS. 2005. Analysis of the isothiocyanates present in cabbage leaves extract and their potential application to control *Alternaria* rot in bell peppers. *Food Res Int* 38: 701–708.

Tsao R, Zhou T. 2000. Interaction of monoterpenoids, methyl jasmonate, and Ca^{2+} in controlling postharvest brown rot of sweet cherries. *HortScience* 35: 1304–1307.

Ugolini L, Martini C, Lazzeri L, D'Avino L, Mari M. 2014. Control of postharvest gray mould (*Botrytis cinerea* Per.: Fr.) on strawberries by glucosinolate-derived allyl-isothiocyanate treatments. *Postharvest Biol Technol* 90: 34–39.

USEPA. 2008. National ambient air quality standards for ozone; final rule. *Code Federal Regulat* 73: 16436–16514.

Utama S, Wills RBH, Ben-Yehoshua S, Kuek C. 2002. *In vitro* efficacy of plant volatiles for inhibiting the growth of fruit and vegetable decay microorganism. *J Agric Food Chem* 50: 6371–6377.

Utto W, Mawson AJ, Bronlund JE. 2008. Hexanal reduces infection of tomatoes by *Botrytis cinerea* whilst maintaining quality. *Postharvest Biol Technol* 47: 434–437.

Vardar C, Ilhan K, Karabulut OA. 2012. The application of various disinfectants by fogging for decreasing postharvest diseases of strawberries. *Postharvest Biol Technol* 66: 30–34.

Vaughn SF, Spencer GF, Shasha BS. 1993. Volatile compounds from raspberry and strawberry fruit inhibit postharvest decay fungi. *J Food Sci* 58: 793–796.

Venditti T, Molinu MG, Dore A, Agabbio M, D'hallewin G. 2005. Sodium carbonate treatment induces scoparone accumulation, structural changes, and alkalinization in the albedo of wounded citrus fruits. *J Agric Food Chem* 53: 3510–3518.

Vigneault C, Leblanc DI, Goyette B, Jenni S. 2012. Invited review: engineering aspects of physical treatments to increase fruit and vegetable phytochemical content. *Can J Plant Sci* 92: 373–397.

Wan M, Li G, Zhang J, Jiang D, Huang HC. 2008. Effect of volatile substances of *Streptomyces platensis* F-1 on control of plant fungal diseases. *Biol Control* 46: 552–559.

Wang SY, Gao H. 2013. Effect of chitosan-based edible coating on antioxidants, antioxidant enzyme system, and postharvest fruit quality of strawberries (*Fragaria × ananassa* Duch.). *LWT Food Sci Technol* 52: 71–79.

Whangchal K, Saengnil K, Singkamanee C, Uthaibutra J. 2010. Effect of electrolyzed oxidizing water and continuous ozone exposure on the control of *Penicillium digitatum* on tangerine cv. "Sai Nam Pung" during storage. *Crop Prot* 29: 386–389.

Wilson CL, Wisniewski ME. 1989. Biological control of postharvest diseases of fruits and vegetables an emerging technology. *Ann Rev Phytopathol* 27: 425441.

Wood EM, Miles TD, Wharton PS. 2013. The use of natural plant volatile compounds for the control of the potato postharvest diseases, black dot, silver scurf and soft rot. *Biol Control* 64: 152–159.

Wuryatmo E, Klieber A, Scott ES. 2003. Inhibition of citrus postharvest pathogens by vapor of citral and related compounds in culture. *J Agric Food Chem* 51: 2637–2640.

Xu X, Tian S. 2008. Salicylic acid alleviated pathogen-induced oxidative stress in harvested sweet cherry fruit. *Postharvest Biol Technol* 49: 379–385.

Yu L, Reitmeir CA, Love MH. 1996. Strawberry texture and pectin content as affected by electron beam irradiation. *J Food Sci* 61: 844–850.

Youssef K, Ligorio A, Sanzani AM, Nigro F, Ippolito A. 2012. Control of storage diseases of citrus by pre- and post-harvest application of salts. *Postharvest Biol Technol* 72: 57–63.

Yun Z, Gao H, Liu P. 2013. Comparative proteomic and metabolomics profiling of citrus fruit with enhancement of disease resistance by postharvest heat treatment. *BMC Plant Biol* 13: 44.

Zhang J, Swingle PP. 2005. Effects of curing on green mold and stem-end rot of citrus fruit and its potential application under Florida packing system. *Plant Dis* 89: 834–840.

Zhang HY, Zheng XD, Yu T. 2007. Biological control of postharvest diseases of peach with *Cryptococcus laurentii*. *Food Control* 18: 287–291.

Zhang H, Ma L, Wang L, Jiang S, Dong Y, Zheng X. 2008. Biocontrol of gray mold decay in peach fruit by integration of antagonistic yeast with salicylic acid and their effects on postharvest quality parameters. *Biol Control* 47: 60–65.

Zhang X, Shen L, Li F, Meng D, Sheng J. 2013. Hot air treatment- induced arginine catabolism is associated with elevated polyamines and proline levels and alleviates chilling injury in postharvest tomato fruit. *J Sci Food Agric* 93: 3245–3251.

Zhao M, Moy J, Paull RE. 1996. Effect of gamma-irradiation on ripening papaya pectin. *Postharvest Biol Technol* 8: 209–222.

Zhou H, Tao N, Jia L. 2014. Antifungal activity of citral, octanal and terpineol against *Geotrichum citri-aurantii*. *Food Control* 37: 277–283.

6 Advances in the Use of 1-MCP

Chris B. Watkins

CONTENTS

6.1 INTRODUCTION

It is almost 20 years since 1-methylcyclopropene (1-MCP) was discovered and patented by Ed Sisler and Sylvia Blankenship (Sisler and Blankenship, 1996). 1-MCP is a cyclopropene that is a competitive inhibitor of ethylene perception that acts by binding irreversibly to ethylene-binding sites, thereby preventing ethylene binding and the eliciting of subsequent signal transduction and translation. The process of discovery of the effects of cyclopropenes and their proposed method of action have been described (Sisler and Serek, 2003; Sisler, 2006). 1-MCP is extremely active but unstable in the liquid phase, but is stabilized in a process whereby 1-MCP is complexed with α-cyclodextrin (Daly and Kourelis, 2000). Floralife Inc. obtained regulatory approval from the United States Environmental Protection Agency (EPA) in 1999 for use of 1-MCP on floriculture and ornamental products.

The 1-MCP formulation was first marketed under the name EthylBloc®. The rights to 1-MCP were subsequently obtained by AgroFresh, formerly a subsidiary of Rohm and Haas, and now part of Dow Chemicals. EPA approval for use of 1-MCP on edible food products was obtained in 2002. 1-MCP has several characteristics that have resulted in rapid approval by regulatory authorities around the world: it is a gaseous molecule that is easily applied, has an excellent safety profile, leaves no residues in treated produce, and is active at very low concentrations normally in the parts per billion range.

The commercialization of 1-MCP as the SmartFresh[SM] Quality System has led to rapid adoption of 1-MCP based technologies for many horticultural industries. By 2011, regulatory approval for use of 1-MCP had been obtained in over 40 countries. 1-MCP is registered for use on a wide variety of fruits and vegetables, including apple, avocado, banana, broccoli, cucumber, date, kiwifruit, mango, melon, nectarine, papaya, peach, pear, pepper, persimmon, pineapple, plantain, plum, squash, and tomato. The specific products that are registered within each country vary greatly, relating to the importance of the crop in that country.

As is the case for controlled atmosphere (CA) storage, most use of SmartFresh[TM] technology is for apples (Mattheis, 2008; Watkins, 2008). The focus on apples is, in large part, due to the large volumes of fruit that are kept in CA storage for periods up to 12 months, depending on the cultivar and growing region. Also, the apple has been a very suitable fruit for 1-MCP because the ideal product in the marketplace is one that is close to that at harvest—one with a crisp fracturable texture, and an acid to sugar ratio appropriate to each cultivar. In contrast, many other climacteric fruits, such as the avocado, banana, pear, and tomato, require a delay, not an inhibition of ripening, to ensure that the consumer receives high quality products with the expected characteristics of color, texture, and flavor. Research continues to seek commercially useful solutions to managing application of 1-MCP on these fruits. Even for the "apple," commercial experience has highlighted that each cultivar is unique, often with specific challenges, and that these can be variable according to growing region and industry size and marketing plans. Also, 1-MCP has provided a tool that has greatly increased understanding of the involvement of ethylene in ripening and senescence processes.

Consequently, a vast international literature has been developed about 1-MCP and its effects on fruits and vegetables, and several reviews are available—for example, Blankenship and Dole (2003), Huber (2008), Sisler (2006), Sisler and Serek (2003), Toivonen (2008), Watkins (2006, 2008, 2010). A comprehensive review on 1-MCP is beyond the scope of a single chapter and this update is therefore focused on three areas of research where new progress is being made, either on understanding the effects of 1-MCP or on development of new handling protocols and technologies:

1. Effects of 1-MCP on ripening and senescence of horticultural produce.
2. Development of handling protocols to maximize the benefits of 1-MCP.
3. Development of novel methods for application of 1-MCP.

6.2 EFFECTS OF 1-MCP ON RIPENING AND SENESCENCE OF HORTICULTURAL PRODUCE

Publications on the effects of 1-MCP on different fruits and vegetables include at least 34 climacteric fruit and fruit vegetables, 17 nonclimacteric fruit and fruit vegetables, and 31 vegetables (Table 6.1). Sometimes, no effects on ripening and

TABLE 6.1
**A Summary of Climacteric and Nonclimacteric Fruit and Vegetables That
Have Been Treated with 1-MCP with Selected References**

Produce Type	Reference
Fruit and Fruit Vegetable—Climacteric	
Apple [*Malus sylvestris* (L) Mill. var.*domestica* (Borkh.) Mansf.] (*Malus x domestica* L.)	Fan et al. (1999), Watkins and Nock (2012), Watkins et al. (2000)
Apricot (*Prunus armeniaca* L.)	Cao et al. (2009), Dong et al. (2002)
Avocado (*Persea Americana* Mill.)	Adkins et al. (2005), Dauny et al. (2003), Hayama et al. (2008), Meyer and Terry (2010)
Banana (*Musa acuminata*, AAA Group)	Golding et al. (1998), Ketsa et al. (2013), Pinheiro et al. (2010)
Blueberry, highbush (*Vaccinium corymbosum* L.)	Chiabrando and Giacalone (2011), DeLong et al. (2003), MacLean and NeSmith (2011)
Cherimoya (*Annona cherimola* Mill.)	Hofman et al. (2001), Li et al. (2009)
Chinese bayberry (*Myrica rubra* Siebold and Zuccarni)	Li et al. (2006)
Custard apple (*Annona squamosa* L.)	Benassi et al. (2003), de Lima et al. (2010b)
Feijoa (*Feijoa sellowiana* Berg.)	Amarante et al. (2008), Rupavatharam et al. (2015)
Fig (*Ficus carica* L.)	Sozzi et al. (2005)
Goldenberry (*Physalis peruviana* L.)	Gutierrez et al. (2008)
Guava (*Psidium guajava* L.)	Hong et al. (2013), Porat et al. (2009), Singh and Pal (2008)
Jujube (*Zizyphus jujube* M.)	Jiang et al. (2004), Zhang et al. (2012b), Zhong and Xia (2007)
Kiwifruit (*Actinidia deliciosa* (A. Chev) C.F. Liang et A.R. Ferguson var. *deliciosa*)	Boquete et al. (2004), Mworia et al. (2011), Vieira et al. (2012)
Lychee (*Litchi chinensis* Sonn.)	De Reuck et al. (2009), Pang et al. (2001), Sivakumar and Korsten (2010)
Mamey sapote (*Pouteria sapote* (Jacq.) H.E. Moore & Stearn)	Ergun et al. (2005)
Mei (*Prunus mume* Sieb.)	Ergun et al. (2005)
Mango (*Mangifera indica* L.)	Sivakumar et al. (2012),Wang et al. (2006, 2009)
Melon (*Cucumis melo* L.)	Ma et al. (2012), Wang et al. (2006, 2009)
Mexican plum (*Spondias purpurea* L.)	Garcia et al. (2011)
Mountain papaya (*Vasconcellea pubescens*)	Balbontin et al. (2007), Moya-Leon et al. (2004)
Muskmelon, Cantaloupe (*Cucumis melo* L. var. *reticulatus*)	de Souza et al. (2008), Jeong et al. (2008, 2007)
Nectarine (*Prunus persica* Lindl.)	Bregoli et al. (2005), DeEll et al. (2008b), Ziliotto et al. (2008)
Papaya (*Carica papaya* L.)	Ahmad et al. (2013), Ergun et al. (2011), Hofman et al. (2001), Manenoi et al. (2007)
Peach (*Prunus persica* L. Batsch)	Jin et al. (2011), Mahajan and Sharma (2012)
Pear, European (*Pyrus communis* L.)	Chiriboga et al. (2013a), DeEll and Ehsani-Moghaddam (2011), Gamrasni et al. (2010), Villalobos-Acuna et al. (2011a)

(Continued)

TABLE 6.1 (*Continued*)
A Summary of Climacteric and Nonclimacteric Fruit and Vegetables That Have Been Treated with 1-MCP with Selected References

Produce Type	Reference
Pear, Asian, or Japanese (*Pyrus pyrifolia* Nakai)	Lee et al. (2012b), Li and Wang (2009), Szczerbanik et al. (2005)
Pear, Laiyang, Yali (*Pyrus bretschneideri* Rehd.)	Fu et al. (2007), Li et al. (2013), Liu et al. (2013)
Persimmon (*Diospyros khaki* L.)	Ahn and Choi (2010), Choi et al. (2013), Harima et al. (2003), Oz (2011)
Plum (*Prunus salicina* L.; *Prunus x domestica* L.)	Candan et al. (2006, 2011), Sharma et al. (2012), Steffens et al. (2013)
Plumcot (*Prunus salicina* Lindl. *x Prunus armeniaca* L.)	Lim et al. (2013)
Sapota (*Manilkara achras* Mill.)	Bhutia et al. (2011)
Soursop (*Annona muricata* L.)	de Lima et al. (2010a), Espinosa et al. (2013)
Tomato (*Solanum lycopersicum* L.)	Baldwin et al. (2011), Sabir and Agar (2011), Su and Gubler (2012); Zhang et al. (2009b)

Fruit and Fruit Vegetable—Nonclimacteric

Cherry (*Prunus avium* L.)	Gong et al. (2002), Mozetic et al. (2006)
Chayote (*Sechium edule* (Jacq.) Sw.)	Cadena-Iniguez et al. (2006)
Cucumber (*Cucumis sativus* L.)	Lima et al. (2005), Nilsson (2005)
Grape (*Vitis vinifera* L.)	Chervin and Deluc (2010), Chervin et al. (2008), Tesniere et al. (2004)
Grapefuit (*Citrus paradisi* Macf.)	Dou et al. (2005), McCollum and Maul (2007)
Lime (*Citrus latifolia* Tanaka)	Kluge et al. (2003)
Loquat (*Eriobotrya japonica* Lindl)	Cao et al. (2010), Edagi et al. (2011), Cai et al. (2006)
Mandarin (*Citrus reticulata* L.)	Li et al. (2012), Salvador et al. (2006)
Orange (*Citrus sinensis* L. Osbeck)	Porat et al. (1999)
Pepper, hot (*Capsicum frutescens* L.)	Huang et al. (2003)
Pineapple (*Ananas comosus* L.)	Dantas et al. (2009), Selvarajah et al. (2001)
Pitaya (*Selenicerus megalanthus* Haw)	Serna et al. (2012)
Rose apple (*Syzygium jambos* Alston)	Plainsirichai et al. (2010)
Strawberry (*Fragaria x ananassa* Duch.)	Villarreal et al. (2010), Bower et al. (2003), Jiang et al. (2001), Ku et al. (1999)
Sweet red and green bell pepper (*Capsicum annuum* L.)	Cao et al. (2012), Fernandez-Trujillo et al. (2009), Ilic et al. (2012)
Watermelon (*Citrullus lanatus* Thunb.)	Mao et al. (2004)
West Indian lime (*Citrus aurantifolia* Swingle)	Win et al. (2006)

Vegetable—Leafy, Root, and Herbs

Asparagus (*Asparagus officinalis* L.)	Liu and Jiang (2006), Zhang et al. (2012a)
Bamboo shoot (*Phyllostachys praecox f. prevernalis*)	Song et al. (2011)
Basil (*Ocimum basilicum* L.)	Aharoni et al. (2010), Hassan and Mahfouz (2010), Kenigsbuch et al. (2009)

(Continued)

TABLE 6.1 (*Continued*)
A Summary of Climacteric and Nonclimacteric Fruit and Vegetables That Have Been Treated with 1-MCP with Selected References

Produce Type	Reference
Bean (*Phaseolus vulgaris* L.)	Cho et al. (2008)
Broccoli (*Brassica oleracea* L.)	Fernandez-Leon et al. (2013), Ku et al. (2013), Ku and Wills (1999)
Cauliflower (*Brassica oleracea*, L. var. *botrytis*)	Ma et al. (2011)
Carrot (*Daucus carota* L.)	Fan and Mattheis (2000), Forney et al. (2007), Kramer et al. (2012)
Chinese cabbage (*Brassica campestris* L. spp. *Pekinensis* (Lour) Olsson)	Porter et al. (2005)
Chinese chive scapes (*Allium tuberosum* Rottler ex Sprengel)	Wu et al. (2009)
Chinese flowering cabbage, Choysum (*Brassica rapa* var. *parachinensis*)	Able et al. (2003)
Chinese kale (*Brassica alboglabra* Bailey)	Sun et al. (2012)
Chinese mustard (*Brassica juncea* var. *foliosa*)	Able et al. (2003)
Chrysanthemum, Garland (*Chrysanthemum coronarium*)	Able et al. (2003)
Coriander (*Coriandrum sativum* L.)	Hassan and Mahfouz (2012), Jiang et al. (2002)
Eggplant (*Solanum melongena* L.)	Massolo et al. (2011)
Endive (*Cichorium intybus* L.)	Salman et al. (2009)
Ginseng (*Panax ginseng* C. A. Mey)	Park et al. (2013)
Lettuce (*Lactuca sativa* L.)	Fan and Mattheis (2000), Saltveit (2004), Tay and Perera (2004)
Mibuna (*Brassica rapa* var. *nipposinica*)	Able et al. (2003)
Mint (*Mentha longifolia* L.)	Kenigsbuch et al. (2007)
Mizuna (*Brassica rapa* var. *nipposinica*)	Able et al. (2003)
Okra (*Hibiscus esculentus* L.)	Huang et al. (2012)
Onion (*Allium cepa* L.)	Chope et al. (2007), Cools et al. (2011), Downes et al. (2010)
Pak choy (*Brassica rapa* var. *italica* L.)	Able et al. (2002, 2003, 2005)
Parsley (*Petroselinum crispum* Mill.)	Lomaniec et al. (2003), Ouzounidou et al. (2013)
Potato (*Solanum tuberosum* L.)	Cheema et al. (2013), Lulai and Suttle (2004), Prange et al. (2005)
Rocket leaves (Eruca sativa (syn. E. *vesicaria* subsp. *sativa* (Miller) Thell., *Brassica eruca* L.))	Koukounaras et al. (2006)
Spinach (*Spinacia oleracea* L.)	Grozeff et al. (2010)
Summer squash (*Cucurbita maxima* var. *Zapallito* (Carr.) Millan)	Massolo et al. (2013)
Tatsoi (*Brassica rapa* var. *rosularis* (L.) Makin.)	Able et al. (2003)
Tsai Tai (*Brassica chinensis*)	Wang et al. (2014), Zhang et al. (2010a)
Water bamboo shoot (*Zizania caduciflora* L.)	Song et al. (2011)

senescence have been detected, but an absence of effects is not common even for nonclimacteric fruits and vegetables.

Several generalizations can be made about responses of fruits and vegetables to 1-MCP, based on the aforementioned reviews, especially those of Huber (2008) and Watkins (2006, 2010), together with more recent research:

1. The primary features of ripening in climacteric fruits such as softening, color development, and volatile production, are inexorably linked to ethylene production, but the specific effects of 1-MCP treatment are closely linked to species, cultivar, and maturity. The capacity to interrupt the progression of ripening, once initiated, has been found to vary by fruit and the attributes that have been studied.
2. Nonclimacteric fruit can be affected by 1-MCP, and are providing insights about the occurrence of ethylene-dependent and ethylene-independent events during ripening, including changes of gene expression (up- and down-regulation). Common benefits of 1-MCP treatment on nonclimacteric products include delayed chlorophyll and protein losses.
3. While the effects of 1-MCP on specific physiological changes can be highly variable, responses to 1-MCP are typically dependent on "concentration × exposure time." If 1-MCP concentrations and exposure periods are appropriate for the product, the final quality attained in the ripened fruit is similar to the untreated produce.
4. Losses of health-promoting compounds such as vitamin C are usually slower in 1-MCP treated produce, whereas effects on phenolic compounds are often less. Levels of health-promoting compounds in treated fruit are usually close to those of untreated fruit if they attain full ripening after 1-MCP-induced delays in ripening.
5. Physiological disorders that are associated with senescence or induced by ethylene (endogenous and exogenous) are inhibited, while those associated with carbon dioxide (CO_2) in the storage environment are often increased. Chilling injury is increased or decreased depending on whether inhibition of ethylene production is related to enhancing or alleviating the specific disorder.
6. Pathological disorders can be increased, decreased, or unaffected by 1-MCP treatment, the effects likely associated with the interaction of the specific pathogen with the product (e.g., species and its maturity or ripeness stage) and the environment.

A major effort has gone into understanding the basis of variation in responses of species and cultivars to 1-MCP. Peaches show relatively limited inhibition in ripening after treatment (Dal Cin et al., 2006; Hayama et al., 2008; Ortiz et al., 2011), while tomatoes respond to treatment even after ripening has started, albeit to a reduced extent (Hoeberichts et al., 2002; Hurr et al., 2005; Tassoni et al., 2006; Wills and Ku, 2002; Zhang et al., 2009a, 2010b). Avocados and bananas are recalcitrant to inhibition by treatment, once ripening has been initiated (Adkins et al., 2005; Golding et al., 1999; Zhang et al., 2011b). In contrast, others such as

European pears may not recover from the inhibition in ripening even when treated with low 1-MCP concentrations (Bai et al., 2006; Chiriboga et al., 2013a; Ekman et al., 2004).

Differences in 1-MCP responses for cultivars within species have been described for many fruit, including apples (Bai et al., 2005; Watkins et al., 2000), avocados (Pereira et al., 2013), guava (Porat et al., 2009), pears (Bai et al., 2006), and tomato (Guillén et al., 2006). In general, early-season fruit with faster rates of metabolism are less responsive to 1-MCP, but the effect of harvest maturity is also critical. If fruits are treated with 1-MCP when immature, ripening can be completely inhibited or abnormal, for example, guava (Bassetto et al., 2005), tomato (Hurr et al., 2005), and banana (Bagnato et al., 2003; Pelayo et al., 2003). For apples, the effects of 1-MCP on quality factors such as firmness generally decline faster in later-harvested than in earlier-harvested fruit (Bulens et al., 2012; Lu et al., 2013; Mir et al., 2001; Toivonen and Lu, 2005) or where internal ethylene concentrations (IECs) are high in fruit at the time of treatment (Jung and Watkins, 2014; Nock and Watkins, 2013; Watkins and Nock, 2012). Inhibition in the ripening of pear by 1-MCP may be less pronounced in more mature pears (Chiriboga et al., 2013b).

The reasons for these differences may include ethylene receptors and their metabolism, 1-MCP diffusivity and reversible and/or irreversible nontarget binding, 1-MCP metabolism, as well as off-gassing. Although information about ethylene receptors has increased and is reviewed earlier, little is known about the receptors in relation to 1-MCP responses across species. Ethylene production of the fruit has been used as an indirect measure of receptivity of ethylene receptors to 1-MCP. Zhang et al. (2009a, 2010b, 2011b) have hypothesized that the IEC modulates the sensitivity of climacteric fruit to 1-MCP, although increased efficacy of 1-MCP applied under hypoxic hypobaria appears more a function of greater 1-MCP ingress into the fruit than lower IEC (Dong et al., 2013a). Overall, such a model is consistent with continued sensitivity to 1-MCP during ripening of fruit such as tomato, which have comparatively low IECs, and reduced sensitivity of fruit such as avocado, which have high IECs and markedly reduced sensitivity to 1-MCP.

Uptake of any gas should be affected by factors such as tissue morphology and cuticular resistance, but notwithstanding issues related to receptor function, most species respond to 1-MCP if given at an appropriate concentration and time. 1-MCP concentrations as low as $0.1 \ \mu L \cdot L^{-1}$ slowed the ripening of mangoes when applied under hypobaria, although normal atmospheric treatment was applied for comparison (Wang et al., 2006). However, comparison of hypobaria and normal atmospheric pressure application demonstrated greater efficacy of infiltration into apple, Chinese pear, and tomato (Chen et al., 2010; Dong et al., 2013a; Kashimura et al., 2010) and, to a lesser extent, in Japanese pear (Kashimura et al., 2010). However, the response of peach to 1-MCP was weak even when 1-MCP was applied under low pressure, indicating that diffusion of 1-MCP through flesh was not a limiting factor (Hayama et al., 2005). Also, while differences between influx and efflux of 1-MCP for avocado and apple have been demonstrated (Dauny et al., 2003), there is little evidence that diffusivity of 1-MCP into horticultural products is restricted.

However, the depletion of 1-MCP in static systems containing fruits and vegetables suggests that 1-MCP is interacting with nontarget binding sites or is being metabolized. The reversible sorption shown by off-gassing of 1-MCP from avocado fruit, which is probably associated with high oil content (Dauny et al., 2003), is not free diffusion based on comparisons with ethane efflux (Dong et al., 2013b). Nanthachai et al. (2007) investigated 1-MCP absorption in a range of produce (apple, asparagus, ginger, green bean, key lime, leaf lettuce, mango, melon, parsnip, plantain, potato, and tangerine) and found that the initial rate of sorption for each correlated with fresh weight, dry matter, insoluble dry matter, and water weight, but not soluble dry matter. The correlation between absorption rate and insoluble dry matter suggests a strong affinity for 1-MCP by cellulosic materials. High rates and capacities for 1-MCP sorption by asparagus spears and plantain peel are associated with high lignin concentrations (Choi and Huber, 2009). It seems unlikely that nontarget binding would affect efficacy of 1-MCP because there is a huge excess of 1-MCP molecules compared with calculated binding sites -0.5 to 4.3×10^6 molecules of 1-MCP available per physiological binding site (Nanthachai et al., 2007). That said, it is possible that diffusion within some tissues within a fruit, for example, locular tissues of tomatoes, can be limiting (Dong et al., 2013b), and at low 1-MCP concentrations required to modulate 1-MCP responses may be more important. The effects of 1-MCP binding to nontarget materials such as wood are discussed in the next section.

The role of 1-MCP metabolism has not received adequate attention and more research is warranted. 1-MCP sorption by fresh cut tissue is due to oxidative metabolism in response to wound-induced production of reactive species (Lee et al., 2012a) and maybe less important for whole fruit. However, evidence that 1-MCP cell free homogenates metabolize 1-MCP exists, and 1-MCP sorption to asparagus spears was reduced by more than 60% under anoxia 0.06 kPa oxygen (Huber et al., 2010).

6.3 DEVELOPMENT OF HANDLING PROTOCOLS TO MAXIMIZE BENEFITS OF 1-MCP

As a new technology available for horticultural industries, SmartFresh™ has led to the development of handling protocols that maximize the beneficial responses of products to 1-MCP or minimize undesirable ones. Most research has been on apple, which is used as a primary example here, but European pear is used as an example of a fruit that requires postharvest ripening to occur in order to meet consumer expectations for color, flavor, and texture.

6.3.1 APPLE

SmartFresh™ is used widely on apple fruit due to its positive effects on a commodity that is grown widely around the world, stored in large volumes in air and CA storage, and then shipped and distributed through marketing chains that can be extensive. Positive responses of apple fruit 1-MCP include reduced ethylene production and respiration rates, maintenance of firmness and acidity, and reduced incidence of many physiological disorders and greasiness (Watkins, 2006). An important aspect of the apple is that the qualities desired by the consumer, such as a crisp fractuable texture and sugar/acid balance, are similar to those at harvest; therefore, in contrast to many other fruit types, postharvest ripening is not usually necessary. The effectiveness of SmartFresh™ in maintaining quality of fruit in the marketplace has been especially important as texture of fruit can deteriorate rapidly after removal from storage and during marketing. Integration of SmartFresh™ into the apple industry has affected use of postharvest chemical treatments, storage temperature regimes, storage atmospheres, storage duration based on fruit maturity at harvest, and poststorage handling and packing procedures (Mattheis, 2008). Decisions about use of SmartFresh™ are influenced by growing region, size of the industry, and cultivar characteristics (Watkins, 2008), including use at the farm-stand level (McArtney et al., 2011) or without refrigeration (Jung and Lee, 2009).

The efficacy of ripening inhibition by 1-MCP is affected by cultivar (DeEll et al., 2008a; DeLong et al., 2004; Fan et al., 1999; Jung and Watkins, 2014; Nock and Watkins, 2013; Rupasinghe et al., 2000; Watkins et al., 2000) and storage method–air versus controlled atmosphere storage (Bai et al., 2005; Watkins and Nock, 2005; Watkins et al., 2000). Research has been carried out to address factors that affect success of 1-MCP treatment, such as its concentration, effects of nontarget materials, and timing of 1-MCP application and how these factors interact with cultivar, harvest maturity, and storage treatments.

6.3.1.1   1-MCP Concentrations, Exposure Time, and Application Temperature

Most research has focused on the effects of 1-MCP concentrations and exposure times in air. 1-MCP is more effective at higher concentrations but affected by cultivar and storage type; for example, fruit of "Queen Cox" were firmer when treated with $10 \ \mu L \cdot L^{-1}$ than with 0.5 or $1 \ \mu L \cdot L^{-1}$ 1-MCP but there was little effect of concentration for "Bramley" (Dauny and Joyce, 2002). Inhibition of ripening of "McIntosh" was greater in fruit treated with $1 \ \mu L \cdot L^{-1}$ than $0.635 \ \mu L \cdot L^{-1}$ (DeEll et al., 2008a). Also, dose responses were found for "McIntosh" but not for three other cultivars, and only in air and not CA storage (Watkins et al., 2000). The full label rate for 1-MCP is $1 \ \mu L \cdot L^{-1}$ in the United States, Canada, and the European Union (EU).

It has been suggested that 1-MCP is better able to reach and attach to the ethylene receptor sites at the higher temperature (Sisler and Serek, 1997), and for apples, lower treatment temperatures can require longer treatment times to get maximum benefit. Temperature is important when application times are shorter, that is, less than 9–12 h, and depends on the cultivar (DeEll et al., 2002; Kostansek and Pereira, 2003). No effect of treatment of warm fruit on the day of harvest compared with overnight cooling was found for four cultivars when treated for 24 h (Watkins and Nock, 2005). It is recognized that 1-MCP applications can be at any temperature, and that a 24 h treatment time will take into account variations in factors such as temperature and room leakiness (Kostansek and Pereira, 2003). While infiltration of 1-MCP into apple fruit greatly reduces the exposure time required to inhibit ripening (Kashimura et al., 2010), such applications are unlikely to have any commercial significance.

6.3.1.2 Nontarget Materials

1-MCP is easily absorbed by nontarget materials that are present in commercial settings. These include wooden and cardboard bins and bin-liner material, and absorption can be further increased if these materials are wet; in contrast, absorption by plastic bins or wall-surface materials is low (Ambaw et al., 2011, 2013a; Rodriguez-Perez et al., 2009; Vallejo and Beaudry, 2006). These findings suggest that the efficacy and uniformity of 1-MCP treatment could be compromised by the presence of nontarget materials. Possible examples include cases where cultivars such as "McIntosh," which recover more easily from 1-MCP inhibition of ripening, and should be treated in plastic bins to maximize the available 1-MCP. Properly designed stack arrangement, sufficient ventilation and appropriately placed air-circulating units are important for large treatment facilities (Ambaw et al., 2013b).

6.3.1.3 Timing of 1-MCP Application

The effect of delays between harvest and 1-MCP treatment can also vary greatly among cultivars, reflecting differences of metabolic rates and associated ethylene production. For example, delays of more than a week between harvest and 1-MCP treatment did not compromise the retention of firmness and the control of superficial scald, brown core, and decay in late-maturing cultivars such as "Delicious" and "Fuji" (Amarante et al., 2010; Lu et al., 2009; Tatsuki et al., 2007). However, 1-MCP treatment delays have a greater influence on rapidly ripening cultivars such as "Anna," "Cortland," "McIntosh," and "Orin" (DeEll et al., 2008a; Lu et al., 2013; Tatsuki et al., 2007). Effects of delayed 1-MCP treatment on quality were more pronounced in air than in CA storage (Watkins and Nock, 2005). The suppliers of SmartFresh™ provide recommendations of maximum delays of seven days between harvest and 1-MCP treatment for most apple cultivars (AgroFresh, 2014). However, the recommended timing is much shorter for "McIntosh," with the Canadian registration requirements and the U.S. recommendations being application of 1-MCP within three days of harvest (AgroFresh, 2014; DeEll et al., 2008a). Rapid CA treatment of fruit with 1-MCP can reduce the need to apply CA for up to two weeks (Watkins and Nock, 2012) or even as long as two months (DeEll and Ehsani-Moghaddam, 2012). The importance of rapid treatment of 1-MCP can cause logistical issues for storage operators. Strategies that have been developed by the industry have included the use of purpose-built rooms, tents, and shipping containers that can be used to treat fruit quickly after harvest.

The SmartFresh™ label was modified in the United States in 2009 and Canada in 2011 to allow multiple 1-MCP treatments. Therefore, 1-MCP can be applied to fruit more than once while storage rooms are being loaded, or during storage. Early studies showed little effect of multiple 1-MCP treatments (DeLong et al., 2004; Mir et al., 2001) but recent research indicates that the repeated treatments of fruit with 1-MCP can increase maintenance of firmness and acidity (DeEll and Ehsani-Moghaddam, 2013; Lu et al., 2013; Nock and Watkins, 2013). However, while control of superficial scald, internal browning, decay and senescent breakdown was maximized with early single and multiple 1-MCP treatments within the first week of cold storage, the incidences of a flesh browning disorder on "Empire" apples and external CO_2 injury could be increased (DeEll and Ehsani-Moghaddam, 2013; Lu et al., 2013; Nock and Watkins, 2013). In general, incidences of both external and internal CO_2 injuries (Argenta et al., 2010; DeEll et al., 2003; Fawbush et al., 2008) and firm-flesh browning of "Empire" apples (Jung and Watkins, 2011) are increased by 1-MCP treatment. CO_2 injuries can be effectively controlled by the antioxidant diphenylamine (DPA) used for control of superficial scald, but DPA has been banned in the EU. Alternative control methods, such as delaying CA storage and employing lower CO_2 concentrations in the early stages of storage, are being investigated.

6.3.2 European Pear

European pear fruit are categorized as summer, for example, "Bartlett," or winter pears, for example, "D'Anjou," "Bosc," and "Conference." In general, both types require a period of cold storage or ethylene exposure to ripen and attain the buttery and juicy texture expected by the consumer (Villalobos-Acuna and Mitcham, 2008). 1-MCP decreases ethylene production and respiration rates, and delays softening of the fruit as well as reduces incidences of superficial scald and internal breakdown, but it is challenging to obtain a delay and not complete inhibition of ripening (Argenta et al., 2003; Bai et al., 2006; Calvo and Sozzi, 2004, 2009; Kubo et al., 2003; Villalobos-Acuna et al., 2011a). The best combination of harvest maturity, 1-MCP concentration, application temperature and time, and storage time to adequately control ripening without developing physiological disorders is still uncertain (Villalobos-Acuna and Mitcham, 2008).

Cultivar and maturity effects are significant factors affecting the responses of pears to 1-MCP. Bai et al. (2006) found that "Bartlett" pears ripened when treated with 0.3 µL · L^{-1} and kept at temperatures ranging 10–20°C for 10–20 days after storage; the specific regime was related to length of storage time and whether fruit were air- or CA-stored. In contrast, ripening of "d'Anjou" pears could not be reinitiated under these conditions and only 1-MCP concentrations as low as 0.05 µL · L^{-1} were satisfactory. In "Bartlett" pears, the same concentration (0.3 µL · L^{-1}) and treatment conditions resulted in diverse responses among pears of different maturities and cold storage periods (Villalobos-Acuna et al., 2011a). Ripening of "Conference" pears is blocked by 0.6 µL · L^{-1} 1-MCP, whereas 0.3 µL · L^{-1} 1-MCP resulted in softening in only some pears (Chiriboga et al., 2011). 1-MCP concentrations (0.025 and 0.050 µL · L^{-1}) that did not prevent ripening of "Conference" pears (Rizzolo et al., 2005) cannot be applied commercially at such low concentrations. A key to attaining high quality fruit may be the application of 1-MCP at a more advanced maturity stage; identifying this stage may be difficult among different orchards but is associated with ethylene production by the fruit (Chiriboga et al., 2013a,b). A similar relationship to ethylene production has been shown for "Spadona," a summer pear with no chilling requirement (Gamrasni et al., 2010).

In commercial settings, another variable is treatment time after harvest. Environmental factors may also be a factor as responses of the same cultivar can vary between growing regions (DeEll and Ehsani-Moghaddam, 2011; Villalobos-Acuna et al., 2011a). When "Bartlett" pears in Ontario, Canada were treated with 0.3 µL · L^{-1} 1-MCP one day after harvest, softening was compromised compared with treatment three and seven days after harvest; since treatments after seven days provided less control of disorders, three days represented a compromise recommendation (DeEll and Ehsani-Moghaddam, 2012). In contrast, "Bartlett" pears grown in California in the United States showed little response to a range of treatment times and temperatures at the same 1-MCP concentration (Villalobos-Acuna et al., 2011a).

Simultaneous application of 1-MCP and ethylene did not have significant effects on inhibition of "Bartlett" pear ripening caused by 1-MCP treatment (Trinchero et al., 2004), whereas positive effects were found for "Bartlett" and "Conference" pears (Chiriboga et al., 2011; Villalobos-Acuna et al., 2011b).

Also, factors such as the cooling method and bin materials that are not so critical for apples, may be more important with use of lower 1-MCP concentrations required to delay ripening of pears (Calvo and Sozzi, 2009).

6.4 DEVELOPMENT OF NOVEL METHODS FOR APPLICATION OF 1-MCP

6.4.1 PREHARVEST

Premature drop of fruit is a phenomenon that is especially severe in some apple cultivars such as "Delicious," "Honeycrisp," and "McIntosh" and pears such as "Bartlett," resulting in loss of fruit before harvest can be completed. Where permitted by registration, drop of apples and pears is often controlled by preharvest sprays of aminoethyvinylglycine (AVG; ReTain), which inhibits activity of a key enzyme in ethylene biosynthesis, ACC synthase, (Bangerth, 1978; Schupp and Greene, 2004), or by naphthaleneacetic acid (NAA), which inhibits abscission but can increase the ripening rates of the fruit (Marini et al., 1993; Smock and Gross, 1947; Yuan and Carbaugh, 2007). AVG applications in the field improve the responses of apples and other fruit to postharvest 1-MCP treatment (Hayama et al., 2008; Moran, 2006). The availability of 1-MCP has opened up the possibility of an additional technology for control of such premature drops.

Early attempts at affecting premature fruit-drop by the use of 1-MCP sprays or as a gas using enclosed plastic bags were not successful (Byers et al., 2005), but AgroFresh has developed formulations of 1-MCP for preharvest application that are registered under the name of Harvista™. The technology had been registered for use in Chile, New Zealand, South Africa, and the United States by 2014. Preharvest 1-MCP applications may also have the benefit of delaying ripening and senescence of fruit where undesirable degrees of inhibition occur when postharvest 1-MCP is applied.

Fruit from Harvista™ treated apple and pear trees show delayed fruit-drop, lower ethylene production rates and IECs than in untreated fruit, firmer fruit, and lower starch pattern indices (McArtney et al., 2008, 2009; Varanasi et al., 2013; Villalobos-Acuna et al., 2010; Watkins et al., 2010; Yuan and Carbaugh, 2007; Yuan and Li, 2008), although effects on maturity indicators are sometimes inconsistent (DeEll and Ehsani-Moghaddam, 2010; McArtney et al., 2008, 2009; Villalobos-Acuna et al., 2010). In addition to effects on slowing the rate of fruit-drop, other benefits of Harvista™ include postharvest benefits of decreasing the rates of softening and loss of acidity (DeEll and Ehsani-Moghaddam, 2010; Varanasi et al., 2013; Villalobos-Acuna et al., 2010; Watkins et al., 2010). Effects of Harvista™ on fruit responses are dependent on the concentration (Elfving et al., 2007; McArtney et al., 2008) and timing of the application before harvest (Elfving et al., 2007; Varanasi et al., 2013; Yuan and Li, 2008). Harvista™ also has the benefits of decreasing water-core development in apples before harvest and development of superficial scald, soft scald, soggy breakdown and internal breakdown, of apples and pears after harvest (DeEll and Ehsani-Moghaddam, 2010; McArtney et al., 2008, 2009; Villalobos-Acuna et al., 2010).

1-MCP has also been applied prior to harvest to other fruit types, usually, but not always with the Harvista™ formulation. Ripening and senescence of figs, mangosteen, plums, and yellow pitahaya fruit were delayed by 1-MCP sprays or fumigation on the tree (Cock et al., 2013; Freiman et al., 2012; Lerslerwong et al., 2013; Steffens et al., 2009). Applications of 1-MCP to sweet cherry trees within three days of ethephon treatment had inconsistent effects on reducing ethephon-induced loss in flesh firmness without inhibiting desirable effects of ethephon on the fruit-removal force (Elfving et al., 2009).

In general, the effects of Harvista™ decrease with longer intervals between treatment and harvest and with increasing time after harvest (Elfving et al., 2007; McArtney et al., 2009; Varanasi et al., 2013; Villalobos-Acuna et al., 2010; Watkins et al., 2010). Combined applications of Harvista™ and NAA or AVG can be more effective than any of them alone (Yuan and Carbaugh, 2007). Also, combinations of Harvista™ and SmartFresh treatments result in more sustained effects on ripening after harvest than either do alone (McArtney et al., 2008, 2009; Varanasi et al., 2013; Watkins et al., 2010). Using tomato plants as a model system, MacKinnon et al. (2009) found that surfactant type influenced 1-MCP effects and that greater spray volume was more effective; but no effect of the spray nozzle type used was detected.

Few biochemical studies on the preharvest effects of 1-MCP are available. Inhibited abscission of cotton and citrus leaves by 1-MCP was associated with reduced gene expression and enzyme activities of endo-β-gluconase and 1-aminocyclopropane-1-carboxylic acid synthase (ACS) and oxidase (ACO) (John-Karuppiah and Burns, 2010; Mishra et al., 2008; Zhong et al., 2001). 1-MCP treatment resulted in lower expression of *MdPG2* and *MdEG1* genes in the abscission zone of "Delicious" apples compared with AVG treatment, whereas expression of *MdACS5A* and *MdACO1* were affected similarly by both treatments (Li and Yuan, 2008; Yuan and Li, 2008). Delay of ethylene production and fruit-softening by 1-MCP was associated with reduced expression of MdACS1, MdACO1, MdETR2, MdERS1, and MdPG1 (Yuan and Li, 2008). Varanasi et al. (2013) suggested that variations in responses of apple fruit to Harvista™ were associated with different regulation patterns of receptor genes (Zhu et al., 2010).

6.4.2   AQUEOUS AND GASEOUS POSTHARVEST APPLICATION

Aqueous application of 1-MCP would provide flexibility when sealed rooms are not available, with selective treatment of fruit to be stored in a room, and a dip or line-spray while fruit are being packed. Also, it may be easier to apply lower 1-MCP concentrations more effectively in the case of fruit such as pear where recovery from 1-MCP is difficult to achieve under the conditions required for gaseous application. 1-MCP applied to apples in water resulted in similar physiological responses (IEC, firmness, and acidity) as the gas form, but required 700-fold higher amounts of 1-MCP (Argenta et al., 2007). In contrast, the efficacy of an aqueous application of 1-MCP for 1 min was as high as gaseous application for 9 h for both avocado and tomato (Choi et al., 2008). Aqueous application of 1-MCP delayed the storage life of plums even at the ripe stage (Manganaris et al., 2007, 2008) and resulted in lower anthocyanin concentrations, PAL activity and CI symptoms such as flesh reddening (Manganaris et al., 2007).

Thus, ripening and associated events were delayed in proportion to the amount of 1-MCP applied to the different fruit types, indicating that aqueous 1-MCP shows similar responses as those treated with gaseous 1-MCP. Ripening factors such as activity of cell-wall associated enzymes, such as polygalacturonase, lycopene, antioxidants, and volatiles of avocado are delayed but recover to reach levels similar to those of untreated fruit (Choi and Huber, 2008; Pereira et al., 2013; Zhang et al., 2013). The application of aqueous 1-MCP also raises issues of interaction with other technologies. Sodium hypochlorite solutions and aqueous 1-MCP are fully compatible if 1-MCP is applied first, but chlorine can result in loss of 1-MCP activity in multiple-use scenario situations (Choi et al., 2009).

Alternative encapsulation systems to α-cyclodextrin complexes include cucurbit[n]urils (CB[n](where n is the number of glycoluril units), which are barrel-like macrocyclic molecules, which are prepared from glycoluril and formalin (Zhang et al., 2011a).

6.4.3   OTHER APPLICATION METHODS

Polyvinyl acetate sachets provide a method to slowly release 1-MCP into the atmosphere around produce (Lee et al., 2006). 1-MCP can also be incorporated into polymer films and released from the film matrix by high relative humidity (Hotchkiss et al., 2007). Both concepts provide interesting prospects for application of 1-MCP to packaged produce but require optimizing of many factors including film characteristics, sachet size, environmental factors, and specific produce items. The use of 1-MCP in cyclodextrin-based nanosponges tested for floriculture systmes (Seglie et al., 2011) has not been explored for fruits and vegetable storage but soy protein 1-MCP-releasing pads have been shown to delay tomato ripening and could be useful for postharvest "in package" treatments (Ortiz et al., 2013).

6.4.4   RELATED COMPOUNDS

Despite the commercialization of Harvista™, claims have been made that 1-MCP application cannot be made outside of a sealed room environment (Goren et al., 2011; Seglie et al., 2010; Sisler et al., 2009). A range of compounds related to 1-MCP that may be used to inhibit ethylene responses of plant tissues continue to be identified, including 1-alkenes (Sisler, 2008), 1-substituted cyclopropenes (Apelbaum et al., 2008), N,N-dipropyl(1-cyclopropenylmethyl)amine, and N,N-di-(1-cyclopropenylmethyl) amine (Seglie et al., 2010; Sisler et al., 2009), and 3-cyclopropyl-1-enlylpropanoic acid sodium salt (Goren et al., 2011). Whether these compounds will be registered for use of food products is uncertain.

REFERENCES

Able AJ, Wong LS, Prasad A, O'Hare TJ, 2002, 1-MCP is more effective on a floral brassica (*Brassica oleracea* var. *italica* L.) than a leafy brassica (*Brassica rapa* var. *chinensis*). *Postharv Biol Technol* 26: 147–155.

Able AJ, Wong LS, Prasad A, O'Hare TJ, 2003, The effects of 1-methylcyclopropene on the shelf-life of minimally processed leafy Asian vegetables. *Postharv Biol Technol* 27: 157–161.

Able AJ, Wong LS, Prasad A, O'Hare TJ, 2005, The physiology of senescence in detached pak choy leaves (*Brassica rapa* var. *chinensis*) during storage at different temperatures. *Postharv Biol Technol* 35: 271–278.

Adkins ME, Hofman PJ, Stubbings BA, Macnish AJ, 2005, Manipulating avocado fruit ripening with 1-methylcyclopropene. *Postharv Biol Technol* 35: 33–42.

AgroFresh, 2014, SmartFresh™ Quality System, Apple Use Recommendations, AgroFresh PA.

Aharoni N, Kenigsbuch D, Chalupowicz D, Faura-Mlinski M, Aharon Z, Maurer D, Ovadia A, Lers A, 2010, Reducing chilling injury and decay in stored sweet basil. *Israel J Plant Sci* 58: 167–181.

Ahmad A, Ali ZM, Zainal Z, 2013, Effect of 1-methylcyclopropene (1-MCP) treatment on firmness and softening related enzymes of 'Sekaki' papaya fruit during ripening at ambient. *Sains Malaysiana* 42: 903–909.

Ahn GH, Choi SJ, 2010, The practical 1-methylcyclopropene treatment method for preventing poststorage softening of 'Fuyu' persimmon fruits. *Korean J Hortic Sci Technol* 28: 254–258.

Amarante CVT do, Argenta LC, Vieira MJ, Steffens CA, 2010, Changes of 1-MCP efficiency by delaying its postharvest application on 'Fuji Suprema' apples. *Revista Brasil Fruticult* 32: 984–992.

Amarante CVT do, Steffens CA, Ducroquet JPHJ, Sasso A, 2008, Fruit quality of feijoas in response to storage temperature and treatment with 1-methylcyclopropene. *Pesq Agropec Bras* 43: 1683–1689.

Ambaw A, Beaudry R, Bulens I, Delele MA, Ho QT, Schenk A, Nicolai BM, Verboven P, 2011, Modeling the diffusion-adsorption kinetics of 1-methylcyclopropene (1-MCP) in apple fruit and non-target materials in storage rooms. *J Food Eng* 102: 257–265.

Ambaw A, Verboven P, Defraeye T, Tijskens E, Schenk A, Opara UL, Nicolai BM, 2013a, Effect of box materials on the distribution of 1-MCP gas during cold storage: A CFD study. *J Food Eng* 119: 150–158.

Ambaw A, Verboven P, Delele MA, Defraeye T, Tijskens E, Schenk A, Nicolai BM, 2013b, CFD modelling of the 3D spatial and temporal distribution of 1-methylcyclopropene in a fruit storage container. *Food Bioprocess Technol* 6: 2235–2250.

Apelbaum A, Sisler EC, Feng XQ, Goren R, 2008, Assessment of the potency of 1-substituted cyclopropenes to counteract ethylene-induced processes in plants. *Plant Growth Regul* 55: 101–113.

Argenta LC, Fan XT, Mattheis JP, 2003, Influence of 1-methylcyclopropene on ripening, storage life, and volatile production by d'Anjou cv. pear fruit. *J Agric Food Chem* 51: 3858–3864.

Argenta LC, Fan XT, Mattheis JP, 2007, Responses of 'Golden Delicious' apples to 1-MCP applied in air or water. *HortScience* 42: 1651–1655.

Argenta LC, Mattheis JP, Fan X, 2010, Interactive effects of CA storage, 1-methylcyclopropene and methyl jasmonate on quality of apple fruit. *Acta Hortic* 857: 259–265.

Bagnato N, Barrett R, Sedgley M, Klieber A, 2003, The effects on the quality of Cavendish bananas, which have been treated with ethylene, of exposure to 1-methylcyclopropene. *Int J Food Sci Technol* 38: 745–750.

Bai J, Mattheis JP, Reed N, 2006, Re-initiating softening ability of 1-methylcyclopropene-treated 'Bartlett' and 'd'Anjou' pears after regular air or controlled atmosphere storage. *J Hortic Sci Biotech* 81: 959–964.

Bai JH, Baldwin EA, Goodner KL, Mattheis JP, Brecht JK, 2005, Response of four apple cultivars to 1-methylcyclopropene treatment and controlled atmosphere storage. *HortScience* 40: 1534–1538.

Balbontin C, Gaete-Eastman C, Vergara M, Herrera R, Moya-Leon MA, 2007, Treatment with 1-MCP and the role of ethylene in aroma development of mountain papaya fruit. *Postharv Biol Technol* 43: 67–77.

Baldwin E, Plotto A, Narciso J, Bai JH, 2011, Effect of 1-methylcyclopropene on tomato flavour components, shelf-life and decay as influenced by harvest maturity and storage temperature. *J Sci Food Agric* 91: 969–980.

Bangerth F, 1978, Effect of a substituted amino-acid on ethylene biosynthesis, respiration, ripening and preharvest drop of apple fruits. *J Am Soc Hortic Sci* 103: 401–404.

Bassetto E, Jacomino AP, Pinheiro AL, Kluge RA, 2005, Delay of ripening of 'Pedro Sato' guava with 1-methylcyclopropene. *Postharv Biol Technol* 35: 303–308.

Benassi G, Correa G, Kluge RA, Jacomino AP, 2003, Shelf-life of custard apple treated with 1-methylciclopropene–an antagonist to the ethylene action. *Brazilian Arch Biol Technol* 46: 115–119.

Bhutia W, Pal RK, Sen S, Jha SK, 2011, Response of different maturity stages of sapota (*Manilkara achras* Mill.) cv. Kallipatti to in-package ethylene absorbent. *J Food Sci Technol-Mysore* 48: 763–768.

Blankenship SM, Dole JM, 2003, 1-Methylcyclopropene: A review. *Postharv Biol Technol* 28: 1–25.

Boquete EJ, Trinchero GD, Fraschina AA, Vilella F, Sozzli GO, 2004, Ripening of 'Hayward' kiwifruit treated with 1-methylcyclopropene after cold storage. *Postharv Biol Technol* 32: 57–65.

Bower JH, Blasi WV, Mitcham EJ, 2003, Effects of ethylene and 1-MCP on the quality and storage life of strawberries. *Postharv Biol Technol* 28: 417–423.

Bregoli AM, Ziosi V, Biondi S, Rasori A, Ciccioni M, Costa G, Torrigiani P, 2005, Postharvest 1-methylcyclopropene application in ripening control of 'Stark Red Gold' nectarines: Temperature-dependent effects on ethylene production and biosynthetic gene expression, fruit quality, and polyamine levels. *Postharv Biol Technol* 37: 111–121.

Bulens I, Van de Poel B, Hertog MLATM, de Proft MP, Geeraerd AH, Nicolai BM, 2012, Influence of harvest time and 1-MCP application on postharvest ripening and ethylene biosynthesis of 'Jonagold' apple. *Postharv Biol Technol* 72: 11–19.

Byers RE, Carbaugh DH, Combs LD, 2005, Ethylene inhibitors delay fruit drop, maturity, and increase fruit size of 'Arlet' apples. *HortScience* 40: 2061–2065.

Cadena-Iniguez J, Arevalo-Galarza L, Ruiz-Posadas LM, Aguirre-Medina JF, Soto-Hernandez M, Luna-Cavazos M, Zavaleta-Mancera HA, 2006, Quality evaluation and influence of 1-MCP on *Sechium edule* (Jacq.) Sw. fruit during postharvest. *Postharv Biol Technol* 40: 170–176.

Calvo G, Sozzi GO, 2004, Improvement of postharvest storage quality of 'Red Clapp's' pears by treatment with 1-methylcyclopropene at low temperature. *J Hortic Sci Biotech* 79: 930–934.

Calvo G, Sozzi GO, 2009, Effectiveness of 1-MCP treatments on 'Bartlett' pears as influenced by the cooling method and the bin material. *Postharv Biol Technol* 51: 49–55.

Candan AP, Graell J, Crisosto C, Larrigaudiere C, 2006, Improvement of storability and shelf-life of 'Blackamber' plums treated with 1-methylcyclopropene. *Food Sci Technol Int* 12: 437–443.

Candan AP, Graell J, Larrigaudiere C, 2011, Postharvest quality and chilling injury of plums: Benefits of 1-methylcyclopropene. *Span J Agric Res* 9: 554–564.

Cao JK, Zhao YM, Wang M, Lu HY, Jiang WB, 2009, Effects of 1-methylcyclopropene on apricot fruit quality, decay, and on physiological and biochemical metabolism during shelf-life following long-term cold storage. *J Hortic Sci Biotech* 84: 672–676.

Cao SF, Yang ZF, Zheng YH, 2012, Effect of 1-methylcyclopene on senescence and quality maintenance of green bell pepper fruit during storage at 20 degrees C. *Postharv Biol Technol* 70: 1–6.

Cao SF, Zheng YH, Wang KT, Rui HJ, Shang HT, Tang SS, 2010, The effects of 1-methyl-cyclopropene on chilling and cell wall metabolism in loquat fruit. *J Hortic Sci Biotech* 85: 147–153.

Cai C, Chen KS, Xu WP, Zhang WS, Li X, Ferguson I, 2006, Effect of 1-MCP on postharvest quality of loquat fruit. *Postharv Biol Technol* 40: 155–162.

Cheema MUA, Rees D, Colgan RJ, Taylor M, Westby A, 2013, The effects of ethylene, 1-MCP and AVG on sprouting in sweetpotato roots. *Postharv Biol Technol* 85: 89–93.

Chen SJ, Zhang M, Wang SJ, 2010, Physiological and quality responses of Chinese 'Suli' pear (*Pyrus bretschneideri* Rehd) to 1-MCP vacuum infiltration treatment. *J Sci Food Agric* 90: 1317–1322.

Chervin C, Deluc L, 2010, Ethylene signalling receptors and transcription factors over the grape berry development: Gene expression profiling. *Vitis* 49: 129–136.

Chervin C, Tira-umphon A, Terrier N, Zouine M, Severac D, Roustan JP, 2008, Stimulation of the grape berry expansion by ethylene and effects on related gene transcripts, over the ripening phase. *Physiol Plant* 134: 534–546.

Chiabrando V, Giacalone G, 2011, Shelf-life extension of highbush blueberry using 1-methyl-cyclopropene stored under air and controlled atmosphere. *Food Chem* 126: 1812–1816.

Chiriboga MA, Saladie M, Bordonaba JG, Recasens I, Garcia-Mas J, Larrigaudiere C, 2013a, Effect of cold storage and 1-MCP treatment on ethylene perception, signalling and synthesis: Influence on the development of the evergreen behaviour in 'Conference' pears. *Postharv Biol Technol* 86: 212–220.

Chiriboga MA, Schotsmans WC, Larrigaudiere C, Dupille E, Recasens I, 2011, How to prevent ripening blockage in 1-MCP-treated 'Conference' pears. *J Sci Food Agric* 91: 1781–1788.

Chiriboga MA, Schotsmans WC, Larrigaudiere C, Dupille E, Recasens I, 2013b, Responsiveness of 'Conference' pears to 1-methylcyclopropene: The role of harvest date, orchard location and year. *J Sci Food Agric* 93: 619–625.

Cho MA, Hurr BM, Jeong J, Lim C, Huber DJ, 2008, Postharvest senescence and deterioration of 'Thoroughbred' and 'Carlo' green beans (*Phaseolus vulgaris* L.) in response to 1-methylcyclopropene. *HortScience* 43: 427–430.

Choi HS, Jung SK, Kim YK, 2013, Storage ability of non-astringent 'Fuyu' persimmon fruit is affected by various concentrations of 1-methylcyclopropene and/or modified atmosphere packaging. *J Hortic Sci Biotech* 88: 195–200.

Choi ST, Huber DJ, 2008, Influence of aqueous 1-methylcyclopropene concentration, immersion duration, and solution longevity on the postharvest ripening of breaker-turning tomato (*Solanum lycopersicum* L.) fruit. *Postharv Biol Technol* 49: 147–154.

Choi ST, Huber DJ, 2009, Differential sorption of 1-methylcyclopropene to fruit and vegetable tissues, storage and cell wall polysaccharides, oils, and lignins. *Postharv Biol Technol* 52: 62–70.

Choi ST, Tsouvaltzis P, Lim CI, Huber DJ, 2008, Suppression of ripening and induction of asynchronous ripening in tomato and avocado fruits subjected to complete or partial exposure to aqueous solutions of 1-methylcyclopropene. *Postharv Biol Technol* 48: 206–214.

Choi ST, Huber DJ, Kim JG, Hong YP, 2009, Influence of chlorine and mode of application on efficacy of aqueous solutions of 1-methylcyclopropene in delaying tomato (*Solanum lycopersicum* L.) fruit ripening. *Postharv Biol Technol* 53: 16–21.

Chope GA, Terry LA, White PJ, 2007, The effect of 1-methylcyclopropene (1-MCP) on the physical and biochemical characteristics of onion cv. SS1 bulbs during storage. *Postharv Biol Technol* 44: 131–140.

Cock LS, Valenzuela LST, Aponte AA, 2013, Physical, chemical and sensory changes of refrigerated yellow pitahaya treated preharvest with 1-MCP. *Dyna-Colombia* 80: 11–20.

Cools K, Chope GA, Hammond JP, Thompson AJ, Terry LA, 2011, Ethylene and 1-methyl-cyclopropene differentially regulate gene expression during onion sprout suppression. *Plant Physiol* 156: 1639–1652.

Dal Cin V, Rizzini FM, Botton A, Tonutti P, 2006, The ethylene biosynthetic and signal trans-duction pathways are differently affected by 1-MCP in apple and peach fruit. *Postharv Biol Technol* 42: 125–133.

Daly J, Kourelis B, 2000, Synthesis methods, complexes and delivery methods for the safe and convenient storage, transport and application of compounds for inhibiting the ethylene response in plants. US Patent 6,017,849.

Dantas OR, Silva SD, Alves RE, Silva ED, 2009, Susceptibility to chilling injury for 'Perola' pineapple treated with 1-methylcyclopropene. *Revista Brasil Fruticult* 31: 135–144.

Dauny PT, Joyce DC, 2002, 1-MCP improves storability of 'Queen Cox' and 'Bramley' apple fruit. *HortScience* 37: 1082–1085.

Dauny PT, Joyce DC, Gamby C, 2003, 1-Methylcyclopropene influx and efflux in 'Cox' apple and 'Hass' avocado fruit. *Postharv Biol Technol* 29: 101–105.

de Lima MAC, Alves RE, Filgueiras HAC, 2010a, Respiratory behavior and softening of sour-sop fruit (*Annona muricata* L.) after postharvest treatments with wax and 1-methylcy-clopropene. *Ciencia E Agrotecnol* 34: 155–162.

de Lima MAC, Mosca JL, da Trindade DCG, 2010b, Delay in ripening of African Pride ate-moya fruits after postharvest treatment with 1-methylcyclopropene. *Ciencia E Tecnol Alimentos* 30: 599–604.

De Reuck K, Sivakumar D, Korsten L, 2009, Integrated application of 1-methylcyclopropene and modified atmosphere packaging to improve quality retention of litchi cultivars dur-ing storage. *Postharv Biol Technol* 52: 71–77.

de Souza PA, Finger FL, Alves RE, Puiatti M, Cecon PR, Menezes JB, 2008, Postharvest con-servation of charentais melons treated with 1-MCP and stored under refrigeration and modified atmosphere. *Hortica Brasil* 26: 464–470.

DeEll JR, Ayres JT, Murr DP, 2008a, 1-Methylcyclopropene concentration and timing of postharvest application alters the ripening of 'McIntosh' apples during storage. *HortTechnology* 18: 624–630.

DeEll JR, Ehsani-Moghaddam B, 2010, Preharvest 1-methylcyclopropene treatment reduces soft scald in 'Honeycrisp' apples during storage. *HortScience* 45: 414–417.

DeEll JR, Ehsani-Moghaddam B, 2011, Timing of postharvest 1-methylcyclopropene treat-ment affects Bartlett pear quality after storage. *Canad J Plant Sci* 91: 853–858.

DeEll JR, Ehsani-Moghaddam B, 2012, Delayed controlled atmosphere storage affects storage disorders of 'Empire' apples. *Postharv Biol Technol* 67: 167–171.

DeEll JR, Ehsani-Moghaddam B, 2013, Effects of rapid consecutive postharvest 1-methylcy-clopropene treatments on fruit quality and storage disorders in apples. *HortScience* 48: 227–232.

DeEll JR, Murr DP, Ehsani-Moghaddam B, 2008b, 1-Methylcyclopropene treatment modifies postharvest behavior of Fantasia nectarines. *Canad J Plant Sci* 88: 753–758.

DeEll JR, Murr DP, Porteous MD, Rupasinghe HPV, 2002, Influence of temperature and duration of 1-methylcyclopropene (1-MCP) treatment on apple quality. *Postharv Biol Technol* 24: 349–353.

DeEll JR, Murr DP, Willey L, Porteous MD, 2003, 1-Methylcyclopropene (1-MCP) increases CO_2 injury in apples. *Acta Hortic* 600: 277–280.

DeLong JM, Prange RK, Bishop C, Harrison PA, Ryan DAJ, 2003, The influence of 1-MCP on shelf-life quality of highbush blueberry. *HortScience* 38: 417–418.

DeLong JM, Prange RK, Harrison PA, 2004, The influence of 1-methylcyclopropene on 'Cortland' and 'McIntosh' apple quality following long-term storage. *HortScience* 39: 1062–1065.

Dong L, Lurie S, Zhou HW, 2002, Effect of 1-methylcyclopropene on ripening of 'Canino' apricots and 'Royal Zee' plums. *Postharv Biol Technol* 24: 135–145.

Dong XQ, Hube DJ, Rao JP, Lee JH, 2013a, Rapid ingress of gaseous 1-MCP and acute suppression of ripening following short-term application to midclimacteric tomato under hypobaria. *Postharv Biol Technol* 86: 285–290.

Dong XQ, Ramirez-Sanchez M, Huber DJ, Rao JP, Zhang ZK, Choi ST, Lee JH, 2013b, Diffusivity of 1-methylcyclopropene in spinach and bok choi leaf tissue, disks of tomato and avocado fruit tissue, and whole tomato fruit. *Postharv Biol Technol* 78: 40–47.

Dou H, Jones S, Ritenour M, 2005, Influence of 1-MCP application and concentration on post-harvest peel disorders and incidence of decay in citrus fruit. *J Hortic Sci Biotech* 80: 786–792.

Downes K, Chope GA, Terry LA, 2010, Postharvest application of ethylene and 1-methylcyclopropene either before or after curing affects onion (*Allium cepa* L.) bulb quality during long term cold storage. *Postharv Biol Technol* 55: 36–44.

Edagi FK, Sasaki FF, Sestari I, Terra FDM, Giro B, Kluge RA, 2011, 1-Methylcyclopropene and methyl salicylate reduce chilling injury of 'Fukuhara' loquat under cold storage. *Ciencia Rural* 41: 910–916.

Ekman JH, Clayton M, Biasi WV, Mitcham EJ, 2004, Interactions between 1-MCP concentration, treatment interval and storage time for 'Bartlett' pears. *Postharv Biol Technol* 31: 127–136.

Elfving DC, Auvil TD, Castillo F, Drake SR, Kunzel H, Kupferman EM, Lorenz B, McFerson JR, Reed AN, Sater C, Schmidt TR, Visser, DB, 2009, Effects of preharvest applications of ethephon and 1-MCP to 'Bing' sweet cherry on fruit removal force and fruit quality. *J Am Pomol Soc* 63: 84–100.

Elfving DC, Drake SR, Reed AN, Visser DB, 2007, Preharvest applications of sprayable 1-methylcyclopropene in the orchard for management of apple harvest and postharvest condition. *HortScience* 42: 1192–1199.

Ergun M, Karakurt Y, Huber DJ, 2011, Cell-wall modification in 1-methylcyclopropene-treated post-climacteric fresh-cut and intact papaya fruit. *Plant Growth Regulat* 65: 485–494.

Ergun M, Sargent ST, Fox AJ, Crane JH, Huber DJ, 2005, Ripening and quality responses of mamey sapote fruit to postharvest wax and 1-methylcyclopropene treatments. *Postharv Biol Technol* 36: 127–134.

Espinosa I, Ortiz RI, Tovar B, Mata M, Montalvo E, 2013, Physiological and physicochemical behavior of soursop fruits refrigerated with 1-methylcyclopropene. *J Food Quality* 36: 10–20.

Fan X, Mattheis JP, 2000, Reduction of ethylene-induced physiological disorders of carrots and iceberg lettuce by 1-methylcyclopropene. *HortScience* 35: 1312–1314.

Fan XT, Blankenship SM, Mattheis JP, 1999, 1-Methylcyclopropene inhibits apple ripening. *J Am Soc Hortic Sci* 124: 690–695.

Fawbush F, Nock JF, Watkins CB, 2008, External carbon dioxide injury and 1-methylcyclopropene. *Postharv Biol Technol* 48: 92–98.

Fernandez-Leon MF, Fernandez-Leon AM, Lozano M, Ayuso MC, Gonzalez-Gomez D, 2013, Different postharvest strategies to preserve broccoli quality during storage and shelf life: Controlled atmosphere and 1-MCP. *Food Chem* 138: 564–573.

Fernandez-Trujillo JP, Serrano JM, Martinez JA, 2009, Quality of red sweet pepper fruit treated with 1-MCP during a simulated post-harvest handling chain. *Food Sci Technol Int* 15: 23–30.

Forney CF, Song J, Hildebrand PD, Fan L, McRae KB, 2007, Interactive effects of ozone and 1-methylcyclopropene on decay resistance and quality of stored carrots. *Postharv Biol Technol* 45: 341–348.

Freiman ZE, Rodov V, Yablovitz Z, Horev B, Flaishman MA, 2012, Preharvest application of 1-methylcyclopropene inhibits ripening and improves keeping quality of 'Brown Turkey' figs (*Ficus carica* L.). *Scient Hortic* 138: 266–272.

Fu L, Cao J, Li Q, Lin L, Jiang W, 2007, Effect of 1-methylcyclopropene on fruit quality and physiological disorders in Yali pear (*Pyrus bretschneideri* Rehd.) during storage. *Food Sci Technol Int* 13: 49–54.

Gamrasni D, Ben-Arie R, Goldway M, 2010, 1-Methylcyclopropene (1-MCP) application to *Spadona* pears at different stages of ripening to maximize fruit quality after storage. *Postharv Biol Technol* 58: 104–112.

Garcia JAO, Barraza MHP, Valdivia VV, Jaimez RG, 2011, Application of 1-methylcyclopropene (1-MCP) on Mexican plum (*Spondias purpurea* L.). *Revista Fitotecnia Mexic* 34: 197–204.

Golding JB, Shearer D, McGlasson WB, Wyllie SG, 1999, Relationships between respiration, ethylene, and aroma production in ripening banana. *J Agric Food Chem* 47: 1646–1651.

Golding JB, Shearer D, Wyllie SG, McGlasson WB, 1998, Application of 1-MCP and propylene to identify ethylene-dependent ripening processes in mature banana fruit. *Postharv Biol Technol* 14: 87–98.

Gong YP, Fan XT, Mattheis JP, 2002, Responses of 'Bing' and 'Rainier' sweet cherries to ethylene and 1-methylcyclopropene. *J Am Soc Hortic Sci* 127: 831–835.

Goren R, Huberman M, Riov J, Goldschmidt EE, Sisler EC, Apelbaum A, 2011, Effect of 3-cyclopropyl-1-enyl-propanoic acid sodium salt, a novel water soluble antagonist of ethylene action, on plant responses to ethylene. *Plant Growth Regulat* 65: 327–334.

Grozeff GG, Micieli ME, Gomez F, Fernandez L, Guiamet JJ, Chaves AR, Bartoli CG, 2010, 1-Methylcyclopropene extends postharvest life of spinach leaves. *Postharv Biol Technol* 55: 182–185.

Guillén F, Castillo S, Zapata PJ, Martínez-Romero D, Valero D, Serrano M, 2006, Efficacy of 1-MCP treatment in tomato fruit. 2. Effect of cultivar and ripening stage at harvest. *Postharv Biol Technol* 42: 235–242.

Gutierrez MS, Trinchero GD, Cerri AM, Vilella F, Sozzi GO, 2008, Different responses of goldenberry fruit treated at four maturity stages with the ethylene antagonist 1-methylcyclopropene. *Postharv Biol Technol* 48: 199–205.

Harima S, Nakano R, Yamauchi S, Kitano Y, Yamamoto Y, Inaba A, Kubo Y, 2003, Extending shelf-life of astringent persimmon (*Diospyros kaki* Thunb.) fruit by 1-MCP. *Postharv Biol Technol* 29: 319–324.

Hassan FAS, Mahfouz SA, 2010, Effect of 1-methylcyclopropene (1-MCP) treatment on sweet basil leaf senescence and ethylene production during shelf-life. *Postharv Biol Technol* 55: 61–65.

Hassan FAS, Mahfouz SA, 2012, Effect of 1-methylcyclopropene (1-MCP) on the postharvest senescence of coriander leaves during storage and its relation to antioxidant enzyme activity. *Scient Hortic* 141: 69–75.

Hayama H, Ito A, Kashimura Y, 2005, Effect of 1-methylcyclopropene (1-MCP) Treatment under sub-atmospheric pressure on the softening of 'Akatsuki' peach. *J Jap Soc Hortic Sci* 74: 398–400.

Hayama H, Tatsuki M, Nakamura Y, 2008, Combined treatment of aminoethoxyvinylglycine (AVG) and 1-methylcyclopropene (1-MCP) reduces melting-flesh peach fruit softening. *Postharv Biol Technol* 50: 228–230.

Hoeberichts FA, Van der Plas LHW, Woltering EJ, 2002, Ethylene perception is required for the expression of tomato ripening-related genes and associated physiological changes even at advanced stages of ripening. *Postharv Biol Technol* 26: 125–133.

Hofman PJ, Jobin-Decor M, Meiburg GF, Macnish AJ, Joyce DC, 2001, Ripening and quality responses of avocado, custard apple, mango and papaya fruit to 1-methylcyclopropene. *Aust J Expt Agric* 41: 567–572.

Hong KQ, He QG, Xu HB, Xie JH, Hu HG, Gu H, Gong DQ, 2013, Effects of 1-MCP on oxidative parameters and quality in 'Pearl' guava (*Psidium guajava* L.) fruit. *J Hortic Sci Biotech* 88: 117–122.

Hotchkiss JH, Watkins CB, Sanchez DG, 2007, Release of 1-methylcyclopropene from heat-pressed polymer films. *J Food Sci* 72: E330–E334.

Huang SL, Li TT, Jiang GX, Xie WP, Chang SD, Jiang YM, Duan XW, 2012, 1-Methylcyclopropene reduces chilling injury of harvested okra (*Hibiscus esculentus* L.) pods. *Scient Hortic* 141: 42–46.

Huang XM, Zhang ZQ, Duan XW, 2003, Effects of 1-methylcyclopropene on storage quality of hot pepper (*Capsicum frutescens*) at room temperature. *China Veget* 1: 9–11.

Huber DJ, 2008, Suppression of ethylene responses through application of 1-methylcyclopropene: A powerful tool for elucidating ripening and senescence mechanisms in climacteric and nonclimacteric fruits and vegetables. *HortScience* 43: 106–111.

Huber DJ, Hurr BM, Lee JS, Lee JH, 2010, 1-Methylcyclopropene sorption by tissues and cell-free extracts from fruits and vegetables: Evidence for enzymic 1-MCP metabolism. *Postharv Biol Technol* 56: 123–130.

Hurr BM, Huber DJ, Lee JH, 2005, Differential responses in colour changes and softening of 'Florida 47' tomato fruit treated at green and advanced ripening stages with the ethylene antagonist 1-methylcyclopropene. *HortTechnology* 15: 617–622.

Ilic ZS, Trajkovic R, Perzelan Y, Alkalai-Tuvia S, Fallik E, 2012, Influence of 1-methylcyclopropene (1-MCP) on postharvest storage quality in green bell pepper fruit. *Food Bioprocess Technol* 5: 2758–2767.

Jeong J, Brecht JK, Huber DJ, Sargent SA, 2008, Storage life and deterioration of intact cantaloupe (*Cucumis melo* L. var. *reticulatus*) fruit treated with 1-methylcyclopropene and fresh-cut cantaloupe prepared from fruit treated with 1-methylcyclopropene before processing. *HortScience* 43: 435–438.

Jeong J, Lee J, Huber DJ, 2007, Softening and ripening of 'Athena' cantaloupe (*Cucumis melo* L. var. *reticulatus*) fruit at three harvest maturities in response to the ethylene antagonist 1-methylcyclopropene. *HortScience* 42: 1231–1236.

Jiang WB, Sheng Q, Jiang YM, Zhou XJ, 2004, Effects of 1-methylcyclopropene and gibberellic acid on ripening of Chinese jujube (*Zizyphus jujuba* M.) in relation to quality. *J Sci Food Agric* 84: 31–35.

Jiang WB, Sheng Q, Zhou XJ, Zhang MJ, Liu XJ, 2002, Regulation of detached coriander leaf senescence by 1-methylcyclopropene and ethylene. *Postharv Biol Technol* 26: 339–345.

Jiang YM, Joyce DC, Terry LA, 2001, 1-Methylcyclopropene treatment affects strawberry fruit decay. *Postharv Biol Technol* 23: 227–232.

Jin P, Shang HT, Chen JJ, Zhu H, Zhao YY, Zheng YH, 2011, Effect of 1-methylcyclopropene on chilling injury and quality of peach fruit during cold storage. *J Food Sci* 76: S485–S491.

John-Karuppiah KJ, Burns JK, 2010, Expression of ethylene biosynthesis and signaling genes during differential abscission responses of sweet orange leaves and mature fruit. *J Am Soc Hortic Sci* 135: 456–464.

Jung SK, Lee JM, 2009, Effects of 1-methylcyclopropene (1-MCP) on ripening of apple fruit without cold storage. *J Hortic Sci Biotech* 84: 102–106.

Jung SK, Watkins CB, 2011, Involvement of ethylene in browning development of controlled atmosphere-stored 'Empire' apple fruit. *Postharv Biol Technol* 59: 219–226.

Jung SK, Watkins CB, 2014, Internal ethylene concentrations in apple fruit at harvest affect sensitivity of fruit to 1-methylcyclopropene. *Postharvest Biol Technol* 96: 1–6.

Kashimura Y, Hayama H, Ito A, 2010, Infiltration of 1-methylcyclopropene under low pressure can reduce the treatment time required to maintain apple and Japanese pear quality during storage. *Postharv Biol Technol* 57: 14–18.

Kenigsbuch D, Chalupowicz D, Aharon Z, Maurer D, Aharoni N, 2007, The effect of CO_2 and 1-methylcyclopropene on the regulation of postharvest senescence of mint, *Mentha longifolia* L. *Postharv Biol Technol* 43: 165–173.

Kenigsbuch D, Ovadia A, Chalupowicz D, Maurer D, Aharon Z, Aharoni N, 2009, Post-harvest leaf abscission in summer-grown basil (*Ocimum basilicum* L.) may be controlled by combining a pre-treatment with 1-MCP and moderately raised CO_2. *J Hortic Sci Biotech* 84: 291–294.

Ketsa S, Wisutiamonkul A, van Doorn WG, 2013, Apparent synergism between the positive effects of 1-MCP and modified atmosphere on storage life of banana fruit. *Postharv Biol Technol* 85: 173–178.

Kluge RA, Jomori MLL, Jacomino AP, Vitti MCD, Padula M, 2003, Intermittent warming in 'Tahiti' lime treated with an ethylene inhibitor. *Postharv Biol Technol* 29: 195–203.

Kostansek E, Pereira W, 2003, Successful application of 1-MCP in commercial storage facilities. *Acta Hortic* 628: 213–219.

Koukounaras A, Siomos AS, Sfakiotakis E, 2006, 1-Methylcyclopropene prevents yellowing of rocket ethylene induced leaves. *Postharv Biol Technol* 41: 109–111.

Kramer M, Bufler G, Ulrich D, Leitenberger M, Conrad J, Carle R, Kammerer DR, 2012, Effect of ethylene and 1-methylcyclopropene on bitter compounds in carrots (*Daucus carota* L.). *Postharv Biol Technol* 73: 28–36.

Ku KM, Choi JH, Kim HS, Kushad MM, Jeffery EH, Juvik JA, 2013, Methyl jasmonate and 1-methylcyclopropene treatment effects on quinone reductase inducing activity and post-harvest quality of broccoli. PLoS ONE 8(10).

Ku VVV, Wills RBH, 1999, Effect of 1-methylcyclopropene on the storage life of broccoli. *Postharv Biol Technol* 17: 127–132.

Ku VVV, Wills RBH, Ben-Yehoshua S, 1999, 1-Methylcyclopropene can differentially affect the postharvest life of strawberries exposed to ethylene. *HortScience* 34: 119–120.

Kubo Y, Hiwasa K, Owino WO, Nakano R, Inaba A, 2003, Influence of time and concentration of 1-MCP application on the shelf life of pear 'La France' fruit. *HortScience* 38: 1414–1416.

Lee JS, Huber DJ, Watkins CB, Hurr BM, 2012, Influence of wounding and aging on 1-MCP sorption and metabolism in fresh-cut tissue and cell-free homogenates from apple fruit. *Postharv Biol Technol* 67: 52–58.

Lee UY, Oh KY, Moon SJ, Hwang YS, Chun JP, 2012, Effects of 1-methylcyclopropene (1-MCP) on fruit quality and occurrence of physiological disorders of asian pear (*Pyrus pyrifolia*), 'Wonhwang' and 'Whasan', duning shelf-life. *Korean J Hortic Sci Technol* 30: 534–542.

Lee YS, Beaudry R, Kim JN, Harte BR, 2006, Development of a 1-methylcyclopropene (1-MCP) sachet release system. *J Food Sci* 71: C1–C6.

Lerslerwong L, Rugkong A, Imsabai W, Ketsa S, 2013, The harvest period of mangosteen fruit can be extended by chemical control of ripening-A proof of concept study. *Scient Hortic* 157: 13–18.

Li CR, Shen WB, Lu WJ, Jiang YM, Xie JH, Chen JY, 2009, 1-MCP delayed softening and affected expression of XET and EXP genes in harvested cherimoya fruit. *Postharv Biol Technol* 52: 254–259.

Li FJ, Zhang XH, Song BC, Li JZ, Shang ZL, Guan JF, 2013, Combined effects of 1-MCP and MAP on the fruit quality of pear (*Pyrus bretschneideri* Reld cv. Laiyang) during cold storage. *Scient Hortic* 164: 544–551.

Li JG, Yuan RC, 2008, NAA and ethylene regulate expression of genes related to ethylene biosynthesis, perception, and cell wall degradation during fruit abscission and ripening in 'Delicious' apples. *J Plant Growth Regulat* 27: 283–295.

Li N, Wang XQ, Ma SJ, Cao SF, Yang ZF, Wang XX, Zheng YH, 2006, Effects of 1-methylcyclopene on fruit decay and quality in Chinese bayberry. *Acta Hort* 712: 699–703.

Li Q, Wu FW, Li TT, Su XG, Jiang GG, Qu HX, Jiang YM, Duan XW, 2012, 1-Methylcyclopropene extends the shelf-life of 'Shatangju' mandarin (*Citrus reticulate* Blanco) fruit with attached leaves. *Postharv Biol Technol* 67: 92–95.

Li ZQ, Wang LJ, 2009, Effect of 1-methylcyclopropene on ripening and superficial scald of Japanese pear (*Pyrus pyrifolia* Nakai, cv. Akemizu) fruit at two temperatures. *Food Sci Technol Res* 15: 483–490.

Lim BS, Yun SK, Nam EY, Chun JP, Cho MA, Chung DS, 2013, Effects of storage temperature and 1-MCP treatment on postharvest quality in plumcot hybrid cv. Harmony. *Korean J Hortic Sci Technol* 31: 203–210.

Lima LCO, Hurr BM, Huber DJ, 2005, Deterioration of beit alpha and slicing cucumbers (*Cucumis sativus* L.) during storage in ethylene or air: Responses to suppression of ethylene perception and parallels to natural senescence. *Postharv Biol Technol* 37: 265–276.

Liu RL, Lai TF, Xu Y, Tian SP, 2013, Changes in physiology and quality of Laiyang pear in long time storage. *Scient Hortic* 150: 31–36.

Liu ZY, Jiang WB, 2006, Lignin deposition and effect of postharvest treatment on lignification of green asparagus (*Asparagus officinalis* L.). *Plant Growth Regulat* 48: 187–193.

Lomaniec E, Aharoni Z, Aharoni N, Lers A, 2003, Effect of the ethylene action inhibitor 1-methylcyclopropene on parsley leaf senescence and ethylene biosynthesis. *Postharv Biol Technol* 30: 67–74.

Lu CW, Cureatz V, Toivonen PMA, 2009, Improved quality retention of packaged 'Anjou' pear slices using a 1-methylcyclopropene (1-MCP) co-release technology. *Postharv Biol Technol* 51: 378–383.

Lu XG, Nock JF, Ma YP, Liu XH, Watkins CB, 2013, Effects of repeated 1-methylcyclopropene (1-MCP) treatments on ripening and superficial scald of 'Cortland' and 'Delicious' apples. *Postharv Biol Technol* 78: 48–54.

Lulai EC, Suttle JC, 2004, The involvement of ethylene in wound-induced suberization of potato tuber (*Solanum tuberosum* L.): A critical assessment. *Postharv Biol Technol* 34: 105–112.

Ma G, Zhang LC, Kato M, Yamawaki K, Asai T, Nishikawa F, Ikoma Y, Matsumoto H, 2011, Effect of 1-methylcyclopropene on the expression of genes for ascorbate metabolism in postharvest cauliflower. *J Japan Soc Hortic Sci* 80: 512–520.

Ma WP, Cao JK, Ni ZJ, Tian WN, Zhao YM, Jiang WB, 2012, Effects of 1-methylcyclopropene on storage quality and antioxidant activity of harvested "Yujinxiang" melon (*Cucumis melo* L.) fruit. *J Food Biochem* 36: 413–420.

MacKinnon DK, Shaner D, Nissen S, Westra P, 2009, The effects of surfactants, nozzle types, spray volumes, and simulated rain on 1-methylcyclopropene efficacy on tomato plants. *HortScience* 44: 1600–1603.

MacLean DD, NeSmith DS, 2011, Rabbiteye blueberry postharvest fruit quality and stimulation of ethylene production by 1-methylcyclopropene. *HortScience* 46: 1278–1281.

Mahajan BVC, Sharma SR, 2012, Effect of 1-methylcyclopropene on storage life and quality of peaches. *Indian J Hortic* 69: 263–267.

Manenoi A, Bayogan ERV, Thumdee S, Paull RE, 2007, Utility of 1-methylcyclopropene as a papaya postharvest treatment. *Postharv Biol Technol* 44: 55–62.

Manganaris GA, Crisosto CH, Bremer V, Holcroft D 2008 Novel 1-methylcyclopropene immersion formulation extends shelf life of advanced maturity 'Joanna Red' plums (*Prunus salicina* Lindell). *Postharv Biol Technol* 47: 429–433.

Manganaris GA, Vicente AR, Crisosto CH, Labavitch JM, 2007, Effect of dips in a 1-methylcyclopropene-generating solution on 'Harrow Sun' plums stored under different temperature regimes. *J Agric Food Chem* 55: 7015–7020.

Mao LC, Karakurt Y, Huber DJ, 2004, Incidence of water-soaking and phospholipid catabolism in ripe watermelon (*Citrullus lanatus*) fruit: Induction by ethylene and prophylactic effects of 1-methylcyclopropene. *Postharv Biol Technol* 33: 1–9.

Marini RP, Byers RE, Sowers DL, 1993, Repeated applications of NAA control preharvest drop of Delicious apples. *J Hortic Sci* 68: 247–253.

Massolo JF, Concellon A, Chaves AR, Vicente AR, 2011, 1-Methylcyclopropene (1-MCP) delays senescence, maintains quality and reduces browning of non-climacteric eggplant (*Solanum melongena* L.) fruit. *Postharv Biol Technol* 59: 10–15.

Massolo JF, Concellon A, Chaves AR, Vicente AR, 2013, Use of 1-methylcyclopropene to complement refrigeration and ameliorate chilling injury symptoms in summer squash. *CyTA J Food* 11: 19–26.

Mattheis JP, 2008, How 1-methylcyclopropene has altered the Washington state apple industry. *HortScience* 43: 99–101.

McArtney S, Parker M, Obermiller J, Hoyt T, 2011, Effects of 1-methylcyclopropene on firmness loss and the development of rots in apple fruit kept in farm markets or at elevated temperatures. *HortTechnology* 21: 494–499.

McArtney SJ, Obermiller JD, Hoyt T, Parker ML, 2009, 'Law Rome' and 'Golden Delicious' apples differ in their response to preharvest and postharvest 1-methylcyclopropene treatment combinations. *HortScience* 44: 1632–1636.

McArtney SJ, Obermiller JD, Schupp JR, Parker ML, Edgington TB, 2008, Preharvest 1-methylcyclopropene delays fruit maturity and reduces softening and superficial scald of apples during long-term storage. *HortScience* 43: 366–371.

McCollum G, Maul P, 2007, 1-Methylcyclopropene inhibits degreening but stimulates respiration and ethylene biosynthesis in grapefruit. *HortScience* 42: 120–124.

Meyer MD, Terry LA, 2010, Fatty acid and sugar composition of avocado, cv. Hass, in response to treatment with an ethylene scavenger or 1-methylcyclopropene to extend storage life. *Food Chem* 121: 1203–1210.

Mir NA, Curell E, Khan N, Whitaker M, Beaudry RM, 2001, Harvest maturity, storage temperature, and 1-MCP application frequency alter firmness retention and chlorophyll fluorescence of 'Redchief Delicious' apples. *J Am Soc Hortic Sci* 126: 618–624.

Mishra A, Khare S, Trivedi PK, Nath P, 2008, Ethylene induced cotton leaf abscission is associated with higher expression of cellulase (*GhCell*) and increased activities of ethylene biosynthesis enzymes in abscission zone. *Plant Physiol Biochem* 46: 54–63.

Moran RE, 2006, Maintaining fruit firmness of 'McIntosh' and 'Cortland' apples with aminoethoxyvinylglycine and 1-methylcyclopropene during storage. *HortTechnology* 16: 513–516.

Moya-Leon MA, Moya M, Herrera R, 2004, Ripening of mountain papaya (*Vasconcellea pubescens*) and ethylene dependence of some ripening events. *Postharv Biol Technol* 34: 211–218.

Mozetic B, Simcic M, Trebse P, 2006, Anthocyanins and hydroxycinnamic acids of Lambert Compact cherries (*Prunus avium* L.) after cold storage and 1-methylcyclopropene treatment. *Food Chem* 97: 302–309.

Mworia EG, Yoshikawa T, Salikon N, Oda C, Fukuda T, Suezawa K, Asiche WO, Ushijima K, Nakano R, Kubo Y, 2011, Effect of MA storage and 1-MCP on storability and quality of 'Sanuki Gold' kiwifruit harvested at two different maturity stages. *J Japan Soc Hortic Sci* 80: 372–377.

Nanthachai N, Ratanachinakorn B, Kosittrakun M, Beaudry RM, 2007, Absorption of 1-MCP by fresh produce. *Postharv Biol Technol* 43: 291–297.

Nilsson T, 2005, Effects of ethylene and 1-MCP on ripening and senescence of European seedless cucumbers. *Postharv Biol Technol* 36: 113–125.

Nock JF, Watkins CB, 2013, Repeated treatment of apple fruit with 1-methylcyclopropene (1-MCP) prior to controlled atmosphere storage. *Postharvest Biol Technol* 79: 73–79.

Ortiz A, Vendrell M, Lara I, 2011, Softening and cell wall metabolism in late-season peach in response to controlled atmosphere and 1-MCP treatment. *J Hortic Sci Biotech* 86: 175–181.

Ortiz CM, Mauri AN, Vicente AR, 2013, Use of soy protein based 1-methylcyclopropene-releasing pads to extend the shelf life of tomato (*Solanum lycopersicum* L.) fruit. *Innov Food Sci Emerging Technol* 20: 281–287.

Ouzounidou G, Papadopoulou KK, Asfi M, Mirtziou I, Gaitis F, 2013, Efficacy of different chemicals on shelf life extension of parsley stored at two temperatures. *Int J Food Sci Technol* 48: 1610–1617.

Oz AT, 2011, Combined effects of 1-methyl cyclopropene (1-MCP) and modified atmosphere packaging (MAP) on different ripening stages of persimmon fruit during storage. *African J Biotech* 10: 807–814.

Pang X, Zhang Z, Duan X, Ji Z, 2001, The effects of ethylene and 1-methylcyclopropene on pericarp browning of postharvest lychee fruit. *J South China Agric Univ* 22: 11–14.

Park MH, Shin YS, Kim SJ, Kim JG, 2013, Effect of 1-methylcyclopropene treatment on extension of freshness and storage potential of fresh ginseng. *Korean J Hortic Sci Technol* 31: 308–316.

Pelayo C, Vilas-Boas EVD, Benichou M, Kader AA, 2003, Variability in responses of partially ripe bananas to 1-methylcyclopropene. *Postharv Biol Technol* 28:3. Volatile profiles of ripening West Indian and Guatemalan-West Indian avocado cultivars as affected by aqueous 1-methylcyclopropene. *Postharv Biol Technol* 80: 37–46.

Pereira MEC, Tieman DM, Sargent SA, Klee HJ, Huber DJ, 2013, Volatile profiles of ripening West Indian and Guatemalan-West Indian avocado cultivars as affected by aqueous 1-methylcyclopropene. *Postharv Biol Technol* 80: 37–46.

Pinheiro ACM, Boas E, Bolini HMA, 2010, Extension of postharvest life of 'Apple' banana subjected to 1-methylcyclopropene (1-MCP)–sensory and physical quality. *Ciencia Tecnol Alimentos* 30: 132–137.

Plainsirichai M, Trinok U, Turner DW, 2010, 1-Methylcyclopropene (1-MCP) reduces water loss and extends shelf life of fruits of Rose apple (*Syzygium jambos* Alston) cv. Tabtim Chan. *Fruits* 65: 133–140.

Porat R, Weiss B, Cohen L, Daus A, Goren R, Droby S, 1999, Effects of ethylene and 1-methylcyclopropene on the postharvest qualities of 'Shamouti' oranges. *Postharv Biol Technol* 15: 155–163.

Porat R, Weiss B, Zipori I, Dag A, 2009, Postharvest longevity and responsiveness of guava varieties with distinctive climacteric behaviors to 1-methylcyclopropene. *HortTechnology* 19: 580–585.

Porter KL, Collins G, Klieber A, 2005, 1-MCP does not improve the shelf-life of Chinese cabbage. *J Sci Food Agric* 85: 293–296.

Prange RK, Daniels-Lake BJ, Jeong JC, Binns M, 2005, Effects of ethylene and 1-methylcyclopropene on potato tuber sprout control and fry colour. *Am J Potato Res* 82: 123–128.

Rizzolo A, Cambiaghi P, Grassi M, Zerbini PE, 2005, Influence of 1-methylcyclopropene and storage atmosphere on changes in volatile compounds and fruit quality of Conference pears. *J Agric Food Chem* 53: 9781–9789.

Rodriguez-Perez LC, Harte B, Auras R, Burgess G, Beaudry RM, 2009, Measurement and prediction of the concentration of 1-methylcyclopropene in treatment chambers containing different packaging materials. *J Sci Food Agric* 89: 2581–2587.

Rupasinghe HPV, Murr DP, Paliyath G, Skog L, 2000, Inhibitory effect of 1-MCP on ripening and superficial scald development in 'McIntosh' and 'Delicious' apples. *J Hortic Sci Biotech* 75: 271–276.

Rupavatharam S, East AR, Heyes JA, 2015, Re-evaluation of harvest timing in 'Unique' feijoa using 1-MCP and exogenous ethylene treatments. *Postharv Biol Technol* 99: 152–159.

Sabir FK, Agar IT, 2011, Influence of different concentrations of 1-methylcyclopropene on the quality of tomato harvested at different maturity stages. *J Sci Food Agric* 91: 2835–2843.

Salman A, Filgueiras H, Cristescu S, Lopez-Lauri F, Harren F, Sallanon H, 2009, Inhibition of wound-induced ethylene does not prevent red discoloration in fresh-cut endive (*Cichorium intybus* L.). *Eur Food Res Technol* 228: 651–657.

Saltveit ME, 2004, Effect of 1-methylcyclopropene on phenylpropanoid metabolism, the accumulation of phenolic compounds, and browning of whole and fresh-cut 'iceberg' lettuce. *Postharv Biol Technol* 34: 75–80.

Salvador A, Carvalho CP, Monterde A, Martinez-Javega JM, 2006, 1-MCP effect on chilling injury development in 'Nova' and 'Ortanique' mandarins. *Food Sci Technol Int* 12: 165–170.

Schupp JR, Greene DW, 2004, Effect of aminoethoxyvinylglycine (AVG) on preharvest drop, fruit quality, and maturation of 'McIntosh' apples. I. Concentration and timing of dilute applications of AVG. *HortScience* 39: 1030–1035.

Seglie L, Martina K, Devecchi M, Roggero C, Trotta F, Scariot V, 2011, The effects of 1-MCP in cyclodextrin-based nanosponges to improve the vase life of *Dianthus caryophyllus* cut flowers. *Postharv Biol Technol* 59: 200–205.

Seglie L, Sisler EC, Mibus H, Serek M, 2010, Use of a non-volatile 1-MCP formulation, N,N-dipropyl(1-cyclopropenylmethyl)amine, for improvement of postharvest quality of ornamental crops. *Postharv Biol Technol* 56: 117–122.

Selvarajah S, Bauchot AD, John P, 2001, Internal browning in cold-stored pineapples is suppressed by a postharvest application of 1-methylcyclopropene. *Postharv Biol Technol* 23: 167–170.

Serna L, Torres LS, Ayala AA, 2012, Effect of pre- and post-harvest application of 1-methylcyclopropene on the maturation of yellow pitahaya (*Selenicerus megalanthus* Haw). *Vitae-Revista Facultad Quimica Farmaceutica* 19: 49–59.

Sharma S, Sharma RR, Pal RK, Paul V, Dahuja A, 2012, 1-Methylcyclopropene influences biochemical attributes and fruit softening enzymes of 'Santa Rosa' Japanese plum (*Prunus salicina* Lindl.). *J Plant Biochem Biotech* 21: 295–299.

Singh SP, Pal RK, 2008, Response of climacteric-type guava (*Psidium guajava* L.) to postharvest treatment with 1-MCP. *Postharv Biol Technol* 47: 307–314.

Sisler EC, 2006, The discovery and development of compounds counteracting ethylene at the receptor level. *Biotech Advances* 24: 357–367.

Sisler EC, 2008, 1-Alkenes: Ethylene action compounds or ethylene competitive inhibitors in plants. *Plant Sci* 175: 145–148.

Sisler EC, Blankenship SM, 1996, Method of counteracting an ethylene response in plants. Washington, DC, 1996. (United States patent 5518988, 1997).

Sisler EC, Goren R, Apelbaum A, Serek M, 2009, The effect of dialkylamine compounds and related derivatives of 1-methylcyclopropene in counteracting ethylene responses in banana fruit. *Postharv Biol Technol* 51: 43–48.

Sisler EC, Serek M, 1997, Inhibitors of ethylene responses in plants at the receptor level: Recent developments. *Physiol Plantar* 100: 577–582.

Sisler EC, Serek M, 2003, Compounds interacting with the ethylene receptor in plants. *Plant Biol* 5: 473–480.

Sivakumar D, Korsten L, 2010, Fruit quality and physiological responses of litchi cultivar McLean's Red to 1-methylcyclopropene pre-treatment and controlled atmosphere storage conditions. *LWT-Food Sci Technol* 43: 942–948.

Sivakumar D, Van Deventer F, Terry LA, Polanta GA, Korsten L, 2012, Combination of 1-methylcyclopropene treatment and controlled atmosphere storage retains overall fruit quality and bioactive compounds in mango. *J Sci Food Agric* 92: 821–830.

Smock RM, Gross CR, 1947, The effect of some hormone materials on the respiration and softening rates of apples. *Proc Am Soc Hortic Sci* 49: 67–77.

Song LL, Gao HY, Chen WX, Chen HJ, Mao JL, Zhou YJ, Duan XW, Joyce DC, 2011, The role of 1-methylcyclopropene in lignification and expansin gene expression in peeled water bamboo shoot (*Zizania caduciflora* L.). *J Sci Food Agric* 91: 2679–2683.

Sozzi GO, Abrajan-Villasenor MA, Trinchero GD, Fraschina AA, 2005, Postharvest response of 'Brown Turkey' figs (*Ficus carica* L.) to the inhibition of ethylene perception. *J Sci Food Agric* 85: 2503–2508.

Steffens CA, do Amarante CVT, Chechi R, Silveira JPG, Brackmann A, 2009, Preharvest spraying with plant regulators aiming fruit maturity delay of 'Laetitia' plums. *Ciencia Rural* 39: 1369–1373.

Steffens CA, Tanaka H, do Amarante CVT, Brackmann A, Stanger MC, Hendges MV, 2013, Conditions of a controlled atmosphere for the storage of 'Laetitia' plums treated with 1-methylcyclopropene. *Revista Ciencia Agron* 44: 750–756.

Su H, Gubler WD, 2012, Effect of 1-methylcyclopropene (1-MCP) on reducing postharvest decay in tomatoes (*Solanum lycopersicum* L.). *Postharv Biol Technol* 64: 133–137.

Sun B, Yan HZ, Liu N, Wei J, Wang QM, 2012, Effect of 1-MCP treatment on postharvest quality characters, antioxidants and glucosinolates of Chinese kale. *Food Chem* 131: 519–526.

Szczerbanik MJ, Scott KJ, Paton JE, Best DJ, 2005, Effects of polyethylene bags, ethylene absorbent and 1-methylcyclopropene on the storage of Japanese pears. *J Hortic Sci Biotech* 80: 162–166.

Tassoni A, Watkins CB, Davies PJ, 2006, Inhibition of the ethylene response by 1-MCP in tomato suggests that polyamines are not involved in delaying ripening, but may moderate the rate of ripening or over-ripening. *J Expt Bot* 57: 3313–3325.

Tatsuki M, Endo A, Ohkawa H, 2007, Influence of time from harvest to 1-MCP treatment on apple fruit quality and expression of genes for ethylene biosynthesis enzymes and ethylene receptors. *Postharv Biol Technol* 43: 28–35.

Tay SL, Perera CO, 2004, Effect of 1-methylcyclopropene treatment and edible coatings on the quality of minimally processed lettuce. *J Food Sci* 69: C131–C135.

Tesniere C, Pradal M, El-Kereamy A, Torregrosa L, Chatelet P, Roustan JP, Chervin C, 2004, Involvement of ethylene signalling in a non-climacteric fruit: New elements regarding the regulation of ADH expression in grapevine. *J Expt Bot* 55: 2235–2240.

Toivonen PMA, 2008, Application of 1-methylcyclopropene in fresh-cut/minimal processing systems. *HortScience* 43: 102–105.

Toivonen PMA, Lu CW, 2005, Studies on elevated temperature, short-term storage of 'Sunrise' Summer apples using 1-MCP to maintain quality. *J Hortic Sci Biotech* 80: 439–446.

Trinchero GD, Sozzi GO, Covatta F, Fraschina AA, 2004, Inhibition of ethylene action by 1-methylcyclopropene extends postharvest life of "Bartlett" pears. *Postharv Biol Technol* 32: 193–204.

Vallejo F, Beaudry R, 2006, Depletion of 1-MCP by 'non-target' materials from fruit storage facilities. *Postharv Biol Technol* 40: 177–182.

Varanasi V, Shin S, Johnson F, Mattheis JP, Zhu YM, 2013, Differential suppression of ethylene biosynthesis and receptor genes in 'Golden Delicious' apple by preharvest and postharvest 1-MCP treatments. *J Plant Growth Regulat* 32: 585–595.

Vieira MJ, Argenta LC, do Amarante CVT, Vieira A, Steffens CA, 2012, Postharvest quality of 'Hayward' kiwifruit treated with 1-MCP and stored under different atmospheres. *Revista Brasil Fruticult* 34: 400–408.

Villalobos-Acuna M, Mitcham EJ, 2008, Ripening of European pears: The chilling dilemma. *Postharv Biol Technol* 49: 187–200.

Villalobos-Acuna MG, Biasi WV, Flores S, Jiang CZ, Reid MS, Willits NH, Mitcham EJ, 2011a, Effect of maturity and cold storage on ethylene biosynthesis and ripening in 'Bartlett' pears treated after harvest with 1-MCP. *Postharv Biol Technol* 59: 1–9.

Villalobos-Acuna MG, Biasi WV, Flores S, Mitcham EJ, Elkins RB, Willits NH, 2010, Preharvest application of 1-methylcyclopropene influences fruit drop and storage potential of 'Bartlett' pears. *HortScience* 45: 610–616.

Villalobos-Acuna MG, Biasi WV, Mitcham EJ, Holcroft D, 2011b, Fruit temperature and ethylene modulate 1-MCP response in 'Bartlett' pears. *Postharv Biol Technol* 60: 17–23.

Villarreal NM, Bustamante CA, Civello PM, Martinez GA, 2010, Effect of ethylene and 1-MCP treatments on strawberry fruit ripening. *J Sci Food Agric* 90: 683–689.

Wang BG, Jiang WB, Liu HX, Lin L, Wang JH, 2006, Enhancing the post-harvest qualities of mango fruit by vacuum infiltration treatment with 1-methylcyclopropene. *J Hortic Sci Biotech* 81: 163–167.

Wang BG, Wang JH, Feng XY, Lin L, Zhao YM, Jiang WB, 2009, Effects of 1-MCP and exogenous ethylene on fruit ripening and antioxidants in stored mango. *Plant Growth Regulat* 57: 185–192.

Wang YQ, Zhang LJ, Zhu SJ, 2014, 1-Methylcyclopropene (1-MCP)-induced protein expression associated with changes in Tsai Tai (*Brassica chinensis*) leaves during low temperature storage. *Postharv Biol Technol* 87: 120–125.

Watkins CB, 2006, The use of 1-methylcyclopropene (1-MCP) on fruits and vegetables. *Biotech Adv* 24: 389–409.

Watkins CB, 2008, Overview of 1-methylcyclopropene trials and uses for edible horticultural crops. *HortScience* 43: 86–94.

Watkins CB, 2010, Managing physiological processes in fruits and vegetables with inhibitors of ethylene biosynthesis and perception. *Acta Hortic* 880: 301–310.

Watkins CB, James H, Nock JF, Reed N, Oakes RL, 2010, Preharvest application of 1-methylcyclopropene (1-MCP) to control fruit drop of apples, and its effects on postharvest quality. *Acta Hortic* 877: 365–373.

Watkins CB, Nock JF, 2005, Effects of delays between harvest and 1-methylcyclopropene treatment, and temperature during treatment, on ripening of air-stored and controlled-atmosphere-stored apples. *HortScience* 40: 2096–2101.

Watkins CB, Nock JF, 2012, Rapid 1-methylcyclopropene (1-MCP) treatment and delayed controlled atmosphere storage of apples. *Postharv Biol Technol* 69: 24–31.

Watkins CB, Nock JF, Whitaker BD, 2000, Responses of early, mid and late season apple cultivars to postharvest application of 1-methylcyclopropene (1-MCP) under air and controlled atmosphere storage conditions. *Postharv Biol Technol* 19: 17–32.

Wills RBH, Ku VVV, 2002, Use of 1-MCP to extend the time to ripen of green tomatoes and postharvest life of ripe tomatoes. *Postharv Biol Technol* 26: 85–90.

Win TO, Srilaong V, Heyes J, Kyu KL, Kanlayanarat S, 2006, Effects of different concentrations of 1-MCP on the yellowing of West Indian lime (*Citrus aurantifolia*, Swingle) fruit. *Postharv Biol Technol* 42: 23–30.

Wu CW, Du XF, Wang LZ, Wang WZ, Zhou Q, Tian XJ, 2009, Effect of 1-methylcyclopropene on postharvest quality of Chinese chive scapes. *Postharv Biol Technol* 51: 431–433.

Yuan RC, Carbaugh DH, 2007, Effects of NAA, AVG, and 1-MCP on ethylene biosynthesis, preharvest fruit drop, fruit maturity, and quality of 'Golden Supreme' and 'Golden Delicious' apples. *HortScience* 42: 101–105.

Yuan RC, Li JG, 2008, Effect of sprayable 1-MCP, AVG, and NAA on ethylene biosynthesis, preharvest fruit drop, fruit maturity, and quality of 'Delicious' apples. *HortScience* 43: 1454–1460.

Zhang LB, Wang GA, Chang JM, Liu JS, Cai JH, Rao XW, Zhang LJ, Zhong JJ, Xie JH, Zhu SJ, 2010a, Effects of 1-MCP and ethylene on expression of three CAD genes and lignification in stems of harvested Tsai Tai (*Brassica chinensis*). *Food Chem* 123: 32–40.

Zhang P, Zhang M, Wang SJ, Wu ZS, 2012a, Effect of 1-methylcyclopropene treatment on green asparagus quality during cold storage. *Int Agrophys* 26: 407–411.

Zhang Q, Zhen Z, Jiang H, Li XG, Liu JA, 2011a, Encapsulation of the ethylene inhibitor 1-methylcyclopropene by Cucurbit 6 uril. *J Agric Food Chem* 59: 10539–10545.

Zhang Z, Huber DJ, Hurr BM, Rao J, 2009a, Delay of tomato fruit ripening in response to 1-methylcyclopropene is influenced by internal ethylene levels. *Postharv Biol Technol* 54: 1–8.

Zhang Z, Huber DJ, Rao J, 2011b, Ripening delay of mid-climacteric avocado fruit in response to elevated doses of 1-methylcyclopropene and hypoxia-mediated reduction in internal ethylene concentration. *Postharv Biol Technol* 60: 83–91.

Zhang ZK, Huber DJ, Hurr BM, Rao J, 2009b, Delay of tomato fruit ripening in response to 1-methylcyclopropene is influenced by internal ethylene levels. *Postharv Biol Technol* 54: 1–8.

Zhang ZK, Huber DJ, Hurr BM, Rao JP, 2010b, Short-term hypoxic hypobaria transiently decreases internal ethylene levels and increases sensitivity of tomato fruit to subsequent 1-methylcyclopropene treatments. *Postharv Biol Technol* 56: 131–137.

Zhang ZK, Huber DJ, Rao JP, 2013, Antioxidant systems of ripening avocado (*Persea americana* Mill.) fruit following treatment at the preclimacteric stage with aqueous 1-methylcyclopropene. *Postharv Biol Technol* 76: 58–64.

Zhang ZQ, Tian SP, Zhu Z, Xu Y, Qin GZ, 2012b, Effects of 1-methylcyclopropene (1-MCP) on ripening and resistance of jujube (*Zizyphus jujuba* cv. Huping) fruit against postharvest disease. *LWT-Food Sci Technol* 45: 13–19.

Zhong GY, Huberman M, Feng XQ, Sisler EC, Holland D, Goren R, 2001, Effect of 1-methylcyclopropene on ethylene-induced abscission in citrus. *Physiol Plant* 113: 134–141.

Zhong QP, Xia WS, 2007, Effect of 1-methylcyclopropene and/or chitosan coating treatments on storage life and quality maintenance of Indian jujube fruit. *LWT-Food Sci Technol* 40: 404–411.

Zhu H, Yuan RC, Greene DW, Beers EP, 2010, Effects of 1-methylcyclopropene and naphthaleneacetic acid on fruit set and expression of genes related to ethylene biosynthesis and perception and cell-wall degradation in apple. *J Am Soc Hortic Sci* 135: 402–409.

Ziliotto F, Begheldo M, Rasori A, Bonghi C, Tonutti P, 2008, Transcriptome profiling of ripening nectarine (*Prunus persica* L. Batsch) fruit treated with 1-MCP. *J Expt Bot* 59: 2781–2791.

7 Advances in Edible Coatings

María Serrano, Domingo Martínez-Romero, Pedro J. Zapata, Fabián Guillén, Juan M. Valverde, Huertas M. Díaz-Mula, Salvador Castillo, and Daniel Valero

CONTENTS

7.1 CONCEPT AND TYPES OF EDIBLE COATINGS

Edible coating can be defined as a thin layer of edible material formed as a coating on a food product and is usually applied by immersing the product in a solution of the coating. This is in contrast to an edible film, which is a preformed, thin layer of edible material that is placed as a wrapping on the food product (Falguera et al., 2011). Food industries and packaging manufacturers have joined in efforts to reduce the amount of food packaging materials, mainly due to environmental and consumer concerns. The development of edible coatings has received much attention in recent years due to advantages such as consumption with the food product, sometimes increasing its organoleptic properties and being produced from agricultural and marine renewable sources, as well as by using several fungal species (ED, 1995, 1998; FDA, 2006; Bourtoom, 2008).

Components of edible coatings can be divided into three categories: hydrocolloids, lipids, and composites. Composites generally contain both hydrocolloid components and lipids and represent a good strategy for enhancing coating properties by taking advantage of the properties of both types of components.

The application of an edible coating onto the fruit surface modifies the internal atmosphere in the same way that do plastic films (Valero and Serrano, 2010), by increasing the carbon dioxide and lowering the oxygen concentrations. Then, the effects of edible coatings on internal gas composition and their interactions on quality parameters must be determined specifically for each fresh produce. The success of edible coatings for fruits depends mainly on selecting the appropriate coating that can give a desirable internal gas composition for each specific product (Cisneros-Zevallos and Krochta, 2002, 2003).

In this chapter, recent trends in edible coatings are summarized with emphasis on their applications on fresh fruit commodities and their effects on physiological behavior, organoleptic quality, nutritive aspects, microbial growth, and levels of bioactive compounds with antioxidant activity.

7.1.1 POLYSACCHARIDES

Polysaccharides are good materials for the formation of edible coating since they show excellent mechanical and structural properties but they are hydrophilic and thus have a poor capacity as a barrier against water vapor and gas diffusion. The principal polysaccharides of interest for edible coatings are cellulose, starch, gums, pectins, alginate, and chitosan. The linear structure of some of these polysaccharides, for example, cellulose, amylose, and chitosan, renders their films tough, flexible, transparent, and resistant to fats and oils (Dhall, 2013). In the future, other complex polysaccharides produced by fungi and bacteria such as xanthan, curdlan, pullan, and hyaluronic acid, will receive more interest.

7.1.1.1 Cellulose

Cellulose is a polymer of D-glucose monomers linked through ß-$(1 \rightarrow 4)$ glycosidic bonds. It has low water solubility in nature but water solubility can be increased by treating cellulose with alkali to swell the structure, followed by reaction with chloroacetic acid, methyl chloride, or propylene oxide to yield carboxymethyl cellulose (CMC), methyl cellulose (MC), hydroxypropylmethylcellulose (HPMC), or hydroxypropyl cellulose (HPC). All these cellulose derivatives are water-soluble, odorless, tasteless, flexible, and transparent, and exhibit higher barrier capabilities to moisture and oxygen transmission than cellulose itself (Krochta and Mulder-Johnston, 1997).

7.1.1.2 Pectin

Pectins are linear or branched polymers with a high content of galacturonic acid and may contain as many as 17 different monosaccharides. Four main types of pectins have been structurally characterized: homogalacturonan, rhamnogalacturonan I, rhamnogalacturonan II, and xylogalacturonan, which differ in both the structure of the macromolecule backbone and the presence and diversity of side chains. Pectin is a class of complex water-soluble polysaccharides used to form coatings. It is a purified carbohydrate product obtained by aqueous extraction of some edible plant material, usually citrus fruits or apples. Under certain circumstances, pectins form gels, which have made them a very important additive in jellies, jams, marmalades and confectionaries, as well as in edible coatings. Pectin is a high-volume and potentially important food ingredient available in high percentages in agricultural waste. Pectin coatings have been also studied for their ability to retard lipid migration and moisture loss, and to improve appearance and handling of foods (Moalemiyan et al., 2011). Due to their hydrophilic nature, pectin edible coatings have low effectiveness as a water-vapor barrier but good properties as barrier to oxygen and carbon dioxide. Low methoxyl pectins are often used as edible coatings because of their ability to form strong gels upon reactions with multivalent metal cations such as calcium. The incorporation of calcium in polysaccharide edible coatings reduces their water vapor permeability, making the coatings water-insoluble (Ferrari et al., 2013).

7.1.1.3 Chitosan

Chitin is the second most abundant natural biopolymer after cellulose. It is the major structural component of the exoskeleton of arthropods and the cell walls of fungi. The chemical structure of chitin is similar to that of cellulose with 2-acetamido-2-deoxy-ß-D-glucose monomers attached via ß-$(1 \rightarrow 4)$ linkages. Chitosan is the deacetylated form of chitin which is soluble in acidic solutions, in contrast to chitin (Shahidi et al., 1999). Thus, chitosan is the low acetyl substituted form of chitin and is composed primarily of glucosamine, 2-amino-2-deoxy-ß-D-glucose, known as $(1 \rightarrow 4)$-2-amino-2-deoxy-D-glucose. Chitosan, is derived from chitin of marine invertebrates and has been used as an edible coating for its ability to form a good film on the commodity surface and control microbial growth. Chitosan is now widely produced commercially from crab and shrimp shell wastes with different deacetylation grades and molecular weights leading to different functional properties (No et al., 2007). Chitosan is water-insoluble but soluble in weak organic acid solutions. Chitosan derivatives in the form of acetate, ascorbate, lactate, and malate are water-soluble. Water-soluble chitosan can also be produced in the form of oligosaccharide by enzymatic or chemical hydrolysis (No et al., 2007). To date, chitosan has attracted considerable interest due to its antimicrobial activity (Dutta et al., 2009) and has widely been used in antimicrobial films. Chitosan has exhibited antimicrobial activity against a wide variety of pathogenic and spoilage microorganisms, including fungi, and Gram-positive and Gram-negative bacteria.

7.1.1.4 Starch

Starch is in abundance in the plant kingdom and has thermoplastic properties upon disruption of its molecular structure (Tharanathan, 2003). Starch granules contain two types of polymeric molecules: amylose, a linear chain of $(1 \rightarrow 4)$-α-D-glucopyranosyl units, and amylopectin, a larger molecule with a backbone of amylose and highly branched side units of D-glucopyranosyl linked by α-(1 $\rightarrow$ 6)-glycosidic bonds. Starch can form films by the interaction of hydroxyl groups through hydrogen bonds. Since these interactions are weak, films are brittle with poor mechanical properties, although a higher proportion of amylose will improve the film characteristics (Campos et al., 2011). A concentration of amylose over 70% as in amylomaize starches gives stronger and more flexible films. The branched structure of amylopectin generally leads to films with poor mechanical properties (decreased tensile strength and elongation). The substitution of the hydroxyl groups in the molecule weakens the hydrogen binding ability and thereby improves freeze-thaw stability and solution clarity. An ether linkage tends to be more stable than an ester linkage (Tharanathan, 2003).

7.1.1.5   Alginate

Alginate is a polymer isolated from brown seaweed (*Phaeophyceae*) and can form translucent, glossy, and strong films with low water vapor and oxygen permeability and high tensile strength. Alginate is a salt of alginic acid and is composed of D-mannuronic acid and L-guluronic acid, and has the ability to crosslink with divalent ions such as calcium to form strong films (Sime, 1990). Alginate films are poor moisture-barriers as they are hydrophilic films, however, the incorporation of calcium reduces water vapor permeability making alginate films water insoluble. The capacity of hydrocolloid-based films as water vapor barriers increases as their solubility in water decreases.

7.1.1.6   Aloe Gel

Aloe vera gel has been identified as a novel coating agent to extend the shelf-life of perishable food crops with good antimicrobial properties, especially as natural antifungal compound (Valverde et al., 2005). *Aloe* spp. are perennial succulents plants characterized by stemless large, thick, fleshy leaves that are lance-shaped and have a sharp apex and a spiny margin. *Aloe* leaves have yellow latex, which is referred to as *Aloe* juice or sap and has a bitter taste. The leaf pulp is the innermost portion of the leaf and is composed of the parenchyma cells that contain the gel (Steenkamp and Stewart, 2007). *Aloe* gel contains polysaccharides, primarily of ß–(1,4)-linked, polydispersed, highly acetylated mannans (acemannan). Many of the medicinal effects of *Aloe* leaf extracts have been attributed to the polysaccharides found in the inner leaf parenchymatous tissue but it is believed that these biological activities should be assigned to a synergistic action of the compounds contained therein rather than a single chemical substance (Eshun and He, 2004; Hamman, 2008). Recently, the leaf characteristics and gel chemical composition of eight *Aloe* species as well as their possible use as edible coatings have been described by Zapata et al. (2013).

7.1.1.7   Gum Arabic

Gum Arabic or gum acacia is a dried, gummy exudate from the stems or branches of the *Acacia* species. It is the least viscous and most soluble of the hydrocolloids (Nisperos-Carriedo, 1994) and is used extensively in the industrial sector because of its emulsification, film-forming, and encapsulation properties. More than half of the world supply is used in confectionary to retard sugar crystallization and to thicken candies, jellies, glazes, and chewing gums, although evidences exist as edible coating for fruits. The main gum to be used commercially is derived from *Acacia senegal* because of its good emulsification properties (Elmanan et al., 2008).

7.1.2 PROTEINS

Proteins have received great attention in edible coating research because of their abundance as agricultural byproducts and food processing residuals. The presence of reactive amino acid residuals enables proteins to be modified and cross-linked through physical and chemical treatments to produce novel polymeric structures (Gennadios, 2002). Protein-based coatings have more interesting mechanical and barrier properties than polysaccharides. Many protein materials have been tested including collagen, corn zein, wheat gluten, SPI, fish proteins, ovalbumin, whey protein isolate and casein (Khwaldia et al., 2004).

7.1.2.1 Whey Protein

Whey proteins from bovine milk have been studied to a great extent because of their ability to form transparent and flexible coatings that exhibit good barrier and mechanical properties (Krochta, 2002). Whey proteins are globular proteins that remain soluble after precipitation of casein at pH 4.6 during cheese making. In bovine milk, these thermolabile proteins consist of mainly α-lactalbumin, β-lactoglobulin, and other proteins present in smaller fractions (e.g., bovine serum albumin, immunoglobulins, and proteasepeptones). Whey proteins are commercially available as whey protein isolates or whey protein concentrates, which have protein content of >90 and 20%–85%, respectively (Reinoso et al., 2008). However, whey protein has a hydrophilic nature and lipids need to be added to the film-forming solution to reduce the water-sensitivity of films.

7.1.2.2 Gelatine

Gelatine is obtained by controlled hydrolysis of the fibrous insoluble protein, collagen, which is the major constituent of animal skin, bones, and connective tissue. The characteristic features of gelatine are high content of amino-acids, such as glycine, proline, and hydroxyproline (Dhall, 2013). Gelatine films can be formed from 20%–30% gelatine, 10%–30% plasticizer (glycerine or sorbitol), and 40%–70% water followed by drying the gelatine gel.

7.1.2.3 Zein

Zein includes a group of alcohol-soluble proteins (prolamines) found in corn endosperm and accounts for 50%+ of the total endosperm protein. Zein has been used intermittently in a number of industrial applications since becoming commercially available in 1938 but is currently limited to formulations of coating agents for the food and pharmaceutical industries. Zein has long been recognized for its film-forming ability and its use as a bioplastic material is of interest because of its environmental and renewable qualities. Zein is a mixture of several peptides of different molecular weight, solubility, and charge that are named as zein fractions and classified according to their relative mass and solubility as α, γ, ß, and δ-zein. α-Zein, the major fraction (85% of total zein), is soluble in 50%–95% isopropyl alcohol (Wang et al., 2005). The utilization of corn zein as a structural polymer has been actively investigated in the last decades (Park and Chinnan, 1995). Zein films cast from aqueous ethanol solutions were rated as moderately good with respect to mechanical properties and moisture and oxygen barrier properties. Zein films plasticized with oleic acid have exhibited tensile and moisture-barrier properties that make them potentially useful as biodegradable packaging materials (Rakotonirainy et al., 2001).

7.1.2.4 Soy Protein

Since the 1960s, soy protein products have been used as nutritional and functional food ingredients in every food category available to the consumer. Recently, soy protein is being used as an ingredient for elaborating edible coatings. The content of protein from soybeans (38%–44%) is much higher than the protein content of cereal grain (8%–15%). Most of the protein in soybeans is insoluble in water but soluble in dilute-neutral salt solutions (Dhall, 2013). Soy protein consists of two major protein fractions referred to as the 7S (conglycinin, 35%) and 11S (glycinin, 52%) fraction. Edible coatings based on soy protein can be produced in either of two ways: surface film-formation on heated soy-milk or film-formation from solutions of soy protein isolates (SPIs) (Gennadios, 2002).

7.1.3 LIPIDS

Because of their apolar nature, hydrophobic lipidic substances are used in fruit coatings mainly as a barrier against moisture migration and to improve surface appearance (Lin and Zhao, 2007). Lipid components commonly used in coatings include natural waxes (e.g., carnauba wax, beeswax, candelilla wax), acylglycerols, and fatty acids. Additionally, some authors include shellac, which is a natural resin, as an ingredient of natural coatings for fruits that are not consumed with peel such as citrus fruits, even though it is not included in the GRAS ingredient list, to provide gloss to food surfaces. Each hydrophobic substance has its own physicochemical properties, and, thus, edible films based on lipids have variable behavior against moisture transfer (Morillon et al., 2002). Lipid compounds include neutral lipids of glycerides, which are esters of glycerol and fatty acids, and the waxes, which are esters of long-chain monohydric alcohols and fatty acids.

Wax was the first edible coating used on fruits, with the Chinese applying wax coatings to oranges and lemons in the 12th and 13th centuries. Although the Chinese did not realize that the full function of edible coatings was to slow down respiratory gas exchange, they found that wax-coated fruits could be stored longer than non-waxed fruits (Park, 1999).

7.2 EFFECTS OF EDIBLE COATINGS ON FRUIT PROPERTIES

7.2.1 EFFECT ON FRUIT PHYSIOLOGY

The quality of fruit is determined by a wide range of characteristics such as nutritional value, organoleptic quality, processing, and shelf-life. Fruits are classified as climacteric and nonclimacteric, with climacteric fruits characterized by an increased rate of respiration and ethylene production early in the ripening process, while in nonclimacteric fruits those changes do not occur. However, in both types of fruits, parameters related to fruit quality change during postharvest storage and marketing, leading to limited storability and shelf-life. The main postharvest changes are in taste, aroma, skin color, firmness, as well as weight-loss due to transpiration. The physiological and biochemical activities involved in fruit-ripening and senescence can be delayed by a range of postharvest treatments (Valero and Serrano, 2010), including the use of edible coatings. Edible coatings maintain the quality of fruit and vegetables by forming a film over the produce, which then serves as a partial barrier to gas transmission and, thereby, creates a modified atmosphere around the commodity that affects fruit physiology and biochemistry.

The data in Table 7.1 summarizes the findings from a wide range of published studies on the effect of various edible coatings on ethylene production, respiration

TABLE 7.1

Effects of Edible Coatings on Ethylene Production, Respiration Rate and Weight Loss in Climacteric Fruit

Climacteric Fruit	Edible Coating	C_2H_4	Respir	Wt. Loss	Reference
Sapote	Wax	↑	↑	↓	Ergun et al. (2005)
Apple–Fuji	Shellac	↓	↓	↓	Hagenmaier (2005)
Red delicious	Shellac	o	↓	↓	Hagenmaier (2005)
Gala	Alginate	ND	ND	↓	Olivas et al. (2007)
Avocado	Methylcellulose	ND	↓	↓	Maftoonazad and Ramaswamy (2005)
Plum	HPMC—Lipid	ND	ND	o	Pérez-Gago et al. (2002)
	Whey Protein	ND	ND	↓	Reinoso et al. (2008)
	Aloe spp.	↓	↓	↓	Guillén et al. (2013)
	Versasheen™	↓	↓	↓	Eum et al. (2009)
	Alginate	↓	↓	↓	Valero et al. (2013)
	A. vera + Rosehip oil	↓	↓	↓	Paladines et al. (2014)
Mango	Semperfresh™	↓	o	↓	Dang et al. (2008)
	A. vera	↓	o	o	Dang et al. (2008)
	Carnauba	↓	↓	↓	Dang et al. (2008)
Tomato	Zein, Alginate	↓	↓	↓	Zapata et al. (2008)
	Gum Arabic	ND	ND	↓	Ali et al. (2010)
	Gum Arabic	↓	↓	↓	Ali et al. (2013)
	A. vera	ND	ND	↓	Athmaselvi et al. (2013)
	A. vera, Shellac	↓	↓	↓	Chauhan et al. (2013)
Nectarine	*A. vera*	↓	↓	↓	Ahmed et al. (2009)
	A. vera	↓	↓	↓	Navarro et al. (2011)
	A. vera + Rosehip oil	↓	↓	↓	Paladines et al. (2014)
Peach	*Aloe* spp.	↓	↓	↓	Guillén et al. (2013)
	A. vera + Rosehip oil	↓	↓	↓	Paladines et al. (2014)

Note: Attribute was decreased (↓), increased (↑), unaffected (o), or was not determined (ND) in each study.

TABLE 7.2

Effects of Edible Coatings on Respiration Rate and Weight Loss in Nonclimacteric Fruit

Nonclimacteric Fruit	Edible Coating	Respir	Wt. Loss	Reference
Strawberry	Wheat gluten	ND	↓	Tanada-Palmu and Grosso (2005)
	Starch	ND	↓	García et al. (1998), Mali and Grossmann (2003)
	Chitosan	ND	↓	Han et al. (2004), Gol et al. (2013)
	CMC, HPMC	ND	↓	Gol et al. (2013)
Bell pepper	Candelilla wax	o	↓	Hagenmaier (2005)
Raspberry	Chitosan	↓	↓	Han et al. (2004)
Sweet cherry	Chitosan acetate	↓	↓	Dang et al. (2010)
	Alginate	↓	↓	Díaz-Mula et al. (2012)
	A. vera	↓	↓	Martínez-Romero et al. (2006)
	A. vera + Rosehip oil	↓	↓	Paladines et al. (2014)
Sour cherry	Aloe vera	↓	↓	Ravanfar et al. (2014)
Mandarin	HPMC - Lipid	↓	↓	Pérez-Gago et al. (2002)
Pomegranate	Starch	↓	↓	Oz and Ulukanli (2012)
Table grape	A. vera	↓	↓	Valverde et al. (2005), Castillo et al. (2010)
	Chitosan	↓	↓	Shiri et al. (2013)

Note: Attribute was decreased (↓), unaffected (o), or was not determined (ND) in each study.

rate and weight loss in a wide range of climacteric fruit. It is clear that the different coatings with polysaccharide, protein, or lipid constituents induced a reduction in ethylene production and respiration rate, as well as a delay in ripening during storage. In addition, the edible coatings were effective in reducing weight loss, resulting in net benefit from an economic perspective. Similarly, Table 7.2 shows the effect of different coatings on reducing respiration rate and weight loss in nonclimacteric fruits, which also achieved maintenance of quality attributes during storage.

7.2.2 EFFECT ON ORGANOLEPTIC QUALITY

In recent years, there is an increasing consciousness of quality, particularly in relation to the effect of eating fruit on the health of consumers; this greatly demands research activities with regard to the production of defined quality, the preservation of quality during marketing, as well as the evaluation of quality parameters and integrating this into the production processes (Valero and Serrano, 2010). The term "quality" is related to the degree of excellence and absence of defects of a fresh produce, which implies either sensory attributes (appearance, color, texture, flavor, and aroma), nutritive (chemical components used to obtain energy), and functional properties (vitamins and other nonnutrient phytochemicals). Shewfelt (1999) suggested that the inherent produce characteristics determine quality, but consumer acceptability is determined by their perception and satisfaction. Thus, quality can be oriented to the produce or to the consumer point of view. It is well documented that during postharvest storage, there is deterioration in fruit quality, primarily affecting the following traits: color, firmness, content of total soluble solids (TSS), and total acidity (TA). The application of edible coatings could modulate changes in these above mentioned parameters and, in turn, extend the marketability and shelf-life of these perishable commodities.

With respect to firmness, Figure 7.1 shows published data on the effect of different edible coatings on a range of fruit species stored at different temperatures. In general, fruit with edible coatings showed higher retention of firmness as compared with control fruit, although the effect was dependent on fruit type, storage temperature, and type of coating. For example, in strawberries that were stored at 11°C for eight days, hydroxypropylmethylcellulose (HPMC) was more effective than carboxymethyl cellulose (CMC) on retaining firmness (Gol et al., 2013). Similarly, in tomato, the most effective coating was gum Arabic followed by zein, while smaller differences were observed when *A. vera* and starch were used as coatings (Zapata et al., 2008; Ali et al., 2010, 2013; Chauhan et al., 2013). However, *A. vera* was very effective in reducing loss of firmness in nectarine, sweet cherry, and table grape (Valverde et al., 2005; Martínez-Romero et al., 2006; Ahmed et al., 2009; Navarro

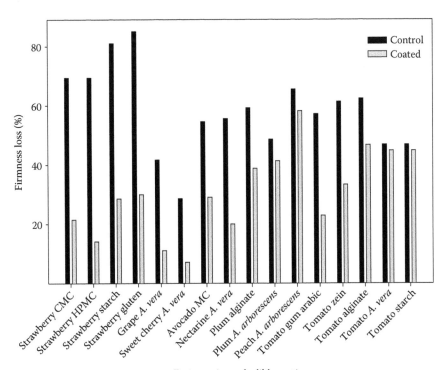

FIGURE 7.1 Firmness loss in fruit with an edible coating and control and after storage. (Charts were prepared from data given in references cited in Tables 7.1 and 7.2.)

et al., 2011). It is well-known that cell-wall hydrolytic enzymes cause dramatic loss of firmness in fruit tissues, the most important being polygalacturonase (PG), cellulase (CL), pectinmethylesterase (PME), and α and β-galactosidases (GAL), among others (Valero and Serrano, 2010), and, thus, reduction in activity of these enzymes would lead to reduced postharvest softening. For example, chitosan, as an edible coating, becomes bound to pectin and thus prevents access of PG to the substrate and maintains firmness in papaya (González-Aguilar et al., 2009). Cellulase activity has been also reduced in carambola fruit treated with chitosan, gum Arabic, and alginate, as well as PME and β-GAL activities (Gol et al., 2013). On the other hand, edible coatings can be used to modify the internal atmosphere of fruits and, in turn, delay senescence (Rojas-Grau et al., 2009). Edible coatings create a passive modified atmosphere that can influence various changes in fresh fruits, such as firmness, and, in the case of climacteric fruit, inhibit ethylene production (Falguera et al., 2011).

Color change is another important parameter related to organoleptic quality; and loss of quality can be measured objectively by increases in HunterLab a/b parameter and decreases in b, Chroma, and Hue angle. Figure 7.2 shows some examples of change in these color parameters and its relation to coating type and fruit species during postharvest storage. From Figure 7.2 it can be inferred that, in general, the application of

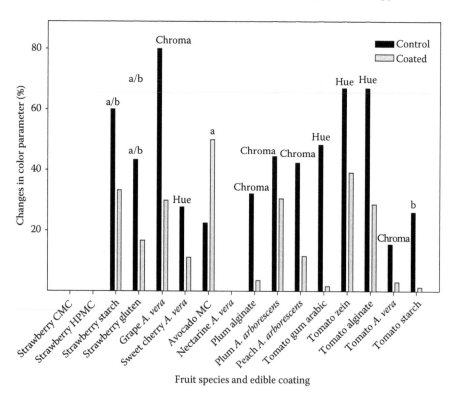

FIGURE 7.2 Color changes as Hunter Lab parameters in fruit with an edible coating and control and after storage. (Charts were prepared from data given in references cited in Tables 7.1 and 7.2.)

different edible coatings led to less changes in the color parameter, although efficacy depended on type of coating, fruit species, and storage conditions. For instance, the Chroma index increased a 80% in control table grapes, and ≅40% in plum and peach and only 18% in tomato, while these increases were much lower in fruits treated with different coatings such as *Aloe vera* or *A. arborescens* gel or alginate (Valverde et al., 2005; Athmaselvi et al., 2013; Guillén et al., 2013; Valero et al., 2013).

The levels of sugars and organic acids are important in determining the taste of ripe fleshy fruit, and the relative content of these constituents depends on the activity and the interaction of sugar and acid metabolism (Valero and Serrano, 2010). TA usually decreases in the fruit flesh during postharvest storage; this is attributed to organic acids being substrates for the respiratory metabolism in detached produce. As can be seen in Figure 7.3, all fruit experienced acidity losses during storage, the magnitude being dependent on fruit species, ranging from 10% in tomato to 70% in peach. However, in general, the different coatings led to reductions in acidity losses, the higher effect being found in tomato and plum coated with alginate and also in plum and peach coated with *A. arborescens* (Zapata et al., 2008; Guillén et al., 2013; Valero et al., 2013).

During postharvest, there is also a general increase in the content of TSS, as has been reported for nectarines, apricots, kiwifruits, and strawberries (Valero and

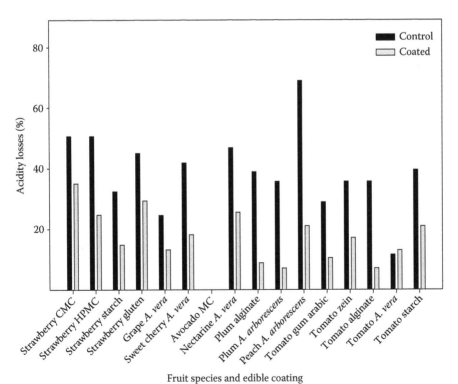

FIGURE 7.3 Acidity loss in fruit with an edible coating and control and after storage. (Charts were prepared from data given in references cited in Tables 7.1 and 7.2.)

Serrano, 2010). This increase in soluble solids is much higher in fruits that accumulate larger amounts of starch during development on the plant, such as mango or bananas. The application of edible coatings on fruit generally leads to lower increases in TSS, such as in strawberry coated with starch, CMC or HPMC (García et al., 1998; Mali and Grossmann, 2003; Gol et al., 2013) and in tomato coated with gum Arabic or starch (Ali et al., 2010; Das et al., 2013), as a consequence of a delay in the postharvest ripening process. However, in other reports, higher increases in TSS have been found in coated fruit than in controls, such as in wax-coated sapote fruits (Ergun et al., 2005) and in zein, alginate or *A. vera* gel-coated tomatoes (Zapata et al., 2008; Athmaselvi et al., 2013).

7.2.3 Effect on Fruit Bioactive Compounds and Antioxidant Activity

Fruits contain hundreds of nonnutrient constituents with significant biological activity, generally called "bioactive compounds" or phytochemicals, which have antioxidant activity and protective effects against several chronic diseases associated to aging, including atherosclerosis, cardiovascular diseases, cancer, cataracts, increased blood pressure, ulcerous, neurodegenerative diseases, brain and immune dysfunction, and even against bacterial and viral diseases. These bioactive compounds vary widely in chemical structure and function in plant tissues and are grouped in vitamins (C and E), carotenoids, phenolics, and thiols (Asensi-Fabado and Munné-Bosch, 2010; Baldrick et al., 2011; Brewer, 2011; Serrano et al., 2011; Li et al., 2012; Valero and Serrano, 2013).

No general tendency has been found for the changes in bioactive compounds during fruit storage. Thus, loss of compounds beneficial to health, such as phenolics and ascorbic acid, has been found in apples, table grapes, and pomegranates, while increases in phytochemicals have been observed in sweet cherry and plum cultivars (Díaz-Mula et al., 2009; Serrano et al., 2009, 2011). Figure 7.4 shows published examples in which the content of total phenolics decreased during storage, and the beneficial effects of the different coatings reduced these phenolic losses. Interestingly, tomato coated with gum Arabic showed increases in total phenolics while they decreased in control fruit (Ali et al., 2010, 2013).

Similar behavior was found for total antioxidant activity (Figure 7.4) with lower losses in chitosan-coated pomegranate, pear and blueberry than in controls (Duan et al., 2011; Ghasemnezhad et al., 2013; Kou et al., 2014) and even TAA increased in sweet cherry and tomato coated with alginate and gum Arabic, respectively, more than in control fruits (Díaz-Mula et al., 2012; Ali et al., 2013).

7.3   EDIBLE COATINGS WITH NATURAL ANTIMICROBIAL COMPOUNDS

In recent years, new edible films and coatings are being formulated with the addition of natural antimicrobial compounds for application onto fresh and minimally processed fruit commodities. This system constitutes an environment-friendly technology and improves the mechanical handling properties that may enhance food quality, safety, stability, by providing a semi-permeable barrier to water vapor, oxygen, and

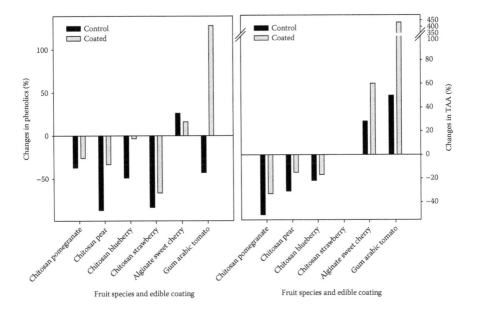

FIGURE 7.4   Changes in total phenolics or total antioxidant activity (TAA) in fruit with an edible coating and control, and after storage. (Charts were prepared from data given in references cited in Tables 7.1 and 7.2.)

carbon dioxide, between the fruit and the surrounding atmosphere, with increased antimicrobial properties (Valencia-Chamorro et al., 2011). There is a wide range of naturally-occurring compounds that exhibit antimicrobial activity, including chitosan, polypetides, and essential oils or spice extracts. As already stated, chitosan is a polysaccharide that shows antimicrobial activity, which has been attributed to its positive charges that would interfere with the negatively charged residues of macromolecules on the cell surface, rendering membrane leakage (Sebti et al., 2007). Most of the antimicrobials proposed to be used in the formulation of coatings must inhibit the spoilage microorganisms (bacteria and fungi) and reduce the food-borne pathogens. In recent years, there is a trend to select the antimicrobials from natural sources and to use generally GRAS compounds, in order to satisfy consumer demands for healthy foods, free of chemical additives (Campos et al., 2011).

The essential oils (EOs) have these characteristics. EOs or the so called volatile or ethereal oils are aromatic oily liquids obtained from plant organs: flower, bud, seed, leave, twig, bark, herb wood fruit, and root (Serrano et al., 2008). Natural compounds that also possess antioxidant effects have been extracted from plants that belong to genus *Thymus, Origanum, Syzygium, Mentha,* and *Eucalyptus* (Burt, 2004). Chemical composition of EOs is complex and strongly dependent on the part of the plant considered (e.g., seed vs. leaves), the moment of harvest (before, during, or after flowering), the harvesting season and the geographical sources. Major components in EOs are phenolic substances, which are thought to be responsible for the antimicrobial properties, and many of them are classified as GRAS (Campos et al., 2011). The antimicrobial activity of the EOs can be attributed to their content of monoterpenes that, due to their lipophilic character, act by disrupting the integrity of microbial cytoplasmic membrane. Lipophilic compounds accumulate in the lipid bilayer according to its specific partition coefficient, leading to disruption of the membrane structure (Liolios et al., 2009).

Table 7.3 gives some examples of different coatings on several whole fruits, in which improvement of coating efficacy was achieved by the incorporation of some natural antimicrobial compounds, such as EOs from different plant origin. The most studied parameter was the contamination of different microorganisms such as bacteria (*E.coli*, *Salmonella* spp. and *S. aureus*) and fungal species (*Penicillium*, *Rhizopus* and *Botrytis*). Apart from the antimicrobial activity, the combined use of the edible coatings with natural antimicrobials was also effective in improving the parameters related to organoleptic and functional quality. As an example, tomato coated with zein at 10% plus the addition of EOs (thymol, carvacrol, and eugenol at 75 μL/L) exhibited a reduced rate of color-change than control tomatoes coated with zein alone after nine days of storage at 10°C (Figure 7.5). The effect was attributed to the reported antioxidant properties of EOs leading to a delay of the postharvest ripening process (Serrano et al., 2008). Accordingly, the addition

TABLE 7.3

Improvement of Coating Efficacy with the Addition of Natural Antimicrobial Compounds

Fruit	Edible Coating	Natural Antimicrobial	Effects	Reference
Blueberry	Chitosan 2%	Phenolics from blueberry extracts	Reduced decay and increased phenolics	Yang et al. (2014)
Pepper, Apple	Pullulan 10%	Summer savory herb	Inhibited growth of Gram-positive and Gram-negative bacteria and *P. expansum*	Kraśniewska et al. (2014)
Strawberry	Alginate 2%	Carvacrol, Methyl Cinnamate	Inhibited *E. coli* and *B. cinerea*	Peretto et al. (2014)
Apple	Cassava starch 2%	Cinnamom or fennel	Inhibited *S. aureus* and *Salmonella*	Oriani et al. (2014)
Plum	Carnauba wax	Lemongrass oil	Inhibited *Salmonella* and *E. coli*, reduced ethylene and improved quality	Kim et al. (2013)
Orange	Chitosan	Tea tree oil	Reduced *P. italicum* growth	Cháfer et al. (2012)
Tomato	Chitosan (1%)	Lime essential oil	Inhibited *Rhizopus stolonifer* and *E. coli*	Ramos-García et al. (2012)
Table grape	Chitosan or HPMC	Bergamot oil	Reduced microbial counts	Sánchez-González et al. (2011)
Lemon	Wax	Carvacrol or thymol	Reduced *P. digitatum*, respiration and acidity loss	Pérez-Alfonso et al. (2012), Castillo et al. (2014)

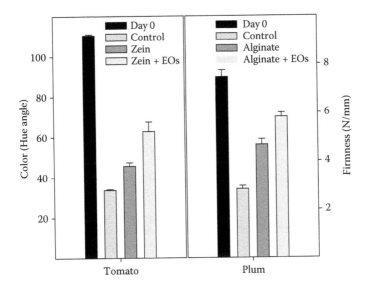

FIGURE 7.5 Color (Hue angle) after nine days' storage of tomatoes coated with zein at 10% or zein + essential oils (EOs), and fruit firmness of plums coated with 3% or alginate + EOs (Serrano and Valero, unpublished data). Data are the mean ± SE.

of these EOs to alginate led to reduced softening, since firmness of plums after 15 days of storage at 2°C was significantly higher in alginate + EOs coated plums compared with alginate alone or controls, for which an accelerated softening process occurred (Figure 7.5).

7.4 CONCLUDING REMARKS

The use of edible coatings for preservation of whole fruit is a matter of high interest taking into account the increasing number of research reports on this issue. For each particular fruit, the design of an appropriate coating formulation is essential for assuring the quality and safety during postharvest storage. The proper selection of an edible coating will depend on the respiration and transpiration rates of the commodity and on the environmental conditions of the storage area. Edible coatings can protect perishable fruits from deterioration during storage by retarding weight loss, reducing respiration rate and ethylene production, improving texture and other quality parameters, and reducing microbial contamination. Many of the polysaccharide-based and protein-based coatings, especially those with inherent antimicrobial or antifungal activities such as chitosan, are attracting more interest as substitutes for traditional lipid coatings.

Edible coatings are effective as a barrier to respiratory gas exchange and water vapor. The efficacy of edible coatings depends on the coating, type and characteristics of the fruit, type of coating, and storage conditions (temperature and duration). More research is needed in order to get a better understanding of the relationship between the internal atmosphere produced by the edible coating and the physiological changes in fruits during storage that will influence the final quality of the coated product. The coatings used for one fruit cultivar may not be appropriate for another since each fruit is different in peel resistance, gas diffusion, and fruit respiration rate among other attributes. One advantage of using edible coatings is that several active compounds can be incorporated into the polymer matrix and consumed with the food, such as the use of natural antimicrobial compounds. In this sense, a new generation of edible coatings is being currently developed allowing the incorporation of EOs for controlling spoiling microorganism and thus enhancing the safety of coated fresh fruits. Finally, sensory evaluation and consumer acceptability tests need to be conducted during the storage of coated fruit.

REFERENCES

Ahmed MJ, Singh Z, Khan AS. 2009. Postharvest *Aloe vera* gel-coating modulates fruit ripening and quality of 'Arctic Snow' nectarine kept in ambient and cold storage. *Int J Food Sci Technol* 44: 1024–1033.

Ali A, Maqbool M, Alderson PG, Zahid N. 2013. Effect of gum Arabic as an edible coating on antioxidant capacity of tomato (*Solanum lycopersicum* L.) fruit during storage. *Postharv Biol Technol* 76: 119–124.

Ali A, Maqbool M, Ramachandran S, Alderson PG. 2010. Gum Arabic as a novel edible coating for enhancing shelf-life and improving postharvest quality of tomato (*Solanum lycopersicum* L.) fruit. *Postharv Biol Technol* 58: 42–47.

Asensi-Fabado MA, Munné-Bosch S. 2010. Vitamins in plants: Occurrence, biosynthesis and antioxidant function. *Trends Plant Sci* 15: 582–592.

Athmaselvi KA, Sumitha P, Revathy B. 2013. Development of *Aloe vera*-based edible coating for tomato. *Int Agrophys* 27: 369–375.

Baldrick FR, Woodside JV, Elborn S, Young IS, Mckinley MC. 2011. Biomarkers of fruit and vegetable intake in human intervention studies: A systematic review. *Crit Rev Food Sci Nutrit* 51: 795–815.

Bourtoom T. 2008. Edible films and coatings: Characteristics and properties. *Int Food Res J* 15: 237–248.

Brewer MS. 2011. Natural antioxidants: Sources, compounds, mechanisms of action, and potential applications. *Comprehensive Rev Food Sci Food Safety* 10: 221–247.

Burt S. 2004. Essential oils: Their antibacterial properties and potential applications in foods—A review. *Int J Food Micro* 94: 223–253.

Campos CA, Gerschenson LN, Flores SK. 2011. Development of edible films and coatings with antimicrobial activity. *Food Bioprocess Technol* 4: 849–875.

Castillo S, Navarro D, Zapata PJ, Guillén F, Valero D, Serrano M, Martínez-Romero D. 2010. Antifungal efficacy of *Aloe vera in vitro* and its use as a preharvest treatment to maintain postharvest table grape quality. *Postharv Biol Technol* 57: 183–188.

Castillo S, Pérez-Alfonso CO, Martínez-Romero D, Guillén F, Serrano M, Valero D. 2014. The essential oils thymol and carvacrol applied in the packing lines avoid lemon spoilage and maintain quality during storage. *Food Control* 35: 132–136.

Cháfer M, Sánchez-González L, González-Martínez C, Chiralt A. 2012. Fungal decay and shelf-life of oranges coated with chitosan and bergamot, thyme, and tea tree essential oils. *J Food Sci* 77: E182–E187.

Chauhan OP, Nanjappa C, Ashok N, Ravi N, Roopa N, Raju PS. 2013. Shellac and *Aloe vera* gel-based surface coating for shelf-life extension of tomatoes. *J Food Sci Technol* (in press).

Cisneros-Zevallos L, Krochta JM. 2002. Internal modified atmospheres of coated fresh fruits and vegetables: Understanding relative humidity effects. *J Food Sci* 67: 1990–1995.

Cisneros-Zevallos L, Krochta JM. 2003. Dependence of coating thickness on viscosity of coating solution applied to fruits and vegetables by dipping method. *J Food Sci* 68: 503–510.

Dang KTH, Singh Z, Swinny EE. 2008. Edible coatings influence fruit ripening, quality, and aroma biosynthesis in mango fruit. *J Agric Food Chem* 56: 1361–1370.

Dang QF, Yan JQ, Li Y, Cheng XJ, Liu CS, Chen XG. 2010. Chitosan acetate as an active coating material and its effects on the storing of *Prunus avium* L. *J Food Sci* 75: S125–S131.

Das DK, Dutta H, Mahanta CL. 2013. Development of a rice starch-based coating with antioxidant and microbe-barrier properties and study of its effect on tomatoes stored at room temperature. *LWT Food Sci Technol* 50: 272–278.

Dhall RK 2013 Advances in edible coatings for fresh fruits and vegetables: A Review. *Crit Rev Food Sci Nutrit* 53: 435–450.

Díaz-Mula HM, Serrano M, Valero D. 2012. Alginate coatings preserve fruit quality and bioactive compounds during storage of sweet cherry fruit. *Food Bioprocess Technol* 5: 2990–2997.

Díaz-Mula HM, Zapata PJ, Guillén F, Martínez-Romero D, Castillo S, Serrano M, Valero D. 2009. Changes in hydrophilic and lipophilic antioxidant activity and related bioactive compounds during postharvest storage of yellow and purple plum cultivars. *Postharv Biol Technol* 51: 354–363.

Duan J, Wu R, Strik BC, Zhao Y. 2011. Effect of edible coatings on the quality of fresh blueberries (Duke and Elliott) under commercial storage conditions. *Postharv Biol Technol* 59: 71–79.

Dutta PK, Tripathi S, Mehrotra GK, Dutta J. 2009. Perspectives for chitosan-based antimicrobial films in food applications. *Food Chem* 114: 1173–1182.

ED. 1995. Directiva 95/2/CE del Parlamento Europeo y del Consejo de 20 de febrero de 1995 relativa a aditivos alimentarios distintos de los colorantes y edulcorantes (http://ec.europa.eu/food/fs/sfp/addit_flavor/flav11_en.pdf, Accessed 06-10-2014).

ED. 1998. Directiva 98/72/CE del Parlamento Europeo y del Consejo de 15 de octubre de 1998 relativa a aditivos alimentarios distintos de los colorantes y edulcorantes (http://eur-lex.europa.eu/LexUriServ/LexUriServ.do?uri=OJ:L:1998:295:0018:0030:EN:PDF, Accessed 06-10-2014).

Elmanan M, Al-Assaf S, Philips GO, Williams PA. 2008. Studies of Acacia exudates gums: Part IV. Interfacial rheology of *Acacia senegal* and *Acacia seyal*. *Food Hydrocolloids* 22: 682–689.

Ergun M, Sargent SA, Fox AJ, Crane JH, Huber DJ. 2005. Ripening and quality responses of mamey sapote fruit to postharvest wax and 1-methylcyclopropene treatments. *Postharv Biol Technol* 36: 127–134.

Eshun K, He Q. 2004. Aloe vera: A valuable ingredient for the food, pharmaceutical and cosmetic industries: A review. *Critic Rev Food Sci Nutrit* 44: 91–96.

Eum HL, Hwang DK, Linke M, Lee SK, Zude M. 2009. Influence of edible coating on quality of plum (*Prunus salicina* Lindl. cv. 'Sapphire'). *Eur Food Res Technol* 229: 427–434.

Falguera V, Quintero JP, Jiménez A, Aldemar Muñoz J, Ibarz A. 2011. Edible films and coatings: Structures, active functions and trends in their use. *Trends Food Res Technol* 22: 292–303.

FDA 21CFR172. 2006. Food additives permitted for direct addition to food for human consumption, Subpart C. Coatings, Films and Related Substances (http://www.accessdata.fda.gov/scripts/cdrh/cfdocs/cfCFR/CFRSearch.cfm?CFRPart=172, Accessed 06-10-2014).

Ferrari CC, Sarantópoulos CIGL, Carmello-Guerreiro SM, Hubinger MD. 2013. Effect of osmotic dehydration and pectin edible coatings on quality and shelf-life of fresh-cut melon. *Food Bioprocess Technol* 6: 80–91.

García MA, Martino MN, Zaritzky NE. 1998. Starch-based coatings: Effect on refrigerated strawberry (*Fragaria ananassa*) quality. *J Sci Food Agric* 76: 411–420.

Gennadios A. 2002. *Protein-based Films and Coatings.* CRC, Boca Raton FL.

Ghasemnezhad M, Zareh S, Rassa M, Sajedi RH. 2013. Effect of chitosan coating on maintenance of aril quality, microbial population and PPO activity of pomegranate (*Punica granatum* L. cv. Tarom) at cold storage temperature. *J Sci Food Agric* 93: 368–374.

Gol NB, Patel PR, Rao TVR. 2013. Improvement of quality and shelf-life of strawberries with edible coatings enriched with chitosan. *Postharv Biol Technol* 85: 185–195.

González-Aguilar GA, Valenzuela-Soto E, Lizardi-Mendoza J, Goicoolea F, Martínez-Telez MA, Villegas-Ochoa MA, Monory-García I, Ayala-Zabala JF. 2009. Effect of chitosan coating in preserving deterioration and preserving quality of fresh cut papaya "Maradol". *J Sci Food Agric* 89: 15–23.

Guillén F, Díaz-Mula HM, Zapata PJ, Valero D, Serrano M, Castillo S, Martínez-Romero D. 2013. *Aloe arborescens* and *Aloe vera* gels as coatings in delaying postharvest ripening in peach and plum fruit. *Postharv Biol Technol* 83: 54–57.

Hagenmaier RD. 2005. A comparison of ethane, ethylene and CO_2 peel permeance for fruit with different coatings. *Postharv Biol Technol* 37: 56–64.

Hamman JH. 2008. Composition and applications of *Aloe vera* leaf gel. *Molecules* 13: 1599–1616.

Han C, Zhao Y, Leonard SW, Traber MG. 2004. Edible coatings to improve storability and enhance nutritional value of fresh and frozen strawberries (*Fragaria x ananassa*) and raspberries (*Rubus ideaus*). *Postharv Biol Technol* 33: 67–78.

Khwaldia K, Pérez C, Banon S, Desobry S, Hardy JS. 2004. Milk proteins for edible films and coatings. *Crit Rev Food Sci Nutrit* 44: 239–251.

Kim IH, Lee H, Kim JE, Song KB, Lee YS, Chung DS, Min SC. 2013. Plum coatings of lemongrass oil-incorporating carnauba wax-based nanoemulsion. *J Food Sci* 78: E1551–E1559.

Kou XH, Guo WL, Guo RZ, Li XY, Xue ZH. 2014. Effects of chitosan, calcium chloride, and pullulan coating treatments on antioxidant activity in pear cv. "Huang Guan" during storage. *Food Bioprocess Technol* 7: 671–681.

Kraśniewska K, Gniewosz M, Synowiec A, Przybył JL, Baczek K, Weglarz Z. 2014. The use of pullulan coating enriched with plant extracts from *Satureja hortensis* L. to maintain pepper and apple quality and safety. *Postharv Biol Technol* 90: 63–72.

Krochta JM. 2002. Proteins as raw materials for films and coatings: Definitions, current status and opportunity. In: *Protein-Based Films and Coatings* ed. A Gennadios. CRC, Boca Raton, FL, pp. 1–42.

Krochta JM, Mulder-Johnston CD. 1997. Edible and biodegradable polymer films: Challenges and opportunities. *Food Technol* 51: 61–74.

Li H, Tsao R, Deng Z. 2012. Factors affecting the antioxidant potential and health benefits of plant foods. *Can J Plant Sci* 92: 1101–1111.

Lin D, Zhao Y. 2007. Innovations in the development and application of edible coatings for fresh and minimally processed fruits and vegetables. *Comprehensive Rev Food Sci Food Safety* 6: 60–75.

Liolios CC, Gortzi O, Lalas S, Tsaknis J, Chinou I. 2009. Liposomal incorporation of carvacrol and thymol isolated from the essential oil of *Origanum dictamnus* L. and *in vitro* antimicrobial activity. *Food Chem* 112: 77–83.

Maftoonazad N, Ramaswamy HS. 2005. Postharvest shelf-life extension of avocados using methyl cellulose-based coating. *LWT Food Sci Technol* 38: 617–624.

Mali S, Grossmann MVE. 2003. Effects of yam starch films on storability and quality of fresh strawberries (*Fragaria ananassa*). *J Agric Food Chem* 51: 7005–7011.

Martínez-Romero D, Alburquerque N, Valverde JM, Guillén F, Castillo S, Valero D, Serrano M. 2006. Postharvest sweet cherry quality and safety maintenance by *Aloe vera* treatment: A new edible coating. *Postharv Biol Technol* 39: 93–100.

Moalemiyan M, Ramaswamy HS, Maftoonazad N. 2011. Pectin-based edible coating for shelf-life extension of Ataulfo mango. *J Food Process Eng* 35: 572–600.

Morillon V, Debeaufort F, Blond G, Capelle M, Voilley A. 2002. Factors affecting the moisture permeability of lipid-based edible films: A Review. *Crit Rev Food Sci Nutrit* 42: 67–89.

Navarro D, Díaz-Mula HM, Guillén F, Zapata PJ, Castill S, Serrano M, Valero D, Martínez-Romero D. 2011. Reduction of nectarine decay caused by *Rhizopus stolonifer*, *Botrytis cinerea* and *Penicillium digitatum* with *Aloe vera* gel alone or with the addition of thymol. *Int J Food Micro* 151: 241–246.

Nisperos-Carriedo MO. 1994. Edible coatings and films based on polysaccharides. In: *Edible Coatings and Films to Improve Food Quality* eds. JM Krochta, EA Baldwin, MO Nisperos-Carriedo, Technomic Publishing, Lancaster, PA, pp. 322–323.

No HK, Meyers SP, Prinyawiwatkul W, Xu Z. 2007. Applications of chitosan for improvement of quality and shelf-life of foods: A review. *J Food Sci* 72: R87–R100.

Olivas GI, Mattinson DS, Barbosa-Cánovas GV. 2007. Alginate coatings for preservation of minimally processed 'Gala' apples. *Postharv Biol Technol* 45: 89–96.

Oriani VB, Molina G, Chiumarelli M, Pastore GM, Hubinger M. 2014. Properties of cassava starch-based edible coating containing essential oils. *J Food Sci* 79: E189–E194.

Oz AT, Ulukanli Z. 2012. Application of edible starch-based coating including glycerol plus oleum Nigella on arils from long-stored whole pomegranate fruits. *J Food Process Preserv* 36: 81–95.

Paladines D, Valero D, Valverde JM, Díaz-Mula HM, Serrano M, Martínez-Romero D. 2014. The addition of rosehip oil improves the beneficial effect of *Aloe vera* gel on delaying ripening and maintaining postharvest quality of several stone fruit. *Postharv Biol Technol* 92: 23–28.

Park HJ. 1999. Development of advanced edible coatings for fruits. *Trends Food Res Technol* 10: 254–260.

Park HJ, Chinnan MS. 1995. Gas and water vapor barrier properties of edible films from protein and cellulosic materials. *J Food Eng* 25: 497–507.

Peretto G, Du WX, Avena-Bustillos RJ, Berrios JDJ, Sambo P, McHugh TH. 2014. Optimization of antimicrobial and physical properties of alginate coatings containing carvacrol and methyl cinnamate for strawberry application. *J Agric Food Chem* 62: 984–990.

Pérez-Alfonso CO, Martínez-Romero D, Zapata PJ, Serrano M, Valero D, Castillo S. 2012. The effects of essential oils carvacrol and thymol on growth of *Penicillium digitatum* and *P. italicum* involved in lemon decay. *Int J Food Micro* 158: 101–106.

Pérez-Gago MB, Rojas C, Del Río MA. 2002. Effect of lipid type and amount of edible hydroxypropyl methylcellulose-lipid composite coatings used to protect postharvest quality of mandarins cv. fortune. *J Food Sci* 67: 2903–2910.

Rakotonirainy AM, Wang Q, Padua GW. 2001. Evaluation of zein films as modified atmosphere packaging for fresh broccoli. *J Food Sci* 66: 1108–1111.

Ramos-García M, Bosquez-Molina E, Hernández-Romano J, Zavala-Padilla G, Terrés-Rojas E, Alia-Tejacal I, Barrera-Necha L, Hernández-López M, Bautista-Baños S. 2012. Use of chitosan-based edible coatings in combination with other natural compounds, to control *Rhizopus stolonifer* and *Escherichia coli* DH5α in fresh tomatoes. *Crop Prot* 38: 1–6.

Ravanfar R, Niakousari M, Maftoonazad N. 2014. Postharvest sour cherry quality and safety maintenance by exposure to Hot-water or treatment with fresh Aloe vera gel. *J Food Sci Technol* (in press)

Reinoso E, Mittal GS, Lim LT. 2008. Influence of whey protein composite coatings on plum (*Prunus domestica* L.) fruit quality. *Food Bioprocess Technol* 1: 314–325.

Rojas-Grau MA, Soliva-Fortuny R, Martín-Belloso O. 2009. Edible coatings to incorporate active ingredients to fresh cut fruits: A review. *Trends Food Res Technol* 20: 438–447.

Sánchez-González L, Pastor C, Vargas M, Chiralt A, González-Martinez C, Chafer M. 2011. Effect of hydroxypropylmethylcellulose and chitosan coatings with and without bergamot essential oil on quality and safety of cold-stored grapes. *Postharv Biol Technol* 60: 57–63.

Sebti I, Chollet E, Degraeve P, Noel C, Peyrol E. 2007. Water sensitivity, antimicrobial, and physicochemical analyses of edible films based on HPMC and/or chitosan. *J Agric Food Chem* 55: 693–699.

Serrano M, Díaz-Mula HM, Valero D. 2011. Antioxidant compounds in fruits and vegetables and changes during postharvest storage and processing. *Stewart Postharv Rev* 2011: 1:1.

Serrano M, Díaz-Mula HM, Zapata PJ, Castillo S, Guillén F, Martínez-Romero D, Valverde JM, Valero D. 2009. Maturity stage at harvest determines the fruit quality and antioxidant potential after storage of sweet cherry cultivars. *J Agric Food Chem* 57: 3240–3246.

Serrano M, Martínez-Romero D, Guillén F, Valverde JM, Zapata PJ, Castillo S, Valero D. 2008. The addition of essential oils to MAP as a tool to maintain the overall quality of fruits. *Trends Food Res Technol* 19: 464–471.

Shahidi F, Arachchi JKV, Jeon YJ. 1999. Food applications of chitin and chitosans. *Trends Food Res Technol* 10: 37–51.

Shiri MA, Ghasemnezhad M, Bakhshi D, Sarikhani H. 2013. Effect of postharvest putrescine application and chitosan coating on maintaining quality of table grape cv. "shahroudi" during long-term storage. *J Food Process Preserv* 37: 999–1007.

Shewfelt RL. 1999. What is quality? *Postharv Biol Technol* 15: 197–200.

Sime WJ. 1990. Alginates. In: *Food Gels* ed. P Harris, Elsevier, London pp. 53–58.

Steenkamp V, Stewart MJ. 2007. Medicinal applications and toxicological activities of *Aloe* products. *Pharmaceut Biol* 45: 411–420.

Tanada-Palmu PS, Grosso CRF. 2005. Effect of edible wheat gluten-based films and coatings on refrigerated strawberry (*Fragaria ananassa*) quality. *Postharv Biol Technol* 36: 199–208.

Tharanathan RN. 2003. Biodegradable films and composite coatings: Past, present and future. *Trends Food Res Technol* 14: 71–78.

Valencia-Chamorro SA, Palou L, Del Río MA, Pérez-Gago MB. 2011. Antimicrobial edible films and coatings for fresh and minimally processed fruits and vegetables: A Review. *Crit Rev Food Sci Nutrit* 51: 872–900.

Valero D, Díaz-Mula M, Zapata PJ, Guillén F, Martínez-Romero D, Castillo S, Serrano M. 2013. Effects of alginate edible coating on preserving fruit quality in four plum cultivars during postharvest storage. *Postharv Biol Technol* 77: 1–6.

Valero D, Serrano M. 2010. *Postharvest Biology and Technology for Preserving Fruit Quality.* CRC-Taylor & Francis, Boca Raton, FL.

Valero D, Serrano M. 2013. Growth and ripening stage at harvest modulates postharvest quality and bioactive compounds with antioxidant activity. *Stewart Postharv Rev* 3: 7.

Valverde JM, Valero D, Martínez-Romero D, Guillén F, Castillo S, Serrano M. 2005. Novel edible coating based on *Aloe vera* gel to maintain table grape quality and safety. *J Agric Food Chem* 53: 7807–7813.

Wang Y, Lopes Filho F, Geil P, Padua GW. 2005. Effects of processing on the structure of zein/oleic acid films investigated by X-ray diffraction. *Mol Biosci* 5: 1200–1208.

Yang G, Yue J, Gong X, Qian B, Wang H, Deng Y, Zhao Y. 2014. Blueberry leaf extracts incorporated chitosan coatings for preserving postharvest quality of fresh blueberries. *Postharv Biol Technol* 92: 46–53.

Zapata PJ, Guillén F, Martínez-Romero D, Castillo S, Valero D, Serrano M. 2008. Use of alginate or zein as edible coatings to delay postharvest ripening process and to maintain tomato (*Solanum lycopersicon* Mill) quality. *J Sci Food Agric* 88: 1287–1293.

Zapata PJ, Navarro D, Guillén F, Castillo S, Martínez-Romero D, Valero D, Serrano M. 2013. Characterisation of gels from different *Aloe* spp. as antifungal treatment: Potential crops for industrial application. *Indust Crops Products* 42: 223–230.

8 Low Ethylene Technology in Non-Optimal Storage Temperatures

Ron B.H. Wills

CONTENTS

8.1 INTRODUCTION

Ethylene ($CH_2 = CH_2$) is notable as the defining metabolic difference between plant and animals systems. Ethylene is only synthesized by plants and has a major regulatory role in the growth, development, and senescence of all plants. This is in contrast to the lack of any known synthesis or metabolic effect on animal systems. In addition, ethylene was the only known endogenous gas to act as a plant-growth regulator, but the relatively recent discovery of the synthesis and metabolic effects of the previously considered toxic gases, nitric oxide (Leshem, 2000), and hydrogen sulphide (Wagner, 2009) in animals, and even more recently in plants, has underscored the limitation of our knowledge of metabolism (see Chapter 9 for postharvest effects of nitric oxide).

Effects of ethylene on fruit and vegetables have been known for many centuries, even if its identity remained unknown due to inadequate analytical techniques. It is believed that the ancient Egyptians and Chinese unknowingly stimulated ethylene production by damaging plant tissues in a confined environment to accelerate the ripening of figs and pears, respectively. In the intervening years, a wide range of other effects of ethylene have been recorded, including early in the twentieth century, when heaters burning kerosene were used to degreen lemons, although the effect was attributed to the heat rather than the ethylene generated as a byproduct of the incomplete combustion of kerosene. It was not until the 1930s that ethylene was identified as the causative agent with Gane (1934) reporting ethylene synthesis by plants and Crocker et al. (1935) proposing ethylene as the agent responsible for fruit-ripening and senescence of vegetative tissues.

The availability of gas chromatography in the 1960s greatly expanded research efforts into the biosynthesis of ethylene. A major contributor to elucidation of the metabolic pathway of ethylene was Shang Fa Yang, who had 30 prolific years at UC Davis on postharvest metabolic studies; the initial stages in the ethylene synthesis pathway is now commonly referred to as the Yang cycle (Bradford, 2008). The essential elements in ethylene synthesis start with conversion of the amino acid methionine → S-adenosyl methionine (SAM) → 1-aminocyclopropane-1-carboxylic acid (ACC) → ethylene. The enzymes catalyzing these reactions are, respectively, methionine adenosyltransferase (MAT), ACC synthase, and ACC-oxidase. The activity of ACC synthase (ACS) is considered the rate-controlling step for ethylene production. The final step requires oxygen to release ethylene (Yang, 1985).

It is of considerable importance that while ethylene is synthesized as part of the natural plant development cycle, it is also induced as a response to a range of external factors, such as mechanical wounding, pathogen attack, and abiotic environmental stress. In this respect, production of ethylene is considered to trigger a range of protective actions by plants, and is thus a beneficial defensive agent (Morgan and Drew, 1997). In addition, ethylene in the ambient air can arise from a range of industrial sources, such as motor vehicle exhaust or industrial effluent, or from senescing vegetation. It can also arise in mixed-load storage or transport where a range of different produces are held in the same chamber, and ethylene evolved from one produce will impinge on other lower ethylene emitters in the same chamber. The rapid diffusivity of ethylene into plant tissues means that exogenous sources of ethylene are as effective as endogenous ethylene in modifying metabolic and physiological behavior.

The tenet of this chapter is that reducing exposure of fruit and vegetables to ethylene can be beneficial in extending postharvest life. An obvious treatment is one that inhibits metabolism of ethylene by produce, but the diffusivity of exogenous ethylene means that any low ethylene technology must be able to counteract all sources of ethylene that the produce might encounter during storage and handling.

Any intervention technology needs to recognize the differing role of ethylene in: (1) initiating ripening in climacteric fruit; and (2) accelerating senescence of nonclimacteric fruit and vegetables (Abeles et al., 1992; Wills et al., 2007a). Climacteric produce are fruits with a distinct ripening pattern, with the defining characteristic being a pronounced increase in respiration to a peak value or climacteric. They are generally harvested at a mature but unripe state, with the aim of preventing ripening during storage and transport to markets, by inhibiting initiation of the full ripening process by ethylene. However, such fruit need to be ripened before consumption so any technological intervention needs to be either reversible or temporary.

Nonclimacteric fruits are harvested at or close to desirable eating quality and do not exhibit dramatic postharvest changes and show only relatively small changes in respiration. Horticultural commodities that are not fruits (i.e., nonseed-bearing vessels) are classed as vegetables, and are nonclimacteric in behavior, as they are harvested at a commercial maturity stage when desirable eating quality is present, although the physiological maturity of vegetables can vary widely (Wills et al., 2007a). For most nonclimacteric produce, the postharvest regime aims to minimize any rate of change, and their response to ethylene is an acceleration of the rate of normal senescence.

8.2 COOL CHAIN MANAGEMENT: THE PREEMINENT TECHNOLOGY

Over the centuries, the seasonality and perishability of most horticultural crops have generated interest in methods that would extend the period over which produce are available in any specific location. The benefit of storing produce in reduced temperatures has long been realized in countries with a cold winter. For example, storage of Chinese cabbages in China and apples in North America and Europe in cellars or caves, sometimes ventilated with cool outside air, was practiced at village and farm level, and provided a much needed food source during winter. The advent of mechanical refrigeration in the 1850s transformed the use of low-temperature storage from an art into a reliable technology. The first mechanical ice-making machine was built in 1851 by James Harrison in Geelong, Australia, and was followed by a commercial refrigeration unit in 1854 that used an ether vapor compression system. However, interest was in producing frozen meat for export to England rather than the cool-storage of fruit and vegetables. The first commercial food cold storage facility in the United States was installed in Boston in 1881, and was followed over the next 10 years by many urban and farm-based facilities (Ryall and Pentzer, 1974; Farrer, 1980). Thus, low-temperature refrigerated storage of food was well established at the beginning of the twentieth century, with apples and pears being the major horticultural crops to be stored.

In the 1950s, apart from the major commodities, there was still a strong seasonality of most produce, with consumption generally close to production areas, and where quality was secondary to availability. In the intervening 60+ years, the horticultural industry has moved to a virtual 12-month availability for many commodities, which is supported by a substantial international trade of a greater diversity and quantity of produce. While many factors have had a role in making this change possible, the major technological impact is considered to be cool-chain management, where a controlled temperature environment is maintained from farm-storage to domestic consumption, to inhibit ripening and senescence. Numerous studies identified the optimum low temperature at which individual produce could be held to achieve the maximum postharvest life. The reduction in metabolism achieved by reducing the temperature should be optimally just above the freezing temperature of about $-1°C$. However, for many produce, the optimum temperature is well above freezing due to abnormal metabolism, leading to chilling-related injury, physiological disorders, or failure to fully ripen when transferred to ambient temperature.

While cool-chain management has become the technology of first resort, its use is energy-intensive and it utilizes costly infrastructure. Refrigeration became *de rigeur* when energy was cheap and greenhouse gas emissions were of no concern. The world has now moved to an energy minimization mode and horticulture, in common with many other industries, is aiming to reduce its carbon footprint from an economic perspective, due to the increasing cost of energy, and to be seen by the community as a good corporate citizen.

East (2010) reviewed postharvest energy usage of horticultural commodities and situations, and cites studies that estimated:

1. Cooling and storage of fresh horticultural produce in California uses about 1 billion kWh of energy, which was 5.5% of the electricity used by agriculture in the state (Thompson and Singh, 2008)
2. Domestic and retail refrigeration account for about 2.5% of Britain's carbon dioxide emissions (Garnett, 2006)
3. The Australian vegetable industry accounts for 0.7% of national emissions (O'Halloran et al., 2008)

In addition, refrigeration in America's supermarkets is the main use of energy, accounting for 36% of all energy costs (E-source, 2013).

In an energy-conscious world, it is appropriate to ask: is cool-chain technology always necessary? Can energy consumption be reduced by storing and transporting produce at higher temperatures than that currently utilized by employing other technologies to inhibit ripening and senescence? One such technology is to reduce activity of ethylene or the concentration around produce or the receptivity of produce to ethylene action. The following sections in this chapter will examine how technology that either reduces exposure to ethylene, or inhibits ethylene action or synthesis, can reduce or even eliminate the need for refrigeration.

8.3 EFFECT OF ETHYLENE CONCENTRATION ON POSTHARVEST BEHAVIOR

For many years, a concentration of 0.1 µL/L was generally cited as the threshold level for the activity of ethylene on postharvest behavior (Burg and Burg, 1962; Knee et al., 1985). Establishment of such a threshold level was probably related as much to the ability to measure very low ethylene levels than any empirical evidence. Peacock (1972) examined the effect on the ripening of green bananas exposed to short bursts of ethylene, and found that while ripening was not immediately initiated, the green life was shortened by any exposure to ethylene. He advocated that for bananas there was no threshold level for ethylene action, and any ethylene level will advance ripening.

The existence of a much lower threshold level for ethylene action was examined in a series of papers by Wills and colleagues (Table 8.1). They determined the

TABLE 8.1
Postharvest Life of Climacteric and Nonclimacteric Produce Continuously Exposed to Ethylene Concentrations of <0.005–10 µL/L in Published Reports by Wills and Colleagues

Produce	Temp (°C)	Quality Issue	Reference
		Climacteric	
Avocado	20, 10, 0	Ripening, chilling (0)	Wills and Gibbons (1998)
Banana	20	Ripening	Wills et al. (1999a)
Mango	20	Ripening	Wills et al. (2001)
Peach	20	Ripening	Wills et al. (2001)
Custard apple	20, 14	Ripening	Wills et al. (2001)
Kiwifruit	20, 0	Ripening	Wills et al. (2001)
Tomato	20	Ripening	Wills et al. (2001)
		Nonclimacteric	
Strawberry	20, 0	Rotting, senescence	Wills and Kim (1995)
Lettuce	20, 0	Leaf tip browning	Kim and Wills (1995)
Green bean	20, 5, 0	Yellowing (20), chilling (5 & 0)	Wills and Kim (1996)
Pak choi, choi sum, gai lan	20, 0	Yellowing	Wills and Wong (1996)
Chinese cabbage	20, 0	Yellowing	Wills et al. (1999b)
Orange	2.5	Chilling	Wills et al. (1999b)
Broccoli, parsley, chive	20, 5	Yellowing	Wills et al. (1999b)
Fancy lettuces (13 types)	5	Leaf-tip browning	Wills et al. (1999b)

postharvest life of a range of climacteric and nonclimacteric produce that were continuously exposed to air containing a logarithmic decrease in ethylene concentration from 10 to 0.001 µL/L (actually <0.005 µL/L, their limit of detection) at ambient and reduced temperatures. They found that for all produce examined, the postharvest life was extended linearly with logarithmic decrease in ethylene regardless of whether postharvest life was due to ripening, rotting, yellowing, browning, chilling-related injury, or general senescence. Examples of this relationship with ethylene concentration are illustrated in Figure 8.1 using data generated for chilling-related injury of "Valencia" orange stored at 2.5°C and ripening of "Hayward" kiwifruit stored at 20°C.

Pranamornkith et al. (2012) showed a similar relationship for "Hort 16A" kiwifruit (*Actinidia chinensis*) that were stored at 1.5°C. Fruit-ripening, as determined by a range of attributes, was progressively delayed as the ethylene concentration in the atmosphere around fruit was decreased in a logarithmic manner from 1 to 0.001 µL/L. The lowest concentration of 0.001 µL/L was confirmed as the actual concentration, using the ultra-sensitive laser-based detection system developed by Harren and colleagues in the Netherlands (Cristescu et al., 2013).

While it can now be categorically stated that the threshold level for a physiological effect of ethylene on horticultural produce is much lower than 0.1 µL/L, it would seem that the actual threshold level cannot yet be categorized. Since the

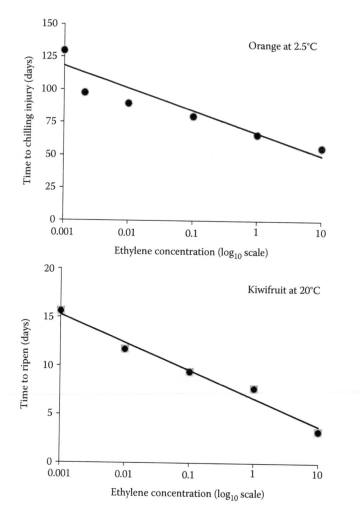

FIGURE 8.1   Effect of ethylene concentration on the postharvest life of oranges at 2.5°C and kiwifruit at 20°C as determined by chilling-related injury and ripening, respectively. (Data from Wills RBH et al. 1999b. *Aust J Expt Agric* 39: 221–224 for "Valencia" orange, and Wills RBH et al. 2001. *Aust J Expt Agric* 41: 89–92 for "Hayward" kiwifruit.)

longest postharvest life in the above studies occurs with the lowest ethylene concentration, the threshold level is likely to be lower than 0.001 μL/L. The difficulty in accurately measuring such low ethylene levels is a barrier to elucidation—it is possible that ethylene may still have some activity at one molecule per cell. However, for commercial situations where ethylene emissions are derived from a range of sources, any reduction in the ethylene concentration in the atmosphere around produce will be beneficial in extending postharvest life. It is immaterial whether the source of ethylene in the atmosphere is from the produce itself, or from exogenous sources. Wills et al. (2000) determined the ethylene level in the atmosphere during marketing, and found that the air in wholesale markets and distribution centers contained on average 0.06 µL/L ethylene, but was lower in supermarket retail outlets at about 0.02 µL/L. However, the level of ethylene in all premises was well above any threshold level, and hence there was some decrease in postharvest life throughout the marketing chain.

The importance of "background" ethylene levels, even in laboratory trials, is illustrated with the study by Bower et al. (2003) who stored Bartlett pears in a range of ethylene concentrations, including a nominal "0 µL/L" treatment in input air. However, on analysis, the ethylene concentration was found to be 0.10 µL/L in chambers of fruit stored at 1°C, and 0.32 µL/L in chambers at 2.5°C. The increase was attributed to endogenous ethylene emissions from the fruit. This also raises the importance of measuring the actual level of ethylene in containers in studies purporting to examine the effect of ethylene on produce. While numerous studies have been reported on the effect of ethylene, many have accepted that the input level of ethylene is the actual level in the produce container. The efficacy of such findings must be considered as suspect.

8.4 INTERACTION OF ETHYLENE AND TEMPERATURE ON POSTHARVEST BEHAVIOR

Exposure of horticultural produce to low temperature and reduced ethylene concentrations independently result in extensions to postharvest life. There are, however, few published studies that quantify the interaction of temperature and ethylene over the ranges likely to be encountered in commercial situations. The most comprehensive study was reported by Wills et al. (2014) on inhibition of the ripening of green bananas. They determined the green life (i.e., time to ripen) of Australian "Cavendish" bananas at all combinations of ethylene at 0.001 (actual <0.002), 0.01, 0.1, and 1.0 µL/L, and temperatures of 15°C, 20°C, and 25°C. The relationship between green life and ethylene and temperature is given in Figure 8.2 and, as expected, shows that green life increased as the temperature and ethylene concentration were reduced, with the greatest increases occurring at the lower end of both the ethylene and temperature range. The value in the data is that it gives various combinations of ethylene concentration and temperature that generate the same green life. Thus, for any specified period required for transport or storage, it is possible to manage the marketing scenario to retain produce quality but minimize cost. Wills et al. (2014) also used the regression equation generated by the data to determine the ethylene concentrations required to allow bananas shipped commercially (by road within Australia and by sea from Central America to southern Europe) to be carried at ambient temperatures, that is, without the need for refrigeration. They concluded that such transport was feasible, although the longer international route may require the use of some other ancillary technology (to be discussed in later sections).

Wills and Kim (1996) also reported a similar relationship for the postharvest life of green beans, where the maximum storage life was achieved at <0.005 µL/L ethylene of 29, 20, and 11 days at 0°C, 5°C, and 20°C, respectively. For each temperature, the postharvest life increased by 2–3 times as the ethylene concentration

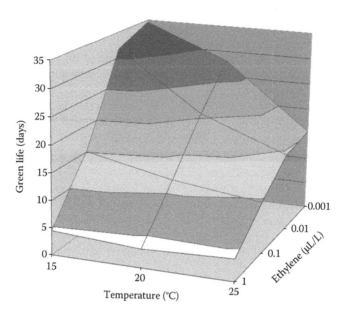

FIGURE 8.2 Interaction of ethylene concentration and storage temperature on green life of "Cavendish" bananas. (Data from Wills RBH et al. 2014. *Postharv Biol Technol* 89: 7–10.)

was increased to 0.01, 0.1, 1, and 10 μL/L. They found that ethylene levels around beans in commercial markets were in the range 0.17–1.17 μL/L. Other reported studies examined too few temperatures and ethylene concentrations to make meaningful assessment of the ethylene/temperature interaction, but it would not seem unreasonable to expect that holding many produce at low ethylene would give a similar postharvest life as at low temperatures in current commercial situations where ethylene levels can readily accumulate.

8.5   METHODS FOR INHIBITING ETHYLENE ACTION

Understanding the metabolism of ethylene has led to the development of various methods to control ripening and senescence of fruits and vegetables. For example, aminoethyoxyvinylglycine (AVG) and aminooxyacetic acid (AOA) inhibit ACC synthase and, therefore, block the conversion of SAM to ACC. They act by reducing ethylene production and are, therefore, ineffective against the action of exogenous ethylene. While they have useful effects with some ornamentals, their development with fruits and vegetables was effectively terminated due to potential toxic effects on humans (Abeles et al., 1992). Other compounds that have some effects on ethylene production or action, but have not found any commercial application, include 3,5-diiodo-4-hydroxybenzoic acid (DIHB) (Robert et al., 1975), ethylene oxide (Lieberman et al., 1964), free radical scavengers such as benzoate and propyl gallate (Apelbaum et al., 1989), and polyamines such as spermidine (Wang, 1987), and 2,4-dinitrophenol (Yu et al., 1980).

Considerable research has focused on the binding of ethylene to a specific enzyme receptor to form an active complex that triggers the postharvest changes. The binding of ethylene to receptors is considered to take place reversibly at a site containing copper, and the affinity of the receptor for ethylene is increased by the presence of oxygen and decreased by carbon dioxide (Yang, 1985; Kanellis et al., 1989). Ethylene action can then be modified by interfering with the binding of ethylene to receptors or changing the quantity of receptors.

Blocking the metal receptor for ethylene was utilized by Beyer (1976) who reported that a foliar spray of silver nitrate prevented a wide range of ethylene-induced responses, including growth inhibition, abscission, and senescence in a range of intact plants. Commercial use of silver as the thiosulphate salt was introduced for a wide range of ornamentals to extend vase life. However, while treatment of fruit and vegetables with silver ion has been shown to inhibit the action of ethylene, silver is not a feasible option due to its toxicity. Even the use of silver on ornamentals is being gradually restricted due to environmental concerns relating to disposal of treatment solution, and of treated plant material at the end of the postharvest life.

Various cyclic olefins have been found to interact with the ethylene receptors and, thereby, block tissue response to ethylene. Among them, 2,5-norbornadiene was found to inhibit binding to receptor sites (Sisler et al., 1986), but the trans-cyclooctenes were more effective (Sisler et al., 1990). However, both have an unpleasant odor and rapidly diffuse from the receptor, which severely limits commercial use (Sisler and Serek, 1999).

8.5.1   1-METHYLCYCLOPROPENE

In recent years, the postharvest research community has focused on the cyclopropenes, which were recognized by Sisler and colleagues at North Carolina State University, and, in particular, with 1-methylcyclopropene (1-MCP). The first publication on 1-MCP was by Serek et al. (1994), who showed that 1-MCP inhibited the ethylene-induced abscission and wilting in potted flowering plants following a relatively short prestorage treatment at nL/L concentrations for a few hours. Numerous studies have since shown a beneficial effect of 1-MCP on a wide range of fruits and vegetables, and it is now in commercial use in many countries for various produce, with apples being the major beneficiary. The effectiveness of 1-MCP is considered due to a physical similarity to ethylene that allows 1-MCP to bind to the metal in the ethylene receptor where it remains bound for a long period.

While 1-MCP is an important consideration for this chapter, which is focusing on low ethylene technologies that will allow produce to be stored above optimal cool-chain temperatures, a comprehensive review of 1-MCP is discussed in Chapter 6. Suffice it in this chapter to restate that common effects of 1-MCP include a reduction in ethylene production, respiration, volatile synthesis, rate of softening, loss of acidity, and loss of chlorophyll. These are typical of delayed ripening and senescence that can be effected by inhibiting ethylene action. The list of effects can also include a reduced incidence of various physiological disorders of apples and chilling-related injury of various fruits, which would also seem to be related to interference with the action of ethylene. However, 1-MCP is not a universal panacea, as issues on uneven ripening can occur in produce such as bananas (Golding et al., 1998; Harris et al., 2000).

The commercial success of 1-MCP can be attributed in large part to the existence of patent protection and the absence of any prior manufacture of 1-MCP. The current patent-holder, Rohm and Hass (Spring House, PA), through its subsidiary, Agrofresh, markets 1-MCP under the trade name SmartFresh™. This provided the funding needed to support international research to demonstrate the effects of 1-MCP, conduct toxicological studies and obtain registration for usage in many countries. In the current complex regulatory environment and community resistance to new chemicals, the lack of patent protection is a significant impediment to commercialization of any new chemical based technology.

Of direct relevance to this chapter is a commercial trial reported by McCormick et al. (2012), in which "Gala" apples treated with 1-MCP were successfully stored at a higher temperature than a room of untreated fruit. 1-MCP was applied at the current commercial rate in Germany (0.625 µL/L for 24 h), then stored in a controlled atmosphere room (2% CO_2, 1.6% O_2), but the temperature was maintained at 4°C, which was 2.5°C higher than the control room. Fruit removed from storage at various times and evaluated for quality after seven days at ambient temperature to simulate marketing showed that 1-MCP-treated fruit at the higher storage temperature were preferred by a consumer sensory panel. An energy audit after six and eight months showed a 35% reduction in energy was achieved by storage at the higher temperature. Thus, a more acceptable product was marketed, but with considerable energy saving.

8.5.2 Nitric Oxide

Nitric oxide (NO) is a highly reactive gaseous free radical that, traditionally, was considered as an industrial byproduct toxic to plants and animals. However, extensive research over the last 20 years has found NO to be metabolized by mammals, and having a regulatory role in many human physiological systems. This has led to the therapeutic use of NO for various conditions, although high levels result in detrimental effects.

The existence of NO in higher plants was first reported by Leshem and Haramaty (1996) who found emissions from pea seedlings. Of particular interest to postharvest horticulture was their finding of an inverse relationship between emitted NO and ethylene, and that the addition of an NO-releasing compound decreased ethylene production. The link between ethylene and NO was furthered by Leshem et al. (1998) who reported that an NO-releasing compound reduced the rate of ethylene production and delayed postharvest changes in a range of climacteric and nonclimacteric produce. In addition, endogenous NO levels were higher in unripe fruits than in ripe and in freshly-cut, compared to senescing flowers. Thus, endogenous NO and ethylene emissions were inversely related and the addition of NO counteracted the senescence-promoting effect of ethylene.

Considerable research on the postharvest effects of NO has been conducted in many countries over the last 10 years, and the findings from these studies have been reviewed in Chapter 9. A summary of the information, presented in Chapter 9, shows that exogenous NO, by fumigation or dipping in an NO-donor solution, inhibits ripening, senescence, chilling-related injury, and development of disease on a wide range of intact and fresh-cut fruit and vegetables. In addition to inhibiting the production of ethylene, NO has been demonstrated to reduce respiration, ion leakage and oxidative stress, enhance the activity of a range of antioxidant enzymes, and inhibit polyphenol oxidase (PPO) activity. The key factor through which NO is affecting metabolism is not known, but inhibition of ethylene production and/or action is a likely important component.

Postharvest application of NO is thus a potential new technology to reduce losses of horticultural produce. NO appears to be synthesized by all plants and can, therefore, be considered as a naturally occurring substance. It would then be analogous to ethylene which, while it has toxicity and handling issues, is an approved postharvest treatment. The human synthesis of NO and its therapeutic use should also be positive in securing regulatory approval. However, plants and mammals do not have totally identical pathways for metabolism of NO and it will need to be proved that NO in fruit and vegetables does not generate toxic byproducts such as nitrosamines. An impediment to commercialization of an NO treatment is the lack of current patent protection. Some organization needs to expend considerable time and resources to develop the safety and efficacy information to gain regulatory approval in many countries, but lack of exclusivity for the treatment would be a major deterrent to commercial operators.

Use of NO gas would require gas-tight fumigation chambers and the monitoring of atmospheric discharges. A more convenient delivery system of a solid that degrades to release NO gas in the presence of produce, as proposed by Wills et al. (2007b), offers greater ease of application. Only a few of the many solid NO-donor compounds have been evaluated, but, currently, sodium nitroprusside (SNP) seems to offer potential for commercial use due to its stability in water and a long history of safe clinical use. A new potentially enticing treatment option is to apply a non-NO-containing compound that stimulates endogenous NO production. The limited research on this aspect has revealed a divergent group of benign compounds. This could be particularly appealing for regulatory purposes if the compound was either naturally occurring or had GRAS or similar other safe status for foods.

8.5.3 Nitrous Oxide

Nitrous oxide (N_2O) is a naturally occurring atmospheric gas mostly originating from soil through the action of aerobic bacteria (Anderson and Levine, 1986). It is chemically inert and has had extensive use for anesthesia in dentistry as it is generally considered to be nontoxic with an exposure limit of up to 50 $\mu L/L$ over a prolonged period (Quarnstrom, 2013). It has a linear structure similar to carbon dioxide with which it has similar physical properties, including high solubility in aqueous media. There are no reports of endogenous nitrous oxide in plants.

Interest in the postharvest effects of nitrous oxide was initiated by the Air Liquide laboratories in France, and Fath et al. (1990) filed a patent for treating horticultural produce with nitrous oxide to extend postharvest life based on its antiethylene properties. Published data by the Air Liquide group (Gouble et al., 1995) showed that nitrous oxide delayed the ripening of tomato and avocado fruits by extending the lag period before ethylene synthesis increased, and lowering the ethylene production rate. The level of nitrous oxide used was quite high at 80% as the effect was achieved by essentially displacing nitrogen from air atmosphere while retaining 20% oxygen. The mode of action was considered to be similar to carbon dioxide, but as nitrous oxide is not toxic to tissues, a higher concentration could be safely maintained around produce and presumably induce a stronger effect. Since the nitrous oxide atmosphere was maintained around produce throughout storage, the technology is limited to controlled-atmosphere storage situations.

Short-term exposure to nitrous oxide was reported by Benkeblia et al. (2001) and Benkeblia and Varoquaux (2003). They found that for onions there was a limited response that required high levels of nitrous oxide up to 100% for 4–15 days, with a reduced incidence of rotting during subsequent storage but there was no effect on the incidence of sprouting. At the end of the exposure period there was a small transient reduction in respiration and a similar increase in soluble solids and acidity. Qadir and Hashinaga (2001) examined short-term exposure of two, four, and six days to 80% nitrous oxide in 20% oxygen followed by ambient air storage at 20°C, and found inhibited development of postharvest decay in apple, guava, persimmon, mandarin, strawberry, and tomato that had been inoculated with seven fungal species. They also prestored strawberries in 10%–80% nitrous oxide with 20% oxygen for two, four, and six days, followed by ambient storage at 2°C, and found nitrous oxide reduced the incidence of *B. cinerea* with the magnitude of the effect being dose- and time-dependent. Thus the most effective treatment was 80% nitrous oxide for six days. Lichanporn and Techavuthiporn (2013) treated longkong fruit with 90% nitrous oxide vapor for 3 h, followed by storage at 13°C and 90% RH, and found delayed development of pericarp browning with a higher level of phenolic compounds but lower levels of the browning associated enzymes, phenylalanine ammonia lyase, polyphenoloxidase, and peroxidase. Even though the effect of nitrous oxide is considered to be similar to carbon dioxide, there have not been any published studies comparing the effects of the two gases.

8.6   METHODS FOR REMOVING ETHYLENE FROM THE ATMOSPHERE

8.6.1   Ventilation

The simplest technology to reduce the concentration of ethylene in the atmosphere around horticultural produce in an enclosed space is to pass ambient air through the container. Ambient air in rural areas contains a low level of ethylene at about 0.001 µL/L (Alberta Environment, 2003). While the trend in modern research is to seek high-technology solutions to postharvest problems, the value of simple systems should not be overlooked. Ventilation is an old physical method but has the potential to be resurrected as a low-cost, environmentally friendly technology.

Ventilating storage chambers with external ambient air was one of the earliest postharvest technologies utilized, although its primary function was for temperature maintenance in cave- or cellar-storage. Cooler external air was drawn into the chamber to maintain a low temperature, but it would have provided a bonus benefit through reducing the accumulation of ethylene. The technology often relied on convection, where the more dense cool external air flowed down into the chamber and the hotter chamber air rose out of the chamber. The technique is still used in less-developed regions with a cold winter, and where refrigerated facilities are expensive and not widely available.

The ventilated transport of bananas by rail in Australia was advanced by the designs of E.W. Hicks, a physicist at the Division of Food Preservation and Transport of CSIRO (Commonwealth Scientific and Industrial Research Organisaton; the Division is now the CSIRO Division of Animal, Food and Health Sciences) (Hicks and Holmes, 1935). The standard covered-rail-van was modified by inclusion of louvered panels into the van walls (Figure 8.3). Forward movement of the van drew external air across boxes of bananas in the van. This minimized the warming of bananas and also prevented the accumulation of ethylene. The overnight transport of bananas from tropical and subtropical growing areas some 1000 km to temperate markets was in such vans until the 1960s when they were gradually superseded by refrigerated road transport, which was perceived to be more flexible in operation but are now a considerable cost to the industry.

The efficient application of ventilation to a commercial situation requires knowledge of the rate of ethylene production at the temperature of the storage chamber

FIGURE 8.3 One type of louvered rail vans that was used for long distance transport of bananas in Australia.

or that encountered during a transport route, the desired storage and/or transport period, and the required ethylene concentration that will allow produce to retain acceptable quality during the postharvest period. Such a study was conducted by Wills et al. (2012) who modeled the effect of rates of ventilation on removal of a continuously generated ethylene supply from a chamber that simulated a stack of banana packages. The modeling also examined the effect of ventilation on water loss from fruit, as weight loss is an undesirable byproduct of excessive ventilation. Their data indicated that ventilation with ambient air in conjunction with a small water tank coupled to inexpensive computer-controlled data loggers on road vehicles can be a cost-effective method to prevent ripening of bananas during long-distance transport. Ventilation could be a valuable technology in developing countries as a reliable, low-cost technology for fixed storage situations where refrigeration is either not available or the energy cost is prohibitive.

8.6.2 Absorption and Oxidation

Ethylene is chemically quite reactive with a range of compounds able to react with the double bond. Early attempts to reduce ethylene levels in the atmosphere around produce was by adsorption onto activated charcoal and variants such as brominated charcoal, as well as a range of diatomaceous earths, clays, and minerals. Most of these did not bring the desired beneficial effect on produce due to insufficient ethylene adsorption capacity and/or extent of reactivity of the adsorbed ethylene. Reid and Dodge (1995) evaluated a number of adsorption products and found limited reduction in ethylene, with the exception of potassium permanganate (discussed in Section 8.6.2.1). There are, however, a number of recently released commercial products that do not disclose the mode of action or the material used, but they appear to act by adsorption—their efficacy is yet to be determined independently of the manufacturer or marketing agents.

In recent years, a range of scrubber units have been developed that use metal catalysts to improve the efficiency of carbon-based reactivity. For example, Martínez-Romero et al. (2009) improved reactivity by incorporating 1% palladium into activated carbon and regenerated its activity by heat pulsing. Such devices are suitable for storage rooms rather than individual packages.

8.6.2.1 Potassium Permanganate

Potassium permanganate ($KMnO_4$) is a strong oxidizing agent that reacts with ethylene and generates carbon dioxide and water as the main end-products. It has a long history of human use as a mild antiseptic to purify water or disinfect skin lesions. An early use with vegetables was as a disinfectant while washing (in a weak solution of potassium permanganate). Its first reported use to reduce the ethylene concentration in the atmosphere around postharvest produce was by Forsyth et al. (1967), on apples. Subsequent studies have reported that the inclusion of potassium permanganate decreased the ethylene concentration and delayed the ripening of a range of climacteric produce.

Of particular interest is the research program in Australia led by Kevin Scott on the ambient storage of bananas in sealed polyethylene bags. The initial report by Scott and Roberts (1966) found that bananas in sealed bags generated a modified atmosphere and resulted in a delay in ripening. The low-oxygen and high-carbon dioxide atmosphere reduces the sensitivity of fruit to ethylene (Liu, 1976), so a much higher ethylene level will be tolerated before ripening is induced. The subsequent study (Scott et al., 1970) included potassium permanganate that had been adsorbed onto an inert porous solid (vermiculite), which reduced the ethylene concentration inside the sealed bag and further delayed ripening; fruit ripened after 21 days, compared to 14 days in a sealed bag without ethylene absorbent and seven days for air-stored fruit. A series of semi-commercial trials between Australia and New Zealand (Scott et al., 1971a,b) showed all treated fruit arrived in the international market still unripe.

Scott and Gandanegara (1974) examined the storage of green bananas over the temperature range of 10–37.5°C in sealed polyethylene bags, with and without inclusion of potassium permanganate, and found that the time to ripen at all temperatures was extended by a modified atmosphere with a further extension due to reduction in ethylene from the absorbent. Of relevance to the storage of produce in a nonrefrigerated environment, the study found that in the presence of potassium permanganate the time to ripen at 25°C was 38 days and at 30°C it was 28 days. Their data suggests that it would be feasible for sealed-bag transport of bananas with ethylene removal to cope with any of the current world trading routes from tropical to temperate countries, to be conducted without refrigeration.

The ability of potassium permanganate to oxidize ethylene, and thereby delay ripening, has been extended to a range of other climacteric fruit such as avocado (Hatton and Reeder, 1972), mango (Esguerra et al., 1978), and kiwifruit (Scott et al., 1984). Fewer studies have been conducted with nonclimacteric produce, but retardation in loss of green color and rotting in lemons (Wild et al., 1976) and lettuce (Kim and Wills, 1995) has been shown, while Kim and Wills (1998) have demonstrated inhibited rotting in strawberries.

Potassium permanganate is now commercially available in pellet form from a diverse range of manufacturers in many countries, although the major market appears to be for removal of odors from air-conditioning systems in buildings. It would seem that a barrier to better commercial uptake is the lack of definitive studies determining the actual amount of potassium permanganate to keep ethylene at a sufficiently low concentration to maintain quality in individual produce for the desired marketing period under specific storage conditions. Wills and Warton (2004), working with a laboratory prepared product on alumina beads, showed that temperature, humidity, and the amount of absorbent all have an impact on the efficiency of ethylene removal. They also found that the efficiency of ethylene removal had declined to 10% of incoming ethylene when close to 50% of the original potassium permanganate was still present in the alumina beads. They made calculations from known rates of ethylene produced by various produce, and showed that at 20°C and 90% RH, the amount of absorbent required in a unit package was manageable, but such use was questionable for produce with higher levels of ethylene production or for very long storage periods.

8.6.2.2 Ozone and Active Oxygen

Ozone is well known to the food industry as a strong oxidizing agent that can remove pathogens and a range of undesirable contaminants from plant and animal products.

However, in such systems, the level of ozone in the atmosphere must be controlled, as it has toxic properties in plants and animals and a level of 0.1 µL is a common permitted limit of exposure in the workplace.

The reaction of ethylene and ozone is well-documented (Dickson et al., 1992). The first reported use of ozone on postharvest produce was reported by Gane (1935) who claimed success in reducing ethylene in the atmosphere of chambers of bananas, but found it difficult to control the level of ozone, which caused severe fruit injury. Scott et al. (1971a,b) used a UV lamp that emits mainly at 254 nm but with trace emissions at 185 nm among other wavelengths and found rapid loss of ethylene in the atmosphere around bananas in a sealed chamber but without any fruit injury. They speculated that while ozone was produced at 185 nm, it was rapidly degraded to atomic oxygen, which was more effective in oxidizing ethylene than ozone. Scott and Wills (1973) then developed a system in which the atmosphere from a chamber was circulated through a scrubber containing UV lamps emitting at the required ratio of 185 and 254 nm where ethylene was degraded but ozone did not emerged from the scrubber. Absoger of France (2013) now manufactures an ethylene scrubber that uses a catalyst of titanium oxide that is activated by a UV lamp to react ethylene with oxygen at elevated temperature. A heat-exchanger is incorporated to allow the heat to be removed before the air is returned to the storage chamber.

A range of ozone generators are commercially available and Skog and Chu (2001) evaluated one such unit in a storage room and found that ozone at a concentration of 0.04 µL/L prevented the accumulation of ethylene in apple and pear storage rooms. They also found that mushrooms, broccoli, and cucumbers could tolerate this ozone level without damage. They suggested that a potential use for ozone generators is in wholesale storage rooms where ethylene-producing produce such as apples are cos-tored with ethylene-sensitive produce. Similarly, Lavigne et al. (2008) reported on a study conducted by the United States army in conjunction with equipment manu-facturer Primaira, to find alternate cost-effective methods for storing fruit and veg-etables on military vessels and camps in a low ethylene environment. They found an ozone generator developed by Primaira was able to maintain a low ethylene level at a cost that was acceptable. The commercially available ozone generating systems come in a range of sizes and prices, and would need to be assessed for individual situations for cost and technical efficacy.

8.6.2.3 Prestorage Modified Atmospheres

An interesting alternative to maintaining a modified atmosphere during storage or transport is the use of a short-term exposure to a low-oxygen or high-carbon dioxide atmosphere followed by storage in ambient air. Wills et al. (1982) showed that hold-ing green bananas in reduced oxygen for three days delayed ripening by about 40% compared to air stored fruit—the ripening times being 19 and 14 days, respectively. The authors suggested that the primary effect was due to inhibition of some aspect of endogenous ethylene production, as the internal ethylene concentration was lower following low-oxygen treatment, and ripening in all treatments coincided with rapid increase in ethylene production. There was, however, no added benefit when carbon dioxide was added into the atmosphere. Wills et al. (1990) examined bananas from more than 20 plantations over three seasons and confirmed that a three-day premarketing holding in nitrogen delayed ripening by about 40%. However, some injury occurred on 4% of fruit but this developed on areas with existing mechanical damage. Klieber et al. (2002) examined shorter exposure periods but found no increase in green life for bananas stored in nitrogen for up to 24 h at 22°C. It would, therefore, seem that a three-day exposure was required to modify the ethylene metabolic system.

There have been a range of other reports of either prestorage in low oxygen or high carbon dioxide, extending postharvest life through decreased respiration or better maintenance of appearance. For example, Couey and Olsen (1975) stored "Golden Delicious" apples in high carbon dioxide (10%–25%) for 7–14 days and obtained better retention of quality during subsequent storage in air, and Couey and Wright (1975) delayed ripening of "d'Anjou" pears after prestorage in >10% CO_2 followed by air or controlled atmosphere storage. Wills et al. (1979) found that the respiration at 20°C of carrot, potato tubers, and zucchini was reduced by about 40% after exposure to either high carbon dioxide (>10%) or low oxygen (<2%) for a few days, and a reduced respiration was maintained for at least two weeks when returned to storage in air. While these studies did not examine the effect on ethylene, it is suspected that inhibition of ethylene synthesis could be at least a contributing factor to the extended postharvest life in subsequent ambient air storage. However, regardless of the mechanism involved, short-term holding in a low-oxygen or high-carbon dioxide atmosphere does offer a nonrefrigerated technology to extend postharvest life.

8.7 ENVIRONMENTAL BALANCE

While the storage of horticultural produce at a reduced temperature is very effective at extending the postharvest life through delay in ripening, senescence, and microbial growth, temperature control through refrigeration comes at an increasing cost and greenhouse gas emissions. An important effect of reducing the storage temperature is to reduce the endogenous production of ethylene, and render produce less sensitive to ethylene, so that a higher concentration is required for the same physiological action. It is then axiomatic that reducing the concentration of ethylene around produce will also generate an extension in postharvest life, and could replace or reduce the need for refrigeration.

There is, however, a general lack of appreciation by industry of the benefit to be derived from reducing the ethylene level around produce. This is in large part due to the relatively small number of studies that include ethylene concentrations well below the previously perceived threshold level of 0.1 μL/L, and which accurately monitor the actual ethylene level in treatments during storage. Published reports that include ethylene levels <0.01 μL/L all show maximum postharvest life was attained at the lowest concentration examined. This implies that a greater postharvest life would be achieved if produce were exposed to even lower ethylene concentrations. The limitation in most studies is the limit of detection of ethylene, which, for gas chromatography, is 0.002–0.005 μL/L, while 0.001 μL/L can be detected by photoacoustic laser-based detection systems. It can be speculated that ethylene concentrations of <0.001 μL/L may give larger extensions in postharvest life but are logistically difficult to attain with any certainty.

Wills et al. (2000) developed a rating scale for nonclimacteric produce stored at 20°C (that is, without refrigeration) based on 0.005 µL/L as the lowest possible ethylene concentration that can be achieved, and hence was designated as retaining 100% of possible postharvest life. From calculations of published data on postharvest life and ethylene concentrations for a range of produce (see Table 8.1), a loss of 10% was ascribed as the limit for an acceptable postharvest environment while 30% was the limit for an unacceptable environment. The ethylene levels associated with <10% loss was ≤0.015 µL/L with >30% loss occurring in >0.1 µL/L, which is the previously considered threshold level. Wills and Warton (2001) attempted to extend the rating scale to climacteric fruits, but concluded that a rating scale would need to segregate individual produce into categories of low, medium, and high ethylene sensitivity with each category having a different set of ethylene standards.

To maintain produce quality throughout any specific marketing situation by controlling the ethylene concentration, rather than through refrigeration, the question then becomes what method of ethylene control should be utilized. The technology used must not only be effective in maintaining quality but also needs to be less costly than the system being replaced. Thus a cost-benefit analysis needs to define a positive economic outcome. McCormick et al. (2012), in their assessment of storing "Gala" apples treated with 1-MCP at 2.5°C higher than untreated fruit, showed that the 1-MCP-treated fruit were more acceptable to consumers and delivered a 35% saving in energy. While 1-MCP is a relatively low-cost option, there is a limited range of produce that can be successfully treated with 1-MCP. Numerous fruit have been shown to subsequently exhibit abnormal ripening, or cannot be easily ripened on demand. Use of nitric oxide is analogous to 1-MCP with a short prestorage treatment at a low concentration. A potential advantage of nitric oxide is that it is naturally produced by plants and animals. However, there is a range of technical and regulatory hurdles that need to be overcome before commercialization on foods becomes possible (these issues are discussed in Chapter 9).

The most environmentally friendly method to minimize the accumulation of ethylene around fruit and vegetables is by ventilation with atmospheric air—no synthetic chemicals are needed, and the air is free. Ventilation can be achieved during transport by directing the natural movement or air across a load of produce or in room or container storage by the inclusion of a fan. The ability to manage water loss is the major challenge but, with modern sensors, computer chips, and small heat exchangers, this should be manageable. For large storage rooms, the use of active oxygen scrubbers is similarly environmentally friendly as excess ozone or active oxygen are degraded before the ethylene-free air is returned to the storage chamber. However, they come at some cost, which needs to be considered in a cost-benefit evaluation.

The use of ethylene absorbents such as potassium permanganate is a relatively low-cost option that is suitable for individual packages or containers. Disposal of the absorbent on arrival at the market destination obviates the need for returnability, an issue that needs to be resolved for more sophisticated technology. While ethylene absorbents are effective in nonsealed packages, they are most efficient when included in a sealed modified atmosphere unit. The system then gains the benefit of reduced ethylene, and inhibition of metabolism from the reduced oxygen and elevated carbon dioxide.

There are, thus, a range of technologies that can reduce the ethylene concentration in the atmosphere around produce, or inhibit the action of ethylene and, thereby, extend postharvest life. For produce with a relatively short marketing chain, the increased postharvest life may be sufficient to store and transport without refrigeration. For longer duration postharvest situations, the reduced ethylene would allow storage at a higher temperature. It should be possible for most produce to use a technology that provides the required postharvest life at a reduced cost, and with some reduction in greenhouse gas emissions.

Finally, the obsession with cool-chain technology has undoubtedly led to its use in situations where produce are held for short periods, and marginal benefit, if any, is obtained. An example could be in supermarket refrigerated display areas. Wills et al. (2000) found the ethylene level in such areas to be quite low at 0.02 μL/L and, when combined with the high rate of produce turnover, it is suggested that there would be minimal change in produce quality if display cabinets were maintained at ambient temperature.

REFERENCES

Abeles FB, Morgan PW, Saltveit ME. 1992. *Ethylene in Plant Biology.* 2nd ed. Academic Press, New York.

Absoger. 2013. Absoger Controlled Atmosphere. www.absoger-controlled-atmosphere-nitrogen-generator.com/fruit/fruit_en_ethylene_scrubber.php (accessed on Dec 10, 2013).

Alberta Environment. 2003. *Assessment Report on Ethylene for Developing Ambient Air Quality Objectives.* Alberta Environment, Edmonton.

Anderson IC, Levine JS. 1986. Relative rates of nitric oxide and nitrous oxide production by nitrifiers, denitrifiers, and nitrate respirers. *Appl Environ Microbiol* 51: 938–945.

Apelbaum A, Wang SY, Burgoon AC, Baker JE, Lieberman M. 1989. Inhibition of the conversion of 1-aminocyclopropane-1-carboxylic acid to ethylene by structural analogs, inhibitors of electron transfer, uncouplers of oxidative phosphorylation, and free radical scavengers. *Plant Physiol* 67: 74–79.

Benkeblia N, Varoquaux P. 2003. Effect of nitrous oxide (N_2O) on respiration rate, soluble sugars and quality attributes of onion bulbs *Allium cepa* "cv. Rouge Amposta" during storage. *Postharv Biol Technol* 30: 161–168.

Benkeblia N, Varoquaux P, Gouble B. 2001. Effect of nitrous oxide (N_2O) shocks on sprouting and rotting of onion bulbs (*Allium cepa* L.). *Sciences Aliment* 21: 193–198.

Beyer EM. 1976. A potent inhibitor of ethylene action in plants. *Plant Physiol* 58: 268–271.

Bower JH, Biasi WV, Mitcham EJ. 2003. Effects of ethylene in the storage environment on quality of "Bartlett" pears. *Postharv Biol Technol* 28: 371–379.

Bradford KJ. 2008. Shang Fa Yang: Pioneer in plant ethylene biochemistry. *Plant Sci* 175: 2–7.

Burg SP, Burg EA. 1962. Role of ethylene in fruit ripening. *Plant Physiol* 37: 179–189.

Couey M, Olsen KL. 1975. Storage response of "Golden Delicious" apples after high carbon dioxide treatment. *J Am Soc Hort Sci* 100: 148–150.

Couey M, Wright TH. 1975. Effect of prestorage carbon dioxide treatment on the quality of "d'Anjou" pears after regular or controlled atmosphere storage. *HortScience* 12: 244–245.

Cristescu SM, Mandon J, Arslanov D, De Pessemier J, Hermans C, Harren FJM. 2013. Current methods for detecting ethylene in plants. *Ann Botany* 111: 347–360.

Crocker W, Hitchcock AE, Zimmerman PW. 1935. Similarities in the effects of ethylene and the plant auxins. *Contrib Boyce Thompson Instit* 7: 231–248.

Dickson RG, Law SE, Kays SJ, Eiteman MA. 1992. Abatement of ethylene by ozone treatment in controlled atmosphere storage of fruits and vegetables. *Proc Int Winter Meeting, Am Soc Agric Eng* 1–9.

East AR. 2010. Energy efficient postharvest storage and handling. *CAB Reviews: Perspectives Agric Veterin Sci Nutrit Natural Resource* 5(61). DOI: 10.1079/PAVSNNR20105061.

Esguerra EB, Mendoza JR, Pantastico EB. 1978. Regulation of fruit ripening II use of perlite-$KMnO_4$ insert as an ethylene absorbent. *Philippine J Sci* 107: 23–31.

E-source. 2013. Managing Energy Costs in Grocery Stores. www.esource.com/BEA/demo/PDF/CEA_groceries.pdf (accessed June 06, 2013).

Farrer KTH. 1980. *A Settlement Amply Supplied: Food Technology in Nineteenth Century Australia*. Melbourne University Press, Carlton Vic.

Fath D, Soudain P, Bordes M. 1990. Procede de traitement de conservation de produits alimentaires vegetaux frais. Eur Patent 90402748.9.

Forsyth FR, Eaves CA, Lockhard CL. 1967. Controlling ethylene levels in the atmosphere of small containers of apples. *Can J Plant Sci* 47: 717–718.

Gane R. 1934. Production of ethylene by some fruits. *Nature* 134: 1008.

Gane R. 1935. Ozone in relation to storage of bananas. *Rep Food Investigation Board for 1934, DSIR. Great Britain*, 128–130.

Garnett T. 2006. Fruit and vegetables and UK greenhouse gas emissions: exploring the relationship. www.fcrn.org.uk/sites/default/files/fruitveg_paper_final.pdf (accessed Nov 04, 2013).

Golding JB, Shearer D, Wylie SG, McGlasson WB. 1998. Application of 1-MCP and propylene to identify ethylene-dependent ripening processes in mature banana fruit. *Postharv Biol Technol* 14: 87–98.

Gouble B, Fath D, Soudain P. 1995. Nitrous oxide inhibition of ethylene production in ripening and senescing climacteric fruits. *Postharv Biol Technol* 5: 311–326.

Harris DR, Seberry JA, Wills RBH, Spohr LJ. 2000. Effect of fruit maturity on efficacy of 1-methylcyclopropene to delay the ripening of bananas. *Postharv Biol Technol* 20: 303–308.

Hatton TT, Reeder WF. 1972. Quality of "Lula" avocados stored in controlled atmospheres with or without ethylene. *J Am Soc Hortic Sci* 97: 339–343.

Hicks EW, Holmes NE. 1935. Further investigations into the transport of bananas in Australia. *CSIR Aust Res Bull* 91, 35pp.

Kanellis AK, Solomos T, Mattoo AK. 1989. Hydrolytic enzyme activities and protein pattern of avocado fruit ripened in air and in low oxygen, with and without ethylene. *Plant Physiol* 90: 259–266.

Kim GH, Wills RBH. 1995. Effect of ethylene on storage life of lettuce. *J Sci Food Agric* 69: 197–201.

Kim GH, Wills RBH. 1998. Interaction of enhanced carbon dioxide and reduced ethylene on the storage life of strawberries. *J. Hortic Sci Biotechnol* 73: 181–184.

Klieber A, Bagnaot N, Barrett R, Sedgley M. 2002. Effect of post-ripening nitrogen atmosphere storage on banana shelf-life, visual appearance and aroma. *Postharv Biol Technol* 25: 15–24.

Knee M, Proctor FJ, Dover CJ. 1985. The technology of ethylene control: Use and removal in post-harvest handling of horticultural commodities. *Ann Appl Biol* 107: 581–595.

Lavigne P, Patterson Z, Chandra S, Affonce D, Benedek K, Carbone P. 2008. Controlling ethylene for extended preservation of fresh fruits and vegetables. Proc 26th Army Sci Conf, Orlando, FL (Accessed at www.dtic.mil/cgi-bin/GetTRDoc?AD=ADA504474‎).

Leshem YY. 2000. *Nitric Oxide in Plants: Occurrence, Function and Use*. Kluwer Academic, Dordrecht, the Netherlands.

Leshem YY, Haramaty E. 1996. The characterization and contrasting effects of the nitric oxide free radical in vegetative stress and senescence of *Pisum sativum* Linn. foliage. *J Plant Physiol* 148: 258–263.

Leshem YY, Wills RBH, Ku VVV. 1998 Evidence for the function of the free radical gas–nitric oxide (NO*) as an endogenous maturation and senescence regulating factor in higher plants. *Plant Physiol Biochem* 36: 825–855.

Lichanporn I, Techavuthiporn C. 2013. The effects of nitric oxide and nitrous oxide on enzymatic browning in longkong (*Aglaia dookkoo* Griff.) *Postharv Biol Technol* 86: 62–65.

Lieberman M, Asen S, Mapson LW. 1964. Ethylene oxide and antagonist of ethylene in metabolism. *Nature* 204: 756.

Liu FW. 1976. Banana response to low concentrations of ethylene. *J Am Soc Hortic Sci* 101: 222–224.

Martínez-Romero D, Guillén F, Castillo S, Zapata PJ, Serrano M, Valero D. 2009. Development of a carbon–heat hybrid ethylene scrubber for fresh horticultural produce storage purposes. *Postharv Biol Technol* 51: 200–205.

McCormick R, Neuwald DA, Streif J. 2012. Commercial apple CA storage temperature regimes with 1-MCP (SmartFresh™): Benefits and risks. *Proc XXVIIIth IHC-IS on Postharvest Technology in the Global Market* (eds. MI Cantwell, DPF Almeida) *Acta Hortic* no 934, 263–270.

Morgan PW, Drew MC. 1997. Ethylene and plant responses to stress. *Physiol Plantarum* 100: 620–630.

O'Halloran N, Fisher P, Rab A. 2008. *Preliminary Estimation of the Carbon Footprint of the Australian Vegetable Industry*. Horticulture Australia Ltd, Sydney, www.vgavic.org.au/research_and_development/Researchers_PDFs/vg08107___carbon_footprint_part_4___estimate.htm (accessed Nov 10, 2013).

Peacock BC. 1972. Role of ethylene in the initiation of fruit ripening. *Queensland J Agric Anim Sci* 29: 137–145.

Pranamornkith T, East A. Heyes J. 2012. Influence of exogenous ethylene during refrigerated storage on storability and quality of *Actinidia chinensis* (cv. Hort16A). *Postharv Biol Technol* 64: 1–8.

Qadir A, Hashinaga F. 2001. Inhibition of postharvest decay of fruits by nitrous oxide. *Postharv Biol Technol* 22: 279–283.

Quarnstrom F. 2013. Nitrous oxide safety: how safe is it for staff? What can be done to make it safer? http://faculty.washington.edu/quarn/nitrousexp.html (accessed Nov 04, 2013).

Reid M, Dodge L. 1995. New ethylene absorbents: No miracle cure. *Perishables Handling Newsletter UC Davis*, 83, 8.

Robert ML, Taylor HF, Wain RL. 1975. Ethylene production by cress roots and excised root segments and its inhibition by 3,5-diiodo-4-hydroxybenzoic acid. *Planta* 126: 273–284.

Ryall AL, Pentzer WT. 1974. *Handling, Transportation and Storage of Fruits and Vegetables*. AVI Publishing, Westport, CT.

Scott KJ, Blake JR, Strachan G, Tugwell BL, McGlasson WB. 1971a. Transport of bananas at ambient temperatures using polyethylene bags. *Tropic Agric (Trin)* 48: 163–165.

Scott KJ, Gandanegara S. 1974. Effect of temperature on the storage life of bananas held in polyethylene bags with an ethylene absorbent. *Tropic Agric (Trin)* 51: 23–26.

Scott KJ, Guigni J, Bailey W. 1984. The use of polyethylene bags and ethylene absorbent to extend the life of kiwifruit (*Actinidia chinensis* Planch) during cool storage. *J Hortic Sci* 50: 563–566.

Scott KJ, McGlasson WB, Roberts EA. 1970. Potassium permanganate as an ethylene absorbent in polyethylene bags to delay the ripening of bananas during storage. *Aust J Exper Agric Anim Husb* 10: 237–240.

Scott KJ, Roberts EJ. 1966. Polyethylene bags to delay ripening of bananas during transport and storage. *Aust J. Exper Agric Anim Husb* 6: 197–199.

Scott KJ, Wills RBH. 1973. Atmospheric pollutants destroyed in an ultra-violet scrubber. *Lab Practice* 22: 103–106.

Scott KJ, Wills RBH, Patterson BD. 1971b. Removal by ultra-violet lamp of ethylene and other hydrocarbons produced by bananas. *J Sci Food Agric* 22: 496–497.

Serek M, Sisler EC, Reid MS. 1994. Novel gaseous ethylene binding inhibitor prevents ethylene effects in potted flowering plants. *J Am Soc Hortic Sci* 119: 1230–1233.

Sisler EC, Blankenship SM, Guest M. 1990. Competition of cyclooctenes and cyclooctadienes for ethylene binding and activity in plants. *Plant Growth Regul* 9: 157–164.

Sisler EC, Reid MS, Yang SF. 1986. Effect of antagonists of ethylene action on binding of ethylene in cut carnations. *Plant Growth Regul* 4: 213–218.

Sisler EC, Serek M. 1999. Compounds controlling the ethylene receptor. *Botan Bull Acad Sinica* 40: 1–7.

Skog J, Chu CL. 2001. Effect of ozone on qualities of fruits and vegetables in cold storage. *Can J Plant Sci* 81: 773–778.

Thompson J, Singh P. 2013. Status of energy use and conservation technologies used in fruit and vegetable cooling operations in California. California Energy Commission, PIER Program 2008. Available from: URL: http://postharvest.ucdavis.edu/datastore-files/234–1165.pdf (accessed Nov 04, 2013).

Wagner CA. 2009. Hydrogen sulfide: a new gaseous signal molecule and blood pressure regulator. *J Nephrol* 22: 173–176.

Wang CY. 1987. Use of ethylene biosynthesis inhibitors in horticulture. *Acta Hortic* 201: 187–194.

Wild BL, McGlasson WB, Lee TH. 1976. Effect of reduced ethylene levels in storage atmospheres on lemon keeping quality. *HortScience* 11: 114–115.

Wills R, McGlasson B, Graham D, Joyce D. 2007a. *Postharvest: An Introduction to the Physiology and Handling of Fruit, Vegetables and Ornamentals*. 5th ed. UNSW Press, Sydney.

Wills RBH, Gibbons SL. 1998. Use of very low ethylene levels to extend the postharvest life of "Hass" avocado fruit. *Int J Food Properties* 1: 71–76.

Wills RBH, Harris DR, Seberry JA. 1999a. Delayed ripening of bananas through minimisation of ethylene. *Tropic Agric (Trin)* 76: 279–282.

Wills RBH, Harris DR, Seberry JA. 2012. Use of ventilation with ambient air to inhibit ripening of banana during long-distance transport. *Food Aust* 64(5): 38–44.

Wills RBH, Harris DR, Spohr LJ, Golding JB. 2014. Reduction of energy usage during storage and transport of bananas by management of exogenous ethylene levels. *Postharv Biol Technol* 89: 7–10.

Wills RBH, Kim GH. 1995. Effect of ethylene on postharvest life of strawberries. *Postharv Biol Technol* 6: 249–255.

Wills RBH, Kim GH. 1996. Effect of ethylene on postharvest quality of green beans. *Aust J Expt Agric* 36: 335–357.

Wills RBH, Klieber A, David R, Siridhata, M. 1990. Effect of brief pre-marketing holding of bananas in nitrogen on time to ripen. *Aust J Expt Agric* 30: 579–581.

Wills RBH, Ku VVV, Shohet D, Kim GH. 1999b. Importance of low ethylene levels to delay senescence of non-climacteric fruit and vegetables. *Aust J Expt Agric* 39: 221–224.

Wills RBH, Pitakserikul S, Scott KJ. 1982. Effects of pre-storage in low oxygen or high carbon dioxide concentrations on delaying the ripening of bananas. *Aust J Agric Res* 33: 1029–1036.

Wills RBH, Soegiarto L, Bowyer MC. 2007b. Use of a solid mixture containing diethylenetriamine/nitric oxide (DETANO) to liberate nitric oxide gas in the presence of horticultural produce to extend postharvest life. *Nitric Oxide Chem Biol* 17: 44–48.

Wills RBH, Warton MA. 2001. A new rating for ethylene action on postharvest fruit and vegetables. In: *Improving Postharvest Technologies for Fruits, Vegetables and Ornamentals*. Vol 1 Improving quality (eds F Artes MI Gill MA Conesa) Internat Instit Refrig, Novograf, Spain, 43–47.

Wills RBH, Warton MA. 2004. Efficacy of potassium permanganate impregnated into alumina beads to reduce atmospheric ethylene. *J Am Soc Hortic Sci* 129: 433–438.

Wills RBH, Warton MA, Ku VVV. 2000. Ethylene levels associated with fruit and vegetables during marketing. *Aust J Expt Agric* 40: 485–470.

Wills RBH, Warton MA, Mussa DMDN, Chew LP. 2001. Ripening of climacteric fruits initiated at low ethylene levels. *Aust J Expt Agric* 41: 89–92.

Wills RBH, Wimalasiri P, Scott KJ. 1979. Short pre-storage exposure to high carbon dioxide or low oxygen atmospheres for the storage of some vegetables. *HortScience* 14: 528–530.

Wills RBH, Wong T. 1996. Effect of low ethylene levels on the storage life of the Asian leafy vegetables, bak choi (*Brassica chinensis*), choi sum (*Brassica parachinensis*) and gai lan (*Brassica alboglabra*). *ASEAN Food J* 11: 145–147.

Yang SF 1985. Biosynthesis and action of ethylene. *HortScience* 20: 41–45.

Yu YB, Adams DO, Yang SF. 1980. Inhibition of ethylene production by 2,4-dinitrophenol and high temperature. *Plant Physiol* 66: 286–290.

9 Potential of Nitric Oxide as a Postharvest Technology

Ron B.H. Wills

CONTENTS

9.1 RISE OF NO FROM VILLAIN TO HERO

Nitric oxide (NO) is a colorless, highly reactive free radical gas. Its production as an industrial byproduct or vehicle pollutant has long been associated with harmful effects on animals, plants, and the environment. However, NO is also synthesized naturally with major sources being through soil microbial activity and lightning discharges, with a meaningful proportion from burning vegetation, which includes discharges from cigarette smoke (Leshem 2000). An important rapid reaction of NO is with atmospheric oxygen and ozone. The reaction has adverse environmental effects through reducing ozone in the stratosphere and through the condensation of the resulting nitrogen oxides to generate acid rain (Casiday and Frey 1998).

Interest in NO was revolutionized when NO was identified as the endothelium–derived relaxing factor that caused dilation of human arteries and was thus implicated in controlling blood pressure and blood distribution. As with many scientific breakthroughs, there is considerable conjecture as to who should be credited with the discovery. Robert Furchgott, Louis Ignarro, and Ferad Murad working independently at universities in New York, California and Texas in the United States published papers within a short period in 1987–1988 and were subsequently recognized as codiscoverers when the trio were awarded the 1998 Nobel Prize for Physiology or Medicine "for their discoveries concerning nitric oxide as a signaling molecule in the cardiovascular system" (Nobel Prize 2013). However, many eminent scientists were convinced that a British team led by Salvador Moncada, who reported the same finding in *Nature* in 1987 (Palmer et al. 1987) should have been included with the Nobel recipients (Berrazuela de 1999).

Extensive research over the last 25 years has found NO to be metabolized in mammals by nitric oxide synthase (NOS, EC 1.14.13.39) which catalyzes the oxidation of the amino acid, L-arginine to L-citrulline with the release of NO (Figure 9.1). It is now known to have a regulatory role in many human physiological systems and is used as a therapeutic adjunct in the treatment of a range of medical conditions. The multiple actions of NO in mammals arise from its ability to readily diffuse through cell membranes into hydrophilic regions of the cell, such as the cytoplasm, as well as the lipid phase of membranes (Tuteja et al. 2004). NO has been identified as a signaling molecule for functions such as stimulating host defenses in the immune system, regulating development of arteriosclerosis and regulating neutral transmission in the brain (Jaffrey and Snyder 1995; Lloyd-Jones and Bloch 1996; Bogdan 2001; Karpuzoglu and Ahmed 2006). NO is also active in peripheral nervous systems, including the respiratory tract, the digestive system, the urinary tract and cerebral vasculature (Feldman et al. 1993; Rand and Li 1995).

Probably the most publicly recognized action of NO is its involvement as a vasodilator in the process of penile erection. In the early 1990s, researchers at Pfizer in the UK developed a new drug called UK-92,480 to treat angina by relaxing coronary arteries. Early trials showed it was not very effective at easing angina but many of the men reported more frequent and longer-lasting erections. In 1988, the compound, sildenafil citrate, marketed as Viagra, was released and, as they say, the rest is

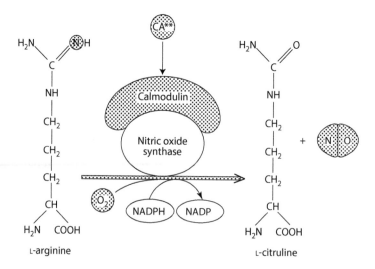

FIGURE 9.1 Biosynthesis of NO. (Reprinted from Leshem YY, Wills RBH. 1998. *Biologia Plantarum* 41: 1–10. With permission.)

history. Viagra works by inhibiting an enzyme that reduces the dilating effect of NO on penile arteries, thus prolonging an erection (Fisch and Braun 2005; Viagra 2013).

The effect of NO on plants was first studied in the 1970s but in the context of whether NO-polluted air had deleterious effects on plant growth. Anderson and Mansfield (1979) reported that NO can either enhance or reduce growth of tomato plants, depending on the concentration of NO, but the critical NO concentration varied between cultivars and levels of soil nutrients. Neighbour et al. (1990) proposed that ambient NO was a determining factor of ozone toxicity in plants, as leaf injury caused by ozone only occurred when NO was added at $>0.002\ \mu L \cdot L^{-1}$. They also suggested that NO at $>0.10\ \mu L \cdot L^{-1}$ might inhibit the generation of stress ethylene.

The foundation for numerous recent studies on the potential benefit of NO action on postharvest commodities can be attributed to the late Ya'acov Leshem from Bar-Ilan University in Israel. Leshem was aware of medical interest in the action and therapeutic benefits of NO and sought to determine if NO had a role in the metabolism of higher plants. His first report (Leshem and Haramaty 1996) on pea seedlings examined a role for NO in plant signaling. They added NO in the form of the NO-donor compound S-nitroso-*N*-acetylpenicillamine (SNAP) to pea leaves and found a greater emission of NO than ethylene and that both NO and ethylene emissions increased when the ethylene precursor 1-aminocyclopropane-1-carboxylic acid (ACC) was added. This led to the suggestion that NO could regulate ethylene production in growing plants. Leshem et al. (1998) then reported an inverse relationship between endogenous NO and ethylene released from four unripe and ripe fruits (Table 9.1) and two cut flowers, while Leshem and Pinchasov (2000) found that increased ethylene production, during ripening of banana and avocado, coincided with reduced NO emission. Leshem et al. (1998) also found alfalfa sprouts that were heat stressed at 37°C showed increased production of NO and decreased production of ethylene, suggesting a stress-coping role for NO. Establishment of the inverse link between endogenous NO and ethylene was the

TABLE 9.1
Endogenous Concentration of NO in Unripe and Ripe Fruit Flesh

Fruit	NO Emission ($\mu M \cdot h^{-1} \cdot g^{-1}$ Fresh Wt)	
	Unripe	Ripe
Avocado	8.4	0.8
Banana	17.5	3.6
Cherry tomato	72.0	34.5
Persimmon	12.6	4.2
Kiwifruit	6.8	4.2
Orange	4.9	2.4

Source: Adapted from Leshem YY, Wills RBH, Ku VVV. 1998. *Plant Physiol Biochem* 36: 825–833.

trigger for substantial research over the last 15 years on the effect of exogenous NO to extend postharvest life to intact produce and fresh-cut horticultural produce and to understand its mode of action.

9.2   POSTHARVEST APPLICATION OF NO

The application of NO to horticultural produce presents a number of technical and logistical problems. These stem from NO, first, as a gas at operating temperatures, and second, as a highly reactive free radical. A major consideration is the rapid oxidation of NO by atmospheric oxygen, with Snyder (1992) reporting a half-life of NO in ambient air of 5 s. This is exacerbated at the sub-ambient temperatures at which horticultural produce are commonly stored, as the rate of oxidation of NO increases with decreasing temperature due to increased intermolecular interaction between NO and oxygen molecules (Tsukahara et al. 1999). Notwithstanding this, the commercial availability of NO gas cylinders makes NO gas fumigation a potential method of application, at least for research.

However, postharvest treatment by dipping in aqueous solutions is a more common and preferred practise in the horticultural industry. The requirements for any treatment are for the NO-generating compound to be quite stable as a solid, relatively stable in solution but to quantitatively release NO when absorbed by produce. Due to the medical use of NO, a wide range of NO-donor compounds have been developed that are relatively stable in solution. Hou et al. (1999) reported that NO-donor compounds can be classified into six categories based on the atom or molecule attached to the NO releasing moiety. These categories are C–NO, N–NO, O–NO, S–NO, heterocyclic-NO, and transition metal-NO compounds. The more important classes for biological purposes are N–NO, S–NO, and transition metal-NO compounds (Hou et al. 1999). While it is not known which NO-donor compounds would be approved for food use, many would seem to be safe as they have been developed for therapeutic purposes but they are certainly useful for research purposes.

9.2.1   Fumigation with NO Gas

The reactivity of NO in air led the early studies to treat horticultural produce with gaseous NO in a nitrogen atmosphere, but the risk of anerobic metabolism at low oxygen limited the treatment time to relatively short periods. However, Soegiarto et al. (2003) examined the rate of loss of NO gas at the relatively low concentrations used to treat produce from atmospheres containing varying levels of oxygen. They found that for 30–40 $\mu L \cdot L^{-1}$ NO, the rate of loss of NO was much lower than expected (Snyder 1992) with the half-life of NO being about 4–6 h in 21% O_2. The rate of loss of NO from the atmosphere was shown to be much greater in the presence of horticultural produce than from air only, indicating a rapid uptake of NO by produce (Table 9.2). They concluded that NO was sufficiently stable at the concentrations and fumigation times utilized and the uptake by produce sufficiently rapid to allow produce to be treated in normal air.

TABLE 9.2
Half-Life of 40 μL · L⁻¹ NO Gas in the Presence of Horticultural Produce

Produce	Half-Life (h)
Carnation	1.7
Strawberry	2.8
Banana	2.2
Air only	6.1

Source: Data from Soegiarto L. et al. 2003. *Postharv Biol Technol* 28: 327–331. With permission.

9.2.2 DIPPING IN SOLUTIONS OF NO-DONOR COMPOUNDS

While a wide range of relatively stable NO-donor compounds have been synthesized, only a few have been used in postharvest horticulture research. The most commonly NO-donor compound utilized is sodium nitroprusside (SNP) (Na$_2$ [Fe (CN)$_5$ NO] 2H$_2$O). SNP is a transition metal NO complex class of donor compound that releases the NO$^+$ cation and which has found clinical use (Pitkanen et al. 1999; Wang et al. 2002). SNP is soluble in water and relatively stable in solution, resisting oxidation at a neutral or slightly acidic pH (Verner 1974) and is widely available and relatively inexpensive.

2,2′-(hydroxynitrosohydrazino)-bisethanamine (diethylenetriamine nitric oxide (DETANO)) was applied by Noritake et al. (1996) to growing potato plants while the first postharvest application of DETANO was by Bowyer et al. (2003) on cut carnation flowers. DETANO is claimed to be the most stable diazeniumdiolate with a half-life in solution of 20 h at 37°C and pH 7.4 and to follow first-order kinetics in degrading to two molar equivalents of the NO• free radical with the reaction being dependent on solution temperature and pH (Lemaire et al. 1999; Fitzhugh and Keefer 2000). The half-life of 0.1 mM DETANO solution markedly decreases at low pH with Davies et al. (2001) reporting it to be 24 s in solutions buffered to pH 2.

The only other NO-donor compounds that have been utilized are: Piloty's acid (N-hydroxybenzenesulfonamide), N-tert-butyl-α-phenylnitrone (PBN) and 3-morpholinosyl-nominine (Sin-1). Each has been examined in one postharvest study.

9.3 EFFECTS OF NO ON POSTHARVEST LIFE

The rationale for most postharvest studies has been to examine whether application of NO as a gas or by dipping in a solution of NO-donor compound inhibited ripening or senescence with some studies focused on control of physiological disorders or diseases. The initial study on NO relating to horticultural produce was by Noritake et al. (1996) who applied a DETANO solution to growing potato plants and found it induced accumulation of the phytoalexin rishitin in tubers. The result suggested a role for NO in stimulating the defence mechanism in plants possibly in response to abiotic stress. The first postharvest report was by Leshem et al. (1998). The study was essentially a preliminary examination of a range of produce to assess the potential of NO gas to extend postharvest life. They reported treatment of strawberry, broccoli, cucumber, Chinese broccoli, kiwifruit and mushroom with a single concentration of NO gas in the range of 0.05–0.25 $\mu mol \cdot L^{-1}$ in a nitrogen atmosphere for 2–16 h followed by storage in air at 20°C and found a 70%–180% extension of postharvest life compared to untreated produce.

9.3.1 EFFECTS ON CLIMACTERIC FRUIT

Studies on a range of climacteric fruit have shown that fumigation with NO gas can regulate ethylene biosynthesis, which is accepted as the trigger for initiation of ripening. Sozzi et al. (2003) fumigated pears with 10 $\mu L \cdot L^{-1}$ NO gas in air for 2 h and showed decreased ethylene production, while fumigation with 10 and 50 $\mu L \cdot L^{-1}$ NO for 12 h delayed yellowing of the skin but did not affect fruit softening. It was concluded that NO had differential effects on pear-ripening and suggested that a time × concentration effect existed for applied NO. Zhu et al. (2006) found that peaches treated with five and 10 $\mu L \cdot L^{-1}$ NO gas in air for 3 h delayed ripening, as measured by the firmness of the fruit, through reduced ethylene biosynthesis. They also reported that NO inhibited lipoxygenase (LOX) activity and suggested that the decrease in LOX activity might be associated with the inhibition of ethylene biosynthesis. Singh et al. (2009) reported that Japanese plums fumigated with NO in air had a reduced rate of respiration and ethylene production during ripening at 21°C. They further showed that NO-fumigation caused a 3–4 day delay in ripening as evidenced by restricted skin-color changes and retarded softening. While Flores et al. (2008) did not report on ethylene, they found that peaches fumigated with 5 $\mu L \cdot L^{-1}$ NO gas in nitrogen for 4 h at 20°C had a reduced respiration rate. They further showed NO treated fruit remained firmer after storage and the degree of electrolyte leakage was also lower. In mango fruit, Zaharah and Singh (2011b) showed that fumigation with 20 $\mu L \cdot L^{-1}$ NO in air for 2 h at 21°C inhibited the activity of 1-aminocyclopropane-1-carboxylic acid synthase (ACS) and 1-aminocyclopropane-1-carboxylic acid oxidase (ACO), which led to a reduced level of ACC. The consequent lower levels of ethylene reduced the activity of fruit-softening enzymes polygalacturonase and pectinesterase but increased the level of endo-1,4-D-glucanase in pulp tissue during ripening and cool storage. Eum et al. (2009b) showed that the ripening of tomato fruit was delayed by fumigation with NO. In addition, Deng et al. (2013) reported that a preharvest spray of SNP onto Golden Delicious apple trees 14 days before harvest delayed postharvest ripening through increasing the level of NO, which delayed the accumulation of ethylene.

Chilling injury is a major storage problem for many produce when stored below a critical threshold low temperature. NO has been implicated in improving chilling tolerance. Singh et al. (2009) reported that Japanese plums fumigated with 10 $\mu L \cdot L^{-1}$ NO in air had reduced incidence of chilling injury during storage at 0°C for 6 weeks. Zaharah and Singh (2011a) fumigated mangoes with 10–40 $\mu L \cdot L^{-1}$ NO gas in air for 2 h and showed that NO treatment not only delayed ripening during storage at 5°C but also reduced the incidence of chilling injury. Zhao et al. (2011) showed that tomatoes dipped in 0.02 mM SNP solution showed reduced chilling injury by inducing NO accumulation and expression of a C-repeat/dehydration-responsive element (CRT/DRE)-binding factors (CBFs), which play an important role in cold-response regulation. They showed SNP protects tomatoes from cold-related injury by inducing expression of LeCBF1 and that NOS activity may be inducing the NO accumulation.

In regard to other low-temperature physiological disorders, Zhu et al. (2010) found that fumigation of peaches with $15 \mu L \cdot L^{-1}$ NO with intermittent warming cycles of one day at 25°C after 14 and 28 days at 5°C was effective in preventing the low-temperature disorder of mealiness. Liu LQ et al. (2012) studied the development of core browning, a low temperature physiological disorder of Yali pear after fumigation with NO gas in nitrogen at 25°C for 3 h followed by storage in air at 2°C. Treatment with $20 \mu L \cdot L^{-1}$ NO was most effective at suppressing core browning. Metabolic effects included an elevated level of NO in the fruit's core tissue and a reduced PPO activity and total phenols content but higher levels of glutathione and ascorbic acid.

Xu et al. (2012) demonstrated that endogenous NO plays a role in chilling injury development. They showed that loquat fruit stored at 1°C induced the accumulation of endogenous NO but pretreatment with the NO scavenger 2-(4-carboxyphenyl)-4,4,5,5-tetramethylimidazoline-1-oxyl-3-oxide (cPTIO) abolished endogenous NO accumulation and enhanced chilling-injury symptoms. The action of NO in alleviating chilling-injury symptoms was ascribed to stimulating the antioxidant defence systems in the fruit. A role for NO in regulating oxidative systems was also attributed by Wu et al. (2012) who found that fumigation of Chinese bayberry (*Myrica rubra*) fruits with $20 \mu L \cdot L^{-1}$ NO gas in nitrogen for 2 h inhibited ethylene production, the incidence of disease and delayed the decrease in firmness, total phenolics and 2, 2-diphenyl-1-picrylhydrazyl (DPPH) radical-scavenging activity. They suggested that NO might enhance the resistance of tissues to decay by maintaining the balance between the formation and detoxification of reactive oxygen species (ROS). Flores et al. (2008) also measured a range of antioxidant enzymes and systems in peaches and showed that fumigation with NO seemed to have a beneficial effect on the oxidation equilibrium and the antioxidant capacity. Zhu et al. (2008) also found that application of NO in solution to kiwifruit reduced the accumulation of malondialdehyde (MDA), superoxide and hydrogen peroxide, delayed the decrease in vitamins C and E, maintained the content of soluble solids, inhibited the activity of LOX and peroxidase (POD), and increased the activity of superoxide dismutase (SOD) and catalase (CAT) during storage.

It is of some interest that Zhang et al. (2013) found an effect when arginine, the metabolic precursor of NO, was applied to tomatoes. They dipped tomatoes in 0.2 mM arginine at reduced pressure (35 kPa) for 0.5 min then stored the fruit at 2°C. Arginine reduced the incidence of chilling injury and enhanced NOS activity and thus the endogenous NO level. They also reported an accumulation of the polyamines putrescine and proline arising from increased activity of the related catabolic enzymes that are often linked to amelioration of chilling injury symptoms.

The use of SNP as the source of NO has been evaluated on a limited number of climacteric fruits with similar findings obtained as with the application of NO gas. In addition to the papers already cited in this section, Zhang et al. (2008) found that plums dipped in 1 mM SNP for 3 min had reduced flesh browning after storage at 2°C but there was no effect on PPO activity. Sis et al. (2012) showed that peaches dipped in 1 mM SNP for 5 min showed an extended postharvest life with a reduced ethylene production rate and increased firmness and antioxidant enzymes activity. Lai et al. (2011) showed that a 1 mM SNP dip applied to tomatoes delayed ripening and reduced the incidence of fungal pathogens such as *Botrytis* spp. They further showed enhanced activity of antioxidant enzymes in NO-treated fruit during storage. The conclusion was that NO suppressed ethylene biosynthesis, stimulated the activity of antioxidant enzymes and regulated the expression of age-related genes. Zheng et al. (2011) inhibited rotting by *Botrytis cinerea* in tomato fruit by application of an extract obtained from *B. cinerea* mycelia. The effect was attributed to stimulation of the plant defense systems with an increase in expression of pathogenesis-related protein 1(PR-1) gene within 2 h followed by enhanced production of NO. The action was considered due to enhanced NOS activity as application of a NOS inhibitor, *N*-nitro-L-arginine, eliminated the beneficial effect of the *Botrytis* extract. Application of a nitrate reductase (NR) inhibitor, sodium azide, (NaN$_3$) did not affect the activity of the elicitor. Activities of phenylalanine ammonia lyase (PAL), chitinase, β-1,3-glucanase, and PPO were also increased by the elicitor treatment as was total phenols content.

9.3.2 EFFECTS ON NONCLIMACTERIC FRUIT AND VEGETABLES

The effect of NO gas on nonclimacteric produce was reported by Wills et al. (2000) who examined strawberries stored at 20°C and 5°C in humidified air, which contained 0.1 μL · L^{-1} ethylene to simulate the ethylene level that might exist in a storage room or transport-container filled with produce. They found that fumigation with NO gas in nitrogen resulted in an increase in postharvest life with the most effective concentration being 5–10 μL · L^{-1} NO at both temperatures. Zhu and Zhou (2007) dipped strawberries in 5 μmol · L^{-1} SNP in water for 2 h at 25°C and found inhibited ethylene production, respiration rate, and reduced ACS activity which led to a lower level of ACC. However, the storage quality of the fruit in this trial was not reported.

Soegiarto and Wills (2004) fumigated broccoli, green beans and bok choy (*Brassica chinensis*) with NO gas in air for 2 h then stored produce at 20°C in air containing 0.1 μL · L^{-1} ethylene. All produce showed an extension in postharvest life through inhibition of yellowing but the optimum concentration of NO differed markedly, being about 50 μL · L^{-1} for green bean and bok choy and 4000 μL · L^{-1} for broccoli. Eum et al. (2009a) have also reported that broccoli florets fumigated with 1000 μL · L^{-1} NO in nitrogen for 5 h delayed the onset of yellowing.

Other studies that used NO gas includes that by Zhu et al. (2009) who found that Chinese winter jujube fumigated with 20 μL · L^{-1} NO gas in nitrogen at 22°C for 3 h inhibited PPO and PAL activity. They also showed that treatment with different NO concentrations from a saturated solution diluted to less than 1 μmol · L^{-1} inhibited *in vitro* activity of PPO and PAL in a dose-dependent manner. Dong J et al. (2012) found that mushrooms fumigated with 10–30 μL · L^{-1} NO gas in air for 2 h had increased antioxidant activity as determined by assays of reducing power, chelating effect on ferrous ions, scavenging effect on hydroxyl free radicals, and DPPH radical scavenging activity. Furthermore, they also reported that NO fumigation enhanced phenolic and flavonoid contents and stimulated activity of PAL and chalcone synthase.

Use of SNP on longan fruit was examined by Duan et al. (2007) who reported that dipping in 1 mM SNP solution for 5 min inhibited pericarp browning and pulp breakdown and reduced PPO activity and the level of MDA, a marker of lipid peroxidation (Mittler 2002), but SNP treatment maintained higher levels of total soluble solids and ascorbic acid during storage. Liu L et al. (2012) dipped "Glorious" oranges in 30–100 μmol $\cdot$ L^{-1} SNP for 10 min and reported a higher level of titratable acidity, soluble protein, ascorbic acid and reducing sugar, a lower weight loss, and lower soluble solids concentration. Ku et al. (2000) reported that rate of water loss was reduced by about 20% from eleven fruits and vegetables and five cut-flowers following a 24 h exposure to NO gas in nitrogen but did not measure the effect beyond this period.

The role of NO in improving the tolerance of nonclimacteric fruit against chilling-related injury was examined by Yang et al. (2011). Cucumbers were fumigated with 25 μL.L^{-1} NO for 12 h at 20°C then stored at 2°C for 15 days. NO reduced the increases in membrane permeability and lipid peroxidation associated with chilling injury and delayed the increases in superoxide anion (O_2^-) and hydrogen peroxide levels. The NO-treated fruit also exhibited higher activity of SOD, CAT, ascorbate peroxidase (APX), POD and higher DPPH-radical scavenging activity. Dong JF et al. (2012) applied a saccharide extract of an unnamed yeast species to cucumbers stored at 4°C and found a reduced incidence of chilling-related injury along with reduced MDA content and ion leakage. The effects were attributed to an associated increase in endogenous NO level. Application of an NO scavenger and NOS inhibitors diminished the rise in NO and suppressed the induced tolerance to cold. The results suggest that NO had an endogenous role in maintaining tolerance to chilling in cucumber fruit by improving the antioxidant defense system.

Yin et al. (2012) examined the effect of NO on wound-healing of sweet potato. Root discs were dipped in SNP or the NO-scavenger cPTIO and stored in the dark at 28°C and 85% RH. 0.5 mM SNP solution generated optimum wound-healing but the effect was reversed by treatment with cPTIO. Other beneficial effects of SNP treatment were enhanced activity of PAL, SOD, CAT, APX, and glutathione reductase. It also maintained higher levels of total phenolics and ascorbic acid and total antioxidative capacity, but peroxide production and MDA accumulation were inhibited. It was suggested that NO promotes wound-healing of sweet potato by improving PAL activity and activating the plant antioxidative defense system.

The interaction of NO and modified atmospheres was examined by Soegiarto and Wills (2006) with strawberries and iceberg lettuce. The effect of low O_2 was examined by fumigating with 10 and 100 μL $\cdot$ L^{-1} NO gas in air and in 2% and 5% O_2 in air followed by storage at 10°C for strawberries and 5°C for lettuce in the same atmospheres. They found an increase in postharvest life due to NO and to storage in low O_2 but there was no additional increase in postharvest life when both treatments were applied concurrently. The effect of elevated CO_2 was examined by placing produce in a sealed polyethylene bag and injecting NO gas at 10 and 20 μL $\cdot$ L^{-1} into the bag. The atmosphere stabilized after seven days at about 20% CO_2 3%

TABLE 9.3

Postharvest Life of Strawberries Held at 10°C in a Modified Atmosphere Package with NO Gas Injected into the Package

Treatment	Postharvest Life (Days)
Air	4.0
MAP	7.2
10 μL · L^{-1} NO in air	6.0
10 μL · L^{-1} NO in MAP	10.7

Source: Data from Soegiarto L, Wills RBH. 2006. *Aust J Expt Agric* 46: 1097–1100. With permission.

O_2 for strawberries and 5% CO_2–13% O_2 for lettuce. The results showed that the postharvest life was enhanced by NO fumigation and was further enhanced by coapplication of the modified atmosphere, presumably due to the CO_2 (Table 9.3). Jiang et al. (2011) examined the effect of modified atmospheres and NO on button mushrooms but used DETANO as the source of NO. They dipped mushrooms for 10 min in DETANO solution and stored produce in sealed modified atmosphere biorientated polypropylene bags at 4°C for up to 16 days. Treatment with 1 mM DETANO maintained a high level of firmness, delayed browning and cap-opening, promoted the accumulation of phenols, ascorbic acid and reduced increases in both respiration and hydrogen peroxide content. The DETANO dip also inhibited PPO activity and increased activity of CAT, SOD, and APX throughout storage period. However, they did not measure the effect of the treatments without modified atmosphere packaging and hence it is not known whether the effects of the NO and MAP treatments were additive.

9.3.3 Effects on Fresh-Cut Produce

Fresh-cut produce is a growing sector of the fresh market with growth driven by consumer demand for ready-to-eat or ready-to-cook fruit and vegetables. However, fresh-cuts are more perishable than intact produce due to stress inflicted during preparation, which involves removal of protective epidermal cells and damage to exposed cells on the cut surface. Apart from the added risk of contamination by plant or human pathogens, shelf-life is compromised due to an increase in respiration, ethylene evolution, water loss and enhanced surface browning (Watada et al. 1996; Artes et al. 1998). Development of surface browning is of particular concern as its appearance is a potent trigger for consumer rejection.

The effect of NO on the development of browning on the cut surfaces of apple slices was examined by Pristijono et al. (2006) who fumigated "Granny Smith" apple slices with NO gas in air. They showed a delay in the onset of browning on the apple surface during storage at 0°C with the most effective treatment being 10 μL · L^{-1} NO for 1 h. They also reported that this treatment was effective in inhibiting surface browning of "Royal Gala," "Golden Delicious," "Sundowner," "Fuji," and "Red

Delicious" apple slices. Pristijono et al. (2008) examined the effect of DETANO on browning and reported that dipping in 10 mg · L^{-1} DETANO in pH 6.5 phosphate buffer for 1 min was the most effective treatment in inhibiting surface-browning of apple slices stored at 0°C. They showed it was necessary to buffer the DETANO solution to be slightly acidic due to vacuolar acids leaking into the dip solution and degrading the DETANO before it was absorbed into the apple slice. They also found that dipping in DETANO solution was more effective than fumigating with NO gas in inhibiting browning. Huque et al. (2013) reported that apple slices dipped in a buffered DETANO solution and fumigated with NO gas had a lower level of total phenols, inhibition of PPO activity, reduced ion leakage, and reduced rate of respiration, but showed no significant effects on ethylene production or lipid peroxide level as measured by MDA and hydrogen peroxide, with DETANO having a greater effect than NO gas. In addition, they found that apple slices dipped in a chlorogenic acid solution enhanced the development of browning but subsequent dipping in DETANO solution negated the effect of chlorogenic acid while fumigation with NO gas gave partial relief. Huque et al. (2013) suggested that an increase in phenols occurs on the apple surface soon after cutting and the effectiveness of NO to inhibit surface-browning may relate to minimizing the activity of phenols on the cut surface, possibly in conjunction with reduced PPO activity.

Huque et al. (2013) also found that the NO-donor compound, Piloty's acid, inhibited browning in apple with the optimum aqueous dip solution concentration being 100 mg · L^{-1}. They compared the optimum concentrations of 500 mg · L^{-1} SNP and 100 mg · L^{-1} Piloty's acid dissolved in water with 10 mg · L^{-1} DETANO in phosphate buffer and fumigation with NO gas to inhibit browning. While all forms of NO extended the postharvest life of apple slices, the order of effectiveness was DETANO > SNP > NO gas = Piloty's acid.

The effect of NO gas and DETANO on iceberg lettuce slices was examined by Wills et al. (2008). They found that the development of browning on the cut surface during storage at 0°C was inhibited, with the optimal treatments being dipping in 500 mg · L^{-1} DETANO in water for 5 min and fumigation with 500 μL · L^{-1} NO for 1 h, with DETANO being more effective than NO gas. The use of a water dip compared to the need for a buffered dip with apple slices was attributed to the higher pH of lettuce tissue not degrading DETANO in solution. Huque et al. (2011) extended the findings to four other types of fresh-cut lettuce (*Lactuca sativa*)—"Green Oak," "Green Coral," "Baby Cos," and "Butter" lettuces—with storage at 5°C. They reported that dipping in a 500 mg · L^{-1} solution of DETANO buffered to pH 6.5 and an aqueous solution of SNP inhibited the development of browning to a greater extent than fumigation with NO gas for 2 h. Use of SNP was considered to be the most feasible commercial option due to the stability of SNP in unbuffered water.

Studies on other fresh-cut produce includes that by Zhu et al. (2009) who fumigated preclimacteric peach slices with a 5 μM NO solution for 10 min followed by storage at 10°C for 10 days. Treatment with NO suppressed the increased rate of leakage and thus maintained compartmentalization between the enzymes and their substrates. They also reported that NO increased PAL activity and total phenol content, and inhibited PPO and POD activity. Cheng et al. (2009) showed that mature green banana slices treated with 5 mM SNP solution had reduced ethylene production, which was associated with reduced ACO activity and expression of the *MA-ACO1* gene. Yang et al. (2010) showed that bamboo shoots dipped in 0.5 mM SNP for 1 h had delayed onset of external browning during storage for 10 days at 10°C. SNP also inhibited PPO, POD, and PAL activity and maintained total phenol content during storage. Furthermore SNP treatment inhibited the synthesis of lignin and cellulose and delayed tissue lignification.

9.4 MODE OF ACTION OF NO ON POSTHARVEST PRODUCE

Postharvest research with NO was stimulated by the findings of Leshem and Haramaty (1996) on pea leaves that NO and ethylene had an antagonistic relationship. The ability of NO to inhibit ethylene production through inhibition of ACS and ACO, resulting in a reduced level of ACC, has now been demonstrated for a range of climacteric and nonclimacteric produce and is generally accepted as one of the important actions of NO in inhibiting ripening and delaying senescence. However, Huque et al. (2013) did not find a reduction in ethylene production in the NO-induced inhibition of browning of apple slices but they suggested the lack of effect of NO was due to the apples being postclimacteric and thus having a substantial production of ethylene. Nonetheless, the data would suggest that ethylene is either not a direct causal factor in the development of surface browning of apples or NO can inhibit browning through other modes of action.

A common finding of many studies is that the applied NO reduced the rate of respiration. This is consistent with findings of Millar and Day (1996) and Zottini et al. (2002) that NO affects the function of mitochondria in plant cells and reduces cell respiration by inhibiting the cytochrome pathway. NO can then be ascribed an anti-senescent action which could arise from a general reduction in the rate of cellular metabolism. Whether this occurs via inhibition of ethylene action or independently, needs to be determined. Reduced ion leakage arising from application of NO has been found in many studies and could be ascribed as a side benefit of better maintenance of cellular integrity through reduced general metabolism.

The role of NO in reducing oxidative stress and its involvement in signaling in growing plants (Besson-Bard et al. 2008) would seem to also apply to postharvest tissue. Reduced lipid oxidation caused by ROS is suggested to be a major cause of membrane deterioration in plant tissues (Mittler 2002). Reduced levels of biomarkers for lipid peroxidation, MDA and hydrogen peroxide, have been found for various postharvest produce, while NO has also been found to stimulate the activity of a range of antioxidant enzymes and enhance PAL activity, which has been implicated in defense mechanisms. Indeed, the stimulation of the phytoalexin, rishitin in growing potato tubers by NO (Noritake et al. 1996) may be an NO-moderated response to either abiotic stress or pathogenic invasion. It would seem that research into whether postharvest NO application can stimulate phytoalexins in other tissues and thus eliminate or reduce the need for chemical biocides to combat pests and diseases could be a worthwhile avenue for study. It is noted that Lazar et al. (2008) in an *in vitro* study found that NO gas inhibited spore germination, sporulation and mycelial growth of three postharvest fungi, *Aspergillus niger, Penicillium italicum* and *Monilinia fructicola*. NO may thus have a dual role in reducing the spread of postharvest diseases.

Inhibition of PPO activity by NO has been shown in numerous intact and fresh-cut produce and was associated with reduced internal- and surface-browning. PPO contains two copper ions in its active center and Zhu et al. (2009) speculated that NO could react with the copper to form copper–nitrosyl complexes, which could change the normal structure of the active site in PPO and thus reduce PPO activity. There may be potential for application of NO with processed fruits and vegetables, where browning of juices and purees during storage is a common problem and treatment with NO to reduce PPO activity may enhance the shelf-life of the processed product. The action of NO on the phenolic substrates, which are substrates for PPO activity, is not clear as the various studies give conflicting data, with some showing a reduced level of phenols while others show an enhanced level.

Alleviation of chilling-related injury by NO is a relatively recent expansion of the mode of action of NO. Endogenous NO production is triggered in response to chilling-stress itself (Xu et al. 2012) or by application of some defense elicitor such as a yeast extract (Dong J et al. 2012; Dong JF et al. 2012). Similarly, application of a fungal extract has been shown to inhibit rotting by stimulation of NO (Zheng et al. 2011). Thus, application of exogenous NO could, therefore, seem to be enhancing a natural defense response to chilling-related injury and disease resistance.

The interaction of NO with modified atmospheres has not been well studied. The only report where NO and a modified atmosphere were simultaneously applied with appropriate air-control treatments (Soegiarto and Wills 2006) showed that exposure to both NO and elevated CO_2 gave an added extension in postharvest life of strawberries and lettuce over that arising from the individual components, whereas exposure to NO and reduced O_2 gave no added benefit. The findings suggest that NO and CO_2 have a different mode of action while that of O_2 and NO is similar and hence not additive. The benefits of modified atmospheres are generally considered to be through an inhibition of respiration via a feedback signal from the respiratory gases.

NO has been applied as a gas and as various NO-donor compounds but all forms have been reported to give a beneficial effect in inhibiting postharvest changes that lead to an extension in postharvest life. However, the different forms of NO degrade to release different NO moieties. DETANO, Piloty's acid and NO gas release the NO$^{\bullet}$ free radical and SNP releases the NO$^+$ cation (Zamora et al. 1995; Hou et al. 1999; Saavedra and Keefer 2002). It might be expected that each NO moiety has different reactivity on produce metabolism, with the moiety more directly linked to the key metabolic action(s) being more effective in inhibiting postharvest change. Since all NO compounds produced some beneficial effect, it would seem likely that the released NO moieties are interconvertible to some extent. Direct comparison of the different forms of NO has only been conducted with fresh-cut apples and lettuces and only in relation to the development of surface-browning. Conclusions drawn from these studies in relation to ripening and general senescence of intact produce must, therefore, be made with some caution. For apple slices, Huque et al. (2013) found all four NO forms inhibited development of browning with DETANO > SNP > NO gas = Piloty's acid, while for the four lettuces evaluated, Huque et al. (2011) found SNP = DETANO > NO gas. As an applied treatment, fumigation with NO gas is therefore less effective than dipping in DETANO and SNP solutions. However, the effect of the solvent must also be considered. In the above studies, DETANO was dissolved in phosphate buffer, and SNP and Piloty's acid in water, while NO gas uses no solvent. Since both phosphate buffer and water alone gave some inhibition of browning, the relative effectiveness of NO gas and DETANO would be, respectively, higher and lower than that found for the actual applied treatments. For the lettuces, Huque et al. (2011) calculated the effectiveness of NO as a percentage of the respective control treatments as SNP > NO gas > DETANO.

While a range of effects of NO on the metabolism of horticultural produce have been identified, it is not clear when NO is acting at the chemical, enzymic, or genetic level. It seems quite possible that NO is active on a range of systems, which could utilize all three modes of action. Also, what is—or are—the key action(s) that generates—or generate—supporting reactions that lead to delayed ripening and senescence? Do these actions differ from those leading to disease-resistance and inhibition of chilling-related injury? Determining the action of endogenous NO in plants would seem to be more complicated than in mammals as, in addition to the synthesis by NOS in mammals, NO in plants can be synthesized by nitrate reductase NR, EC 1.6.6.1-3) which primarily catalyzes the conversion of nitrate to nitrite but under stress conditions can generate NO (Leshem 2000; Yamasaki and Sakihama 2000). However, Zheng et al. (2011) found NOS to be active in NO generated for defense against Botrytis rotting in tomatoes, but NR was not active.

Studies are beginning to examine the effects of NO at the genetic level. The inhibition of rotting in tomato fruit using a *B. cinerea* extract reported by Zheng et al. (2011) was attributed to stimulation of the plant defense systems through expression of the pathogenesis-related protein 1(PR-1) gene followed by enhanced NOS activity and enhanced production of NO. Zhao et al. (2011) demonstrated that an elevated NO content up-regulated the expression of a cold-regulation gene (LeCBF1) and also induced antioxidative processes to scavenge ROS resulting in treatment improved tolerance to cold in tomatoes in cold storage while Eum et al. (2009b) showed that NO can regulate ripening in tomato fruit through genes that regulate ethylene biosynthesis. Kuang et al. (2012) found NO enhanced the expression of the histone deacetylase 2 gene, DlHD2, but suppressed the expression of two ethylene-responsive genes, DlERF1 and DlERF2, of longan fruit during storage at 25°C. It seems that NO may be acting on a range of systems, which have different levels of importance in different produce and in response to differing postharvest stresses.

It is still an open question whether exogenous NO is acting as a foreign agent similar to the inhibition of ethylene by 1-MCP or is accentuating the level—and hence activity—of endogenous NO. Early studies by Leshem (2000) indicated that the level of endogenous NO in mature produce is relatively low. Utilization of NO scavengers as used by Xu et al. (2012) would seem to be useful to investigate these scenarios. The interaction between applied ethylene and NO would appear worth examining, given the central role of ethylene in ripening and general senescence. Based on the early observations of Leshem (2000), it is not unreasonable to suggest that NO has a major role in maintaining cellular integrity and organization during development and maturation of the produce but is replaced by the deteriorative effect of ethylene in mature produce, which have reached the end of their developmental life. Application of exogenous NO could then be analogous to applying a juvenile hormone to prevent aging.

9.5   POTENTIAL FOR COMMERCIAL USAGE OF NO

Since NO seems to be synthesized by all plants it can be considered as a naturally occurring substance. As such, it would be analogous to ethylene which, while it has toxicity and handling issues, is an approved postharvest treatment. The human synthesis of NO and the therapeutic use of various forms of NO should also be positive factors in gaining regulatory approval. However, plants and mammals do not have totally identical pathways for metabolism of NO. Of particular interest is the presence in higher plants of NR which catalyzes the conversion of nitrate to nitrite. Nitrites have the potential to be converted to nitrosamines which are potent carcinogens (Walker 1990). The addition of NO gas or individual donor compounds would need to be shown to not generate nitrosamines. The use of NO with ornamentals would not pose such a regulatory hurdle and regulatory approval should be easier to obtain.

Use of NO gas would require gas-tight fumigation chambers and the monitoring of atmospheric discharges from treatment chambers but these are not insurmountable issues, particularly if the NO was added to air rather than into a nitrogen-rich atmosphere. A more convenient delivery system for NO gas could be along the lines proposed by Wills et al. (2007) who developed a solid mixture that quantitatively released NO gas in the presence of horticultural produce. The mixture of DETANO, citric acid and wheat starch (added as a filler and moisture absorbent) in the ratio of 1:10:20 was found to be stable when stored in dry air. However, in humid air, absorption of moisture from the atmosphere led to reaction of DETANO with citric acid and the evolution of NO gas (Figure 9.2). They found that when the dry mixture was placed in containers of strawberries and mushrooms, the moisture given off by produce activated the mixture and resulted in a similar extension in postharvest life as achieved by direct fumigation with NO gas. The industrial use of such a solid mixture could be through tablets or sachets placed in packages in a packing house or even added into storage rooms. The package would not need to be sealed as the steady release of NO gas over a few days is rapidly absorbed by the surrounding produce.

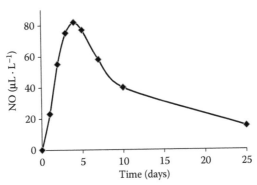

FIGURE 9.2 Atmospheric concentration of NO gas released from a solid mixture of DETANO:citric acid:starch (1:10:20 w/w) held in a sealed container in air at 20°C and 95% RH. (From Wills RBH, Soegiarto L, Bowyer MC. University of Newcastle, Australia, unpublished data.)

Only a few of the many NO-donor compounds have been evaluated and then on only a few produce. To date, SNP would seem to offer potential for commercial use due to its stability in water and resistance to oxidation at a neutral or slightly acidic pH (Verner 1974). It has also had a long history of safe clinical use for hypertensive and cardiac conditions (Ziesche and Franciosa 1977), which could assist in gaining regulatory approval. While DETANO has been shown to be more effective than SNP for some produce, it is unstable at even mildly acidic pH (Hrabie et al. 1993). The quality of water used in commercial operations is variable and would require the use of a buffered solution to ensure a near neutral pH is maintained throughout dipping. For ornamentals, the addition of DETANO or some other solid NO-donor compound to vase water would seem to offer a commercial opportunity.

An alternative treatment option which is even more under-explored is to apply a non-NO containing compound that stimulates endogenous NO production. This could be particularly appealing for regulatory purposes if the compound was either naturally occurring or had GRAS or similar other safe status for foods. A potential research line could be a yeast extract as used by (Dong J et al. 2012; Dong JF et al. 2012) to reduce chilling-related injury of cucumber. The recent paper of Zhang et al. (2013), reporting that application of arginine, the precursor of the NOS catalyzed synthesis of NO (Figure 9.1), stimulated NO production by tomatoes and resulted in reduced chilling-related injury, offers an interesting simpler option. Since L-arginine is an essential amino acid, there should be minimal hurdles in gaining regulatory approval. Another option is to apply some physical condition such as heat or atmosphere that stimulates NO production. An early paper by Leshem et al. (1998) showed that increased production of NO occurred when alfalfa sprouts were heat stressed at 37°C.

ACKNOWLEDGMENT

Tribute is paid to the late Prof. Ya'acov Leshem, Bar-Ilan University, Israel–an innovative biologist for whom scientific curiosity was the finest art. His ability for cross-disciplinary thinking pioneered scientific investigation into the role of NO in postharvest horticulture. The author was a beneficial recipient of Leshem's willingness to share his vision and values in the all too brief time spent in research collaboration.

REFERENCES

Anderson LS, Mansfield TA. 1979. The effects of nitric oxide pollution on the growth of tomato. *Environ Pollut* 20: 113–121.

Artes F, Castaner M, Gil MI. 1998. Enzymatic browning in minimally processed fruit and vegetables. *Food Sci Technol Int* 4: 377–389.

Berrazuela de JR. 1999. The Nobel Prize of nitric oxide. The unjust exclusion of Dr Salvador Moncada. *Rev Esp Cardiol* 52: 221–226.

Besson-Bard A, Pugin A, Wendehenne D. 2008. New insights into nitric oxide signaling in plants. *Annu Rev Plant Biol* 59: 21–39.

Bogdan C. 2001. Nitric oxide and the immune response. *Nat Immunol* 2: 907–916.

Bowyer MC, Wills RBH, Badiyan D, Ku VVV. 2003. Extending the postharvest life of carnations with nitric oxide—comparison of fumigation and *in vivo* delivery. *Postharv Biol Technol* 30: 281–286.

Casiday R, Frey R. 1998. Acid Rain. Inorganic Reactions Experiment http://www.chemistry. wustl.edu/~edudev/LabTutorials/Water/FreshWater/acidrain.html (accessed 25 June 2013).

Cheng G, Yang E, Lu W, Jia Y, Jiang Y, Duan X. 2009. Effect of nitric oxide on ethylene synthesis and softening of banana fruit slice during ripening. *J Agric Food Chem* 57: 5799–5804.

Davies KM, Wink DA, Saavedra JE, Keefer LK. 2001. Chemistry of the diazeniumdiolates. 2. Kinetics and mechanism of dissociation to nitric oxide in aqueous solution. *J Am Chem Soc* 123: 5473–5481.

Deng LL, Pan XQ, Chen L, Shen L, Sheng JP. 2013. Effects of preharvest nitric oxide treatment on ethylene biosynthesis and soluble sugars metabolism in "Golden Delicious" apples. *Postharv Biol Technol* 84: 9–15.

Dong J, Zhang M, Lu L, Sun L, Xu M. 2012. Nitric oxide fumigation stimulates flavonoid and phenolic accumulation and enhances antioxidant activity of mushroom. *Food Chem* 135: 1220–1225.

Dong JF, Yu Q, Lu L, Xu MJ. 2012. Effect of yeast saccharide treatment on nitric oxide accumulation and chilling injury in cucumber fruit during cold storage. *Postharv Biol Technol* 68: 1–7.

Duan X, Su X, You Y, Qu H, Li Y, Jiang Y. 2007. Effect of nitric oxide on pericarp browning of harvested longan fruit in relation to phenolic metabolism. *Food Chem* 104: 571–576.

Eum HL, Hwang DK, Lee SK. 2009a. Nitric oxide reduced chlorophyll degradation in broccoli (*Brassica oleracea* L. var. italica) florets during senescence. *Food Sci Technol Int* 15: 223–228.

Eum HL, Kim HO, Choi SB, Lee SK. 2009b. Regulation of ethylene biosynthesis by nitric oxide in tomato (*Solanum lycopersicum* L.) fruit harvested at different ripening stages. *Eur Food Res Technol* 228: 331–338.

Feldman PL, Griffith OW, Stuehr DJ. 1993. The surprising life of nitric oxide. *Chem Eng News* 20: 26–38.

Fisch H, Braun S. 2005. *The Male Biological Clock*. Simon & Schuster, New York.

Fitzhugh AL, Keefer LK. 2000. Diazeniumdiolates: Pro- and anti-oxidant applications of the "NONOates". *Free Radical Biol Med* 28: 1463–1469.

Flores FB, Sánchez-Bel P, Valdenegro M, Romojaro F, Martínez-Madrid MC, Egea MI. 2008. Effects of a pretreatment with nitric oxide on peach (*Prunus persica* L.) storage at room temperature. *Eur Food Res Technol* 227: 1599–1611.

Hou YC, Janczuk A, Wang PG. 1999. Current trends in the development of nitric oxide donors. *Current Pharm Design* 5: 417–441.

Hrabie JA, Klose JR, Wink DA, Keefer LK. 1993. New nitric oxide releasing zwitterions derived from polyamines. *J Org Chem* 58: 1472–1476.

Huque R, Wills RBH, Golding JB. 2011. Nitric oxide inhibits cut-surface browning in four lettuce types. *J Hortic Sci Biotech* 86: 97–100.

Huque R, Wills RBH, Pristijono P, Golding JB. 2013. Effect of nitric oxide (NO) and associated control treatments on the metabolism of fresh-cut apple slices in relation to development of surface browning. *Postharv Biol Technol* 78: 16–23.

Jaffrey SR, Snyder SH. 1995. Nitric oxide: A neural messenger. *Annu Rev Cell Development Biol* 11: 417–440.

Jiang T, Zheng X, Li J, Jing G, Cai L, Ying T. 2011. Integrated application of nitric oxide and modified atmosphere packaging to improve quality retention of button mushroom (*Agaricus bisporus*). *Food Chem* 126: 1693–1699.

Karpuzoglu E, Ahmed SA. 2006. Estrogen regulation of nitric oxide and inducible nitric oxide synthase (iNOS) in immune cells: Implications for immunity, autoimmune diseases, and apoptosis. *Nitric Oxide* 15: 177–186.

Ku VVV, Wills RBH, Leshem YY. 2000. Use of nitric oxide to reduce postharvest water loss from horticultural produce. *J Hortic Sci Biotech* 75: 268–270.

Kuang JF, Chen JY, Luo M, Wu KQ, Sun W, Jiang YM, Lu WJ. 2012. Histone deacetylase HD2 interacts with ERF1 and is involved in longan fruit senescence. *J Exp Bot* 63: 441–454.

Lai T, Wang Y, Li B, Qina G, Tiana S. 2011. Defense responses of tomato fruit to exogenous nitric oxide during postharvest storage. *PostharvBiol Technol* 62: 127–132.

Lazar EE, Wills RBH, Binh HT, Harris A. 2008. Fungitoxic effect of gaseous nitric oxide (NO) on mycelium growth, sporulation and spore germination of the postharvest horticulture pathogens, *Aspergillus niger, Monilinia fructicola* and *Penicillium italicum*. *Lett Appl Micro* 46: 688–692.

Lemaire G, Alvarez-Pachon FJ, Beuneu C, Lepoivre M, Petit JF. 1999. Differential cyto-static effects of NO donors and NO producing cells. *Free Radical Biol Med* 26: 1274–1283.

Leshem YY. 2000. *Nitric Oxide in Plants: Occurrence, Function and Use*. Kluwer Academic, Dordrecht.

Leshem YY, Haramaty E. 1996. The characterization and contrasting effects of the nitric oxide free radical in vegetative stress and senescence of *Pisum sativum* Linn. foliage. *J Plant Physiol* 148: 258–263.

Leshem YY, Pinchasov Y. 2000. Non-invasive photoacoustic spectroscopic determination of relative endogenous nitric oxide and ethylene content stoichiometry during the ripening of strawberries *Fragaria anannasa* (Dutch.) and avocados *Persea americana* (Mill.). *J Exp Bot* 51: 1471–1473.

Leshem YY, Wills RBH. 1998. Harnessing senescence delaying gases nitric oxide and nitrous oxide: A novel approach to postharvest control of fresh horticultural produce. *Biologia Plantarum* 41: 1–10.

Leshem YY, Wills RBH, Ku VVV. 1998. Evidence for the function of the free radical gas—nitric oxide (NO•)–as an endogenous maturation and senescence regulating factor in higher plants. *Plant Physiol Biochem* 36: 825–833.

Liu L, Wang J, Qu L, Li S, Wu R, Zeng K. 2012. Effect of nitric oxide treatment on storage quality of "Glorious" oranges. *Procedia Eng* 37: 150–154.

Liu LQ, Dong Y, Guan JF. 2012. Effects of nitric oxide fumigation on post-harvest core browning in "Yali" pear (*Pyrus bretschneideri* Rehd.) during cold storage. *J Hortic Sci Biotechnol* 87: 571–576.

Lloyd-Jones DM, Bloch KD. 1996. The vascular biology of nitric oxide and its role in athero-genesis. *Annu Rev Med* 47: 365–375.

Millar AH, Day DA. 1996. Nitric oxide inhibits the cytochrome oxidase but not the alternative oxidase of plant mitochondria. *FEBS Lett* 398: 155–158.

Mittler R. 2002. Oxidative stress, antioxidants and stress tolerance. *Trends Plant Sci* 7: 405–410.

Neighbour EA, Pearson M, Mehlhorn H. 1990. Purafil-filtration prevents the development of ozone-induced frost injury: A potential role for nitric oxide. *Atmospheric Environ* 24: 711–715.

Nobel Prize. 2013. The Nobel Prize in Physiology or Medicine 1998. http://www.nobelprize.org/nobel_prizes/medicine/laureates/1998, (accessed 3 June 2013).

Noritake T, Kawakita K, Doke N. 1996. Nitric oxide induces phytoalexin accumulation in potato tuber tissues. *Plant Cell Physiol* 37: 113–116.

Palmer RMJ, Ferrige AG, Moncada S. 1987. Nitric oxide accounts for the biological activity of endothelium-derived relaxing factor. *Nature* 327: 524–527.

Pitkanen OP, Laine H, Kemppainen J, Eronen E, Alanen A, Raitakari M, Kirvela O. et al. 1999. Sodium nitroprusside increases human skeletal muscle blood-flow, but does not change flow distribution or glucose uptake. *J Physiol* 521: 729–737.

Pristijono P, Wills RBH, Golding JB. 2006. Inhibition of browning on the surface of apple slices by short-term exposure to nitric oxide (NO) gas. *Postharv Biol Technol* 42: 256–259.

Pristijono P, Wills RBH, Golding JB. 2008. Use of the nitric oxide-donor compound, diethyl-enetriamine-nitric oxide (DETANO) as an inhibitor of browning in apple slices. *J Hortic Sci Biotech* 83: 555–558.

Rand MJ, Li CG. 1995. Nitric oxide as a neurotransmitter in peripheral nerves: Nature of transmitter and mechanism of transmission. *Annu Rev Physiol* 57: 659–682.

Saavedra JE, Keefer LK. 2002. Nitrogen-based diazeniumdiolates: Versatile nitric oxide-releasing compounds in biomedical research and potential clinical applications. *J Chem Educ* 79: 1427–1434.

Singh SP, Singh Z, Swinny EE. 2009. Postharvest nitric oxide fumigation delays fruit ripening and alleviates chilling injury during cold storage of Japanese plums (*Prunus salicina* Lindell). *Postharv Biol Technol* 53: 101–108.

Sis SA, Mostofi Y, Boojar MMA, Khalighi A. 2012. Effect of nitric oxide on ethylene bio-synthesis and antioxidant enzymes on Iranian peach (*Prunus persica* cv. Anjiri). *J Food Agric Environ* 10: 125–129.

Snyder SH. 1992. Nitric oxide: First in a new class of neurotransmitters? *Science* 257: 494–496.

Soegiarto L, Wills RBH. 2004. Short-term fumigation with nitric oxide gas in air to extend the postharvest life of broccoli, green bean and bok choy. *HortTechnology* 14: 538–540.

Soegiarto L, Wills RBH. 2006. Effect of nitric oxide, reduced oxygen and elevated carbon dioxide levels on the postharvest life of strawberries and lettuce. *Aust J Expt Agric* 46: 1097–1100.

Soegiarto L, Wills RBH, Seberry JA, Leshem YY. 2003. Nitric oxide degradation in oxy-gen atmospheres and rate of uptake by horticultural produce. *Postharv Biol Technol* 28: 327–331.

Sozzi GO, Trinchero GD, Fraschina AA. 2003. Delayed ripening of "Bartlett" pears treated with nitric oxide. *J Hortic Sci Biotech* 78: 899–903.

Tsukahara H, Ishida T, Mayumi M. 1999. Gas-phase oxidation of nitric oxide: Chemical kinet-ics and rate constant. *Nitric Oxide Chem Biol* 3: 191–198.

Tuteja N, Chandra M, Tuteja R, Misra MK. 2004. Nitric oxide as a unique bioactive signalling messenger in physiology and pathophysiology. *J Biomed Biotechnol* 4: 227–237.

Verner IR. 1974. Sodium nitroprusside: Theory and practice. *Post Med J* 50: 576–581.

Viagra MD. 2013. Viagra: A chronology http://www.about-ed.com/viagra-history (accessed 25 June 2013).

Walker R. 1990. Nitrates, nitrites and N-nitroso compounds: A review of the occurrence in food and diet and the toxicological implications. *Food Addit Contam* 7: 717–768.

Wang PG, Xian M, Tang X, Wu X, Wen Z, Chai T, Janczuk A. 2002. Nitric oxide donors: Chemical activities and biological applications. *Chem Rev* 102: 1091–1134.

Watada AE, Ko NP, Minott DA. 1996. Factors affecting quality of fresh-cut horticultural prod-ucts. *Postharv Biol Technol* 9: 115–125.

Wills RBH, Ku VVV, Leshem YY. 2000. Fumigation with nitric oxide to extend the posthar-vest life of strawberries. *Postharv Biol Technol* 18: 75–79.

Wills RBH, Pristijono P, Golding JB. 2008. Browning on the surface of cut lettuce slices inhib-ited by short-term exposure to nitric oxide (NO). *Food Chem* 107: 1387–1392.

Wills RBH, Soegiarto L, Bowyer MC. 2007. Use of a solid mixture containing diethylenetri-amine/nitric oxide (DETANO) to liberate nitric oxide gas in the presence of horticultural produce to extend postharvest life. *Nitric Oxide Chem Biol* 17: 44–48.

Wu F, Yang H, Chang Y, Cheng J, Bai S, Yin J. 2012. Effects of nitric oxide on reactive oxygen species and antioxidant capacity in Chinese Bayberry during storage. *Scient Hortic* 135: 106–111.

Xu M, Dong J, Zhang M, Xu X, Sun L. 2012. Cold-induced endogenous nitric oxide generation plays a role in chilling tolerance of loquat fruit during postharvest storage. *Postharv Biol Technol* 65: 5–12.

Yamasaki H, Sakihama Y. 2000. Simultaneous production of nitric oxide and peroxynitrite by plant nitrate reductase: *in vitro* evidence for the NR-dependent formation of active nitrogen species. *FEBS Lett* 468: 89–92.

Yang H, Wu F, Cheng J. 2011. Reduced chilling injury in cucumber by nitric oxide and the antioxidant response. *Food Chem* 127: 1237–1242.

Yang H, Zhou C, Wu F, Cheng J. 2010. Effect of nitric oxide on browning and lignification of peeled bamboo shoots. *Postharv Biol Technol* 57: 72–76.

Yin JY, Bai SF, Wu FH, Lu GQ, Yang HQ. 2012. Effect of nitric oxide on the activity of phenylalanine ammonia-lyase and antioxidative response in sweet potato root in relation to wound-healing. *Postharv Biol Technol* 74: 125–131.

Zaharah SS, Singh Z. 2011a. Postharvest nitric oxide fumigation alleviates chilling injury delays fruit ripening and maintains quality in cold-stored "Kensington Pride" mango. *Postharv Biol Technol* 60: 202–210.

Zaharah SS, Singh Z. 2011b. Mode of action of nitric oxide in inhibiting ethylene biosynthesis and fruit softening during ripening and cool storage of "Kensington Pride" mango. *Postharv Biol Technol* 62: 258–266.

Zamora R, Grzesiok A, Webert H, Feelisch M. 1995. Oxidative release of nitric oxide accounts for guanylyl cyclase stimulating, vasodilator and anti-platelet activity of Piloty's acid: A comparison with Angeli's salt. *Biochem J* 312: 333–339.

Zhang OD, Cheng GP, Li J, Yi C, Yang E, Qu HX, Jiang YM, Duan XW. 2008. Effect of nitric oxide on disorder development and quality maintenance of plum fruit stored at low temperature. *Acta Hortic* 804: 549–554.

Zhang XH, Shen L, Li FJ, Meng DM, Sheng JP. 2013. Amelioration of chilling stress by arginine in tomato fruit: Changes in endogenous arginine catabolism. *Postharv Biol Technol* 76: 106–111.

Zhao R, Sheng J, Lv S, Zheng Y, Zhang J, Yu M, Shen L. 2011. Nitric oxide participates in the regulation of LeCBF1 gene expression and improves cold tolerance in harvested tomato fruit. *Postharv Biol Technol* 62: 121–126.

Zheng Y, Shen L, Yu MM, Fan B, Zhao DY, Liu LY, Sheng J. 2011. Nitric oxide synthase as a postharvest response in pathogen resistance of tomato. *Postharv Biol Technol* 60: 38–46.

Zhu L, Zhou J, Zhu S, Guo L. 2009. Inhibition of browning on the surface of peach slices by short-term exposure to nitric oxide and ascorbic acid. *Food Chem* 114: 174–179.

Zhu L, Zhou J, Zhu SH. 2010. Effect of a combination of nitric oxide treatment and intermittent warming on prevention of chilling injury of "Feicheng" peach fruit during storage. *Food Chem* 121: 165–170.

Zhu S, Liu M, Zhou J. 2006. Inhibition by nitric oxide of ethylene biosynthesis and lipoxygenase activity in peach fruit during storage. *Postharv Biol Technol* 42: 41–48.

Zhu S, Sun L, Liu M, Zhou J. 2008. Effect of nitric oxide on reactive oxygen species and antioxidant enzymes in kiwifruit during storage. *J Sci Food Agric* 88: 2324–2331.

Zhu S, Sun L, Zhou J. 2009. Effects of nitric oxide fumigation on phenolic metabolism of postharvest Chinese winter jujube (*Zizyphus jujuba* Mill. cv. Dongzao) in relation to fruit quality. *LWT Food Sci Technol* 42: 1009–1014.

Zhu S, Zhou J. 2007. Effect of nitric oxide on ethylene production in strawberry fruit during storage. *Food Chem* 100: 1517–1522.

Ziesche S, Franciosa JA. 1977. Clinical application of sodium nitroprusside. *Heart Lung* 6: 99–103.

Zottini M, Formentin E, Scattolin M, Carimi F, Schiavo FL, Terzi M. 2002. Nitric oxide affects plant mitochondrial functionality *in vivo*. *FEBS Lett* 515: 75–78.

10 Methyl Jasmonate in Postharvest

Ahmad Sattar Khan, Zora Singh, and Sajid Ali

CONTENTS

10.1 INTRODUCTION

Jasmonic acid (JA) or its methyl ester (methyl jasmonate, MJ), collectively known as jasmonates (JAs), are naturally occurring plant growth regulators (PGRs) (Sembder and Parthier 1993; Creelman and Mullet 1995) that are involved in variety of responses in higher plants including cell division, fruit growth (Kondo and Fukuda 2001), and fruit ripening (Fan et al. 1998b). The level of JAs has been reported to change during fruit development. In apple, the concentration of endogenous MJ is higher at maturity compared to the initial fruit-growing period (Kondo et al. 2000),

211

whereas in strawberry (Gansser et al. 1997), grape (Kondo and Fukuda 2001), and sweet cherry (Kondo et al. 2000), MJ concentration is higher at initial stages of fruit growth followed by a steady decrease during fruit development.

MJ was discovered in floral extracts of *Jasminium grandiflorum* L. (Demole et al. 1962) while JA was isolated from the pathogenic fungus *Lasiodiplodia theobromae* (Aldridge et al. 1971). However, it was not until 10 years later that its biological activity in extracts from *Artemisia absinthium* L. was reported (Ueda and Kato 1980). JAs have now been discovered in numerous other species and are considered as ubiquitous in nature (Hamberg and Gardner 1992; Meyer et al. 1984; Ueda et al. 19991).

Considerable research has been conducted on controlling the rate of ripening of climacteric fruit after harvest. Conventional approaches suitable for the extension of postharvest storage life of such fruit include use of cold storage, but this has limited value with many fruit due to their sensitivity to chilling-related injury. An alternative method is to inhibit ethylene biosynthesis and/or prevent ethylene actions, and this has been achieved with exogenous application of various compounds such as 1-methylcyclopropene (1-MCP) (Khan and Singh 2009), polyamines (PA) (Mattoo and Handa 2008), 1-aminoethoxyvenylglycine (AVG) (Rath et al. 2006), and nitric oxide (NO) (Zaharah and Singh 2013). However, many fruit show uneven ripening following application of these chemicals. Exogenous application of MJ has been found to induce uniform fruit ripening in various climacteric fruit (Fan et al. 1998b); this chapter reviews MJ/JA biosynthesis and actions, interaction of MJ/JA with other PGRs, their role in fruit ripening and color development, and effect on physiological disorders and diseases.

10.2 BIOSYNTHESIS OF JASMONATES

Biosynthesis of jasmonates is initiated by conversion of linolenic acid (LA) to 13-hydroperoxylinolenic acid (HLA) by lipoxygenase (LOX). HLA is converted by hydroperoxide dehydrase or hydroperoxide dehydratase allene oxide synthase (AOS), also known as allene oxide cyclase (AOC), to 12-oxo-phytodienoic acid (12-oxo-PDA), the immediate precursor of JA. JA can be catabolized to form MJ and other conjugates with biological activity (Hamberg and Gardner 1992). Accumulation of JA in reaction to physical wounding, and/or treatment with some other elicitors and systems can effectively be blocked with LOX inhibitors (Baldwin et al. 1996; Doares et al. 1995; Farmer et al. 1994; Pena-Cortes et al. 1993).

10.2.1 LINOLENIC ACID AND LIPASES

Application of LA to fruit plants results in the accumulation of JA, which indicates that the level, availability, and distribution of LA can determine the rate of synthesis of JA (Farmer and Ryan 1992). However, the level of free LA in fruit may increase manifold after wounding (Conconi et al. 1996), which partly results from the release of LA from some phospholipids, and/or utilization of LA present in the fruit tissues prior to wounding. The escalation in JA might result from the triggering of the phospholipases, which then releases LA from membranes (Farmer and

Ryan 1992). Phospholipase D cleaves the head group of the phospholipids and triggers the release of LA. Moreover, phospholipase D has also been reported to play a significant role in plant defense mechanisms (Wang 1993). However, in some cases, fatty acids may also be oxidized by LOX before release from lipids for synthesis to JA. An oxidative burst associated with plant defenses may also stimulate the activity of phospholipases to release oxidized fatty acids for synthesis to JA (Bradley et al. 1992). Observations that enzymes involved in JA biosynthetic pathways are predominantly localized in the plastids (Bell et al. 1995). Blée and Joyard (1996) suggested that some imperative mechanisms must be present to transport released LA to the target plastids.

10.2.2 Lipoxygenase

Treatment with LOX inhibitors (Baldwin et al. 1996; Pena-Cortes et al. 1993) and/or transgenic plants with the reduced activities of LOX (Bell et al. 1995) have reduced capacity to synthesize JA. Plant LOX oxygenates LA at the position of 9 or 13, and while 13-hydroperox-LA is generally the major product, the function of 9-hydroperoxides in the plants is imprecise (Ehret et al. 1994). Changes in the abundance and distribution of LOX during development are due to expression of different LOX gene family members. For instance, two LOX genes, "*AtLox1*" and "*AtLox2*," (Bell and Mullet 1993) have been identified. Because "*AtLox1*" lacks clear targeting sequences, it is almost certainly localized in cytoplasm, while "*AtLox2*" is localized in chloroplasts (Bell et al. 1995). Plastid transit sequences suggest that LOX, which catalyzes HLA, is restricted to chloroplasts (Peng et al. 1994).

10.2.3 Allene Oxide Synthase

Hydroperoxide lyase (HPLS) has been reported to cleave HLA to 6-carbon-aldehydes and 12-oxo-dodecenoic acid volatiles. JAs production requires that HLA must be metabolized by AOS to produce allene oxide. AOS activity has been restricted only to the outer envelope of plastids (Blée and Joyard 1996), but this needs to be confirmed. Moreover, AOS-over-expression has been reported to increase the level of JA. AOS activity has been found to be higher in tissues with higher JA levels, signifying that the differences in JA concentration, which occur during the development phase of plants, may be caused by the variations in AOS-abundance and/or activity (Simpson and Gardner 1995). The enzymes catalyzing reduction of 12-oxo-PDA and/or β-oxidation to JA have been validated *in vitro* but have not been extensively characterized (Vick 1993). In tomato fruit, JL5 gene repressed the translation of the HLA to 12-oxo-PDA (Howe et al. 1996), which, in turn, can be transformed in AOC or AOS activity. An inhibitor of the biosynthesis of JA, diethyldithiocarbamic acid (DIECA), was shown to reduce the levels of HLA to 13-hydroxy-LA (Feys et al. 1994). This suggests that the DIECA inhibits the biosynthesis of JA by dropping the precursor pools leading to allene oxide. Salicylic acid (SA), another facilitator of some the plant defense reactions, impedes conversion of the 13-S-hydroperoxy-LA to 12-oxo-PDA (Farmer et al. 1994; Pena-Cortes et al. 1995).

10.3 METHODS OF MJ APPLICATION

Application of MJ can be done by either vapor or emulsions, but effectiveness varies among cultivars, and the maturity of a produce (Ayala-Zavala et al. 2005; Buta and Moline 1998; Gonzalez-Aguilar et al. 2001, 2003; Meir et al. 1996; Meyer et al. 1991). Wang and Buta (1994) conducted the first application of MJ to alleviate chilling-related injury in *Cucurbita pepo* by dipping in an MJ emulsion, but vapor treatment has been proved more effective for the prevention of chilling-related injury in guava and papaya fruits (Gonzalez-Aguilar et al. 2003, 2004).

In general, vapor application of MJ is accomplished by soaking a filter paper in a solution of MJ and placing the paper in a sealed container of fruit. The amount of MJ applied in a treatment can be quantified by analysis of the headspace in the container (Khan and Singh 2007). The required duration of exposure to MJ differs between produce. For example, tomato fruit need to be treated with 10^{-5} M MJ for 3–4 h, whereas mango requires exposure for 20–24 h at the same concentration (Gonzalez-Aguilar et al. 2001). For fruit treated by immersion, the solution of MJ also needs to include a surfactant at 0.05% concentration. The produce is then immersed in the emulsion for a much shorter time of 1–3 min.

10.4 INTERACTION WITH OTHER PLANT GROWTH REGULATORS

MJ has been found to modulate expression of numerous genes that affect various aspects of fruit development, some responses to abiotic and/or biotic stresses, and regulation of ripening. It might be expected that MJ would interact with other PGRs. The ubiquitin-mediated-protein dilapidation has been identified as a controlling mechanism in numerous PGR pathways (Devoto et al. 2002). MJ signaling utilizes this pathway with three MJ-signaling-genes, CORONATINE INSENSITIVE-1 (COI1), SGT1b/JAI4, and AUXIN RESISTANT-1 (AXR1), identified as part of ubiquitin-proteasome-pathway. COI1 has a principal role in signaling of MJ and is required for all MJ-dependent-responses (Feys et al. 1994; Xie et al. 1998). It has also been found to encode F-box-protein and is the first signal for MJ in the ubiquitin-mediated-protein breakdown (Xie et al. 1998). This hypothesis has further been reinforced by demonstration that COI1 is found in an E3 type ubiquitin-ligase functional complex (Devoto et al. 2002; Feng et al. 2003).

A target of COI1-dependent-proteasome degradation has been identified as RPD3b histone deacetylase (Devoto et al. 2002). However, histone deacetylation is supposed to lessen the accessibility of the chromatin to transcription machinery; therefore, COI1-dependent-proteasome degradation of RPD3b may be a possible mechanism for the MJ-dependent-transcription suppression (Lusser et al. 2001). Conversely, expression of RPD3b-associated histone deacetylase (RPD3a/HD19) has the reverse effect, thus increasing the transcription of the MJ-induced ERF1-transcription factors and ERF1-targets (Zhou et al. 2005).

10.4.1 Ethylene

MJ and ethylene may either collaborate or act antagonistically in the regulation of various stress-related responses, such as disease or pathogen attack and wounding

(Cuello et al. 1990; Devoto et al. 2002; Devoto and Turner 2003; Miszcak et al. 1995; Farmer et al. 2003; Turner et al. 2002). In case of attack by pathogens, both MJ and ethylene cooperate in a synergistic manner for activation of specific defense gene expression (Lorenzo et al. 2003; Xu et al. 1995). Fruit with inhibited synthesis of either MJ or ethylene show reduced induction of defense and enhanced pathogen susceptibility (Knoester et al. 1998; Staswick et al. 1998; Thomma et al. 1998; Vijayan et al. 1998). However, for various wound responses, antagonistic interactions have been reported between the MJ and ethylene for activation of reactions confined in damaged fruit tissues (Rojo et al. 2003). Two MJ transcription factors, ERF1 and AtMYC2/JIN1, have been found to participate in regulation of these defense interactions. ERF1 expression is usually triggered by infection of necrotrophic pathogens and regulates *in vivo* expression of defense-related-genes PDF1.2 and b-CHI (Berrocal-Lobo and Molina 2004). Moreover, expression of ERF1 and/or targets of ERF1 are prompted by MJ and ethylene, but this promotion further depends on synchronization of MJ and ethylene signaling corridors. Numerous other ERFs are also prompted by MJ and ethylene interactions, and are probably also involved in the regulation of the defense mechanism against intruding pathogens, (Brown et al. 2003). Interestingly, ERF1 positively regulates the expression of pathogen-response genes, and, ultimately, precludes the JA-mediated induction of the wound-response gene VSP2 (Lorenzo et al. 2004). Two JA-regulated transcription factors, JAMYC2 and JAMYC10, have been reported in tomato fruit and found to specifically recognize a T/G box in promotion of the leucine aminopeptidase (LAP) JA-induced gene (Boter et al. 2004). JAR1 genes encode an enzyme that has JA adenylation activity, which forms conjugates among the JA and several other amino acids, specifically with isoleucine (Staswick et al. 1992). JAR1 also conjugates JA with the ethylene precursor 1-aminocyclopropane-1-carboxylic acid (ACC).

10.4.2 ABSCISIC ACID

Exogenous MJ inhibits fungal spore germination and shows synergistic interaction with ABA, but this may not reflect endogenous functions (Staswick et al. 1992). Antagonistic interaction between MJ and ABA pathways have been perceived to modulate the pathogen defense reactions, while, positive relations between MJ and ABA have been reported in wound-responses (Anderson et al. 2004). Similarly, JIN1/AtMYC2 can be a major controller of interactions between MJ and ABA. In addition to MJ, expression of AtMYC2 is also stimulated by ABA in a COI1-gene-dependent way, which proposes that ABA leads MJ in AtMYC2-mediated activation wound-responses (Lorenzo et al. 2004). Moreover, the deleterious effects of ABA on MJ regulated-expression of defense genes (Anderson et al. 2004) might not be caused merely by AtMYC2 activation by ABA. It would seem that ABA acts independently of MJ in pathogen defense regulation and AtMYC2 is not be the solitary point of convergence (Boter et al. 2004; Lorenzo et al. 2004).

10.4.3 SALICYLIC ACID

The relationship between salicylic acid (SA) and MJ has been intensively studied (Devoto and Turner 2003; Li et al. 2004; Turner et al. 2002). MPK4 that encodes

MAP-kinase 4 is probably the major gene identified as a key regulator of the negative interaction between JA and SA in defense activation. It has been reported that MPK4 is mandatory for induction of MJ-regulated defense genes, including PDF1.2 and THI2.1. MPK4 loss of the functions constitutively expresses SA-regulated genes, possibly due to the elevated levels of SA. This suggests that a MAP-kinase-cascade encompasses MPK4-regulated JA-SA-interaction by inhibiting biosynthesis of SA and promotes perception of MJ. Likewise, constitutive ERF1 over-expression promotes activation of the MJ and/or ethylene-dependent defenses and increases tolerance of pathogens, which suggests possible positive interaction between MJ and ethylene, as well as SA-signaling pathways (Berrocal-Lobo et al. 2002). It has also been reported that induction of SA after infection of *Pseudomonas* inhibits accumulation of MJ with this negative interface being controlled by NPR1 protein (Spoel et al. 2003). NPR1 stimulation by SA cooperates with participants of TGA transcription factors, and, ultimately, triggers SA-dependent expression of PR genes. Surprisingly, negative effects of NPR1 on the signaling of JA does not need its nuclear localization but appears to be exercised via NPR1 function in the cytosol of fruit (Spoel et al. 2003). Moreover, NPR1 could not work solely, because the NPR1-related gene NPR4 has been reported to be required for the full expression of MJ-induced PDF1.2 genes (Angelini et al. 2005).

MJ induces biosynthesis of anthocyanins by the induction of chalcone synthase and dihydroflavonol-4-reductase gene expression (Saniewski et al. 1998). Therefore, postharvest MJ application increased anthocyanin contents (Rudell et al. 2002) and the level of antioxidants in banana, guava, mango, and papaya (Kondo et al. 2005). Application of MJ also enhances the activity of phenylalanine ammonia lyase (PAL) and, consequently, the contents of various other flavonoids (Gonzalez-Aguilar et al. 2004; Tomas-Barberan et al. 2001). Moreover, MJ promotes biosynthesis of β-carotene in tomato (Saniewski and Czapski 1983; Saniewsky et al. 1997) and apples (Perez et al. 1993). This effect may be due to direct stimulation of the β-carotene or by indirect enhancement of β-carotene synthesis through promotion of ethylene synthesis by MJ (Perez et al. 1993).

10.4.4 POLYAMINES

The polyamines (PA) mainly consist of putrescine, spermidine, and spermine (Langebartels et al. 1991), and play important roles in different characteristics throughout plant development, including stress-responses and senescence (Angelini et al. 2005). MJ treatment has been found to increase the levels of spermidine and spermine but decrease putrescine (Wang and Buta 1994). Both spermidine and spermine have been shown to be involved in lipid peroxidation reduction, degradative enzyme inhibition, and stabilization of the membrane structures (Escribano and Merodio 1994; Smith 1985). The higher levels of PA in MJ-treated tissues appear to be active in the reduction of chilling-related stress (Wang and Buta 1994).

Application of MJ to apple and plum fruit has been reported to increase the activities of the key ethylene synthesising enzymes, aminocyclopropane-1-carboxylic acid synthase (ACS), and ACC oxidase (ACO), and hence stimulates synthesis of ethylene (Khan and Singh 2007; Olias et al. 1992). Biosynthetic conduits for both ethylene and PA share S-adenosyl methionine (SAM) as a metabolite and, hence, plant cells can commit SAM into synthesis of PA and/or ethylene (Cassol and Mattoo 2002). Lower levels of MJ and JA have been reported in mango skin and pulp during early growth stages, and increase during ripening, with ACC levels higher in pulp than skin tissue (Lalel et al. 2003). This proposes that JA or MJ and ACC are related to the ripening of mango.

10.5   MJ AND FRUIT-RIPENING

JA and MJ are usually present in low concentrations in different parts of plants, but the highest levels are reported in fruit (Meyer et al. 1984). In apples, endogenous MJ has been found to be low during early fruit development but increase towards harvest (Kondo et al. 2000). Similarly, Lalel et al. (2003) reported that the level of MJ in mango pulp was significantly higher at harvest but decreased during ripening. On the other hand, levels of MJ in the nonclimacteric fruit such as strawberries (Gansser et al. 1997; Mukkun and Singh 2009), cherries (Kondo et al. 2000), and grapes (Kondo and Fukuda 2001), have been found to be higher at immature stages, and decrease steadily during development. Furthermore, application of MJ to unripe green strawberries enhanced respiration, production of ethylene, and synthesis of anthocyanins, and chlorophyll degradation, thus suggesting a natural role for MJ in strawberry-ripening (Perez et al. 1997).

Application of MJ triggers the ripening of climacteric fruit by increasing ethylene production (Fan et al. 1997; Perez et al. 1997; Saniewski et al. 1987). MJ has been found to also enhance development of aroma (Fan et al. 1997; Lalel et al. 2003; Olias et al. 1992), and various pigment changes (Lalel et al. 2003; Perez et al. 1997, 1993). A transient increase in endogenous MJ was observed in apple and tomato during the onset of fruit-ripening, with promotion of ethylene synthesis (Czapski and Saniewski 1992; Fan et al. 1998a). JAs have been found to enhance ethylene production through stimulation of ACS and ACO activity (Saniewski 1995).

MJ stimulates the production of ethylene, ACC, and ACO activity in the preclimacteric stage of various fruit particularly apple (Fan et al. 1997; Staswick et al. 1992). Responses of fruit to exogenous MJ/JA, suggested that JAs may be involved in the regulation of the early fruit development (Fan et al. 1997). Endogenous concentrations of MJ and JA increased during the commencement of ripening in apple and tomato, increased rapidly prior to an increase in ethylene, and decreased during the later stages of ripening (Fan et al. 1998a). Moreover, activities of ACS and ACO have been reported to be stimulated by low concentrations of MJ (Khan and Singh 2007). Response of color changes to MJ vapor also depended on the fruit developmental stage, and its affect was maximized as fruit began to generate ethylene (Fan et al. 1998a). MJ treatment also caused a higher respiration and ethylene production in white strawberries, while a decrease in carbon dioxide evolution and ethylene has been observed in the ripe and overripe strawberry fruits (Perez et al. 1997).

10.6   MJ AND FRUIT QUALITY

Fruit quality consists of numerous attributes that encompass appearance, texture, flavor, nutrient content, and phytochemicals (Schreiner and Huyskens-Keil 2006).

Higher endogenous levels of MJ during fruit-ripening and exogenous application of MJ have been reported in numerous studies to influence the postharvest fruit quality parameters in both climacteric and nonclimacteric fruit, and these findings are summarized in Table 10.1. Exogenous application of MJ has been found to improve fruit color and increase aroma volatile production, chlorophyll degradation and carotene synthesis, as well as enhance levels of antioxidants of many fruit along with reduced incidence of rotting and chilling injury. Some of the studies reported in Table 10.1 on various effects of MJ on fruit quality are considered in more detail in the sections that follow.

10.6.1 Color Development

A red peel color is an important quality parameter of many ripe fruit. Anthocyanins are the major compounds responsible for red color in apples. MJ has been reported to increase anthocyanin accumulation and, hence, peel color change in peach, strawberry, and apple fruits (Perez et al. 1997; Saniewski et al. 1998; Shafiq et al. 2013; Wang 1999). While this is generally considered due to enhanced synthesis of ethylene by JAs (Fan et al. 1997; Saniewski and Czapski 1983), the effect for apples has been suggested as being at least partly independent of ethylene (Fan and Mattheis 1999a; Shafiq et al. 2013). Moreover, increased sugar content in raspberries following MJ treatment has been attributed to induced anthocyanin accumulation (Pirie and Mullins 1976). Similarly, ripe "Kent" mango that had been treated with MJ developed more intense red and yellow colors and exhibited superior fruit quality compared to control fruit (Gonzalez-Aguilar et al. 2001). Also, the natural peel-color change from green to yellow developed faster and more uniformly in "Manila" mango exposed to MJ (Herrera et al. 2004). Exogenous application of MJ also stimulated ethylene synthesis and aroma volatiles in "Kensington Pride" mango during fruit-ripening. It would also seem that pre- and post-harvest application of MJ can be used to improve fruit color development at harvest and during postharvest handling in a wide range of fruit.

10.6.2 Fruit-Softening

MJ application slightly reduced the firmness in strawberries during storage (Perez et al. 1997), "Kent" mango exhibited a rapid loss of fruit firmness after storage (Gonzalez-Aguilar et al. 2001) and increased softening in Japanese plum fruit, which was attributed to enhanced ethylene production and activity of ACS and ACO enzymes (Khan and Singh 2007). In other studies, MJ did not significantly affect fruit firmness during cold storage (Gonzalez-Aguilar et al. 2000). When MJ was applied in combination with modified atmosphere packaging (MAP) it resulted in reduced loss of firmness in papaya fruit, but reduced firmness appeared to be related to the effects of film-packaging rather than MJ (Gonzalez-Aguilar et al. 2003).

10.6.3 Soluble Solids and Sugars

MJ treatment affects the quality attributes such as soluble solids concentration (SSC), titratable acidity (TA), and pH of a range of fruits. For example, MJ application to

TABLE 10.1

Effect of Exogenous Application of Methyl Jasmonate on Postharvest Quality of Fruit

Fruit	Cultivar	MJ Concentration	Conditions	Inferences	Reference
Apple	"Braeburn"	4480 mg/L	105 & 175 DAFB	Increased red color, phenolic content, and antioxidant capacity	Ozturk et al. (2014)
	"Delicious"	400 mg/L ethephon + 5 μM 1-MCP + 5 mM MJ	12 h at 20°C	Increase internal ethylene, firmness, malic acid, and aroma volatiles	Kondo et al. (2005)
	"Cripps Pink"	10 mM	Preharvest spray 169 DAFB	Increased red blush, anthocyanins, chlorogenic acid, phloridzin, and flavanols	Shafiq et al. (2013)
	"Cripps Pink"	1 or 5 mM	Spray 3 week before harvest	Increased red blush, anthocyanins, and flavonoids	Shafiq et al. (2011)
	—	1000 ppm	20°C for 4 days	Increased PAL activity, phenolics, and antioxidants	Heredia and Cisneros-Zevallos (2009)
	"Fuji"	10 μM 1-MCP + 400 μM MJ	20°C for 40 hr	Increased anthocyanin and carotene; reduced chlorogenic acid	Rudell and Mattheis (2008)
	"Fuji"	2.24 and 4.48 g/L	Preharvest spray 48 and 172 DAFB	Increased fruit size and color development; reduced fruit-softening, and bitter pit	Rudell and Fellman (2005)
	"Fuji"	10 mM 1-MCP + 2 mM MJ	20°C for 15 days	Inhibited ethylene production, alcohols, and esters	Fan and Mattheis (1999b)
	"Fuji"	2.24 g/L	120 h at 20°C	Increased anthocyanin, carotene, and color development	Rudell et al. (2002)
	"Golden Delicious"	0.75 μL 1-MCP + 2.0 mM MJ + 2 mM SA	0°C for 14 week	Increased production of esters	Li et al. (2006)
	"Golden Delicious"	400 mg/L ethephon + 5 μM 1-MCP + MJ	20°C for 12 h	Increased internal ethylene, firmness, malic acid, and aroma volatiles	Kondo et al. (2005)

(Continued)

TABLE 10.1 (Continued)

Effect of Exogenous Application of Methyl Jasmonate on Postharvest Quality of Fruit

Fruit	Cultivar	MJ Concentration	Conditions	Inferences	Reference
	"Golden Delicious"	1 and 10 μM for 12 h	22°C till ripe	Increased ethylene synthesis and improved color development	Fan et al. (1998b)
	"Golden Delicious"	8 ppm	Storage for 2 week	Promoted β-carotene synthesis and chlorophyll degradation	Perez et al. (1993)
Avocado	"Etinger", "Fuerte" "Hass"	2.5 μM	4–10 week at 2°C	Reduced chilling-related injury and wound percentage	Meir et al. (1996)
Banana	"Namwa"	0.39 mM PDJ	6 and 12°C for 4 days	Reduced chilling-related injury and increased antioxidative enzymes	Kondo et al. (2005)
	"Cavendish"	0.1 mM	7°C for 5 days	Increased chilling-tolerance over expression of *ICE-CBF MaCBF1, MaCBF2, MaCOR1, MaKIN2, MaRD2,* and *MaRD5* genes	Zhao et al. (2013)
	"Williams 8818"	1.5 mM	Preharvest	Higher chitinase, β-1,3-glucanase, TPC, SOD, POD, PPO, CAT, and PAL activities	Sun et al. (2013)
Bayberry	"Wumei"	10 μM	0°C for 12 days	Reduced fruit decay, higher PAL activity, increased TPC, flavonoids, and anthocyanins	Wang et al. (2009)
Blackberry	"Chester Thornless", "Hull Thornless"	0.1 mM	4 week	Higher TSS, flavonoids, antioxidants, anthocyanin, and TPC	Wang et al. (2008)
Cherry tomato	—	0.1% chitosan + 500 mM MJ	Incubated 7 days at 28°C	Inhibited mycelial growth, reduced disease, higher activities of PPO, POD, and PAL	Chen et al. (2014)
Fragaria chiloensis	—	100 μM	Stored for 9 days	Increased ethylene, anthocyanin, expression of LOX, AOS, and OPR-3 genes	Concha et al. (2013)

Commodity	Cultivar	Concentration	Treatment	Effects	Reference
Grapefruit	"Marsh seedless"	10 μM	4–10 week	Reduced chilling-related injury and wounds	Meir et al. (1996)
Grape	—	250 ppm	20°C for 4 days	Increased PAL activity, phenolics, and antioxidants	Heredia and Cisneros-Zevallos (2009)
	"Monastrell"	0.3 mM MJ + 10 mM BTH	Preharvest application	Higher anthocyanin and flavonol contents	Ruiz-Garcia et al. (2013)
Guava	White and red flesh	10^{-4} and 10^{-5} M	5°C for 15 days, 20°C for 2 days	Reduced chilling-related injury & ion leakage, increased sugars, PAL, and LOX activities	Gonzalez-Aguilar et al. (2004)
Guava	"Klomsalee"	0.39 mM PDJ	Storage at 6°C and 12°C	Reduced chilling-related injury and increased anti-oxidative enzymes	Kondo et al. (2005)
	White variety	220 μM MJ + 1% CaCl$_2$	6°C for 28 days	Increased shelf-life	Barriga-Tellez et al. (2011)
Loquat	"Fuyang"	10 μM	20°C for 24 h, 1°C for 35 days	Proline accumulation, reduced chilling-related injury, Δ^1-pyrroline-5-carboxylate synthetase, GABA, ornithine δ-aminotransferase, glutamate decarboxylase	Cao et al. (2012)
	"Fuyang"	24 μM	20°C for 24 h, 1°C for 35 days	Reduced chilling-related injury, internal browning, increased firmness, PAL, PPO, POD, and polygalacturonase activities	Cao et al. (2010)
	"Fuyang"	10 μM	20°C for 24 h, 1°C for 35 days	Reduced chilling-related injury, increased fruit firmness, and activities of SOD and CAT	Cao et al. (2009)
	"Jiefangzhong"	10 μM	20°C for 6 days	Reduced decay incidence and higher polyamines	Cao et al. (2014)

(Continued)

TABLE 10.1 (*Continued*)
Effect of Exogenous Application of Methyl Jasmonate on Postharvest Quality of Fruit

Fruit	Cultivar	MJ Concentration	Conditions	Inferences	Reference
	"Jiefangzhong"	10 μM	20°C for 24 h, 1°C for 35 days	Inhibited chilling-related injury, internal browning, monodehydroascorbate reductase, dehydroascorbate reductase, and glutathione reductase activities	Cai et al. (2011)
	"Jiefangzhong"	10 μM	20°C for 6 days	Reduced anthracnose rot and lower disease incidence	Cao et al. (2008)
Mandarin	"Ponkan"	100 μM	20°C for 4 days	Inhibited disease incidence, higher activities of PPO, POD, and CATs	Guo et al. (2014)
Mango	"Kensington Pride"	10^{-3} M	21°C	Increased ethylene, skin color, fatty acids, total aroma volatiles	Lalel et al. (2003)
	"Kent"	10^{-4} M	5°C for 14 days 2°C for 5 days	Reduced chilling-related injury, increased color development, higher sugars, TSS, and ascorbic acid contents	Gonzalez-Aguilar et al. (2001)
	"Tommy Atkins"	10^{-4} M	5°C for 21 days + 5 days shelf	Reduced ion leakage and chilling-related injury; increased shelf-life and TSS	Gonzalez-Aguilar et al. (2000)
Nectarines	—	1000 ppm	20°C for 4 days shelf	Increased PAL activity, phenolics, and antioxidants	Heredia and Cisneros-Zevallos (2009)
Papaya	"Khakdam"	0.39 mM PDJ	6 and 12°C for 4 days	Reduced chilling-related injury, increased antioxidative enzymes	Kondo et al. (2005)
	"Sunrise"	10^{-5} and 10^{-4} M	10°C for 32 days + 4 days shelf	Inhibited fungal decay and reduced chilling-related injury	Gonzalez-Aguilar et al. (2003)

Fruit	Cultivar	Concentration	Storage conditions	Effects	Reference
Peach	"Baifeng"	1 μM	0°C for 5 weeks	Reduced chilling-related injury and inhibited activities of PPO, POD; increased activities of SOD, CAT, and increased phenolics	Jin et al. (2009)
	"Jiubao"	0.1 mM	5°C for 21 days with 3 day shelf	Reduced chilling-related injury; higher activity of POD and increased SSC/TA ratio	Meng et al. (2009)
	"Stark Red Gold"	0.22 mM PDJ	Preharvest application	Reduced ethylene production, softening, increased level of PA	Ziosi et al. (2009)
Pear		1000 ppm	20°C for 4 days	Increased PAL activity, phenolics, AOX, and antioxidants	Heredia and Cisneros-Zevallos (2009)
	"La France"	0.39 mM PDJ	20°C for 9 days	Increased ethylene production, expression of ACS and ACO	Kondo et al. (2007)
Pineapple	"Pattavia"	10^{-5} M	10°C for 21 days	Reduced chilling-related injury, fruit weight loss, and electrolyte leakage	Nilprapruck et al. (2008)
	"Red Spanish"	10^{-4} M	7°C for 12 days	Decreased microbiological growth	Martinez-Ferrer and Harper (2005)
Plum	"Amber Jewel", "Angelino", "Black Amber"	10^{-3} M	During ripening for 13 days	Increased SSC, respiration rates, ethylene production, activities of ACS and ACO enzymes	Khan and Singh (2007)
	"Black Amber"	1500 mL	0°C for 4 weeks	Increased firmness and phenolics	Ozturk et al. (2012)
	"Fortune"	1120 mg/L	0°C for 4 weeks	Increased total phenolics, antioxidant, and chlorogenic acid	Karaman et al. (2013)
	—	1000 ppm	20°C for 4 days	Increased PAL activity, phenolics, and antioxidants	Heredia and Cisneros-Zevallos (2009)
Pomegranate	"Malas Yazdi" "Malas Ashkezar"	0.4 mM MJ, 1%–2% $CaCl_2$, 1–2 mM SA	0.5°C for 2 months	Reduced chilling-related injury, electrolyte leakage, and increased TSS	Mirdehghan and Ghotbi (2014)

(Continued)

TABLE 10.1 *(Continued)*
Effect of Exogenous Application of Methyl Jasmonate on Postharvest Quality of Fruit

Fruit	Cultivar	MJ Concentration	Conditions	Inferences	Reference
	"Mollar de Elche"	0.01 and 0.1 mM	84 days	Increased total phenolics and anthocyanins; reduced chilling-related injury and pitting	Sayyari et al. (2009)
Raspberry (black and red)	"Jewel" "Autumn Bliss"	0.1 mM MJ + 0.01– 0.1 mM MJ (two sprays)	4 weeks	Higher SSC, total sugars, fructose, glucose, sucrose and lower TA, malic acid, citric acid, flavonoids, and antioxidant	Wang and Zheng (2005)
Rose apple	"Kheaw"	0.39 mM PDJ	6 and 12°C for 4 days	Reduced chilling-related injury and increased anti-oxidative enzymes	Kondo et al. (2005)
Strawberry	"Camarosa"	50 μM	Storage at 85% RH	Increased respiration, ethylene production, color development, anthocyanin, and β-caroten synthesis	Perez et al. (1997)
		164 μL	4°C for 7 days	Increased antioxidants, reduced nitrite production	Flores et al. (2013)
	"Pajaro"	50 μM	—	Increased ethylene production and ACO activity	Mukkun and Singh (2009)
	—	1000 ppm	20°C for 4 days	Increased PAL activity, phenolics, AOX, and antioxidants	Heredia and Cisneros-Zevallos (2009)
Sweet Cherry	—	2 mM SA + 0.2 mM MJ	Storage at 25°C	Reduced disease, increased β-1,3-glucanas, PAL, POD activities	Yao and Tian (2005b)

Tomato	"Ailsa Craig"	10 mM for 10 min	25°C for 14 days	Inhibited gray mold, stimulated CAT, APX gene expression, enhanced ASC and GSH contents	Zhu and Tian (2012)
	"Beefsteak"	22.4 µM	0°C for 21 days	Reduced chilling-related injury and increased levels of AOX	Fung et al. (2006)
	"Beefstake"	0.01 mM	5°C for 4 weeks	Alleviated chilling-related injury, increased HSPs, especially class II HSPs	Ding et al. (2001)
	"Castlemart"	0.5 µM	15 days	Increased lycopene biosynthesis and over-expression of lycopene biosynthetic genes	Liu et al. (2012)
Tomato	—	10^{-4} m	With MAP for 9 weeks	Delayed ethylene production, reduced softening and chilling-related injury	Siripatrawan and Assatarakul (2009)
	"Charleston"	220 µL/L	10°C for 27 days	More color development, firmness, reduced CO_2 and ethylene production	Baltazar et al. (2007)

Note: DAFB = days after full bloom, PPO = polyphenol oxidase, SOD = superoxide dismutase, POD = peroxidise, TPC = total phenolic content, CAT = catalase. BTH = benzo(1,2,3)thiadiazole-7-carbothioic acid S-methyl ester, TSS = total soluble solids, TA = titratable acidity, AOX = alternative oxidase, HSPs = heat-shock proteins, PDJ = *n*-propyl-dihydrojasmonate, PAL = phenylalanine ammonia lyase, GABA = gamma-aminobutyric acid, LOX = lysyl oxidase.

"Tommy Atkins" mango increased SSC in ripe fruit over untreated fruit (Fan et al. 1998b; Gonzalez-Aguilar et al. 2000, 2001; Herrera et al. 2004). Similarly, post-harvest MJ-treated Japanese plum fruit exhibited improved fruit quality, including color, ascorbic acid, total carotenoids, and SSC (Khan and Singh 2007).

The effect of MJ application on the sugar content of fruit reported in the literature is controversial. Higher glucose and fructose levels were reported in "Kent" mango fruit treated with 10^{-4} M MJ (Gonzalez-Aguilar et al. 2004) but in another study on "Kent" mango, fruit treated with 10^{-4} M MJ increased the sucrose level without influencing fructose and glucose levels, while 10^{-5} M MJ decreased fructose and glucose without any change in sucrose (Gonzalez-Aguilar et al. 2001).

10.6.4 PHYTOCHEMICALS

Phytochemicals in the fruit can be improved significantly by MJ treatment. Examples of this action are accrual of resveratrol in grapes; and accumulation of anthocya-nins and antioxidants, alkaloids, and saponins; and taxol formation in various fruits (Ahuja et al. 2012; Verpoorte et al. 2000; Wang and Zheng 2005). MJ was found to stimulate the activity of PAL in guava, which led to accumulation of total phenolic compounds, while in tomato and guava fruit, MJ enhanced LOX activity (Gonzalez-Aguilar et al. 2004; Heitz et al. 1997). Postharvest application of MJ vapor increased levels of ascorbic acid, total carotenoids and total antioxidants, in Japanese plums (Khan and Singh 2007).

10.6.5 OXIDATIVE STRESS

MJ treatment has been found to increase antioxidant enzyme activity of many sys-tems, including ascorbate peroxidase, superoxide dismutase, glutathione peroxidase, dehydro ascorbate reductase, guaiacol peroxidase, glutathione reductase, and mono-dehydro ascorbate reductase (Chanjirakul et al. 2006). Treatment with MJ increased the level of polyphenols of guava, and enhanced catechin, chlorogenic acid, epicate-chin, and phloridzin levels in apple (Gonzalez-Aguilar et al. 2004; Rudell et al. 2002).

Ascorbic acid is an antioxidant, and MJ application enhanced its level in various fruit (Gonzalez-Aguilar et al. 2004). Moreover, it has been shown that MJ treatment enhanced the transcription of genes involved in ascorbic acid *de novo* synthesis in plant-cell suspensions. Treatment with MJ vapor before cold storage also enhanced expression of alternative oxidase genes that are related to reduction of ROS forma-tion, which could affect the marked increased tolerance of fruit to chilling-related injury (Fung et al. 2004).

10.7 MJ AND PHYSIOLOGICAL DISORDERS AND DISEASES

Limitation on the use of low-temperature storage is often due to the occurrence of flesh-softening, internal breakdown, or fruit decay during storage or in poststorage handling. Potential applications of MJ in the prevention of chilling-related injury and other postharvest disorders and diseases have been documented in Table 10.1. Some aspects are discussed below.

10.7.1 Chilling-Related Injury

Use of MJ and JA to ameliorate the sensitivity of fruit to develop chilling-related injury was first reported by Wang and Buta (1994), and has since been demonstrated in many commodities. Some examples of this effect are MJ applied as a dip or vapor to avocado and grapefruit has been reported to decrease chilling-related injury during storage (Meir et al. 1996), while banana stored at 5°C for one week was protected from development of chilling-related injury when treated with *n*-propyl-dihydrojasmonate (PDJ) (Chaiprasart et al. 2002). MJ treatments also decreased decay in strawberry stored at 5°C and 10°C (Moline et al. 1997). Use of MJ in combination with hot water is a common treatment protocol that not only reduces ion leakage but also the incidence of chilling-related injury (Gonzalez-Aguilar et al. 2000, 2003, 2004). However, MJ also increased abscission in the harvested cherry tomatoes, which is considered undesirable, as cherry tomatoes are marketed attached to the peduncles.

Meir et al. (1996) proposed that MJ might interact with signal transduction cascades of the chemical changes involved in the tolerance of chilling injury. It is now known that MJ produces specific signals and activates certain enzymatic reactions involved in the synthesis of proteins known to have significant roles in the postharvest life of horticultural crops (Buta and Moline 1998; Droby et al. 1999; Gonzalez-Aguilar et al. 2004).

A role for ethylene is implicated in a study of bananas treated with PDJ before storage at 5°C reduced chilling-related injury after cold storage and at 20°C for one week during ripening. Compared to control, treated bananas exhibited increased ethylene evolution and it can be hypothesized that this might play an important role in inhibiting chilling-related injury, although more evidence is required to support this hypothesis (Ding et al. 2002). Alternately, MJ also induces mRNA accumulation of heat-shock proteins (HSPs) in tomato that can partly explain reduction in chilling-related injury symptoms (Ding et al. 2001). Exposure of the heated fruit to ambient temperature conditions prior to transfer under the low temperatures eliminates the tolerance capability to low temperature or chilling-stress (Sabehat et al. 1996). Thus, biochemical and physiological changes instigated by MJ treatment might vary when the target produce is stored at cold or ambient temperature after treatment.

10.7.2 Inhibition to Disease

Fungal diseases can be controlled by synthetic fungicides, but development of resistance by the pathogens and escalating consumer concerns about chemical use and residues have stimulated the search for alternative measures to control fruit diseases (Fan et al. 1998b). One such approach is to utilize methods that stimulate the natural defense systems of plants to invading fungi (Meir et al. 1998). JAs are involved in various signal transduction systems and result in the induction of defense-responses against pathogen attack (Gundlach et al. 1992; Guo et al. 2014; Nojiri et al. 1992; Yao and Tian 2005b). Application of JA has been reported to induce resistance against a range of organisms in a range of fruit. Some examples are: resistance to *B. cinerea*, *C. acutatum*, and *A. brassicicola*, in papaya, loquat, and tomato, respectively (Cao et al. 2008; Gonzalez-Aguilar et al. 2003), while application of MJ increased resistance against *A. alternata* and *B. cinerea* in strawberry (Moline et al. 1997) and eliminated decay in raspberry fruit (Wang 2003).

MJ treatment in combination with C. *laurentii* also exhibited positive effects for the control of blue mold and brown rot of peaches (Yao and Tian 2005a). This inhibitory mechanism is considered to be due to *C. laurentii* and to MJ-induced resistance. Yao and Tian (2005a) reported that direct inhibition of blue mold growth *in vitro* was due to the application of MJ. A probable mode of action of MJ in the induction of disease resistance might be due to phenol and anthocyanin synthesis and the antagonistic effects of ethylene (Gonzalez-Aguilar et al. 2004; Meir et al. 1998; Rudell et al. 2002).

REFERENCES

Ahuja I, Kissen R, Bones AM. 2012. Phytoalexins in defence against pathogens. *Trends Plant Sci* 17: 73–90.

Aldridge DC, Galt S, Giles D, Turner WB. 1971. Metabolites of *Lasiodiplodia theobromae*. *J Chem Soc C* 1623–1627.

Anderson JP, Badruzsaufari E, Schenk PM, Manners JM, Desmond OJ, Ehlert C, Maclean DJ, Ebert PR, Kazan K. 2004. Antagonistic interaction between abscisic acid and jasmonate ethylene signaling pathways modulates defence gene expression and disease resistance in *Arabidopsis*. *Plant Cell* 16: 3460–3479.

Angelini R, Manes F, Federico R. 2005. Spatial and functional correlation between diamine-oxidase and peroxidase activities and their dependence upon de-etiolation and wounding in chick-pea stems. *Planta* 182: 89–96.

Ayala-Zavala JF, Wang SY, Wang CY, Gonzalez-Aguilar GA. 2005. Methyl jasmonate in conjunction with ethanol treatment increases antioxidant capacity, volatile compounds and postharvest life of strawberry fruit. *Eur Food Res Technol* 221: 731–738.

Baldwin IT, Schmelz EA, Zhang ZP. 1996. Effects of octadecanoid metabolites and inhibitors on induced nicotine accumulation in *Nicotiana sylvestris*. *J Chem Ecol* 22: 61–74.

Baltazar A, Espina-Lucero J, Ramos-Torres I, Gonzalez-Aguilar G. 2007. Effect of methyl jasmonate on properties of intact tomato fruit monitored with destructive and nondestructive tests. *J Food Eng* 80: 1086–1095.

Barriga-Tellez LM, Garnica-Romo MG, Aranda-Sanchez JI, Correa GA, Bartolome-Camacho MC, Martınez-Flores HE. 2011. Nondestructive tests for measuring the firmness of guava fruit stored and treated with methyl jasmonate and calcium chloride. *Int J Food Sci Technol* 46: 1310–1315.

Bell E, Mullet JE. 1993. Characterization of an Arabidopsis lipoxygenase gene responsive to methyl jasmonate and wounding. *Plant Physiol* 103: 1133–1137.

Bell E, Creelman RA, Mullet JE. 1995. A chloroplast lipoxygenase is required for wound-induced jasmonic acid accumulation in *Arabidopsis*. *Proc Nat Acad Sci* 92: 8675–8679.

Berrocal-Lobo M, Molina A. 2004. Ethylene response factor 1 mediates *Arabidopsis* resistance to the soil-borne fungus *Fusarium oxysporum*. *Mol Plant-Microbe Interact* 17: 763–770.

Berrocal-Lobo M, Molina A, Solano R. 2002. Constitutive expression of ethylene-response-factor1 in Arabidopsis confers resistance to several necrotrophic fungi. *Plant J* 299: 23–32.

Blée E, Joyard J. 1996. Envelope membranes from spinach chloroplasts are a site of the metabolism of fatty acid hydroperoxides. *Plant Physiol* 110: 445–455.

Boter M, Ruiz-Rivero O, Abdeen A, Prat S. 2004. Conserved MYC transcription factors play a key role in jasmonate signaling both in tomato and Arabidopsis. *Genes Dev* 18: 1577–1591.

Bradley DJ, Kjellbom P, Lamb CJ. 1992. Elicitor- and wound-induced oxidative cross-linking of a proline-rich plant cell wall protein: A novel, rapid defence response. *Cell* 70: 21–30.

Brown RL, Kazan K, McGrath KC, Maclean DJ, Manners JM. 2003. A role for the GCC-box in jasmonate-mediated activation of the PDF1.2 gene of Arabidopsis. *Plant Physiol* 132: 1020–1032.

Buta JG, Moline HE. 1998. Methyl Jasmonate extends shelf-life and reduce microbial contamination of fresh-cut celery and peppers. *J Agric Food Chem* 46: 1253–1256.

Cai Y, Cao S, Yang Z, Zheng Y. 2011. MeJA regulates enzymes involved in ascorbic acid and glutathione metabolism and improves chilling tolerance in loquat fruit. *Postharv Biol Technol* 59: 324–326.

Cao S, Zheng Y, Wang K, Jin P, Rui H. 2009. Methyl jasmonate reduces chilling injury and enhances antioxidant enzyme activity in postharvest loquat fruit. *Food Chem* 115: 1458–1463.

Cao S, Zheng Y, Yang Z, Tang S, Jin P. 2008. Control of anthracnose rot and quality deterioration in loquat fruit with methyl jasmonate. *J Sci Food Agric* 88: 1598–1602.

Cao, S, Cai Y, Yang Z, Joyce DC, Zheng Y. 2014. Effect of MeJA treatment on polyamine, energy status and anthracnose rot of loquat fruit. *Food Chem* 145: 86–89.

Cao, S, Cai Y, Yang Z, Zheng Y. 2012. MeJA induces chilling tolerance in loquat fruit by regulating proline and γ-aminobutyric acid contents. *Food Chem* 133: 1466–1470.

Cao S, Zheng Y, Wang K, Rui H, Tang S. 2010. Effect of methyl jasmonate on cell wall modification of loquat fruit in relation to chilling injury after harvest. *Food Chem* 118: 641–647.

Cao S, Zheng Y, Yang Z, Tang S, Jin P, Wang K, Wang X. 2008. Effect of methyl jasmonate on the inhibition of *Colletotrichum acutatum* infection in loquat fruit and the possible mechanisms. *Postharv Biol Technol* 49: 301–307.

Cassol T, Mattoo AK. 2002. Do polyamines and ethylene interact to regulate plant growth, development and senescence. In: *Molecular and Cellular Biology: New Trends*, (eds P Nath, A Mattoo S Ranade, J Weil) SMPS Publishers, Dehradun, India.

Chaiprasart P, Gemma H, Iwahori S. 2002. Reduction of chilling injury in stored banana fruits by jasmonic acid derivative and abscisic acid treatment. *Acta Hortic* 575: 689–696.

Chanjirakul K, Wang SY, Wang CY, Siriphanich J. 2006. Effect of natural volatile compounds on antioxidant capacity and antioxidant enzymes in raspberries. *Postharv Biol Technol* 40: 106–115.

Chen J, Zou X, Liu Q, Wang F, Feng W, Wan N. 2014. Combination effect of chitosan and methyl jasmonate on controlling *Alternaria alternata* and enhancing activity of cherry tomato fruit defence mechanisms. *Crop Prot* 56: 31–36.

Concha CM, Figueroa NE, Poblete LA, Oblate FA, Schwab W, Figueroa CR. 2013. Methyl jasmonate treatment induces changes in fruit ripening by modifying the expression of several ripening genes in *Fragaria chiloensis* fruit. *Plant Physiol Biochem* 70: 433–444.

Conconi A, Miquel M, Browse JA, Ryan CA. 1996. Intracellular levels of free linolenic and linoleic acids increase in tomatoes leaves in response to wounding. *Plant Physiol* 111: 797–803.

Creelman RA, Mullet JE. 1995. Jasmonic acid distribution and action in plants: Regulation during development and response to biotic and abiotic stress. *Proc Nat Acad Sci* 92: 4114–4119.

Cuello J, Quiles MJ, Garcia C, Sabater B. 1990. Effect of light and growth substances on senescence of barley leaf segments at different developmental stages. *Botan Bull Acad Sci* 107: 117–112.

Czapski J, Saniewski M. 1992. Stimulation of ethylene production and ethylene forming enzyme in fruits of the non-ripening nor and rin tomato mutants by methyl jasmonate. *J Plant Physiol* 139: 265–268.

Demole E, Lederer E, Mercier DE. 1962. Isolement et détermination de la structure du jasmonate de méthyle, constituant odorant charactéristique de l'essence de jasmin. *Helvet Chim Acta* 45: 675–685.

Devoto A, Nieto-Rostro M, Xie D, Ellis C, Harmston R, Patrick E, Davis J, Sherratt L, Coleman M, Turner JG. 2002. COI1 links jasmonate signalling and fertility to the SCF ubiquitin-ligase complex in Arabidopsis. *Plant J* 32: 457–466.

Devoto A, Turner JG. 2003. Regulation of jasmonate-mediated plant responses in Arabidopsis. *Ann Botan* 92: 329–337.

Ding CK, Wang CY, Gross KC, Smith DL. 2001. Reduction of chilling injury and transcript accumulation of heat shock proteins in tomato fruit by methyl jasmonate and methyl salicylate. *Plant Sci* 161: 1153–1159.

Ding CK, Wang CY, Gross KC, Smith DL. 2002. Jasmonate and salicylate induce the expression of pathogenesis-related-protein genes and increase resistance to chilling injury in tomato fruit. *Planta* 214: 895–901.

Doares SH, Narvaez-Vasquez J, Conconi A, Ryan CA. 1995. Salicylic acid inhibits synthesis of proteinase inhibitors in tomato leaves induced by systemin and jasmonic acid. *Plant Physiol* 108: 1741–1746.

Droby S, Porat R, Cohen L, Weiss B, Shapiro B, Philosoph-Hadas S, Meir S. 1999. Suppressing green mold decay in grapefruit with postharvest jasmonate application. *J Am Soc Hortic Sci* 124: 184–188.

Ehret R, Schab J, Weiler EW. 1994. Lipoxygenases in *Bryonia dioica* Jacq. tendrils and cell cultures. *J Plant Physiol* 144: 175–182.

Escribano MI, Merodio C. 1994. The relevance of polyamine levels in cherimoya (*Annona cherimola* Mill.) fruit ripening. *J Plant Physiol* 143: 207–212.

Fan X, Mattheis JP. 1999a. Methyl jasmonate promotes apple fruit degreening independently of ethylene action. *HortSci* 34: 310–312.

Fan X, Mattheis JP. 1999b. Impact of 1-methylcyclopropene and methyl jasmonate on apple volatile production. *J Agric Food Chem* 47: 2847–2853.

Fan X, Mattheis JP, Fellman JK. 1998a. Responses of apples to postharvest jasmonate treatment. *J Am Soc Hortic Sci* 123: 421–425.

Fan X, Mattheis JP, Fellman JK. 1998b. A role for jasmonates in climacteric fruit ripening. *Planta* 204: 444–449.

Fan X, Mattheis JP, Fellman JK, Patterson ME. 1997. Effect of methyl jasmonate on the ethylene and volatile production by Summer Red apples depends on fruit developmental stage. *J Agric Food Chem* 45: 208–211.

Farmer EE, Ryan CA. 1992. Octadecanoid precursors of jasmonic acid activates the synthesis of wound inducible proteinase inhibitors. *Plant Cell* 4: 129–134.

Farmer EE, Almeras E, Krishnamurthy V. 2003. Jasmonates and related oxylipins in plant responses to pathogenesis and herbivory. *Curr Opinion Plant Biol* 6: 372–378.

Farmer EE, Caldelari D, Pearce G, Walker-Simmons MK, Ryan CA. 1994. Diethyldithiocarbamic acid inhibits the octadecanoid signaling pathway for the wound induction of proteinase inhibitors in tomato leaves. *Plant Physiol* 106: 337–3442.

Feng S, Ma L, Wang X, Xie D, Dinesh-Kumar SP, Wei N, Deng XW. 2003. The COP9 signal interacts physically with SCF COI1 and modulates jasmonate responses. *Plant Cell* 15: 1083–1094.

Feys B, Benedetti CE, Penfold CN, Turner JG. 1994. Arabidopsis mutants selected for resistance to the phytotoxin coronatine are male sterile, insensitive to methyl jasmonate, and resistant to a bacterial pathogen. *Plant Cell* 6: 751–759.

Flores G, Perez C, Gil C, Blanch GP, Ruiz del Castillo ML. 2013. Methyl jasmonate treatment of strawberry fruits enhances antioxidant activity and the inhibition of nitrite production in LPS-stimulated raw 264.7 cells. *J Func Foods* 5: 1803–1809.

Fung RWM, Wang CY, Smith DL, Gross KC, Tian M. 2004. MeSA and MeJA increase steady-state transcript levels of alternative oxidase and resistance against chilling injury in sweet pepper (*Capsicum annuum* L,). *Plant Sci* 166: 711–719.

Fung RWM, Wang CY, Smith DL, Gross KC, Tao Y, Tian M. 2006. Characterization of alternative oxidase (AOX) gene expression in response to methyl salicylate and methyl jasmonate pre-treatment and low temperature in tomatoes. *J Plant Physiol* 163: 1049–1060.

Gansser D, Latza S, Berger RG. 1997. Methyl jasmonates in developing strawberry fruit (*Fragaria ananassa* Duch. cv. Kent). *J Agric Food Chem* 45: 2477–2480.

Gonzalez-Aguilar GA, Buta JG, Wang CY. 2001. Methyl jasmonate reduces chilling injury symptoms and enhances colour development of "Kent" mangoes. *J Sci Food Agric* 81: 1244–1249.

Gonzalez-Aguilar GA, Buta JG, Wang CY. 2003. Methyl jasmonate and modified atmosphere packaging (MAP) reduce decay and maintain postharvest quality of papaya "Sunrise". *Postharv Biol Technol* 28: 361–370.

Gonzalez-Aguilar GA, Fortiz J, Cruz R, Baez R, Wang CY. 2000. Methyl jasmonate reduces chilling injury and maintains postharvest quality of mango fruit. *J Agric Food Chem* 48: 515–519.

Gonzalez-Aguilar GA, Tiznado-Hernandez ME, Zavaleta-Gatica R, Martinez-Tellez MA. 2004. Methyl jasmonate treatments reduce chilling injury and activate the defence response of guava fruits. *Biochem Biophys Res Commun* 313: 694–701.

Gundlach H, Miiller MJ, Kutchan TM, Zenk MH. 1992. Jasmonic acid is a signal transducer in elicitor-induced plant cell cultures. *Proc Nat Acad Sci* 89: 2389–2393.

Guo J, Fang W, Lu H, Zhu R, Lu L, Zheng X, Yu T. 2014. Inhibition of green mold disease in mandarins by preventive applications of methyl jasmonate and antagonistic yeast Cryptococcus laurentii. *Postharv Biol Technol* 88: 72–78.

Hamberg M, Gardner HW. 1992. Oxylipin pathway to jasmonates: Biochemistry and biological significance. *Biochim Biophys Acta* 1165: 1–18.

Heitz T, Bergey DR, Ryan CA. 1997. A gene encoding a chloroplast-targeted lipoxygenase in tomato leaves is transiently induced by wounding, systemin and methyl jasmonate. *Plant Physiol* 114: 1085–1093.

Heredia JB, Cisneros-Zevallos L. 2009. The effects of exogenous ethylene and methyl jasmonate on the accumulation of phenolic antioxidants in selected whole and wounded fresh produce. *Food Chem* 115: 1500–1508.

Herrera MR, De la Cruz J, Garcia HS. 2004. Methyl jasmonate improved sensory attributes and reduced severity of chilling injury in hydrothermally treated Manila mangoes. *IFT Annual Meeting*, LasVegas, NV, USA.

Howe GA, Lightner J, Browse J, Ryan CA. 1996. An octadecanoid pathway mutant (JL5) of tomato is compromised in signal-ling for defence against insect attack. *Plant Cell* 11: 2067–2077.

Jin P, Zheng Y, Tang S, Rui H, Wang CY. 2009. A combination of hot air and methyl jasmonate vapor treatment alleviates chilling injury of peach fruit. *Postharv Biol Technol* 52: 24–29.

Karaman S, Ozturk B, Genc N, Celik SM. 2013. Effect of preharvest application of methyl jasmonate on fruit quality of plum (*Prunus salicina* Lindell cv. "Fortune") at harvest and during cold storage. *J Food Process Preserv* 37: 1049–1059.

Khan AS, Singh Z. 2007. Methyl jasmonate promotes fruit ripening and improves fruit quality in Japanese plum. *J Hortic Sci Biotechnol* 82: 695–706.

Khan AS, Singh Z. 2009. 1-MCP application suppresses ethylene biosynthesis and retards fruit softening during cold storage of "Tegan Blue" Japanese plum. *Plant Sci* 176: 539–544.

Knoester M, van Loon LC, van den Heuvel J, Henning J, Bol JF, Linthorst HJM. 1998. Ethylene-insensitive tobacco lacks non-host resistance against soil-borne fungi. *Proc Nat Acad Sci* 95: 1933–1937.

Kondo S, Tomiyama A, Seto H. 2000. Changes of endogenous jasmonic acid and methyl jasmonate in apples and sweet cherries during fruit development. *J Am Soc HorticSci* 125: 282–287.

Kondo S, Kittikorn M, Kanlayanarat S. 2005. Preharvest antioxidant activities of tropical fruit and the effect of low temperature storage on antioxidants and jasmonates. *Postharv Biol Technol* 36: 309–318.

Kondo S, Setha S, Rudell DR, Buchanan DA, Mattheis JP. 2005. Aroma volatile biosynthesis in apples affected by 1-MCP and methyl jasmonate. *Postharv Biol Technol* 36: 61–68.

Kondo S, Yamada H, Setha S. 2007. Effect of jasmonates differed at fruit ripening stages on 1-aminocyclopropane-1-carboxylate (ACC) synthase and ACC oxidase gene expression in pears. *J Amer Soc HorticSci* 132: 120–125.

Kondo S, Fukuda K. 2001. Changes of jasmonates in grape berries and their possible roles in fruit development. *Scientia Hortic* 91: 275–288.

Kondo S, Posuya P, Kanlayanarat S. 2001. Changes in physical characteristics and polyamines during maturation and storage of rambutans. *Scientia Hortic* 91: 101–109.

Lalel HJD, Singh Z, Tan SC. 2003. The role of methyl jasmonate in mango ripening and biosynthesis of aroma volatile compounds. *J Hortic Sci Biotechnol* 78: 470–484.

Langebartels CI, Kerner K, Leonardi S, Schraudner M, Trost M, Heller W, Sandermann HJ. 1991. Biochemical plant responses to ozone I, Differential induction of polyamine and ethylene biosynthesis in tobacco. *Plant Physiol* 95: 882–889.

Li DP, Xu YF, Sun LP, Liu LX, Hu XL, Li DQ, Shu HR. 2006. Salicylic acid, ethephon, and methyl jasmonate enhance ester regeneration in 1-MCP-treated apple fruit after long-term cold storage. *J Agric Food Chem* 54: 3887–3895.

Li J, Brader G, Palva ET. 2004. The WRKY70 transcription factor: A node of convergence for jasmonate-mediated and salicylate-mediated signals in plant defence. *Plant Cell* 16: 319–331.

Liu L, Wei J, Zhang M, Zhang L, Li C, Wang Q. 2012. Ethylene independent induction of lycopene biosynthesis in tomato fruits by jasmonates. *J Exp Bot* 16: 5751–5761.

Lorenzo O, Chico JM, Sanchez-Serrano JJ, Solano R. 2004. Jasmonate-insensitive1 encodes a MYC transcription factor essential to discriminate between different jasmonate-regulated defence responses in Arabidopsis. *Plant Cell* 16: 1938–1950.

Lorenzo O, Piqueras R, Sánchez-Serrano JJ, Solano R. 2003. Ethylene response factor-1 integrates signals from ethylene and jasmonate pathways in plant defence. *Plant Cell Environ* 15: 165–178.

Lusser A, Kolle D, Loidl P. 2001. Histone acetylation: Lessons from the plant kingdom. *Trend Plant Sci* 6: 59–65.

Martinez-Ferrer M, Harper C. 2005. Reduction of microbial growth and improvement of storage quality in fresh-cut pineapple after methyl jasmonate treatment. *J Food Quality* 28: 3–12.

Mattoo AK, Handa AK. 2008. Higher polyamines restore and enhance metabolic memory in ripening fruit. *Plant Sci* 174: 386–393.

Meir S, Droby S, Davidson H, Alsevia S, Cohen L, Horev B, Philosoph-Hadas S. 1998. Suppression of Botrytis rot in cut rose flowers by postharvest application of methyl jasmonate. *Postharv Biol Technol* 3: 235–243.

Meir S, Philosoph-Hadas S, Lurie S, Droby S, Akerman M, Zauberman G, Shapiro B, Cohen E, Fuchs Y. 1996. Reduction in the chilling injury in stored avocado, grapefruit, and bell pepper by methyl jasmonate. *Can J Botan* 74: 870–874.

Meng X, Han J, Wang Q, Tian S. 2009. Changes in physiology and quality of peach fruits treated by methyl jasmonate under low temperature stress. *Food Chem* 114: 1028–1035.

Meyer A, Gross D, Schmidt J, Jensen E, Vorkefeld S, Sembdner G. 1991. Curcubic acid related metabolites of the plant growth regulator dihydrojasmonic acid in barley (*Hordeum vulgare*). *Biochem Physiol Pflanzen* 187: 401–408.

Meyer A, Miersch O, Büttner C, Dathe W, Sembdner G. 1984. Occurrence of the plant growth regulator jasmonic acid in plants. *J Plant Growth Reg* 3: 1–8.

Mirdehghan SH, Ghotbi F. 2014. Effects of salicylic acid, jasmonic acid, and calcium chloride on reducing chilling injury of pomegranate (*Punica granatum* L.) fruit. *J Agric Sci Technol* 16: 163–173.

Miszcak A, Lange E, Saniewski M, Czapski J. 1995. The effect of methyl jasmonate on ethylene production and CO_2 evolution in Jonagold apple. *Acta Agrobotanica* 48: 121–128.

Moline EH, Buta JG, Safner RA, Maas JL. 1997. Comparison of three volatile natural products for the reduction of postharvest decay in strawberries. *Adv Strawberry Res* 6: 43–48.

Mukkun L, Singh Z. 2009. Methyl jasmonate plays a role in fruit ripening of "Pajaro" strawberry through stimulation of ethylene biosynthesis. *Scientia Hortic* 123: 5–10.

Nilprapruck P, Pradisthakarn N, Authanithee F, Keebjan P. 2008. Effect of exogenous methyl jasmonate on chilling injury and quality of pineapple (*Ananas comosus* L.) cv. Pattavia. *Silpakorn Univ Sci Technol J* 2: 33–42.

Nojiri H, Yamane H, Seto H, Yamaguchi I, Murofushi N, Yoshihara T, Shibaoka H. 1992. Qualitative and quantitative analysis of endogenous jasmonic acid in bulbing and non-bulbing onion plants. *Plant Cell Physiol* 33: 1225–1231.

Olias JM, Sanz LC, Rios JJ, Perez AG. 1992. Inhibitory effect of methyl jasmonate on the volatile ester-forming enzyme system in Golden Delicious apples. *J Agric Food Chem* 40: 266–270.

Ozturk B, Kucuker E, Karaman S, Ozkan Y. 2012. The effects of cold storage and aminoethoxyvinylglycine (AVG) on bioactive compounds of plum fruit (*Prunus salicina* Lindell cv. "Black Amber"). *Postharv Biol Technol* 72: 35–41.

Ozturk B, Ozkan Y, Yildiz K. 2014. Methyl jasmonate treatments influence bioactive compounds and red peel color development of Braeburn apple. *Turk J Agric Forestry* 38: 1–12.

Pena-Cortes H, Fisahn J, Willmitzer L. 1995. Signals involved in wound-induced proteinase inhibitor II gene expression in tomato and potato plants. *Proc Nat Acad Sci* 92: 4106–4113.

Pena-Cortes H, Albrecht T, Prat S, Weiler EW, Willmitzer L. 1993. Aspirin prevents wound-induced gene expression in tomato leaves by blocking jasmonic acid biosynthesis. *Planta* 191: 123–128.

Peng YL, Shirano Y, Ohta H, Hibino T, Tanaka K, Shibata D. 1994. A novel lipoxygenase from rice. *J Biol Chem* 269: 3759–3761.

Perez AG, Sanz C, Richardson DG, Olias JM. 1993. Methyl jasmonates vapour promotes β-carotene synthesis and chlorophyll degradation in "Golden Delicious" apple peel. *J Plant Growth Reg* 12: 163–167.

Perez AG, Sanz C, Olias R, Olias JM. 1997. Effect of methyl jasmonate on *in vitro* strawberry ripening. *J Agric Food Chem* 45: 3733–3737.

Pirie A, Mullins MG. 1976. Changes in anthocyanin and phenolics content of grapevine leaf and fruit tissue treated with sucrose, nitrate and abscisic acid. *Plant Physiol* 58: 468–472.

Rath AC, Kang IK, Park CH, Yoo WJ, Byun JK. 2006. Foliar application of aminoethoxyvinylglycine (AVG) delays fruit ripening and reduces pre-harvest fruit drop and ethylene production of bagged "Kogetsu" apples. *Plant Growth Regulat* 50: 91–100.

Rojo E, Solano R, Sanchez-Serrano JJ. 2003. Interactions between signaling compounds involved in plant defence. *J Plant Growth Regulat* 22: 82–98.

Rudell DR, Mattheis JP, Fan X, Fellman JK. 2002. Methyl jasmonate enhances anthocyanin accumulation and modifies production of phenolics and pigments in "Fuji" apples. *J Am Soc Hortic Sci* 127: 435–441.

Rudell DR, Fellman JK. 2005. Pre-harvest application of methyl jasmonate to "Fuji" apples enhances red coloration and affects fruit size, splitting and bitter pit incidence. *HortSci* 40: 1760–1762.

Rudell DR, Mattheis JP. 2008. Synergism exists between ethylene and methyl jasmonate in artificial light-induced pigment enhancement of "Fuji" apple fruit peel. *Postharv Biol Technol* 47: 136–140.

Ruiz-Garcia Y, Gil-Munoz R, Lopez-Roca JM, Martinez-Cutillas A, Romero-Cascales I, Gomez-Plaza E. 2013. Increasing the phenolic compound content of grapes by preharvest application of abscisic acid and a combination of methyl jasmonate and benzothiadiazole. *J Agric Food Chem* 61: 3978–3983.

Sabehat A, Susan L, Weiss D. 1996. The correlation between heat-shock protein accumulation and persistence and chilling tolerance in tomato fruit. *Plant Physiol* 110: 531–547.

Saniewski M. 1995. Methyl jasmonate in relation to ethylene production and other physiological processes in selected horticultural crops. *Acta Hortic* 394: 85–94.

Saniewski M, Czapski J. 1983. The effect of methyl jasmonate on lycopene and β-carotene accumulation in ripening red tomatoes. *Experientia* 39: 1373–1374.

Saniewski M, Czapski J, Nowacki J, Lange E. 1987. Effect of methyl jasmonate on the ethylene and 1-aminocyclopropane-1-carboxylic acid production in apple fruits. *Biol Plantar* 29: 199–203.

Saniewski M, Miszczak A, Kawa-Miszczak L, Wegrzynowicz-Lesiak E, Miyamoto K, Ueda J. 1998. Effects of methyl jasmonate on anthocyanin accumulation, ethylene production, and CO_2 evolution in uncooled and cooled tulip bulbs. *J Plant Growth Regulat* 17: 33–37.

Saniewski M, Nowacki J, Czapski J. 1987. The effect of methyl jasmonate on ethylene production, chlorophyll degradation and polygalacturonase activity in tomatoes. *J Plant Physiol* 129: 175–180.

Saniewsky M, Urbanek H, Czapsky J. 1997. Effect of methyl jasmonate on ethylene production, chlorophyll degradation and polygalacturonase activity in tomatoes. *J Plant Physiol* 127: 177–181.

Sayyari M, Babalar M, Kalantari S, Serrano M, Valero D. 2009. Effect of salicylic acid treatment on reducing chilling injury in stored pomegranates. *Postharv Biol Technol* 53: 152–154.

Schreiner M, Huyskens-Keil S. 2006. Phytochemicals in fruit and vegetables: Health promotion and postharvest elicitors. *Cri Rev Plant Sci* 25: 267–278.

Sembder G, Parthier B. 1993. The biochemistry and the physiological and molecular actions of jasmonates. *Ann Rev Plant Physiol Plant Molec Biol* 44: 569–589.

Shafiq M, Singh Z, Khan AS. 2011. Pre-harvest application of methyl jasmonate and training system influence the red blush development and quality of "Cripps Pink" apple. *J Hortic Sci Biotechnol* 86: 422–430.

Shafiq M, Singh Z, Khan AS. 2013. Time of methyl jasmonate application influences the development of "Cripps Pink" apple fruit colour. *J Sci Food Agric* 93: 611–618.

Simpson TD, Gardner HW. 1995. Allene oxide synthase and allene oxide cyclase, enzymes of the jasmonic acid pathway, localized in Glycine max tissues. *Plant Physiol* 108: 199–202.

Siripatrawan U, Assatarakul K. 2009. Methyl jasmonate coupled with modified atmosphere packaging to extend shelf life of tomato (*Lycopersicon esculentum* Mill.) during cold storage. *Internat J Food Sci Technol* 44: 1065–1071.

Smith TA. 1985. Polyamines. *Ann Rev Plant Physiol* 36: 117–143.

Spoel SH, Koornneef A, Claessens SM, Korzelius JP, Van Pelt JA, Mueller MJ, Buchala AJ, Metraux JP, Brown R, Kazan K. 2003. NPR1 modulates cross-talk between salicylate- and jasmonate-dependent defence pathways through a novel function in the cytosol. *Plant Cell* 15: 760–770.

Staswick PE, Su W, Howell SH. 1992. Methyl jasmonate inhibition of root growth and induction of a leaf protein are decreased in an *Arabidopsis thaliana* mutant. *Proc Nat Acad Sci* 89: 6837–6840.

Staswick PE, Yuen GY, Lehman CC. 1998. Jasmonate signaling mutants of *Arabidopsis* are susceptible to the soil fungus *Pythium irregulare. Plant J* 15: 747–754.

Sun D, Lu X, Hu Y, Li W, Hong K, Mo Y, Cahill DM, Xie J. 2013. Methyl jasmonate induced defence responses increase resistance to *Fusarium oxysporum* in banana. *Scientia Hortic* 164: 484–491.

Thomma BPHJ, Eggermont K, Penninckx IAMA, Mauch-Mani B, Vogelsang R, Cammue BPA, Broekaert WF. 1998. Separate jasmonate-dependent and salicylate-dependent defence-response pathways in Arabidopsis are essential for resistance to distinct microbial pathogens. *Proc Nat Acad Sci* 95: 15107–15111.

Tomas-Barberan FA, Gil MI, Cremin P, Waterhouse AL, Hess-Pierce B, Kader AA. 2001. HPLC-DAD-ESIMS analysis of phenolic compounds in nectarines, peaches, and plums. *J Agric Food Chem* 49: 4748–4760.

Turner JG, Ellis C, Devoto A. 2002. The jasmonate signal pathway. *Plant Cell* 14: S153–S164.

Ueda J, Kato J. 1980. Isolation and identification of a senescence-promoting substance from wormwood (*Artemisia absinthium* L.). *Plant Physiol* 66: 246–249.

Ueda J, Miyamoto K, Kato J. 1991. Identification of jasmonic acid from *Euglena gracilis* Z as a plant growth regulator. *Agric Biol Chem* 55: 275–276.

Verpoorte R, van der Heijden R, Memelink J. 2000. Engineering the plant cell factory for secondary metabolite production. *Transgenic Res* 9: 323–343.

Vick BA. 1993. Oxygenated fatty acids of the lipoxygenase pathway. In: *Lipid Metabolism in Plants* (ed TS Moore) CRC Press, Boca Raton, FL.

Vijayan P, Shockey J, Levesque CA, Cook RJ, Browse J. 1998. A role for jasmonate in pathogen defence of *Arabidopsis. Proc Nat Acad Sci* 95: 7209–7214.

Wang CY. 1993. Chilling Injury of tropical horticultural commodities. *HortSci* 29: 986–988.

Wang CY. 2003. Maintaining postharvest quality of raspberries with natural volatile compounds. *Internat J Food Sci Technol* 38: 869–875.

Wang CY, Buta JG. 1994. Methyl jasmonate reduces chilling injury in *Cucurbita pepo* through its regulation of abscisic acid and polyamine levels. *Environ Exper Botan* 43: 427–432.

Wang K, Jin P, Cao S, Shang H, Yang Z, Zheng Y. 2009. Methyl jasmonate reduces decay and enhances antioxidant capacity in Chinese bayberries. *J Agric Food Chem* 57: 5809–5815.

Wang SY. 1999. Methyl jasmonate reduces water stress in strawberry. *J Plant Growth Regulat* 18:127–134.

Wang SY, Zheng W. 2005. Preharvest application of methyl jasmonate increases fruit quality and antioxidant capacity in raspberries. *Int J Food Sci Tech* 40: 187–195.

Wang SY, Bowman L, Ding M. 2008. Methyl jasmonate enhances antioxidant activity and flavonoid content in blackberries (*Rubus* sp.) and promotes antiproliferation of human cancer cells. *Food Chem* 107: 1261–1269.

Xie DX, Feys BF, James S, Nieto-Rostro M, Turner JG. 1998. COI1: An Arabidopsis gene required for jasmonate-regulated defence and fertility. *Science* 280: 1091–1094.

Xu Y, Chang PF, Liu D, Narasimhan ML, Raghothama KG, Hasegawa PM, Bressan RA. 1995. Plant defence genes are synergistically induced by ethylene and methyl jasmonate. *Plant Cell* 6: 1077–1085.

Yao HJ, Tian SP. 2005a. Effects of a biocontrol agent and methyl jasmonate on postharvest diseases of peach fruit and the possible mechanisms involved. *J Appl Microbiol* 98: 941–950.

Yao HJ, Tian SP. 2005b. Effects of pre- and post-harvest application of salicylic acid or methyl jasmonate on inducing disease resistance of sweet cherry fruit in storage. *Postharv Biol Technol* 35: 253–262.

Zaharah SS, Singh Z. 2013. Nitric oxide fumigation delays mango fruit ripening. *Acta Hortic* 992: 543–550.

Zhao ML, Wang JN, Shan W, Fan JG, Kuang JF, Wu KQ, Li XP, Chen WX, He FY, Chen JY, Lu WJ. 2013. Induction of jasmonate signalling regulators MaMYC2s and their physical interactions with MaICE1 in methyl jasmonate-induced chilling tolerance in banana fruit. *Plant Cell Environ* 36: 30–51.

Zhou C, Zhang L, Duan J, Miki B, Wu K. 2005. Histone deacetylase19 is involved in jasmonic acid and ethylene signaling of pathogen response in Arabidopsis. *Plant Cell* 17: 1196–1204.

Zhu Z, Tian S. 2012. Resistant responses of tomato fruit treated with exogenous methyl jasmonate to *Botrytis cinerea* infection. *Scientia Hortic* 142: 38–43.

Ziosi V, Bregoli AM, Fregola F, Costa G, Torrigiani P. 2009. Jasmonate-induced ripening delay is associated with up-regulation of polyamine levels in peach fruit. *J Plant Physiol* 166: 938–946.

11 Postharvest Oxidative Stress in Fresh Fruits

Sukhvinder Pal Singh

CONTENTS

11.1 INTRODUCTION

Oxygen (O_2) is the most abundant molecule in the biological system and is often a source of free radicals as its partially reduced species are generated through normal metabolic processes. Most of the O_2 consumed by nonphotosynthetic plant tissues is reduced to water by the terminal oxidase(s) of the respiratory electron transport chain in the mitochondria (Apel and Hirt, 2004). Reactive oxygen species (ROS) are the partially reduced forms of molecular oxygen that result from either the excitation of O_2 to form singlet oxygen or the transfer of one, two, or three electrons to O_2 to form, respectively, superoxide (O_2^-), hydrogen peroxide (H_2O_2) or hydroxyl radical ($OH^\bullet$) (Hodges et al., 2004). Both O_2^- and $OH^\bullet$ are extremely reactive and can cause oxidative injury leading to cell death. The average life span of these ROS varies from nanoseconds (e.g., OH) to milliseconds (e.g., O_2^-, H_2O_2). The $OH^\bullet$ can also be generated by the interaction of O_2^- and H_2O_2 in the presence of transition metal ions, so called "Haber-Weiss reaction." Cells do not possess detoxification mechanism for $OH^\bullet$ due to its very high reactivity, and rely on mechanisms preventing its formation. These mechanisms include the preceding elimination of O_2^- and

H_2O_2 and/or sequestering metal ions that catalyze the Haber-Weiss reaction with metal-binding proteins such as ferritins or metallothioneins (Gechev et al., 2006). In addition to H_2O_2, O_2^- can also react with nitric oxide radical ($NO^\bullet$) to form peroxynitrite ($ONOO^-$), which can rapidly protonate to peroxynitrous acid ($ONOOH$), a powerful oxidizing agent. The reactions among various types of ROS can, therefore, generate intermediates or products that are capable of causing extreme levels of oxidative injury to the cell.

There are multiple sites and sources of ROS production in a plant cell, adding to the complexity of complete understanding of their chemistry and metabolism. As a part of normal metabolism, ROS levels are maintained at basal levels, except during stress conditions and developmental signals. The major areas of aerobic biochemistry (e.g., respiratory and photosynthetic electron transport; oxidation of glycolate, xanthine, and glucose) are involved in ROS production, in addition to their generation by several enzyme systems (e.g., plasmalemma-bound NADPH-dependent superoxide synthase and SOD) (Noctor and Foyer, 1998). Chloroplasts are the major sites of ROS production in photosynthetic tissues. The over-energization of photosynthetic electron transfer chains cause production of O_2^-, mainly by electron leakage from Fe–S centers of photosystem-I or reduced ferrodoxin to O_2 (Mehler reaction) (Apel and Hirt, 2004; Gechev et al., 2006). The O_2^- is rapidly metabolized into H_2O_2 by the action of SOD. Under excess light conditions, the photoinhibition of photosystem-II also causes the drastic increase in singlet oxygen production. The importance of chloroplastic sources may, however, decrease in ripening fruits and commodities placed in dark storage (Hodges et al., 2004). Mitochondria are the major consumers of O_2 and, therefore, contribute to the production of ROS (Møller, 2001; Navrot et al., 2007). During stress conditions, high O_2 concentration may build up in the cell due to reduced respiratory activities. Consequently, the high redox status of the electron transport chain (ETC), and high intra-mitochondrial O_2 concentration favor the leakage of electrons from single-electron-reduced components to molecular oxygen, resulting in the increased production of ROS. The potential sites of the leakage of single electrons to molecular O_2 are complex I and complex III (Navrot et al., 2007). Peroxisomes and glyoxysomes also contribute to the ROS generation in the cell during photorespiration and fatty-acid oxidation, respectively.

11.2   MECHANISMS FOR ROS SCAVENGING

Cellular homeostasis is achieved by a delicate balance among different pathways operating in various organelles. In a plant cell, ROS are produced during normal metabolism as well as stress conditions. The concentration of ROS in the cell is an important factor in determining their role as beneficial molecules in various signal transduction processes or in causing oxidative damage (Suzuki and Mittler, 2006). ROS has the ability to initiate cascade reactions and their products and intermediates result in damage to the lipids, proteins, and DNA. Therefore, the level of ROS is regulated by scavenging them to avoid accumulation to toxic levels in the cell. The scavenging mechanism of these ROS is also as complex and diverse as their generation sites. It involves both enzymatic and nonenzymatic antioxidant systems to act cooperatively to mitigate the ROS levels (Hodges et al., 2004).

11.2.1 Enzymatic Antioxidants

The enzymatic antioxidants include various enzymes such as superoxide dismutase (SOD), catalase (CAT), guaiacol-peroxidase (POD), ascorbate peroxidase (APX), monodehydroascorbate reductase (MDHAR), dehydroascorbate reductase (DHAR), glutathione reductase (GR), glutathione peroxidase (GPX), and glutathione-s-transferase (GT), which are responsible for direct or indirect scavenging of ROS (Noctor and Foyer, 1998). SOD, CAT, and POD are known as primary antioxidant enzymes, which serve as frontline defense antioxidants. The dismutation of O_2^- into H_2O_2 is catalyzed by SOD. The activity of SOD converts a ROS into another type of ROS, H_2O_2, which is subsequently detoxified into water (H_2O) and O_2 by the CAT and APX. CAT is considered to be more efficient as it does not require any reducing power for H_2O_2 detoxification as opposed to guaiacol-POD and APX, but it has low substrate affinity (Apel and Hirt, 2004). The activity of APX requires an ascorbate (AA) and glutathione (GSH) regeneration system, which is also called ascorbate-glutathione cycle. APX catalyzes the conversion of H_2O_2 into H_2O using two molecules of AA as a reducing power with a concomitant production of two molecules of monodehydro-ascorbate (MDHA). MDHA radicals rapidly disproportionate into dehydroascorbate (DHA) and AA; the latter reaction is catalyzed by MDHAR using NADPH as the electron donor. The DHA is reduced back to AA by the action of DHAR, using GSH as the reducing agent. Finally, GR can regenerate GSH from glutathione disulfide (GSSG) using NAD(P)H as a reducing agent. The efficient removal of ROS by the enzymatic antioxidants is therefore dependent on low molecular weight reductants, AA and GSH, because these supply electrons necessary for the activity of enzymes (Noctor and Foyer, 1998).

In addition to other enzymatic antioxidants, plant mitochondria have evolved another system of mitigating the production of ROS during stress conditions (Purvis, 2004). In addition to cytochrome oxidase, plant mitochondria have an alternate oxidase (AOX), which is located on the matrix side of the inner membrane, and oxidizes ubiquinol directly, without producing a proton electrochemical gradient across the inner membrane. Therefore, electron flow through the AOX is not coupled to the synthesis of ATP and causes the reduction in redox potential of the component of ETC and lowers the concentration of O_2 in the mitochondria. This overall process reduces the potential for leakage of electrons from the system, especially under the enhanced respiration experienced both during and after stress (Purvis, 2004). The up-regulation of AOX activity in response to various environmental stimuli can thus contribute to the reduction in ROS production in the cell.

11.2.2 Nonenzymatic Antioxidants

Ascorbate (AA) is one of the most abundant low-molecular-weight antioxidants present in plant tissues (Davey and Keulemans, 2004). AA can donate electrons, which makes it an effective ROS-detoxifying compound in a wide range of enzymatic and nonenzymatic reactions in the aqueous phase. AA serves as an antioxidant to scavenge ROS which oxidizes AA to MDHA and DHA. MDHA and DHA can be converted back into AA to maintain the cellular pool by the action of MDHAR and DHAR, respectively. Therefore, the amount of AA present in the cell is modulated by both its biosynthesis and oxidation loss. The ratio of AA to DHA is an indicator of the redox potential of the cell and can also serve as a marker for the degree of oxidative stress (Davey and Keulemans, 2004). AA also regenerates tocopherols from the tocoperoxyl radical thereby providing membrane protection (Blokhina et al., 2003). It also plays important roles in resistance to a number of environmental stresses, such as pathogen infection, hypoxia stress, high light and UV-B radiation (Noctor and Foyer, 1998). It has been implicated in the regulatory role in some fundamental cellular processes such as photo-protection, cell-cycle, and cell expansion (Blokhina et al., 2003).

Glutathione (GSH) is a nonprotein sulfur-containing tripeptide (Glu-Cys-Gly) and acts as a storage and transport form of reduced sulfur (Tausz et al., 2004). GSH is related to the sequestration of xenobiotics and heavy metals, and is also an essential component of the cellular antioxidative defense system, which keeps ROS under control. Being an integral component of the AA-GSH cycle, it is involved in regeneration of AA during the removal of excess H_2O_2 from the cell. Its role in regeneration of another antioxidant, AA, provides it a central role in the antioxidant defense system. In addition to AA-GSH cycle, GSH is possibly involved in degradation of H_2O_2 in a reaction catalyzed by GPX, but the role of GSH as a substrate of GPX is still questionable (Szalai et al., 2009). GSH plays a very important role in the removal of lipid peroxides through the activity of glutathione-s-transferases (GT). The accumulation of GSSG, the oxidized form of GSH, leads to a decrease in GSH/GSSG ratio, which symbolizes the oxidized environment and the reduction of antioxidant capacity of the glutathione system (Noctor and Foyer, 1998).

Polyphenolic compounds, a group of plant secondary metabolites of >9000 individual molecules, are involved in an array of processes, including plant–pathogen interactions, pollination, light-screening, seed development, and allellopathy (Hernández et al., 2009). Many biosynthetic genes for polyphenolic compounds are induced under stress conditions and, accordingly, their levels increase during exposure to biotic and abiotic stresses, such as wounding, drought, metal toxicity, and nutrient deprivation. *In vitro* antioxidant tests reveal that the antioxidant capacities of phenolic compounds are several-fold higher than those of ascorbate (vitamin C) or α-tocopherol (vitamin E), two well known *in planta* antioxidants. Phenolic compounds have been suggested to act as antioxidants, protecting plants from oxidative stress. Phenolic compounds and peroxidases are generally localized in vacuoles of plant cells. When H_2O_2 is formed in vacuoles or tonoplasts, or when H_2O_2 formed outside of vacuoles is diffused into the organelles, vacuolar peroxidases oxidize phenolics, especially flavonols and hydroxycinnamic acids, to form phenoxyl radicals (Takahama, 2004). Phenoxyl radicals formed by peroxidase-dependent reactions are reduced by AA in vacuoles. This suggests that ascorbate/phenolics/peroxidase systems in the vacuoles can scavenge H_2O_2.

Tocopherols are a group of closely related phenolics benzochroman derivatives having extensive ring alkylation (Lurie, 2003). There are four tocopherols isomers (α-, β-, γ-, and δ-); α-tocopherol has the highest antioxidant activity among all. The molecule is amphipathic in nature; the hydrophobic tail is located in a membrane and is associated with the acyl chains of fatty acids or their residues, while the polar chromanol head group lies at the membrane–cytosol interface. It acts as an antioxidant either by chemical scavenging of free-radicals or by physical deactivation, thereby preventing the proliferation of oxidative chain reactions. The major role of α-tocopherol is to protect the biological membrane by acting as a chain-reaction terminator for the removal of polyunsaturated fatty acid (PUFA) radicals generated during lipid peroxidation (Blokhina et al., 2003).

Carotenoids are among the most common natural pigments in plants and play a major role in the protection against photooxidative processes (Lurie, 2003). These are very efficient antioxidants scavenging singlet molecular oxygen and peroxyl radicals and, thus, prevent oxidative damage to the tissue. In plant tissues, carotenoids mostly occur in the chloroplast membranes and help prevent oxidative injury from ROS produced in the photosynthetic electron transport chain. The richness of chloroplast membranes with PUFA linolenate makes them more prone to peroxidation. Therefore, the presence of carotenoids prevents the membrane damage to the chloroplast by ROS (Lurie, 2003).

11.3 POSTHARVEST OXIDATIVE STRESS

Oxidative stress occurs when critical balance is disrupted by excess of ROS, reduction in antioxidants, or both. The important homeostatic parameters must be set to counteract the oxidant effects and to restore redox balance by the cell. There are several factors that augment the development of oxidative stress in plants and include, but are not limited to, temperature extremes, drought and flooding, salinity, ozone exposure, UV irradiation, heavy metal toxicity, herbicides, and environmental pollutants (Apel and Hirt, 2004). These environmental perturbations enhance the ROS production in the plant or plant organs. If ROS production exceeds the antioxidant capacity of the tissue, it causes oxidative damage to the cell, which ultimately leads to cell death.

Fresh fruits are living entities that continue to respire and transpire even after harvest. The postharvest physiological and biochemical changes have both desirable and undesirable effects on fruit quality. The most important desirable changes in climacteric fruit are brought by the ripening process, which involves the changes in skin and flesh color, texture, taste, and emission of aroma-volatile compounds. The advanced stages of ripening culminate into the senescence process, which eventually leads to the death of fruit, and is the most undesirable from economic viewpoint. On the other hand, nonclimacteric fruit are generally harvested at the ripe stage, but there are still some desirable postharvest changes in the development of fruit texture and flavor. Similar to climacteric fruits, the onset of senescence in nonclimacteric fruits is also a natural process that leads to death. Fruit-ripening has been described as an oxidative phenomenon that involves production and removal of ROS from the tissue, and the failure or reduction in the capability of the antioxidant system leads to the oxidative injury coinciding with the onset/advancement of senescence (Masia, 1998; Singh et al., 2012).

The objective of a postharvest technology is to slow down the ripening and/or delay the onset of senescence in fruit and to deliver a high-quality fruit to the consumer/processor. The postharvest life of a fruit can be manipulated by regulating its ripening and senescence rates. The better understanding of the physiology of ripening and senescence of fruit can, therefore, lead to the development of more effective postharvest practices aimed to minimize qualitative and quantitative losses. Low temperatures have been widely used as a means to retard the biological activity of fruit that enables its storage, transport and distribution over longer periods and distances. Studies have shown that chilling temperatures can potentially cause oxidative injury to the fruit, depending on the storage temperature and length of exposure. Most of the postharvest practices aimed to extend the storage or shelf-life of produce have been known to act as stress factors. It is therefore obvious that the development of oxidative stress in postharvest situations is indicative of the significant stress experienced by a commodity (Toivonen, 2004).

Some of the postharvest physiological and biochemical changes during the supply chain result in development of certain disorders that affect the consumer acceptability of fruit. The commercially important storage disorders, such as superficial scald in apple, flesh and core browning in apples and pears, and internal browning in plums, have been attributed to the adverse effects of postharvest oxidative stress in fruit. Chilling-related injury is also one of such major physiological disorders that limit the long-term cold-storage of fruits below a critical threshold temperature and adversely affect the consumer experiences, especially when the symptoms are invisible or internal. Fruits of tropical and sub-tropical origin are generally more susceptible to chilling-related injury than their temperate/Mediterranean counterparts. The development of chilling-related injury in temperate/Mediterranean fruits is generally a function of duration of exposure to a particular temperature. There is increasing evidence that chilling-related injury in fruits is also a result of oxidative damage to the membrane, which results in irregularities in the physiology and metabolism of the cell.

A large volume of literature has accumulated on the importance of antioxidant metabolism in relation to fruit quality and durability in postharvest supply chain. The sites of ROS production and their roles in cell physiology are complex and intricate in nature, and so are the antioxidant defense systems evolved by the plant or plant organs in order to cope with the environmental stresses. Postharvest oxidative stress in fruits can be detected directly and indirectly (Toivonen, 2004). The degree of oxidative stress can be directly determined by the measurements of accumulation of ROS, increases in lipid peroxidation products, enhanced membrane disintegration or accumulation of brown pigments. The indirect measurements include the determination of dynamics of nonenzymatic and enzymatic antioxidant systems. The understanding of inter-relationship of oxidative stress parameters and their association with storage disorders offer attractive alternatives to develop the innovative postharvest practices. The comprehensive measurement of various components of antioxidant protection systems is, therefore, central to demonstrating their role in fruit quality with regard to physiological disorders. As a general rule, plants have two primary strategies of coping with oxidative stress: avoidance and tolerance (Hodges et al., 2004). Postharvest fruits are themselves not in situations to adopt the "avoidance" strategy, but the human interventions to mitigate the conditions promoting ROS production can contribute to "avoidance." However, "tolerance" to oxidative stress of postharvest fruits has been associated with their inherent antioxidant potential. But there are several other factors that affect the capability of fruits to encounter postharvest oxidative stress and some of these important factors are reviewed and discussed in the following sections.

11.4 FACTORS AFFECTING POSTHARVEST OXIDATIVE STRESS

11.4.1 GENOTYPE/CULTIVAR

Postharvest behavior of a cultivar is determined by its genetic makeup. Some cultivars with a faster rate of ripening and senescence have poor postharvest storage and shelf-life. Generally, the cultivars with higher antioxidant potential have better stress resistance, nutritional quality, and storage characteristics (Davey and Keulemans, 2004; Łata et al., 2005). Screening of apple cultivars for some enzymatic and non-enzymatic antioxidants revealed a great variation in the activities and concentration of these antioxidants among cultivars at harvest and following cold storage (Davey et al., 2004; Davey and Keulemans, 2004; Łata et al., 2005). As compared to variation at harvest, apple cultivars differed substantially in their ability to maintain L-AA levels during storage. Generally, cultivars that could maintain their AA and GSH pools also had better storage properties (Davey and Keulemans, 2004). For instance, "Sunrise" and "Gravenstein" cultivars showed maximal losses of AA and GSH—by about 80% and 50%—following shelf-life and these cultivars are known to have poor storage qualities, with susceptibility to browning and rot. For cultivars such as "Arlet" and "Angold," there was no significant difference between AA at harvest, cold storage or shelf-life, and could be stored for up to six months at 1°C. Some cultivars such as "Greenstar" and "Braeburn" were able to maintain or actually slightly increase both AA and GSH levels under both cold storage and shelf-life and retain quality for up to six months at 1°C (Davey and Keulemans, 2004). Other researchers have also shown a great variation in the concentration of antioxidants in different cultivars of apples (Łata et al., 2005).

The concentrations of antioxidants also differ with the fruit tissue. AA and GSH levels in the peel tissue were 6.7- and 2.8-fold higher, respectively, than those in the cortical tissue (Davey et al., 2004). Similarly, in 25 cultivars of apples, on average, GR, CAT, AA, GSH, and phenolics were approximately 2.9, 1.5, 4.4, 1.7, 2.1, 2.1, and 2.5 times higher in the apple peel than the whole apple fruit (Łata et al., 2005). A recent study also confirmed that three apple cultivars, "Delicious," "Golden Delicious," and "Gala" showed higher AA content in peel than in cortical tissue as the peel tissue has to play a protective role in response to both abiotic and biotic stresses such as ultraviolet radiation, wind, pathogens, and insects (Felicetti and Mattheis, 2010). Similarly, Japanese plum cultivars have been reported to show variations in their oxidative and antioxidative systems during fruit ripening. Climacteric-type cultivars, "Blackamber" and "Amber Jewel," showed a faster decline in the ability of enzymatic and nonenzymatic antioxidative systems, parallel to the faster rate of ripening and senescence, as compared to the suppressed-climacteric cultivar, "Angeleno" (Singh et al., 2012).

Cultivar is also a major factor in determining the susceptibility of fruit to postharvest physiological disorders. For instance, apple cultivars greatly differ in their susceptibility to superficial scald, a very serious postharvest physiological disorder.

Cultivars such as "Empire," "Gala," "Golden Delicious," and "Crispin" are resistant to scald development, whereas others such as "Delicious," "McIntosh," "Cortland," "Granny Smith," "Idared," "Rome Beauty," and "Fuji" are susceptible to this disorder (Emongor et al., 1994). It is believed that the biosynthesis of α-farnesene and its subsequent degradation to conjugated trienes (CT) may cause the disruption, discoloration and death of surface cells (MacLean et al., 2006). Scald development may also be related to ROS production either through low-temperature stress or through α-farnesene metabolism. No clear relationships between antioxidant enzyme activities and susceptibility to superficial scald development have been observed in different apple cultivars (Du and Bramlage, 1994, 1995; Meir and Bramlage, 1988). The confounding effects of cultivar, maturity, ripeness, and environmental conditions, may be ascribed to obstruct the development of a clear relationship between the scald development and antioxidant systems in peel (Rao et al., 1998). Some studies have shown that activities of the H_2O_2-degrading enzymes, POD and APX, were higher in scald-resistant selections of "White Angel" × and "Rome Beauty" apples, indicating the role of ROS in induction of scald (Rao et al., 1998). The amount of α-farnesene produced during scald development may be less important than the activity of the antioxidant enzyme system (Whitaker et al., 2000). To reduce the incidence of superficial scald in apples, the success of postharvest treatments of fruit with antioxidants and storage under low O_2 atmospheres support the notion that antioxidant metabolites and/or enzymes may influence the resistance to scald (DeLong and Prange, 2003). The studies indicate that scald is characterized by cellular oxidation within the hypodermis, and oxidative stress theory can partially explain the susceptibility to this disorder exhibited by different cultivars of apples.

"Clemenules" and "Clementine" cultivars of mandarins are considered to be chilling-tolerant compared to "Nova" and "Fortune" cultivars that are chilling-sensitive (Sala, 1998). The tolerance to chilling in these cultivars has been attributed to the increased activities of CAT, APX, and GR in response to chilling-related stress. Further studies on the role of antioxidant defense showed that CAT is the major antioxidant enzyme involved in the defense mechanism of mandarin fruits against chilling-related stress (Sala and Lafuente, 2000). In another study on pepper (*Capsicum annuum* L.) fruit, susceptibility to chilling-related injury appeared to be related to the activities of SOD and CAT, which were much higher in "Buchon," a chilling-tolerant cultivar, than "Nockgwang," a chilling-sensitive cultivar (Lim et al., 2009). These studies suggest that cultivars with tolerance to chilling-related stress are generally equipped with a more efficient antioxidative system.

11.4.2 Harvest Maturity

Harvest maturity has been shown to affect the development of postharvest oxidative stress in fruits, thereby influencing the storage potential and susceptibility to oxidative injury (Toivonen, 2003a). The development of postharvest physiological disorders in fruit crops has been related to the antioxidant levels at harvest and the changes in their concentrations during cold storage. The accumulation of lipid-soluble antioxidants in apple peel, due to delayed harvest, has been shown to decrease the incidence of superficial scald during cold storage (Barden and Bramlage, 1994;

Diamantidis et al., 2002), but a clear relationship between antioxidant enzymes and scald-resistance has not been established, due to contradictory reports (Du and Bramlage, 1995; Rao et al., 1998). However, the flesh-related disorders in apples and pears have been shown to increase with advanced maturity. The decrease in activities of SOD and CAT in flesh tissue increased the incidence of Braeburn browning disorder (BBD) in late-harvested "Braeburn" apples (Toivonen et al., 2003). Another study showed that the changes in activities of antioxidant enzymes, SOD, CAT, and POD, in the flesh tissue of "Golden Smoothee" apples were mainly influenced by the climatic conditions during the last phase of on-tree fruit maturation (Molina-Delgado et al., 2009). The cooler season resulted in a higher antioxidant potential of the fruit in terms of higher activities of these enzymes and AA content. In the cooler season, the delay in harvesting caused a significant decrease in the activities of SOD and POD, and AA levels in the flesh tissue. In general, the changes in quality parameters during on-tree ripening were not related to the capability of the fruit to produce ethylene, but rather to endogenous levels of antioxidants, especially CAT and AA, at the earliest picking date (Molina-Delgado et al., 2009).

The decline in AA concentration in the flesh tissue of apple cultivars during the final stages of fruit maturation has been demonstrated by various researchers (Davey et al., 2007; Felicetti and Mattheis, 2010; Molina-Delgado et al., 2009). AA concentrations were also found to be negatively correlated with mean preharvest daytime temperature; however, since preharvest temperature and harvest date themselves are closely linked, it was not possible to definitively separate the relative contributions of genetic and environmental factors to these traits (Davey et al., 2007). The delayed harvesting of "Conference" and "Passa Crassana" pears increased susceptibility to core browning due to a decrease, with advanced maturity, in the ability of the antioxidant system to protect from ROS (Lentheric et al., 1999; Veltman et al., 1999). The behavior of peel tissue and flesh tissue seems to be quite different with regards to maturity and subsequent changes in respective antioxidant systems (Toivonen, 2003a).

It is well-known that sensitivity to chilling-related injury is strongly influenced by the fruit's maturity. Fruit at advanced stages of maturity or ripening are chilling-tolerant compared to less mature or unripe fruit. In mangoes, the activities of SOD, CAT, APX, and PPO in pre-yellow and yellow fruit were reported to be higher than those of the green, from day six to day 12, during cold storage (Zhao et al., 2009). A lower content of MDA, higher levels of AA and GSH were maintained in pre-yellow and yellow fruit than that in green fruit. These results suggested that chilling-tolerance of pre-yellow and yellow mangoes compared to green fruit was due to their higher antioxidant potential (Zhao et al., 2009) and, therefore, cold storage of mangoes at advanced stages of maturity can help alleviating chilling-related injury to some extent. Another typical example is pepper (*Capsicum annuum* L.) fruit. Pepper fruit harvested at mature green and breaker stages were more prone to chilling-related injury than those at red-ripe stage (Lim et al., 2009; Lin et al., 1993). The higher APX and SOD activities in mitochondria from ripe-red pepper fruit might play a role in avoiding the accumulation of ROS generated in the mitochondria (Jiménez et al., 2002). The higher activities of SOD and CAT have been correlated with chilling-tolerance at the red-ripe stage in peppers (Lim et al., 2009). Cherries harvested at advanced stages of maturity, four days later than commercial harvest dates, showed higher levels of phenolics and total antioxidant activity at harvest and also after 16 days of cold storage at 2°C plus two days of shelf-life (Serrano et al., 2009). In "Amber Jewel" plums, the incidence and severity of chilling-related injury was higher in fruit from the delayed harvest compared to commercial harvest (Singh and Singh, 2013c). The status of enzymatic and nonenzymatic antioxidative system during cold storage of Japanese plums was reported to be more important in providing protection against oxidative injury (expressed as chilling-related injury) than the at-harvest antioxidant status. Delayed harvested fruit experienced more oxidative stress during cold storage compared to the fruit harvested at commercial maturity. It has been widely argued that the changes in antioxidant components during cold storage of fruit should be considered more important in providing protection against oxidative injury than their at-harvest antioxidant status.

11.4.3 STORAGE TEMPERATURE AND DURATION

Storage temperature and duration are important factors that govern the development of postharvest oxidative stress in fruits (Hodges et al., 2004; Toivonen, 2004). Despite the inhibitory effects of certain postharvest procedures on fruit-ripening and senescence, the storage potential of each fruit cultivar is definite. Storage at optimum temperature is essential to maintain fruit quality. However, storage at temperatures below and above the optimal limit accelerates the loss of fruit quality, either by inducing chilling-related injury or by a faster rate of ripening and senescence. Even storage life under optimum conditions is also limited to a certain time period, depending on the cultivar and several other factors such as fruit maturity.

Antioxidant metabolism has been shown to play a very important role in determining the storage potential of fruit. The accumulation of ROS continues during postharvest phase of fruit to a variable extent, depending upon the environmental conditions and the antioxidant potential of the fruit. The production and accumulation of ROS beyond the antioxidant capability of fruit tissue causes the oxidative damage, resulting in senescence and the appearance of visible injuries on the fruit. The impact of storage-duration on the development of oxidative stress has been widely studied in fruits, but there is relatively less information about the role of storage-temperature. With the increase in storage-duration, there is an increased accumulation of ROS in the fruit tissue. The increase in concentration of H_2O_2 has been reported to occur during 16 weeks of cold storage at 0.5°C in different apple selections derived from a cross between "White Angel" and "Rome Beauty" (Rao et al., 1998). The higher levels of H_2O_2 have been associated with oxidative damage and the development of flesh-browning in "Pink Lady"™ apples (De Castro et al., 2008), and also related to water core in "Fuji" apples (Kasai and Arakawa, 2010). On the other hand, only a transient increase in the concentration of H_2O_2 was observed in "Golden Smoothee" apples during the first three days of storage, and then it declined continuously during the subsequent 90 days of storage period (Vilaplana et al., 2006). Contrary to H_2O_2, the accumulation of lipid peroxidation products continued during the entire storage-duration, which indicated the development of oxidative stress with the progression of storage (Vilaplana et al., 2006). The cultivar differences do exist in the storage behavior of fruits, especially with regard to accumulation of ROS, depending on tissue type. The increase in ROS concentration and accumulation of lipid peroxidation products as a function of storage duration have been reported in loquats (Cao et al., 2009), peaches (Zheng et al., 2007), oranges (Huang et al., 2008), mangoes (Singh and Dwivedi, 2008; Wang et al., 2009), and plums (Singh and Singh, 2012a,b, 2013a,b,c).

The activities of primary antioxidant enzymes, SOD, CAT, and POD, have been known to change with respect to storage-duration. The activities of CAT and POD increased significantly in the skin tissues of apple cultivars, "Empire," "Cortland," and "Delicious" during the cold storage at 0°C for 12 or 24 weeks, while no significant change in SOD activity was reported (Du and Bramlage, 1995). Similarly, in flesh tissue, CAT activity has been reported to increase during cold storage of "Golden Delicious," "Golden Smoothee" and "Fuji" cultivars, with less impact on SOD activity (Masia, 1998; Vilaplana et al., 2006). CAT activity increased only gradually in "Golden Delicious" while peaking in "Fuji" after 49 days of cold storage. The decrease in CAT after 84 days of storage was coincident with the appearance of water-core disorder in "Fuji" apples (Masia, 1998). Another study has shown the decreases in activities of CAT and POD and increase in SOD activity during 16 weeks of cold storage of apple selections derived from a cross between "White Angel" and "Rome Beauty" (Rao et al., 1998). These studies show mixed results in terms of changes in the activities of various enzymes in even the same fruit tissue; some were contradictory and others were in agreement with one another.

The changes in activities of various antioxidant enzymes have been studied in other fruits such as mandarins (Sala, 1998; Sala and Lafuente, 2000, 2004), mangoes (Wang et al., 2006, 2009), oranges (Huang et al., 2008), pears (Larrigaudière et al., 2001b, 2004), peaches (Wang et al., 2004, 2005, 2006), plums (Singh et al., 2012; Singh and Singh, 2013a), and raspberries (Chanjirakul et al., 2006). These studies indicate that the increased activities of SOD, CAT, and POD enzymes are important to regulate the ROS levels in the fruit tissue. The decreases in the activities of these enzymes were generally coincident with the failure of the overall antioxidant system, resulting in loss of membrane integrity, appearance of physiological disorders, and termination of storage life.

AA concentrations generally decrease during cold storage of fruits. The decrease in concentration of AA in apples has been reported to be a function of storage-duration (Barden and Bramlage, 1994; Davey et al., 2004, 2007; Davey and Keulemans, 2004; De Castro et al., 2008; Fawbush et al., 2009; Felicetti and Mattheis, 2010; Łata et al., 2005) and a subsequent increase in the number of physiological disorders (Davey et al., 2004; Davey and Keulemans, 2004; De Castro et al., 2008; Kasai and Arakawa, 2010). The role of AA in better postharvest storage performance of pears has been widely studied and accepted (Franck et al., 2003; Lentheric et al., 1999; Veltman et al., 1999, 2000). The factors that contribute to the core breakdown in pears have been closely related to the loss of AA in the fruit (Franck et al., 2003). In general, a significant decrease in GSH levels has been reported to occur during long-term cold storage of fruits such as mangoes (Zhao et al., 2009), oranges (Huang et al., 2008), and pawpaws (Galli et al., 2009). High AA levels in some apple cultivars have been linked to the high GSH levels, resulting in better storage properties

(Davey and Keulemans, 2004). Apple cultivars such as "Greenstar" and "Braeburn" were able to maintain or actually slightly increase GSH levels under both cold storage and during shelf-life, and could retain quality for up to six months at 1°C (Davey and Keulemans, 2004). A comprehensive study on the changes in different antioxidants in apple fruit during cold storage revealed that GSH levels were significantly higher in the fruit skin after 45 days of storage in air and then declined after 90 days to harvest levels (Łata, 2008). The study supports the notion that concentration of GSH increases in response to chilling-related stress as a mechanism for acclimatization, but prolonged storage duration may result in decrease in its concentration to original levels or below. At the same time, the increase in GSSG concentration has been reported to cause a decrease in the ratio of GSH:GSSG, which has been considered to be a marker of oxidative stress in the tissue (Tausz et al., 2004). The decrease in glutathione concentration has been associated with internal browning in methyl bromide-fumigated "Thompson Seedless" grapes (Liyanage et al., 1993) and methyl-iodide-induced phytotoxicity in lemons (Ryan et al., 2007). These studies also reflect that most changes in different antioxidant compounds, including GSH, are cultivar-specific. The role of GSH as an antioxidant has not been widely studied in fruits. The concentrations of other antioxidant compounds such as phenolics and flavonoids are strongly influenced by the temperature during and duration of storage. Strawberry fruit stored at 10°C or 5°C showed higher antioxidant capacity, total phenolics, and anthocyanins, than those stored at 0°C (Ayala-Zavala et al., 2004). Strawberries stored at 0°C retained an acceptable overall quality for the longest storage duration; however, berries stored at temperatures higher than 0°C showed higher content of aroma compounds and antioxidant capacity during the postharvest period.

11.4.4 STORAGE ATMOSPHERES

Storage atmospheres containing low O_2 and/or high CO_2 can retard the fruit metabolism through reduced rates of respiration and ethylene production. The modification of storage atmospheres also provides other benefits such as alleviation of certain physiological disorders, and suppression of microbial and insect activity. The presence of low O_2 atmospheres surrounding the fruit can decrease inter- and intra-cellular O_2 levels in the flesh tissue (Whitaker, 2004). Therefore, there is a potential for reducing the rates of oxidative processes in the fruit. Both beneficial and detrimental effects can be expected from storage under controlled/modified atmosphere (CA/MA) conditions, depending on the concentrations of gases, the physiological maturity of fruit and the preharvest environmental conditions. Control of superficial scald in apples is a classic example of the suppression of oxidative reactions in the fruit by CA/MA storage. Low O_2 atmospheres have been known to reduce the incidence and severity of superficial scald in apples (Patterson and Workman, 1962). The storage of "Starkrimson Delicious" apples in atmospheres containing 0.7% O_2 for 6–8 months, markedly suppressed incidence of scald compared with air-storage (Lau et al., 1998). "Braeburn" fruit held in atmospheres containing 1.2 or 1.5% O_2 + 1.0 or 1.2% CO_2 for six months had significantly less core-browning and superficial scald than fruit held in air for the same period (Lau, 1998). However, CA-stored fruit were highly susceptible to Braeburn browning disorder (BBD) and internal cavities (IC) after cool growing seasons. And the susceptibility of fruit to BBD and IC was greatest in late-harvested fruit stored in 3.0% CO_2 and 1.5% O_2. Therefore, seasonal environmental conditions, coupled with the stage of fruit maturity, can override the benefits of low O_2 storage atmospheres in providing protection against superficial scald (Emongor et al., 1994; Lau, 1998). CA storage in combination with postharvest treatment with an antioxidant such as diphenylamine (DPA) or ethylene action inhibitor, 1-MCP, can reduce the incidence of scald in apples and pears (Rizzolo et al., 2005; Watkins et al., 2000). These studies suggest that CA storage alone cannot provide protection against the incidence of scald in apple and pear, without the supplementation of a postharvest treatment with an antioxidant or 1-MCP.

The role of high CO_2 during CA storage in development of postharvest oxidative stress has also been studied in fruits. "Pink Lady"™ apples held in CA containing 1.5 kPa O_2 and 5 kPa CO_2 accumulated more H_2O_2 than those stored in air, indicating stress from the high-CO_2 concentrations in storage and causing flesh browning symptoms (De Castro et al., 2008). A direct comparison of the effects of CO_2 concentration in CA on the oxidative stress during storage of "Blanquilla" pears was investigated (Larrigaudière et al., 2001b). During six months of storage in regular air or CA containing either 2% O_2 + 0.7% CO_2, or 2% O_2 + 5% CO_2, the incidence of core-browning only occurred in pears stored at 5% CO_2 which was associated with the accumulation of lipid peroxidation products (ethane and TBARS) and lowest concentration of AA after six months of storage. The effectiveness of the antioxidative defense system might have decreased, causing lipid peroxidation and finally browning. The study suggested that oxidative damage was involved in the CO_2-related physiological damage occurring during CA-storage in pears (Larrigaudière et al., 2001b).

It is believed that significant changes in the antioxidant metabolites and enzymes occur during the initial stages of CA-storage of apples and pears, which subsequently affect the long-term storage behavior of fruit. Some of these changes may have detrimental effects on fruit quality as reported in "Conference" pears. "Conference" pears, subjected to CA containing 2% O_2 and 5% CO_2 or regular air for 21 days at −1°C, showed a rapid decrease in total ascorbate content and increase in DHA under CA conditions (Larrigaudière et al., 2001a). A sharp increase in the activities of SOD, APX, GR, and LOX was coupled with the accumulation of H_2O_2 and lipid peroxidation products, when the fruit were exposed to CA. These changes marked a significant amount of oxidative stress in "Conference" pears exposed to CA containing 5% CO_2, which can further lead to development of core-browning during extended storage (Larrigaudière et al., 2001a). It is generally accepted that decrease in AA concentration caused by late-harvesting and 5% CO_2 concentration in the storage atmosphere are mainly associated with the appearance of brown heart in "Conference" pears (Zerbini et al., 2002).

A study on apple cultivars, "Šampion," "Jonagold," "Topaz," "Sawa" and the clone U 633, showed that the pool of antioxidants such as AA, thiols, and phenolic compounds, in the skin tissues of fruit, showed a pronounced increase under CA conditions (1.5% O_2 + 1.5% CO_2) during the initial 45 days of storage and should be considered as an acclimation response (Łata, 2008). After 90 days of storage, the antioxidant status was kept more efficiently in the CA as compared to storage in regular air. Cranberries (*Vaccinium macrocarpon* Aiton) could be stored for two months at 3°C under storage atmosphere containing 30% CO_2 + 21% O_2 (Gunes et al., 2002). The storage atmosphere did not affect the content of total phenolics or flavonoids. However, the total antioxidant activity of the fruit increased overall by about 45% in fruits stored in air, while this increase was prevented by storage in atmospheres containing 30% CO_2 + 21% O_2. Longan (*Dimocarpus longan* Lour.) fruit stored in a 5% O_2 atmosphere (balance N_2) for six days at 28°C markedly delayed pericarp browning in association with maintenance of high total phenolic content and reduced activities of polyphenol oxidase (PPO), POD and phenylalanine ammonia lyase (Cheng et al., 2009). Moreover, the fruit stored in a 5% O_2 atmosphere exhibited a lower relative leakage rate and pulp breakdown and higher DPPH radical scavenging activity than fruit stored in air.

The short-term exposure to MA has also been reported to be beneficial for inhibiting lipid peroxidation and enhancing the antioxidant potential of fruit. Postharvest exposure of kiwifruit cv. "Xuxiang" to pure N_2 gas for 6 h maintained a high level of firmness within 14 days of storage at 1 ± 1°C and reduced the decrease in the firmness during shelf-life (Song et al., 2009). A similar study was conducted on loquat, in which fruit exposed to 100% N_2 for 6 h could retain their quality during 35 days storage at 5°C (Gao et al., 2009). In kiwifruit and loquat, the short-term anoxic treatment reduced the increases in membrane permeability and lipid peroxidation, delayed the increases in both O_2^- production rate and H_2O_2 content, increased activities of SOD and POD, but reduced LOX activity throughout storage period (Gao et al., 2009; Song et al., 2009). These studies show that the beneficial effects obtained from anoxic treatment could be due to reduced lipid peroxidation, enhanced antioxidant ability and membrane integrity maintenance.

The nonenzymatic and enzymatic antioxidants in mangoes have also been studied with respect to CA storage (Kim et al., 2007; Niranjana et al., 2009). According to Kim et al. (2007), gallic acid, total hydrolysable tannins, total soluble phenolics, and antioxidant capacity significantly decreased throughout mango fruit ripening from mature-green to full-ripe stages, but CA-storage can possibly delay the decrease through delayed fruit ripening (Kim et al., 2007). Higher concentration of phenolics compounds and activities of CAT and POD and lower levels of carotenoids were found in "Alphonso" mangoes stored in CA containing 5% O_2 and 5% CO_2 compared to those kept in regular air after 45 days storage at 8°C, which could be attributed to the effect of CA on fruit ripening. CA-storage was also effective to alleviate symptoms of chilling-related injuries in mangoes (Niranjana et al., 2009). Peach fruit "cv. Okubao" stored at 0°C for 60 days under CA of 5% O_2 + 5% CO_2, or CA with high O_2 concentration 70% O_2 + 0% CO_2 for 15 days, then in CA with 5% O_2 + 5% CO_2 showed reduced chilling-related injuries compared to those held in regular air (Wang et al., 2005). CA-storage delayed the reduction in activities of SOD, CAT, and POD, which have been associated with the alleviation of chilling-related injuries in peaches. High O_2 treatment also induced the activities of SOD and CAT, but no significant effect on alleviating chilling-related injuries was found compared to CA-storage. The extent of lipid peroxidation was greatly reduced under CA conditions, resulting in the maintenance of membrane integrity (Wang et al., 2005). A recent study showed that storage of Japanese plums "cv. Blackamber" in CA (1 or 2.5% O_2 and 3% CO_2) was favorable for the glutathione pool and the enzymes related to restoration of glutathione in the AA–GSH cycle, but the effects of CA on the ascorbate pool were not favorable (Singh and Singh, 2013a). The enzymatic and nonenzymatic antioxidative systems were efficiently operating under low O_2 atmospheres to scavenge the ROS produced in plums in the response to long-term chilling and gaseous stresses. CA conditions appeared to be limiting the oxidative processes to some extent, resulting in reduced oxidative damage in plums (Singh and Singh, 2013a).

Posthypoxic and/or postanoxic conditions are among the stresses in which ROS are implicated as the principle cause of injury. When aerobic conditions are reestablished, a burst of ROS takes place, resulting in posthypoxic or postanoxic injury to the tissues. A study was conducted on the monitoring of gene expression of enzymes involved in ascorbate biosynthesis, oxidation, and recycling in tomato, in response to hypoxia, and postanoxic stress (Ioannidi et al., 2009). To create hypoxic conditions, mature green tomato fruit were subjected to atmospheres containing 0%, 0.5%, and 3% O_2 (balance N_2) for 1, 3, 6, 12, 24, 48, and 72 h at 22°C. For testing posthypoxic stress, mature green fruit were subjected to 100% N_2 for 48 h and then were exposed to air. Postanoxic stress caused mature green tomatoes to respond by inducing the transcript accumulation of all AA biosynthetic genes as early as 3–6 h after return to air, coinciding with elevated levels of AA. Similarly, enzymes involved in AA recycling responded to postanoxic stress by increasing their mRNA steady-state levels upon return to air. This was an indication of the magnitude of the oxidative damage and the crucial role of AA in scavenging ROS under these conditions. The induction of AA recycling genes suggest that this massive activation of transcription is needed to increase the reduced AA pool in order to compensate for the oxidative stress.

A study on the effects of anaerobic stress (exposure to N_2 atmospheres for 24 h) on the proteome of mandarins and grapefruit revealed that the majority of the identified citrus anaerobic proteins (ANPs) in both mandarins (eight out of 10 proteins) and grapefruit (five out of nine proteins) were stress-related proteins, such as SOD, APX and LOX (Shi et al., 2008). The activation of SOD and APX in response to anaerobiosis might be crucial to preserve the redox status of the cells. Furthermore, LOX is one of the main enzymes known to be involved in the generation of ROS and, thus, the observed suppression of the LOX protein in mandarin flavedo may be part of the cellular efforts to cope with the undesired accumulation of ROS. The increase in the abundance of two oxidoreductases—zinc oxidoreductase and quinine oxidoreductase—in flavedo may contribute to the elevated anaerobic stress tolerance of "StarRuby" grapefruit (Shi et al., 2008).

It is therefore evident that storage of fruit under optimal CA conditions can potentially maintain or slightly increase the antioxidant potential of the fruit. The short-term hypoxia or anoxia can also stimulate the antioxidant defense system in the fruit. But, there is absolute lack of information on the transcript levels of various antioxidant enzymes involved in ROS scavenging and AA–GSH cycle in fruit subjected to CA/MA. Most studies have been restricted to measuring the activities of SOD, CAT, and POD and lipid peroxidation. The emerging tools of proteomics and metabolomics have provided great insight into the development of anaerobic stress in mandarin and grapefruit (Shi et al., 2008). The application of these tools has also been extended to study chilling-related injuries in peaches (Dagar et al., 2010). There is vast scope for research in this area to explicate the processes involved in developing oxidative stress leading to physiological disorders in various fruits.

11.4.5 POSTHARVEST TREATMENTS

Postharvest oxidative stress has been associated with the enhanced rates of ripening and senescence, the development of physiological disorders, and reduction in nutritional quality of fruits (Hodges et al., 2004). Therefore, the regulation of oxidative stress can be important to improving the shelf-life and quality maintenance during postharvest handling, storage and distribution of fruits (Toivonen, 2003b). Postharvest treatments to control oxidative stress-related injury in fruits involve two approaches: (1) the use of antioxidant dips and coatings to directly prevent oxidation reactions, and (2) postharvest treatments, such as low or high temperature, exposure to CA/MA and growth regulators, to enhance endogenous resistance or tolerance to oxidative stress. The postharvest treatments such as low-temperature conditioning (cold shock treatments), heat treatments, and antioxidant dips, etc., also play an important role in strengthening the antioxidant system of fruit (Toivonen, 2003b). The role of storage-temperature, storage-duration and CA/MA in alleviating oxidative stress in fruits has been reviewed in previous sections. In this section, the role of postharvest treatments with ethylene action inhibitor, 1-MCP, and putative ethylene biosynthesis inhibitor, nitric oxide, in relation to oxidative stress in fruits will be discussed.

11.4.5.1   1-Methylcyclopropene

Among different strategies to control ethylene, the postharvest exposure of fresh produce to 1-methylcyclopropene (1-MCP) has emerged as the greatest tool of commercial importance all over the world (Blankenship and Dole, 2003; Watkins, 2006). The major beneficial effects of 1-MCP in fresh horticultural produce are discussed in greater detail in Chapter 6. In this section, the effects of 1-MCP on the oxidative behavior of fruits will be reviewed. Studies have shown that 1-MCP treatment can reduce the lipid peroxidation in different fruits during postharvest storage and shelf-life. The inhibition of accumulation of lipid peroxidation products in the skin and flesh tissues of various fruits such as apples, pears, loquats, and mangoes (Cao et al., 2009; Singh and Dwivedi, 2008; Vilaplana et al., 2006), indicates a positive role for 1-MCP in alleviation of oxidative stress and maintenance of membrane integrity. The maintenance of membrane structural integrity has been demonstrated by the reduced ion leakage from the skin or flesh tissue of 1-MCP-treated fruits such as apricot, lychee (litchi), loquat, and pear (Cao et al., 2009; Egea et al., 2010; Larrigaudière et al., 2004; Sivakumar and Korsten, 2010). 1-MCP treatment has been shown to affect the activities of primary antioxidant enzymes, SOD, CAT, and POD. A significant increase in the activities of SOD, CAT, POD, and APX has been observed in flesh tissues of fruits, such as apricot, loquat, mangoes, and pears, in response to 1-MCP treatment which might have helped to regulate the concentration of ROS in the fruit tissue, resulting in lesser oxidative damage. Studies have also shown that the concentrations of ROS were significantly lower in the 1-MCP-treated fruit compared to untreated control (Cao et al., 2009; Larrigaudière et al., 2008; Singh and

Dwivedi, 2008; Vilaplana et al., 2006; Wang et al., 2009). There are contradictory reports on the effects of 1-MCP on the nonenzymatic antioxidants and antioxidant capacity of fruits. 1-MCP treatment has been shown to reduce the levels of AA in the flesh tissues of apples and pears (Larrigaudière et al., 2004; Vilaplana et al., 2006). Another study showed that there was no consistent effect on the concentrations of AA and GSH in 1-MCP-treated pears (Silva et al., 2010). The antioxidant capacity in apples has been reported to increase in the skin tissue (Larrigaudière et al., 2004; MacLean et al., 2006; Shaham et al., 2003; Vilaplana et al., 2006) and remain unaffected in the flesh tissue (Vilaplana et al., 2006). The beneficial effects of 1-MCP in retention of AA in the fruit have been mainly attributed to the retardation of fruit-ripening and senescence (Egea et al., 2010; Sivakumar and Korsten, 2010). The stimulation of phenylpropanoid pathway by 1-MCP treatment has been attributed to the increase or maintenance of phenolics and flavonoids in the skin or flesh tissue of fruits.

The effects of 1-MCP are influenced by the fruit cultivar, the 1-MCP dose, exposure duration, and postharvest conditions. However, the response of the antioxidant system to 1-MCP treatment has been tissue-specific in fruits. For instance, 1-MCP treatment of "Golden Smoothee" apples increased the susceptibility of skin tissue to browning by increased activity of PPO and POD (Larrigaudière et al., 2008), while it was beneficial for the flesh tissue due to decreased lipid peroxidation and lower accumulation of H_2O_2 (Vilaplana et al., 2006). Total phenolics in "Empire" apples showed a significant increase in the skin, but a decrease in the flesh tissue during cold storage for five months (Fawbush et al., 2009). The control of superficial scald in "Granny Smith" apples has been attributed to the increase in lipid soluble antioxidants in the skin of 1-MCP-treated fruit, while activities of CAT and APX were suppressed by 1-MCP treatment (Shaham et al., 2003).

The exposure of fruit to higher doses of 1-MCP has some detrimental effects on the antioxidant system of fruit. In strawberries, exposure to 1 μL L^{-1} 1-MCP increased the susceptibility to disease through reduced activity of PAL and lower concentration of phenolics (Jiang et al., 2001). Similarly, in limes, the higher dose of 1 μL L^{-1} 1-MCP increased the activity of POD, enhanced the skin-yellowing as compared to lower doses, which had an inhibitory effect on the loss of chlorophyll (Win et al., 2006). The impact of postharvest conditions on the efficacy of 1-MCP in influencing the antioxidant potential of fruit has been demonstrated in "Sunrise" apples (Qiu et al., 2009). The study showed that the phenolics fraction of the flesh tissue digestate showed higher Folin Ciocalteu reaction, reducing capacity in response to 1-MCP treatment, only if the fruit were stored at ≥15°C for three weeks. The effects of 1-MCP treatment on the retention of nonenzymatic antioxidants during storage of "Empire" apples were also influenced by the storage atmosphere, but most of the results were inconsistent (Fawbush et al., 2009). The literature indicates the role of 1-MCP in influencing the oxidative processes involved in ripening, senescence and physiological disorder development.

11.4.5.2 Nitric Oxide

NO is an important signaling molecule, which plays an important role in ROS metabolism during normal and stress conditions in plants (Lamattina et al., 2003). NO is itself a free radical molecule with very high reactivity. The ability of NO to regulate the levels and toxicity of ROS imparts it a "cytoprotective" role. Oxidative stress, which is provoked by increased concentrations of O_2^-, H_2O_2, and alkyl peroxides, can therefore be alleviated through protective effects of NO (Beligni et al., 2002). Additionally, NO itself possesses antioxidant properties (Beligni et al., 2002). The reaction of NO with lipid alcoxyl ($LO^•$) and peroxyl ($LOO^•$) radicals is rapid and beneficial to prevent the propagation of ROS mediated lipid peroxidation. The interaction between NO and other cellular antioxidants provides indirect protection from ROS damage. The toxic effects of NO are generated when it reacts with O_2^- to form peroxynitrites ($ONOO^-$) (Lamattina et al., 2003). Peroxynitrites can further react with thiol groups of proteins and polyunsaturated radicals of fatty acid lipids of membrane, causing serious damage to proteins, lipids, and DNA, and is also capable of generating the most damaging hydroxyl radicals through its protonation. Therefore, NO can be cytotoxic and cytoprotective, depending on the local concentration of NO as an effect of the rate of synthesis, translocation, effectiveness of removal of this reactive nitrogen species, as well as its ability to directly interact with other molecules and signals (Arasimowicz and Floryszak-Wieczorek, 2007; Beligni and Lamattina, 1999).

The role of NO in delaying fruit-ripening and senescence has been established and discussed in more detail in Chapter 9. NO is thought to be involved in interfering ethylene biosynthetic pathway in fruit (Lesham and Wills, 1998; Singh et al., 2009; Zhu et al., 2006, 2008), which helps to inhibit ethylene-dependent responses in the fruit. In the past few years, the role of NO in the alleviation of postharvest oxidative stress has been reported in fruits such as kiwifruit, longan, peach, and tomato (Duan et al., 2007; Zhu et al., 2006, 2008). The concentrations of NO that were higher than optimum increased the oxidative stress in the fruit tissue as shown by increased lipid peroxidation and accumulation of ROS (Zhu et al., 2008). Therefore, the cytoprotective effects of NO can only be obtained at certain concentrations of NO, which are required to be optimized for different fruits. In general, the postharvest treatment of fruit with either donor-compounds of NO or NO gas, resulted in decreased levels of lipid peroxidation, reduced accumulation of superoxides and H_2O_2, and increased activities of antioxidant enzymes favorable for delay in initiation and reduction of oxidative stress (Duan et al., 2007; Zhu et al., 2006, 2008). These studies demonstrate that NO is a powerful tool to regulate the oxidative stress in fresh fruits, depending upon the concentration and stage of application of NO.

11.5 CONCLUSIONS

The process of senescence in a plant or plant organs such as fruit, involves various degenerative changes in DNA, proteins and lipids, caused by oxidative stress. The oxidative stress develops when the well-regulated balance between prooxidants and antioxidants is disturbed, in favor of the prooxidants (Apel and Hirt, 2004). The ROS are produced as byproducts of normal metabolism for their role in signal transduction, but their proliferation in the cell can cause oxidative damage to the macromolecules (DNA, proteins, and lipids). The concept of oxidative stress has emerged to a great extent to explain the anomalies in development of various chronic diseases and ageing of humans and animals. The role of oxidative stress in plants, due to environmental perturbations, became of significant interest to researchers in the middle to late 1980s (Hodges et al., 2004). However, the research on post-harvest stress physiology in relation to oxidative stress gained impetus in the late 1990s. Postharvest procedures adopted to extend the potential storage or market-life of a fruit also act as stress factors (Toivonen, 2004). The development of oxidative stress in the fruit tissue has been demonstrated in response to the prolonged stress conditions and inherent metabolic changes. Postharvest oxidative stress has been associated with processes of fruit-ripening, senescence, and physiological disorders. Several factors have been proposed to influence the occurrence and severity of post-harvest oxidative stress in fruits, including: genotype; harvest maturity; storage conditions; storage duration; and growth regulators, including ethylene and polyamines. Postharvest oxidative stress has been associated with the development of serious physiological disorders such as chilling-related injury in citrus (Sala, 1998), super-ficial scald in apples (Barden and Bramlage, 1994; Whitaker, 2004), and internal browning in apples (Toivonen, 2003b), pears (Veltman et al., 1999, 2000) and plums (Singh and Singh 2012a,b). Several strategies have been developed to delay the devel-opment of oxidative stress, either through enrichment of antioxidants (e.g., antioxi-dant dip treatment in apples to prevent scald) or through activation and sustenance of antioxidative systems in fruits, to cope with the increasing ROS production during postharvest handling and storage. The understanding of oxidative behavior of fruits to various preharvest and postharvest factors has led to minimizing the detrimental effects of oxidative stress on fruit quality and shelf-life. Oxidative stress measure-ments, such as lipid peroxidation and activities of antioxidant enzymes and nonen-zymatic antioxidants, are essential to understand the physiology and biochemistry of postharvest processes. The oxidative stress indicators may be developed as biomark-ers and tools for predicting and monitoring the postharvest behavior of fruits during the supply chain. However, the complexity of oxidative and antioxidative systems in multiple fruit species and/or cultivars is a great challenge to develop and evaluate the robustness of these biomarkers. The oxidative stress theory holds potential to screen germplasm for desirable postharvest traits and also to evaluate the efficacy of various postharvest treatments or procedures aimed to enhance storage stability of fruits.

REFERENCES

Apel K, Hirt H. 2004. Reactive oxygen species: Metabolism, oxidative stress and signal trans-duction. *Annu Rev Plant Biol* 55: 373–399.

Arasimowicz M, Floryszak-Wieczorek J. 2007. Nitric oxide as a bioactive signalling molecule in plant stress responses. *Plant Sci* 172: 876–887.

Ayala-Zavala JF, Wang SY, Wang CY, González-Aguilar GA. 2004. Effect of storage tempera-tures on antioxidant capacity and aroma compounds in strawberry fruit. *LWT-Food Sci Technol* 37: 687–695.

Barden CL, Bramlage WJ. 1994. Relationships of antioxidants in apple peel to changes in ct-farnesene and conjugated trienes during storage, and to superficial scald development after storage. *Postharv Biol Technol* 4: 23–33.

Beligni MV, Fath A, Bethke PC, Lamattina L, Jones RL. 2002. Nitric oxide acts as an antioxidant and delays programmed cell death in barley aleurone layers. *Plant Physiol* 129: 1642–1650.

Beligni MV, Lamattina L. 1999. Is nitric oxide toxic or protective? *Trends Plant Sci* 4: 299–300.

Blankenship SM, Dole JM. 2003. 1-Methylcyclopropene: A review. *Postharv Biol Technol* 28: 1–25.

Blokhina O, Virolainen E, Fagerstedt KV. 2003. Antioxidants, oxidative damage and oxygen deprivation stress: a review. *Ann Bot* 91: 179–194.

Cao S, Zheng Y, Wang K, Rui H, Tang S. 2009. Effects of 1-methylcyclopropene on oxidative damage, phospholipases and chilling injury in loquat fruit. *J Sci Food Agric* 89: 2214–2220.

Chanjirakul K, Wang SY, Wang CY, Siriphanich J. 2006. Effect of natural volatile compounds on antioxidant capacity and antioxidant enzymes in raspberries. *Postharv Biol Technol* 40: 106–115.

Cheng G, Jiang Y, Duan X, Macnish A, You Y, Li Y. 2009. Effect of oxygen concentration on the biochemical and chemical changes of stored longan fruit. *J Food Qual* 32: 2–17.

Dagar A, Friedman I, Lurie S. 2010. Thaumatin-like proteins and their possible role in protection against chilling injury in peach fruit. *Postharv Biol Technol* 57: 77–85.

Davey MW, Auwerkerken A, Keulemans J. 2007. Relationship of apple vitamin C and antioxidant contents to harvest date and postharvest pathogen infection. *J Sci Food Agric* 87: 802–813.

Davey MW, Franck C, Keulemans J. 2004. Distribution, developmental and stress responses of antioxidant metabolism in Malus. *Plant Cell Environ* 27: 1309–1320.

Davey MW, Keulemans J. 2004. Determining the potential to breed for enhanced antioxidant status in Malus: Mean inter- and intravarietal fruit vitamin C and glutathione contents at harvest and their evolution during storage. *J Agric Food Chem* 52: 8031–8038.

De Castro E, Barrett DM, Jobling J, Mitcham EJ. 2008. Biochemical factors associated with a CO2-induced flesh browning disorder of Pink Lady apples. *Postharv Biol Technol* 48: 182–191.

DeLong JM, Prange RK. 2003. Superficial scald–a postharvest oxidative stress disorder. In: Hodges DM (ed.) *Postharvest Oxidative Stress in Horticultural Crops.* Food Products Press, New York, USA, pp. 92–112.

Diamantidis G, Thomai T, Genitsariotis M, Nanos G, Bolla N, Sfakiotakis E. 2002. Scald susceptibility and biochemical/physiological changes in respect to low preharvest temperature in "Starking Delicious" apple fruit. *Scientia Hortic* 92: 361–366.

Du Z, Bramlage WJ. 1994. Superoxide dismutase activities in senescing apple fruit (*Malus domestica* Borkh.). *J Food Sci* 59: 581–584.

Du Z, Bramlage WJ. 1995. Peroxidative activity of apple peel in relation to development of poststorage disorders. *HortSci* 30: 205–208.

Duan X, You YL, Su X, Qu H, Joyce DC, Jiang Y. 2007. Influence of the nitric oxide donor, sodium nitroprusside, on lipid peroxidation and anti-oxidant activity in pericarp tissue of longan fruit. *J Hortic Sci Biotech* 82: 467–473.

Egea I, Flores FB, Martínez-Madrid MC, Romojaro F, Sánchez-Bel P. 2010. 1-Methylcyclopropene affects the antioxidant system of apricots (*Prunus armeniaca* L. cv. Búlida) during storage at low temperature. *J Sci Food Agric* 90: 549–555.

Emongor VE, Murr DP, Lougheed EC. 1994. Preharvest factors that predispose apples to superficial scald. *Postharv Biol Technol* 4: 289–300.

Fawbush F, Nock JF, Watkins CB. 2009. Antioxidant contents and activity of 1-methylcyclopropene (1-MCP)-treated "Empire" apples in air and controlled atmosphere storage. *Postharv Biol Technol* 52: 30–37.

Felicetti E, Mattheis JP. 2010. Quantification and histochemical localization of ascorbic acid in "Delicious," "Golden Delicious," and "Fuji" apple fruit during on-tree development and cold storage. *Postharv Biol Technol* 56: 56–63.

Franck C, Baetens M, Lammertyn J, Verboven P, Davey MW, Nicolai BM. 2003. Ascorbic acid concentration in "cv. Conference" pears during fruit development and postharvest storage. *J Agric Food Chem* 51: 4757–4763.

Galli F, Archbold DD, Pomper KW. 2009. Pawpaw fruit chilling injury and antioxidant protection. *J Am Soc Hortic Sci* 134: 466–471.

Gao H, Tao F, Song L, Chen H, Chen W, Zhou Y, Mao J, Zheng Y. 2009. Effects of short-term N2 treatment on quality and antioxidant ability of loquat fruit during cold storage. *J Sci Food Agric* 89: 1159–1163.

Gechev TS, Breusegem FV, Stone JM, Denev I, Laloi C. 2006. Reactive oxygen species as signals that modulate plant stress responses and programmed cell death. *BioEssays* 28: 1091–1101.

Gunes G, Liu RH, Watkins CB. 2002. Controlled-atmosphere effects on postharvest quality and antioxidant activity of cranberry fruits. *J Agric Food Chem* 50: 5932–5938.

Hernández I, Alegre L, Breusegem FV, Munné-Bosch S. 2009. How relevant are flavonoids as antioxidants in plants? *Trends Plant Sci* 14: 125–132.

Hodges DM, Lester GE, Munro KD, Toivonen PMA. 2004. Oxidative stress: importance for postharvest quality. *HortSci* 39: 924–929.

Huang R, Liu J, Lu Y, Xia R. 2008. Effect of salicylic acid on the antioxidant system in the pulp of "Cara cara" navel orange (*Citrus sinensis* L. Osbeck) at different storage temperatures. *Postharv Biol Technol* 47: 168–175.

Ioannidi E, Kalamaki MS, Engineer C, Pateraki I, Alexandrou D, Mellidou I, Giovannonni J, Kanellis AK. 2009. Expression profiling of ascorbic acid-related genes during tomato fruit development and ripening and in response to stress conditions. *J Exp Bot* 60: 663–678.

Jiang Y, Joyce DC, Terry LA. 2001. 1–Methylcyclopropene treatment affects strawberry fruit. *Postharv Biol Technol* 23: 227–232.

Jiménez A, Gómez JM, Navarro E, Sevilla F. 2002. Changes in the antioxidative systems in mitochondria during ripening of pepper fruit. *Plant Physiol Biochem* 40: 515–520.

Kasai S, Arakawa O. 2010. Antioxidant levels in watercore tissue in "Fuji" apples during storage. *Postharv Biol Technol* 55: 103–107.

Kim Y, Brecht JK, Talcott ST. 2007. Antioxidant phytochemical and fruit quality changes in mango (*Mangifera indica* L.) following hot water immersion and controlled atmosphere storage. *Food Chem* 105: 1327–1334.

Lamattina L, García-Mata C, Graziano M, Pagnussat G. 2003. Nitric oxide: the versatility of an extensive signal molecule. *Annu Rev Plant Biol* 54: 109–136.

Larrigaudière C, Lentheric I, Pinto E, Vendrell M. 2001a. Short-term effects of air and controlled atmosphere storage on antioxidant metabolism in "Conference" pears. *J Plant Physiol* 158: 1015–1022.

Larrigaudière C, Pintó E, Lentheric I, Vendrell M. 2001b. Involvement of oxidative processes in the development of core browning in controlled-atmosphere stored pears. *J Hortic Sci Biotech* 76: 157–162.

Larrigaudière C, Vilaplana R, Soria Y, Recasens I. 2004. Oxidative behaviour of "Blanquilla" pears treated with 1-methylcyclopropene during cold storage. *J Sci Food Agric* 84: 1871–1877.

Larrigaudière C, Ubach D, Soria Y, Recasens I, Chiriboga MA, Cascia G. 2008. Biochemical changes in 1–MCP treated skin tissue during cold storage and their relationship with physiological disorders. *Acta Hortic* 796: 119–124.

Łata B. 2008. Apple peel antioxidant status in relation to genotype, storage type and time. *Sci Hortic* 117: 45–52.

Łata B, Przeradzka M, Bińakowska M. 2005. Great differences in antioxidant properties exist between 56 apple cultivars and vegetation seasons. *J Agric Food Chem* 53: 7970–7978.

Lau OL. 1998. Effect of growing season, harvest maturity, waxing, low O_2 and elevated CO_2 on flesh browning disorders in "Braeburn" apples. *Postharv Biol Technol* 14: 131–141.

Lau OL, Barden CL, Blankenship SM, Chen PM, Curry EA, DeEll JR, Lehman-Salada L, Mitcham EJ, Prange RK, Watkins CB. 1998. A North American cooperative survey of "Starkrimson Delicious" apple responses to 0.7% O2 storage on superficial scald and other disorders. *Postharv Biol Technol* 13: 19–26.

Lentheric I, Pinto E, Vendrell M, Larrigaudiére C. 1999. Harvest date affects the antioxidant systems in pear fruits. *J Hortic Sci Biotech* 74: 791–795.

Lesham YY, Wills RBH. 1998. Harnessing senescence delaying gases nitric oxide and nitrous oxide: A novel approach to postharvest control of fresh horticultural produce. *Biologia Plantarum* 41: 1–10.

Lim CS, Kang SM, Cho JL, Gross KC. 2009. Antioxidizing enzyme activities in chilling-sensitive and chilling-tolerant pepper fruit as affected by stage of ripeness and storage temperature. *J Am Soc Hortic Sci* 134: 156–163.

Lin WC, Hall JW, Saltveit ME. 1993. Ripening stage affects the chilling sensitivity of greenhouse-grown peppers. *J Am Soc Hortic Sci* 118: 791–795.

Liyanage C, Luvisi DA, Adams DO. 1993. The glutathione content of grape berries is reduced by fumigation with methyl bromide or methyl iodide. *Am J Enol Vitic* 44: 8–12.

Lurie S. 2003. Antioxidants, In: Hodges DM ed. *Postharvest Oxidative Stress in Horticultural Crops*. Food Products Press, New York, USA, pp. 131–150.

MacLean DD, Murr DP, DeEll JR, Horvath CR. 2006. Postharvest variation in apple (*Malus domestica* Borkh.) flavonoids following harvest, storage, and 1-MCP treatment. *J Agric Food Chem* 54: 870–878.

Masia A. 1998. Superoxide dismutase and catalase activities in apple fruit during ripening and post-harvest and with special reference to ethylene. *Physiologia Plantarum* 104: 668–672.

Meir S, Bramlage WJ. 1988. Antioxidant activity in Cortland apple peel and susceptibility to superficial scald after storage. *J Am Soc Hortic Sci* 113: 412–418.

Molina-Delgado D, Larrigaudiere C, Recasens I. 2009. Antioxidant activity determines on-tree maturation in "Golden Smoothee" apples. *J Sci Food Agric* 89: 1207–1212.

Møller IM. 2001. Plant mitochondria and oxidative stress. Electron transport, NADPH turnover and metabolism of reactive oxygen species. *Annu Rev Plant Physiol Plant Mol Biol* 52: 561–591.

Navrot N, Rouhier N, Gelhaye E, Jacquot JP. 2007. Reactive oxygen species generation and antioxidant systems in plant mitochondria. *Physiologia Plantarum* 129: 185–195.

Niranjana P, Rao KPG, Rao DVS, Madhusudhan B. 2009. Effect of controlled atmosphere storage (CAS) on antioxidant enzymes and DPPH-radical scavenging activity of mango (*Mangifera indica* L.) cv. Alphonso. *African J Food Agric Nutr Dev* 9: 779–792.

Noctor G, Foyer CH. 1998. Ascorbate and glutathione: Keeping active oxygen under control. *Ann Rev Plant Physiol Plant Mol Biol* 49: 249–279.

Patterson ME, Workman M. 1962. The influence of oxygen and carbon dioxide on the development of apple scald. *Proc Am Soc Hortic Sci* 80: 130–136.

Purvis AC. 2004. Regulation of oxidative stress in horticultural crops. *HortSci* 39: 930–932.

Qiu S, Lu C, Li X, Toivonen PMA. 2009. Effect of 1-MCP on quality and antioxidant capacity of *in vitro* digests from "Sunrise" apples stored at different temperatures. *Food Res Int* 42: 337–342.

Rao MV, Watkins CB, Brown SK, Weeden NF. 1998. Active oxygen species metabolism in "White Angel" × "Rome Beauty" apple selections resistant and susceptible to superficial scald. *J Am Soc Hortic Sci* 123: 299–304.

Ryan FJ, Leesch JG, Palmquist DE, Aung LH. 2007. Glutathione concentration and phytotoxicity after fumigation of lemons with methyl iodide. *Postharv Biol Technol* 45: 141–146.

Sala JM. 1998. Involvement of oxidative stress in chilling injury in cold-stored mandarin fruits. *Postharv Biol Technol* 13: 255–261.

Sala JM, Lafuente MT. 2000. Catalase enzyme activity is related to tolerance of mandarin fruits to chilling. *Postharv Biol Technol* 20: 81–89.

Sala JM, Lafuente MT. 2004. Antioxidant enzymes activities and rindstaining in "Navelina" oranges as affected by storage relative humidity and ethylene conditioning. *Postharv Biol Technol* 31: 277–285.

Serrano M, Díaz-Mula HM, Zapata PJ, Castillo S, Guillén F, Martinez-Romero D, Valverde JM, Valero D. 2009. Maturity stage at harvest determines the fruit quality and anti-oxidant potential after storage of sweet cherry cultivars. *J Agric Food Chem* 57: 3240–3246.

Shaham Z, Lers A, Lurie S. 2003. Effect of heat or 1-methylcyclopropene on antioxidative enzyme activities and antioxidants in apples in relation to superficial scald development. *J Am Soc Hortic Sci* 128: 761–766.

Shi JX, Chen S, Gollop N, Goren R, Goldschmidt EE, Porat R. 2008. Effects of anaerobic stress on the proteome of citrus fruit. *Plant Sci* 175: 478–486.

Silva FJP, Gomes MH, Fidalgo F, Rodrigues JA, Almeida DPF. 2010. Antioxidant properties and fruit quality during long-term storage of "Rocha" pear: Effects of maturity and storage conditions. *J Food Qual* 33: 1–20.

Singh R, Dwivedi UN. 2008. Effect of ethrel and 1-methylcyclopropene (1-MCP) on anti-oxidants in mango (*Mangifera indica* var. Dashehari) during fruit ripening. *Food Chem* 111: 951–956.

Singh SP, Singh Z, Swinny EE. 2009. Postharvest nitric oxide fumigation delays fruit ripening and alleviates chilling injury during cold storage of Japanese plums (*Prunus salicina* Lindell). *Postharv Biol Technol* 53: 101–108.

Singh SP, Singh Z. 2012a. Postharvest oxidative behaviour of 1–methylcyclopropene–treated Japanese plums (*Prunus salicina* Lindell) during storage under controlled and modified atmospheres. *Postharv Biol Technol* 74: 26–35.

Singh SP, Singh Z. 2012b. Role of membrane lipid peroxidation, enzymatic and non-enzymatic antioxidative systems in the development of chilling injury in Japanese plums. *J Am Soc Hort Sci* 137: 473–481.

Singh SP, Singh Z, Swinny EE. 2012. Climacteric level during fruit ripening influences lipid peroxidation and enzymatic and non-enzymatic antioxidative systems in Japanese plums (*Prunus salicina* Lindell). *Postharv Biol Technol* 65: 22–32.

Singh SP, Singh Z. 2013a. Controlled and modified atmospheres influence chilling injury, fruit quality and antioxidative system of Japanese plums (*Prunus salicina* Lindell). *Int J Food Sci Technol* 48: 363–374.

Singh SP, Singh Z. 2013b. Dynamics of enzymatic and non-enzymatic antioxidants in Japanese plums during storage at safe and lethal temperatures. *LWT-Food Sci Technol* 50: 562–568.

Singh SP, Singh Z. 2013c. Postharvest cold storage-induced oxidative stress in Japanese plums in relation to harvest maturity. *Aus J Crop Sci* 7: 391–400.

Sivakumar D, Korsten L. 2010. Fruit quality and physiological responses of litchi cultivar McLean's Red to 1-methylcyclopropene pre-treatment and controlled atmosphere storage conditions. *LWT-Food Sci Technol* 43: 942–948.

Song L, Gao H, Chen H, Mao J, Zhou Y, Chen W, Jiang Y. 2009. Effects of short-term anoxic treatment on antioxidant ability and membrane integrity of postharvest kiwifruit during storage. *Food Chem* 114: 1216–1221.

Suzuki N, Mittler R. 2006. Reactive oxygen species and temperature stresses: A delicate balance between signaling and destruction. *Physiologia Plantarum* 126: 45–51.

Szalai G, Kellős T, Galiba G, Kocsy G. 2009. Glutathione as an antioxidant and regulatory molecule in plants under abiotic stress conditions. *J Plant Growth Reg* 28: 66–80.

Takahama U. 2004. Oxidation of vacuolar and apoplastic phenolic substrates by peroxidase: Physiological significance of the oxidation reactions. *Phytochem Rev* 3: 207–219.

Tausz M, Šircelj H, Grill D. 2004. The glutathione system as a stress marker in plant ecophysiology: Is a stress-response concept valid? *J Exp Bot* 55: 1955–1962.

Toivonen PMA. 2003a. Effects of storage conditions and postharvest procedures on oxidative stress in fruits and vegetables, In: Hodges DM ed. *Postharvest Oxidative Stress in Horticultural Crops.* Food Products Press, New York, USA, pp. 69–90.

Toivonen PMA. 2003b. Postharvest treatments to control oxidative stress in fruits and vegetables, In: Hodges DM ed. *Postharvest Oxidative Stress in Horticultural Crops*. Food Products Press, New York, USA, pp. 225–246.

Toivonen PMA. 2004. Postharvest storage procedures and oxidative stress. *HortSci* 39: 938–942.

Toivonen PMA, Wiersma PA, Gong Y, Lau OL. 2003. Levels of antioxidant enzymes and lipid soluble antioxidants are associated with susceptibility to internal browning in "Braeburn" apples. *Acta Hortic* 600: 57–61.

Veltman RH, Kho RM, van Schaik ACR, Sanders MG, Oosterhaven J. 2000. Ascorbic acid and tissue browning in pears (*Pyrus communis* L. cvs Rocha and Conference) under controlled atmosphere conditions. *Postharv Biol Technol* 19: 129–137.

Veltman RH, Sanders MG, Persijn ST, Peppelenbos HW, Oosterhaven J. 1999. Decreased ascorbic acid levels and brown core development in pears (*Pyrus communis* L. cv. Conference). *Physiologia Plantarum* 107: 39–45.

Vilaplana R, Valentines MC, Toivonen PMA, Larrigaudiére C. 2006. Antioxidant potential and peroxidative state of "Golden Smoothee" apples treated with 1-methylcyclopropene. *J Am Soc Hortic Sci* 131: 104–109.

Wang B, Wang J, Feng X, Lin L, Zhao Y, Jiang Y. 2009. Effects of 1-MCP and exogenous ethylene on fruit ripening and antioxidants in stored mango. *Plant Growth Regul* 57: 185–192.

Wang L, Chen S, Kong W, Li S, Archbold DD. 2006. Salicylic acid pretreatment alleviates chilling injury and affects the antioxidant system and heat shock proteins of peaches during cold storage. *Postharv Biol Technol* 41: 244–251.

Wang YS, Tian SP, Xu Y. 2005. Effects of high oxygen concentration on pro- and anti-oxidant enzymes in peach fruits during postharvest periods. *Food Chem* 91: 99–104.

Wang YS, Tian SP, Xu Y, Qin GZ, Yao H. 2004. Changes in the activities of pro- and anti-oxidant enzymes in peach fruit inoculated with *Cryptococcus laurentii* or *Penicillium expansum* at 0 or 20°C. *Postharv Biol Technol* 34: 21–28.

Watkins CB. 2006. The use of 1-methylcyclopropene (1-MCP) on fruits and vegetables. *Biotechnol Adv* 24: 389–409.

Watkins CB, Nock JF, Whitaker BD. 2000. Responses of early, mid and late season apple cultivars to postharvest application of 1–methylcyclopropene (1–MCP) under air and controlled atmosphere storage conditions. *Postharv Biol Technol* 19: 17–32.

Whitaker BD. 2004. Oxidative stress and superficial scald of apple fruit. *HortSci* 39: 933–937.

Whitaker BD, Nock JF, Watkins CB. 2000. Peel tissue α-farnesene and conjugated trienol concentrations during storage of "White Angel" × "Rome Beauty" hybrid apple selections susceptible and resistant to superficial scald. *Postharv Biol Technol* 20: 231–241.

Win TO, Srilaong V, Heyes J, Kyu KL, Kanlayanarat S. 2006. Effects of different concentrations of 1-MCP on the yellowing of West Indian lime (*Citrus aurantifolia*, Swingle) fruit. *Postharv Biol Technol* 42: 23–30.

Zerbini PE, Rizzolo R, Brambilla A, Grassi M. 2002. Loss of ascorbic acid during storage of "Conference" pears in relation to the appearance of brown heart. *J Sci Food Agric* 82: 1007–1013.

Zhao Z, Cao J, Jiang W, Gu Y, Zhao Y. 2009. Maturity-related chilling tolerance in mango fruit and the antioxidant capacity involved. *J Sci Food Agric* 89: 304–309.

Zheng X, Tian SP, Meng X, Li B. 2007. Physiological and biochemical responses in peach fruit to oxalic acid treatment during storage at room temperature. *Food Chem* 104: 156–162.

Zhu S, Liu M, Zhou J. 2006. Inhibition by nitric oxide of ethylene biosynthesis and lipoxygenase activity in peach fruit during storage. *Postharv Biol Technol* 42: 41–48.

Zhu S, Sun L, Liu M, Zhou J. 2008. Effect of nitric oxide on reactive oxygen species and antioxidant enzymes in kiwifruit during storage. *J Sci Food Agric* 88: 2324–2331.

12 Advances in Postharvest Maintenance of Flavor and Phytochemicals

Jun Song

CONTENTS

12.1 INTRODUCTION

Fruits and vegetables play an important role in the human diet. Consumption of fresh fruit is increasing, as consumers become more aware of the importance of food consumption in relation to human health and its role in disease prevention. In addition to their nutritional benefits, fruits are recognized for their delicious flavor, which is important for enjoyment of food. The quality of fresh fruit and vegetables comprises multiple aspects, including texture, appearance, color, flavor, and nutrition (Defilippi et al., 2004). Among the quality indices, flavor may be one of the most important traits; it is characteristically diverse, and brings enjoyment to the consumption of fruit and vegetables. The growing understanding of flavor as an important quality trait is reflected in the increased research efforts on the control and maintenance of the quality of flavor in many fruit and vegetables. Discussing the trends in research on flavor, Klee (2010) pointed out that most new technology developments have focused more on production, yield, decay resistance, and appearance, rather than on flavor (Klee, 2010). Improving the flavor properties of fresh fruit would add value, increase consumption, add health benefits by increasing consumption, and create new markets for these commodities. It is also needed to satisfy consumer demands (Kader, 2004; Morris and Sands, 2006). Meanwhile, fruits and vegetables offer a wide variety of bioactive compounds such as flavonoids, phenolics, anthocyanins, phenolic acids, stilbenes, and tannins, as well as nutritive compounds such as essential oils, carotenoids, vitamins, and minerals. Many of these compounds have potent antioxidant, anticancer, antimutagenic, antimicrobial, anti-inflammatory, and anti-neurodegenerative properties, both *in vitro* and *in vivo*. There has been a clear recognition for the need to shift from judging quality based on appearance, to judging quality based on flavor and nutritional value, in the postharvest life of fruits and vegetables (Kader, 2004; Klee, 2010). Therefore, it is worthwhile to look at these unique quality characteristics, and assess the literatures from a postharvest perspective, to reveal control and regulatory mechanisms that can be used to maximize the eating quality of fresh produce. As indicated in other sections of this book, the main focus will be on recently reported studies on flavor, and developed technologies and systems that will assist the horticulture industry to be more environmentally sustainable, and economically competitive in minimizing postharvest quality loss, and to generate products that are appealing and acceptable to consumers. It is also important to point out that many factors, from production to consumption, such as preharvest genetics, environment, culture practices, fertilizers, irrigation, and stress conditions, will influence postharvest flavor and phytochemical quality. It is not possible to cover each individual commodity as well as each reported study; therefore, only selected commodities will be discussed.

12.2 FLAVOR

12.2.1 Flavor and Flavor Perception

Flavor has been used as a term to describe the consumer's perception (eating experience) of a product. It integrates multiple indices, including taste (sweetness and acidity), and smell (aroma/volatiles). Sweetness and acidity are important flavor factors that have been long recognized. However, human perception of volatile compounds is much more complicated, and determined by at least three factors: (a) fruit volatile concentration; (b) aroma perception threshold; and (c) human aroma receptors. Great progress has been made to characterize the production of fruit volatiles, in the past twenty years. However, the science around flavor has been found to be more complicated than most people first perceived. On the chemistry side, in addition to identification of sugars and acids, over 300 chemical compounds have been reported and identified as aroma/volatiles compounds in many fruits and vegetables. These compounds include esters, acids, alcohols, aldehydes, ketones, and terpenes. The diverse composition of volatiles among commodities indicates the complex blend of flavor and complex pathways leading to volatile biosynthesis, and regulation in each type of fruit and vegetable. Despite the low concentration of these compounds present in nature, in comparison with sugars and acids, most of these compounds have very different aroma thresholds, and therefore, some may still be important contributors to the overall flavor perception, even at trace concentration. As a typical example, aroma compounds in tomatoes have been intensively studied, but among the hundreds of compounds identified, only 7 compounds are believed to be critical to aroma/flavor perception (Buttery et al., 1989). New research has demonstrated, however, that even the aroma threshold cannot solely be representative of human perception. Some of the most abundant volatiles do not contribute to consumer liking, whereas other, less abundant ones, do. For example, C_6 volatiles (hexanal, cis-3 hexenal, trans-2-hexenal, and hexyl alcohol) had been identified as among the impact volatile compounds with high abundance, contributing to the flavor of the tomato fruit, before it was proved that, even if the production of these compounds was reduced significantly, it has no significant impact on consumer liking or flavor perception (Tieman et al., 2012). New findings also point to the fact that aroma volatiles contribute to perceived sweetness, independent of sugar concentration, suggesting a novel way to increase perception of sweetness without adding sugar (Tieman et al., 2012). A positive correlation of overall flavor with total titratable acidity, and total soluble solids, also indicates that these two components play an important role in determining overall flavor. Subjectively measured traits, including fruity odor and fruity flavor, had positive correlations with overall flavor (Panthee et al., 2013). However, there was no correlation between fruit firmness and overall flavor. This indicated that sweetness can be a target for breeding programs aiming to improve the overall flavor of the tomato fruit. It implies that breeders can simply focus on selecting tomato accessions for a few traits, including sweet taste, juiciness, and juicy texture (Panthee et al., 2013).

For other fruit, depending on the flavor characteristic, there may be different targets. For example, it is well known that many fruits produce significant amount of esters, which have "fruity" and "sweet" flavor characteristics. Due to low aroma thresholds, these compounds contribute significantly to the overall "fruity" flavor of ripe and mature fruit. The divergent blends of these ester compounds result in very different flavor perceptions by consumers of these fruit. Butyl acetate, hexyl acetate, and 2-methylbutyl acetate are the most important volatiles present in ripe fruit—as found in "Golden Delicious" apples (Song and Bangerth, 1996). While, hexyl hexanoate and hexyl butanoate may present a "senescent" and "over-ripe" flavor to the same apples, but at the late stage of fruit-ripening and senescence (Song and Forney, 2008). In a study investigating the effect of modified atmosphere packaging (MAP) using ultra microperforated film (MP), on the sensory quality of apple slices, a significantly higher mean score for fruity aroma, for the MAP packages from the southern hemisphere, was directly attributable to the higher volatile concentrations for straight-chain esters (Isaacson et al. 2006), estragole, total-volatiles, other-volatiles, and terpene compounds (Cliff et al., 2010). Interestingly, there was higher perceived sweetness of the fruit in the MP packages, compared to the steady-state atmosphere packages. While this does not appear to be associated with changes in the soluble-solids content, it may be the result of the psychological interaction and perception of sweetness and fruitiness, as noted in the literature. In a model system with no sugar, samples are perceived sweeter when fruity odors/aromas are present. Such associations are attributed to cross-modality (taste–smell) interactions particular for congruent sensation. While the compounds involved do not physically or physiologically interact, there is a strong and very real learned association, particularly for repeated exposure, sensations typically occurring with regard to maturing fruit (Cliff et al., 2010). This research is the first of its type to document that higher volatile concentrations, associated with the MP films, translates to greater perceived fruitiness, and more flavorful apple slices, as perceived by sensory panelists.

These new findings indicate the complex interaction of human flavor perception and flavor chemistry. The application of aroma thresholds is dependent on different volatile compounds and individual fruit, and there is no universal fit using aroma thresholds to all cases of the flavor perceptions of consumers.

12.2.2 Biosynthesis and Regulations of Volatile Aroma Compounds

Despite intensive research for many years, biosynthetic pathways for most volatile compounds in fruits and vegetables have not been elucidated. Traditionally, volatiles can be classified as "primary" or "secondary," indicating whether they were present in intact fruit tissue, or produced as a result of tissue disruption (Dirinck et al., 1989). For many fruits and vegetables, such as apples, bananas, melons, and strawberry, flavor impact volatiles can be considered as primary compounds; however, the main flavor volatiles in tomatoes, cucumbers, and lemons are secondary volatiles. Therefore, aroma perception of fresh produce can vary, depending on how aroma is generated, and when the product is consumed. It is necessary, in every aroma study, to clearly state the procedure for sample preparation and assessment. Aroma produced from an intact fruit relates to the consumer's perception and judgment of the product for being "ripe," "unripe," or "off-flavor," while aroma volatiles produced after tissue disruption may reflect the consumer's perception of flavor during eating and chewing. It is also important to pay attention to the effects of tissue disruption on flavor generation. Due to enzyme activities, and nonenzyme chemical reactions, volatiles generated after tissue disruption may be unstable, and prone to change, and therefore, it is necessary to stabilize and standardize the volatile analysis procedure for volatile research. Methods to stabilize volatiles involve using buffers and salts, including calcium and sodium, which have been used on tomato samples (Buttery et al., 1989). Although the metabolic pathways for volatile biosynthesis in most fruit are not fully understood, it is known that more than 80% of the volatiles produced by ripe apples are esters (Dirinck et al., 1989). Therefore, production of these compounds has been the major focus of research in the past. The last enzyme that is responsible for ester formation is alcohol acyltransferase (AAT, EC 2.3.1.84), which combines alcohols and acyl CoAs to form esters. Other enzymes, such as lipoxygenase (LOX, EC 1.13.11.12), alcohol dehydrogenase (ADH, EC 1.1.1.1.), and pyruvate decarboxylase (PDC, EC 1.2.4.1), are also believed to be involved in the pathways to provide aldehydes and alcohols (Dixon and Hewett, 2000; Fellman et al., 2000). Using transgenic lines that block ethylene biosynthesis, AAT was found to be regulated by ethylene, while ADH and LOX were unaffected by ethylene modulation. Isoleucine, which is an important precursor for the branched esters of 2-methyl-related compounds, increases in the peel, and is also regulated by ethylene (Defilippi et al., 2004). Exogenous ethylene treatment of immature apple fruit induced fruit-ripening and volatile production, with preference to branched esters (Song, 1994).

It was reported that more than 88 acyltransferase were found in *Arabidopsis*, but a limited number of them have been characterized with a known biochemical function (St-Pierre and De Luca, 2000). Cloning of MpAAT has been conducted in "Royal Gala" apples. It was report that MpAAT is expressed in leaves, flowers, and fruit. The MpAAT gene product (protein) has the AAT function, and can use a wide range of substrates, including both straight chain (C_3–C_{10}) and branched chain alcohols. However, the binding of alcohol substrates is rate-limiting, compared with the binding of CoA substrates. The preference of MpAAT1 for alcohol substrates is dependent on substrate concentration, which determines the aroma profiles of apple fruit (Souleyre et al., 2005). Another AAT gene, MdAAT2, was cloned in "Golden Delicious" apples. It has low sequence identifiers, in comparison with other fruit AATs. In contrast to other apple varieties, the MdAAT2 of "Golden Delicious" was exclusively expressed in the fruit (Li et al., 2006). The MdAAT2 protein is about 47.9 kD, and is localized primarily in the fruit peel. Data also demonstrates that the expression of the MdAAT2 protein is regulated at the transcription level, in the fruit peel. MdAAT2 was inhibited by treatment with 1-methylcyclopropene (1-MCP), and influenced by ethylene (Li et al., 2006). The expression levels of both MdAAT1 and MdAAT2 increased with the progression of fruit-ripening, and correlated with the total amount of esters detected in "Golden Delicious" and "Granny Smith" apple cultivars (Zhu et al., 2008). In particular, MdAAT1 and MdAAT2 coincide with the genes described by Souleyre et al. (2005), and Li et al. (2006), respectively. As with AAT in melons and strawberries, it was also revealed that substrate availability is more important, than AAT activity, in determining ester formation in apples (Beekwilder et al., 2004). Another study conducted on banana and strawberry fruit found similar results—that the substrates determine the characteristic aroma profiles.

It has been reported that both the fatty acid and branched amino acids may function as the precursors of volatile formation (Rowan et al., 1999; Song and Bangerth, 2003). The straight-chain volatile, with C_2, C_4, and C_6 carbons, may be derived from fatty acids metabolism, to aldehyde, which is further converted to alcohols by the ADH. It is widely assumed that lipoxygenases may contribute to the breakdown of long-chain fatty acid to C_6 aldehydes, which are converted to alcohols by aldehyde dehydrogenase. Branched amino acids are important substrates for branched chain volatiles such as 2-methylbutylactate or ethyl-2-methyl-butanoate (Rowan et al., 1996).

The source of alcohols and aldehydes, leading to ester synthesis in apple fruit, is not fully understood. LOX catalyzes the first step in the metabolic pathway to convert the 18:2 and 18:3 fatty acids to C_6 volatiles, including *cis*-3-hexenal, hexanal, and their alcohols. Despite the possible role of LOX in tomatoes, it is interesting to note that there is no close correlation between LOX activity and ester formation in "Golden Delicious" apples at early and middle maturity harvests (Song, 1994). However, a better relationship between LOX and ester formation can be seen in late harvested fruit. Using multivariance analysis of the biosynthesis of volatile compounds, it was concluded that LOX and PDC are responsible for the differential production volatiles that occur in CA and regular air (RA) stored fruit, while no difference in AAT activity was found (Lara et al., 2006). Further research on fatty acid biosynthesis in apple fruit, with different harvest maturities, revealed that newly synthesized free fatty acids may be used as precursors for aroma biosynthesis (Song and Bangerth, 2003).

Strawberry fruit produce more than 300 volatile aroma compounds, the majority being esters, alcohols, aldehydes, terpenes, and acids. Total volatile production significantly increases as fruit-ripening advances from white to pink, to red. This increase of volatile compounds is closely related to total volatiles, esters, and acids, but negatively related to alcohols, which are predominant in white fruit. Using a quantitative proteomic approach, the increase of total volatile compounds and esters is well correlated with the protein profiles associated with aroma production in strawberry. It identified and quantified two AATs, two quinone oxidoreductases and *O*-methyltransferase (Li et al., 2013). The involvement of AAT in ester formation of many fruit is well known (Pérez et al., 2002). AAT is one of the most important enzymes to catalyze the formation of esters in strawberry fruit during ripening (Aharoni et al., 2000). At the proteomic level, two AAT proteins (AAT1 and AAT2) increased more between the pink-to-red stage, than from the white-to-pink stage, in both "Mira" and "Honeoye" strawberry fruits. This change confirmed the results at the transcript level reported by Aharoni, that AAT expression is induced in fruit-ripening (Aharoni et al., 2000). In addition, two pyruvate decarboxylases (PDC) in both "Honeoye" and "Mira," and acyl-CoA synthetase in "Mira," were also found to be significantly upregulated. These increases can be explained by the high demand for acyl-CoA for ester and fatty acid biosynthesis during ripening. Proteomic work provides evidence that PDC and AAT become more abundant concurrent with rapid volatile production, in ripening strawberry fruit.

The furanones—4-hydroxy-2, 5 dimethyl (2,3H) (DMHF, furaneol), and 4-methoxy-2, 5 dimethyl (2,3H) (DMMF, mesifurane)—have been recognized as key volatile compounds in strawberry, and a group of 3(2H)–furanone compounds have been identified in strawberry fruit (Ulrich et al., 1997; Bood and Zabetakis, 2002). Using HS-SPME, DMMF was identified in both "Mira" and "Honeoye" cultivars (Li et al., 2013). DMMF has a signature olfactory characteristic of "sweet" and "burnt sugar" notes, and its methylation is mediated by an *O*-methyltransferase (*FaOMT*), whose activity increases during fruit-ripening (Lavid et al., 2002). The formation of furanones is under the control of FaOMT, a gene that is responsible for the variation of the natural mesifurane content in strawberries (Wein et al., 2002; Zorrilla-Fontanesi et al., 2012). In both "Honeoye" and "Mira," it was found that there was a significant increase in OMF protein, as fruit ripened from white to red, and which coincided with the increase of DMMF concentration. In addition, two quinone oxidoreductases were identified, which both increased during ripening. The quinone oxidoreductase FaQR, has been shown to be strongly induced by ripening, and to be auxin-dependent (Raab et al., 2006). This enzyme functions as an enone oxidoreductase in the biosynthesis of DMMF, and is responsible for the conversion of DMHF to DMMF. In anti-ACO line kiwi fruit, it increased 5-fold in expression, and peaked at 168 h after ethylene treatment (Atkinson et al., 2011). These results reveal the important role of quinone oxidoreductase in volatile biosynthesis of strawberry during fruit-ripening.

In addition to enzymes and substrates, other factors may also be equally important. In early harvested immature fruit, fruit treated with 1-MCP, or fruit subjected to long-term low oxygen storage, metabolic energy, such as ATP, may be limited, and thus constrain volatile production. All of these conditions can reduce fruit respiration, as well as rate of metabolism (Rudell et al., 2002). Saquet et al. (2003) postulated that other cofactors, such as pyridine nucleotides (NADH and NADPH), are not reduced in CA-stored fruit. Low rates of respiration decrease ATP as well as the ATP:ADP ratio, which directly reduces fatty acid biosynthesis (Saquet et al., 2003). Further study is needed to elucidate the control mechanisms of volatile biosynthesis during fruit-ripening and senescence.

The branched amino acids leucine and isoleucine, are important substrates for branched chain volatiles, such as 2-methylbutyl acetate, and ethyl-2-methylbutanoate in apple, or 3-methylbutyl acetate in banana (Tressl and Drawert, 1973; Fellman et al., 2000). Feeding studies show that this pathway may be present in many fruits, such as apple, banana, and strawberry (Tressl and Drawert, 1973; Rowan et al., 1996; Fellman et al., 2000; Perez et al., 2002). In melons, both fatty acid and branched amino acid biosynthesis are important contributors to volatile formation. The amino acids alanine, valine, leucine, iso-leucine, and methionine, increased during fruit-ripening, in close association with volatile production, and are believed to supply the carbon chains for four groups of esters, ethyl acetate, 2-methylpropyl, 2-methylbutyl, and thioether ester, respectively (Wyllie et al., 1995; Wang et al., 1996). When a comparison of amino acid content was made between the highly aromatic melon "Makdimon" and the low-aroma melon "Alice," no significant difference in amino acids concentration was found. Therefore, the difference in volatile concentrations in melon is not due to the availability of amino acid substrates, but rather is dependent on other biosynthetic pathways (Wyllie et al., 1995).

Using tomatoes as another example, the analysis of "secondary" volatile compounds become a more routine procedure (Baldwin et al., 2007). A recent report indicated that several genes related to volatile biosynthesis in tomatoes have been identified through the candidate gene approach, or screening for genes encoding enzymes that might function in a given synthetic pathway (Klee and Tieman, 2013). However, the whole production process from selection of genetic materials, preharvest practice, to postharvest techniques, will influence flavor and postharvest quality.

Molecular biology tools need to be applied to better understand the biosynthesis of aroma volatiles in fruit. Genomic tools, which have been developed on model systems, could be applied to fruit and vegetables. Although few genomes of fruits and vegetables have been completely sequenced, many genomic tools that are available today for *Arabidopsis* (Arabidopsis Information Resource), *Solanaceae* (Solanaceae Genomic Network), and *Rosaceae* (Rosaceae Genome Database), are paving the way for future research and development in genomics and proteomics. Apple has more than 60,000 protein sequences reported in NCBI (http://www.ncbi.nlm.nih. gov/protein, accessed on December 31, 2014). A high-quality draft of the complete genome of pear (*Pyrus × bretschneideri*) reports 42,812 protein coding genes, with ca. 28% encoding multiple isomers (Wu et al., 2013); there were 46,518 protein sequences reported for public use on (http://www.ncbi.nlm.nih.gov/protein, accessed on December 31, 2014). Undoubtedly, the ultimate success of proteomic research is dependent on the completion of genome sequences of fruits and vegetables, which will provide complete information about proteins, and their modifications, that can be identified in biological studies.

Applying an expressed sequence tag (EST), frequency analysis with EST database revealed that EST clusters from fruit-derived tissue show strong sequence homology, with biochemically characterized enzymes, that are involved in ester biosynthesis, including acyl-CoA dehydrogenase, acyl-CoA oxidase, enoyl-CoA hydratase, acyl carrier proteins, malonyl-CoA:ACP transacylase, LOXs, 3-ketoacyl-CoA thiolase, acyl-CoA synthetase, and acyl carrier proteins (Park et al., 2006). Two highly divergent ADH genes (CmADH1 and CmADH2) are expressed in ripening melon fruit, and have been cloned in melon (*Cucumis melo* var. *Cantalipensis*). These enzymes are closely related to fruit ethylene production, and are inhibited by 1-MCP, indicating that they are under control of ethylene. Sequence analysis indicated that CmADH1 has 83% homology with apple Md-ADH (Marinquez et al., 2006). These findings add new clues for enzymes responsible for aroma volatile biosynthesis beyond AAT, LOX, and fatty acid ß-oxidation, and may open new opportunities for aroma volatile research to identify unknown pathways related to fruit volatile biosynthesis.

12.2.3 ADVANCES IN GENETIC INFORMATION ON FLAVOR COMPOUNDS IN FRUIT

In a review paper, Klee and Tieman (2013) discussed these two specific challenges for flavor improvement from a breeder's perspective:

- What are the important chemicals that contribute to consumer preference, either positive or negative?
- What genes control the synthesis of these chemicals, and what are the possible alleles of the most important genes?

The answers to these questions will lead to the introduction of a series of genes into the genome (Klee and Tieman, 2013). With the great progress in genome sequencing of many fruit and vegetable species, full genomic information has been made available, not only to plant breeders, but also to chemists, biochemists, and physiologists, to aid in the improvement of flavor quality. Genetic mapping of quantitative trait loci (QTLs) involves identifying and determining the degree of association between certain traits and a set of genetic markers. A saturated genetic map, covering the entire genome, is essential for accurate QTL identification. The identification of QTLs linked to important traits in apple, such as disease resistance, tree growth, and fruit quality, is still at an initial stage. A quantitative genetic analysis of traits associated with apple fruit-flesh firmness, using a population derived from a "Prima" × "Fiesta" cross, was conducted (King et al., 2000). QTLs accounting for differing degrees of variation for firmness, stiffness, and a number of sensory attributes, were identified on seven linkage groups (LG), with large effects on LG1, LG10, and LG16. Further work extended the range of mechanical measurements to include compression, and wedge fracture tests (King et al., 2000). The wedge fracture tests identified significant QTLs on LG16 and LG1. The QTL on LG16 was located in the same region as the QTL identified for certain sensory textural attributes, such as crispness and juiciness.

Employing linkage mapping, and map calculation using a subset of 57 out of the 86 individuals of the cross between "Fiesta" and "Discovery," volatile organic compounds from ripe apple fruit were analyzed with proton transfer reaction–mass spectrometry (PTR–MS). Abundance of mass 61, which is a common fragment of acetate esters, and the observed ratio between mass 43 and 61, are in excellent agreement with that observed for acetate esters, indicating a possible link between ester production and QTL (Zini et al., 2005). During further study on the same populations, a set of QTL associated with major volatile compounds was identified, and confirmed across three different locations. The volatile organic compound (VOC) groups showing higher values of heritability among locations were related to esters, which is one of the most important compound classes for fruit aroma. In this specific case, several masses enabled the detection of QTL, located in two regions of chromosome 2, while another volatile peak, assigned to ethylene, was mapped on chromosome 15 (Costa et al., 2013). Using the assembled "Golden Delicious" genome database, 17 apple AAT members were targeted and annotated. When the MdAAT1 primers were used, a 468 bp nucleotide sequence was obtained for sequence comparisons among the 102 apple cultivars (Dunemann et al., 2012). High sequence similarity was observed among the 102 cultivars, with only four SNPs observed at nt positions 62, 110, 425, and 459 bp, according to the position in the reference gene sequence MDP0000637737 of "Golden Delicious." These SNPs caused changes in the amino acid sequence at positions 258 (I–T), 274 (C–Y), and 379 (V–A). The association study revealed significant associations of individual SNPs and specific haplotypes, with the trait "ester concentration." These remarkable phenotypic differences observed between apple cultivars were clearly caused by the genotype, with minimal influence of year of harvest, suggesting apple aroma as a highly heritable trait (Dunemann et al., 2012). Even if consumer preference for the flavor of apple varieties is highly subjective, ester-accentuated flavor types are especially successful on markets worldwide. Therefore, the ester content of apple fruit is an important goal in an apple cultivar breeding program. The use of marker-assisted selection using functional markers for fruit flavor traits, including ester content, is assumed to shorten the long traditional breeding activities.

Genetic investigation of LOX genes detected a total of 15 QTLs for eight volatiles (esters and C_6 aldehyde, hexanal) in apples, which were located on chromosomes 2, 7, 9, and 12. At least four genome regions were identified to be associated with LOX candidate genes by QTL mapping. The QTLs associated with the MdLOX5 gene cluster on chromosomes 2 and 7, might be explained by the function of type 2 (13-LOX), while the association of QTL with the MdLOX7 cluster on chromosome 12, is more complex, and needs more functional study at the protein level (Vogt et al., 2013). Unlike AAT, the genetic information with LOX may not link directly to the esters production, since the direct substrates are not directly controlled by the LOX. Ester formation may be regulated at the subsequent enzymatic levels (HPL and ADH), and the availability of fatty acids may affect the production of LOX-derived volatiles, masking other influence enzymes (Vogt et al., 2013).

Sweetness and acidity play an important role in fruit flavor. Fruit acidity is due to the presence of organic acids, and malic and citric acids are the main acids found in most ripe fruit (Seymour et al., 1993). A recent study, conducted using two half-sib populations GMAL 4595 [Royal Gala × PI (Plant Introduction) 613988] and GMAL 4590 of 438 trees, demonstrated that the Ma locus, which is controlling the malic acid content in apples, is the primary genetic factor determining fruit titratable acidity, and/or pH in both "Royal Gala" and the two *Malus sieversii* accessions PI 613988 and PI 613971 (Xu et al., 2012). In the "Golden Delicious" genome, the homologous Ma region, defined by the five flanking markers, is no larger than 150 kb, and contained 44 predicted genes. In addition, there are two minor QTL detected for fruit TA and pH with M2 specific to Royal Gala, and M3 to PI 613988. With the QTL and markers identified/developed, it might be possible to screen apple breeding populations at the seedling stage, to remove or select most of the targeted genotypes (Xu et al., 2012). The eight new simple sequence repeat (SSR) markers developed would be useful in marker-assisted breeding in apple. Construction of the fine map of the Ma locus represents an important step forward in isolating the Ma gene. Those 44 predicted genes are largely hypothetical, and quite diverse, including 19 of hypothetical proteins. Further study is required to confirm and reveal the identity of the Ma gene (Xu et al., 2012).

12.2.4 Postharvest Treatments' Influence on Flavor Quality

The effect of harvest maturity and postharvest treatments on volatile biosynthesis and flavor in many fruits and vegetables, has been well documented (Mehinagic et al., 2006; Song and Forney, 2008; Bennett, 2012).

When fresh-cut cantaloupe cubes were treated with 1.0 µL/L of 1-MCP for 24 h at 5°C, and stored in air at 5°C for nine days, most quality attributes were unaffected by the treatment with 1-MCP (Amaro et al., 2013). It preserved soluble solids, total phenolics, total carotenoids, and β-carotene content, but significant softening occurred. 1-MCP-treated fresh-cut cantaloupe accumulated higher levels of propyl acetate, 2-methylbutyl acetate, methyl butanoate, methyl 2-methyl butanoate, methyl hexanoate, 2-methylbutyl alcohol, and phenethyl alcohol, and lower levels of benzyl alcohol and heptanal, than untreated controls, particularly those derived from the amino acids isoleucine and phenylalanine, but had no significant effect on other phytochemicals or quality attributes.

12.3 BIOACTIVE COMPOUNDS AND PHYTOCHEMICALS

It is well known that fruits and vegetables are major sources of nutrients and phytochemicals for the human diet. Increasingly more epidemiological evidence has become available for possible effects of specific compound(s) on human health and disease prevention. This chapter will emphasize recent research on phytochemicals and bioactive compounds in fruits and vegetables, with specific interest on postharvest aspects. It is not intended to be a comprehensive coverage of all the research on bioactive compounds. The main goal of this section is to provide updated information on the effects of various factors such as genetics, environment, and postharvest storage, on the bioactive compounds and phytochemicals. It is expected that new knowledge of these factors can lead to multidisciplinary strategies to maximize the bioavailability and health potential of fruit and vegetables, during postharvest handling.

For better understanding of the bioactive compounds and phytochemicals in fruits and vegetables, it is necessary to outline the major groups and categories. In general, bioactive compounds and phytochemicals can be divided into three categories: polyphenols, glucosinolates, and carotenoids. Under the polyphenols category, further subgroups can be seen as flavonoids, phenolic acids, stilbenes, and lignans. The well-known flavonoids groups in many fruits can further divided into isoflavones, flavones, flavonols, flavanols, anthocyanins, and flavanones. The abundance of each group of compounds differs significantly in different fruits and vegetables. For example, berry fruit contain significant amounts of flavonoids, which can be seen as distinct blue or red in color. High bush blueberry (*Vaccinium corymbosum*) and strawberry fruit contain large amounts of total phenolics (251–310 mg/100 g and 222–225 mg/100 g, gallic acid equivalent, respectively), and total anthocyanins (92–129 mg/100 g, cyanidin 3-glucoside FW and 35.6 mg/100 g, pelargonidin 3-glucoside FW, respectively) (Aaby et al., 2007). The distribution of phenolic compounds in tissues varies greatly in different parts of fruits and vegetables. Strawberry fruit achenes contain about 14% higher levels of phenolics and 12% higher levels of anthocyanin, when compared to the strawberry flesh (Aaby et al., 2005). The skin of apples has a concentration of 5-caffeoylquinic acid and catechins higher than in the cortex parenchyma. The levels of hydroxycinnamic acid and catechin in apples and pears generally increase for a short time during initial development, and are significantly high in the raw fruits (Boyer and Liu, 2004). In mature cherries, the phenolic acids, such as derivatives of synaptic acid, are often localized in the epicarp. In grape skin, the concentration of hydroxylcinnamoyl tartaric acid is several-fold greater than in the pulp (Li et al., 2012).

Antioxidant capacity has been widely used to describe the biochemical and physical characteristics of the bioactive and phytochemical compounds in plants and food. Several methods have been developed through the years, and widely applied to monitor the antioxidant capacity, including: ORAC (oxygen radical absorbance capacity); TRAP (total radical-reducing antioxidant potential); FRAP (ferric-reducing antioxidant potential); and TEAC (Trolox equivalent antioxidant capacity). The advantages and disadvantages of these methods have been reviewed (Prior et al., 2005). CAA (celluare antioxidant activity) and DPPH assays have also been reported and applied to evaluate the antioxidant capacity of fruits and vegetables (Wolfe et al., 2008). The ORAC assay has been considered by the U.S. Department of Agriculture (USDA) as the official method for antioxidant-capacity analysis, and tables of ORAC values for phytochemicals, foods, and single antioxidants, are available on the USDA website (http://www. usda.gov/wps/portal/usdahome).

A high positive correlation was found between total phenolic and flavonoid content, implying that flavonoids constitute an important group of phenolic compounds in peaches and nectarines (Cantín et al., 2009). Moreover, a linear positive relationship was observed between antioxidant capacity using the DPPH assay, and total phenolics for the flesh of the peach and nectarine genotypes, and has also been observed for apricots and plums.

Antioxidant properties of fruits and vegetables mainly contributed to their polyphenols and vitamin content. A study performed in our laboratory found significant differences in flavonoids, such as cyanidin-3-glucoside and pelargonidin-3-glucoside, and the antioxidant capacities of strawberry fruit, at different development stages. No obvious correlation between total phenolic compounds and ORAC and FRAP were found in strawberries (Li et al., 2013). A study performed on apples and pears found significant differences in antioxidant capacities in the presence or absence of the peel. It appears that on the basis of weight, the peel contribution to phenolic content and the ORAC value is several times higher than that of the pulp.

12.3.1 GENETIC FACTORS AFFECTING BIOACTIVE COMPOUNDS AND PHYTOCHEMICALS

Production and maintenance of bioactive compounds and phytochemicals in fruits and vegetables are affected by any pre- and postharvest factors, including agriculture practices, environmental factors, genetics, harvest maturity, postharvest storage, and processing. Among them, genetics may be the primary factor.

Significant variation in the level of anthocyanins content between five commercial strawberry cultivars and three breeding lines were found (Fredericks et al., 2013). One breeding line (BL 2006-221) was an exceptional source of anthocyanins (~1 g/kg fresh weight), with approximately double the level of current commercial cultivars. Hue angle and anthocyanin concentration also showed a good correlation. This indicates that a great divergence of anthocyanin content exists in strawberry fruit.

By combining the metabolomic data (LC–MS), analysis with genetic linkage maps, 488 mQTLs (metabolite quantitative trait loci) were detected in peel, and 254 mQTLs in apple flesh, using the software MetaNetwork (Khan et al., 2012). Despite the fact that half of the metabolites did not have mQTL detected, procyanidins, phenolic esters, (+)-catechin, (–)-epicatechin, and kaempferol hexose rhamnose, showed similar segregation patterns, apparently being controlled by the same dominant and recessive alleles of LG16 from the cross of "Prima" and "Fiesta." Structural genes involved in the phenylpropanoid biosynthetic pathway were located using the apple genome sequence. The structural gene leucoanthocyanidin reductase was in the mQTL hotspot on LG16, with seven transcription factor genes (Khan et al., 2012).

A single QTL responsible for up to 62% of the variation in the anthocyanin content was mapped on a Syrah X Grenache F1 pseudo-testcross in grapes (Fournier-Level et al., 2009). Among the 68 unigenes identified in the grape genome within the QTL interval, a cluster of four Myb-type genes was selected on the basis of physiological evidence (VvMybA1, VvMybA2, VvMybA3, and VvMybA4). From a core collection of natural resources, 32 polymorphisms revealed significant association, and extended linkage disequilibrium was observed. Using a multivariate regression method, it was demonstrated that five polymorphisms in VvMybA genes, except VvMybA4, accounted for 84% of the observed variation. All these polymorphisms led to either structural changes in the MYB proteins, or differences in the VvMybAs promoters. It was concluded that the continuous variation in anthocyanin content in grape was explained mainly by a single gene cluster of three VvMybA genes. Recent studies on 20 tomato cultivars and breeding lines indicated that the levels of phytochemicals (carotenoids and phenolic compounds), and their antioxidant activity, were significantly dependent on the genetic background (Li et al., 2011). Similar results were reported, which showed that the content and activity of total phenolics, anthocyanins, flavonoids, and vitamin C, were significantly different in genotypes from peach and nectarine breeding progenies. Recently, an integrated fruit quality gene map of *Prunus*, containing 133 genes, putatively involved in the determination of fruit texture, pigmentation, flavor, and chilling injury resistance, was presented (Martínez-García et al., 2013).

These results confirm the importance of genotype on the availability of bioactive compounds and the antioxidant capacity of peach and nectarine fruits and, consequently, on their benefits to health. Therefore, the peach cultivars used as progenitors in the crosses of a breeding program have a vital importance to release new cultivars with high bioactive compounds content. On the other hand, the high number of evaluated genotypes, from different genetic origins, and with a large phenotypic variability, constitutes a considerable contribution to peach species, and especially for breeding purposes.

12.3.2   POSTHARVEST STORAGE EFFECTS ON BIOACTIVE COMPOUNDS AND PHYTOCHEMICALS

Changes in both the quality and phytochemical composition of plants can occur rapidly, depending on postharvest handling, such as storage and processing conditions. Such changes may not always result in reduction of the health-promoting compounds. The concentration of phytochemicals and antioxidant activity in some fruits and vegetables were actually enhanced by postharvest storage and processing parameters (Bengtsson, 2010). The two major chemical changes causing deterioration are lipid oxidation, and nonenzymatic browning, during storage and food processing, which can lead to altered color and flavor. Lipid oxidation is influenced by light, oxygen, temperature, and water activity, and the presence of catalysts, such as transition metals iron and copper. Nonenzymatic browning can occur easily during the storage of dried and concentrated foods. Different phytochemicals are affected by these factors differently. Carotenoids are very sensitive to heat, and can incur significant losses during different vegetable processing steps (Tiwari and Cummins, 2013). The main cause of carotenoid degradation in foods is oxidation.

Significant differences in anthocyanin concentration, as a result of fruit-ripening, was shown in two strawberry cultivars ("Mira" and "Honeoye"), when expressed as total anthocyanins content of total 520 nm area count, equivalent to either cyaniding 3-*O*-glucoside or pelargodin-3-glucoside. Anthocyanin content was significantly higher in red fruit of both cultivars, being 22.9 and 17.07 mg/100 g FW of Cy-3-glc equivalents, or 58.93 and 43.9 mg/100 g FW Pel-3-glc, respectively. In red ripe fruit, "Honeoye" showed higher anthocyanin content than "Mira." Meanwhile, a significant decrease in total flavonoid compounds during ripening was observed, while no significant decrease in total phenolic compounds was found (Li et al., 2013). A quadratic relationship in the antioxidant capacity, measured as ORAC, was found in "Mira" at three ripeness stages, while a significant decrease of 15% of the antioxidant capacity, measured as FRAP, was found in both cultivars, with advanced ripeness at the red—as compared with the white—stage.

Flavonoids and other phenolic compounds are relatively stable at high temperature, and over long storage. Phenolics in plants exist in both free and conjugated forms. Postharvest loss of phenolics is mainly due to enzymatic oxidation by polyphenol oxidase and peroxidases (Wang et al., 2007). Degradation of anthocyanins is pH dependent. Percentage degradation of total anthocyanins, and total antioxidant activity in unblanched blueberry juice (conventional and organic), after five months of storage at 23°C, was reported to be 72%–79% and 21%–43%, respectively (Syamaladevi et al., 2012). MCP treatment was found to maintain the ascorbic acid, carotenoid, total phenolic, and flavonoid content of mango fruit (Sivakumar et al., 2011). Apples treated with 1-MCP had no significant change in overall phenolic content, and antioxidant activity, although the effects on individual phenolic compounds, and on polyphenols, in different parts, varied. Similar results were found in cherry fruit, where cherries treated with 1-MCP, hexanal, or a combination of both, showed enhanced fruit quality and extended shelf-life, but no significant effect on polyphenols, such as anthocyanins and phenolic acids (Sharma et al., 2010). As briefly discussed above, although the effect of postharvest storage of fresh produce on phytochemicals is multifaceted, optimizing the various parameters can lead to good retention of nutritionally important food bioactives, such as those with strong antioxidant activities. Studies have also shown that the phytochemical content of a particular cultivar can vary significantly, due to other factors.

Controlled atmosphere (CA) and MAP have been widely applied on many fresh fruits and vegetables, as they are very effective in maintaining the quality, and extending the marketability of fresh produce. Many studies have also shown that CA or MAP technologies offer the possibility to retard the respiration rate, maintain bioactive compounds, and extend the shelf-life of fruits and vegetables, as compared with conventionally stored or packaged samples. Overall, total phenolic, flavonoid, and anthocyanin concentrations, as well as antioxidant activity, were relatively stable during air and CA storage. In air-stored fruit, total phenolic concentrations were higher in the peel of 1-MCP treated fruit, than in the control fruit, but slightly lower in the flesh of 1-MCP-treated fruit. In CA-stored fruit, interactions between O_2 partial pressures, temperature, and storage duration, were detected, but overall, few consistent trends were observed. However, flavonoid concentrations were higher in the flesh of 1-MCP-treated than untreated fruit kept in 2 kPa O_2 while anthocyanin concentrations, only measured in the peel, were not affected by 1-MCP treatment. There were no correlations found between total phenolics and antioxidant activity. Ascorbic acid concentrations declined in both peel and flesh tissues of untreated and 1-MCP-treated fruit stored in air, while changes of ascorbic acid concentrations in CA-stored fruit were inconsistent (Fawbush et al., 2009). For better phytochemical retention, shelf-life and phytochemicals of broccoli florets seemed to be maintained with polypropylene microperforated film and refrigerated conditions (Nath et al., 2011). Several families of phytochemicals, such as phenolic acids, isoflavones, flavones, flavonols, and glucosinolates, were determined in both fresh and fresh-cut samples, including tomato, carrot, grape, eggplant, and broccoli (Alarcón-Flores et al., 2014). Both samples of produce have potential and similar beneficial properties, regarding their content of phytochemicals, except tomato, which should be consumed as fresh. Processes such as slicing, grating, and dicing, as well as storing conditions (temperature and light), were observed to impact the content in eggplant; the content of phenolic acids is statistically different, depending on the presentation. The content of phytochemicals was higher when fresh-cut carrots were stored at 4°C, regardless of the presence or absence of light.

12.4   FUTURE RESEARCH PERSPECTIVES

From the postharvest perspective, research plays a critical role in both the frontiers of flavor and phytochemicals. Further research is needed on understanding how the agriculture production system affects flavor and phytochemicals, and on aligning research between nutrition and agricultural production. Research efforts on quality evaluation need to shift from yield and quantity to eating-quality factors (such as taste, flavor, and nutritional value), which have positive effects on fruit and vegetable consumption. More research and evidence is also needed on the content and bioavailability of bioactive compounds.

Production of flavor compounds and bioactive compounds is the result of complex metabolic networks involving many pathways and control mechanisms. It is necessary to remember that volatile biosynthesis pathways and bioactive compounds are only a part of the complex network of fruit metabolism during ripening, and is influenced by many factors, such as genetics, production practices, and postharvest handling. The metabolomic diversity is also caused by low enzyme specificity, and is directly related to the availability of substrates. Future work must, therefore, take into account the entire metabolic pathway, rather than a single enzyme or section of the pathway. Flavor/aroma is one of the most important quality indices of fruits and vegetables. It contributes not only to the attractiveness of food, but may also be associated with nutritional quality, due to many commonly shared precursors in secondary metabolism (Klee, 2010). This implies that new research needs to address the whole network of biology, rather than individual—or a group of—metabolites. This system approach employs state-of-the-art metabolomics, genomic, and proteomic tools, to study fundamental metabolism, and its regulation and localization. Combining results of genetic, chemical and sensory properties, will lead to a better understanding of how to optimize and retain fruit flavor quality in the marketplace, for the benefit of both consumers and the fruit industry.

Volatile biosynthesis in both climacteric and nonclimacteric fruit is highly integrated with fruit-ripening and senescence. Therefore, a better understanding of fruit-ripening and its triggers will help in our understanding of fruit volatile production (Alexander and Grierson, 2002). Linkages of flavor biosynthesis with climacteric respiration, endogenous ethylene content, ethylene biosynthesis, and its response to the inhibitors AVG and 1-MCP, will provide new insights into control mechanisms. Using AAT, LOX, ADH, PDC, and other enzymes, as examples, it has been demonstrated how these enzymes influence fruit volatile biosynthesis in fruit. While precursor studies indicate that AAT is not the limiting factor in volatile production, it may also imply that both fatty acids and branched amino acid substrates were physically unavailable to the enzyme. Due to overwhelming evidence that substrates may be the bottleneck of volatile production in most fruit, it becomes very important to clarify the compartmentalization of volatile biosynthesis within the fruit cell, in order to understand the biosynthesis of those substrates or precursors, and their compartmentalization and transport in the cell. Information on aroma threshold levels sheds light on the nature of chemicals that may contribute to human perception of aroma volatiles.

While precursor studies of early harvested or 1-MCP-treated fruit indicate that AAT is not the limiting factor of volatile production, it may also imply that the substrates were physically unavailable to the enzymes. Understanding where substrates are produced, and where enzymes are localized, will improve our understanding of the volatile biosynthesis system. Immunolocalization, immunoblot, and fluorescence imaging techniques, could help to determine protein expression, localization, activity state, and cell compartmentation (Giepmans et al., 2006). Information on aroma thresholds will continue to aid in the identification of compounds that contribute to human perception of fruit flavor. Little information is available about the sensory contribution of many volatiles. Of particular interest is an objective olfactory description of fruit "freshness," "ripeness," "off-flavor" and "over-ripeness." Combining sensory and instrumental analysis should refine our understanding of fruit-ripening and consumer preferences, and help to develop production and postharvest handling technologies to optimize fruit flavor quality (Song and Forney, 2008).

Similar to flavor, bioactive compounds and phytochemicals in fruits and vegetables can provide health benefits beyond their basic nutritional values. Phytochemicals such as carotenoids, phenolics, and glucosinolates are among the most important food bioactives that have positive impact on human health. In addition to the unique phytochemical profile of different plants, the amount and composition of a particular plant or food can also be changed by the various conditions during growth, postharvest storage, and processing. Fully understanding the roles of these factors will help the development of strategies to preserve the bioactive components, and maximize their bioavailability, and ultimately, their potential health benefit. This is an undertaking that requires a multidisciplinary approach, and support from the agricultural and agrifood sectors.

Genetic markers can assist fruit breeders to assess genetic diversity of the germplasm, determine heritability, and predict cultivars. A major advantage of using markers in fruit breeding is the improved efficiency of enabling early selection for any quality traits, simultaneous selection for multiple traits, including resistance gene pyramiding, and selection for traits that are expensive to screen by phenotype. According to published data, markers have been increasingly used for selection in apple breeding. To increase selection efficiency, and to reduce MAS cost in apple breeding, knowledge of the most useful phenotypic characters is essential. The latest advances in apple genetics offer unprecedented opportunities for cultivar improvements. Future apple breeding programs should consider the creation of new cultivars combining fruit quality features such as texture, storability, well-balanced acid/sugar ratio, and desirable aroma (Ulrich and Dunemann, 2012).

Improvement of the flavor and phytochemistry of fruits and vegetables will only be successful through multidisciplinary research. Metabolomics, genomic, proteomics, and integrative multiomics techniques, will provide the essential information and knowledge on the production and regulation of flavor and phytochemical compounds in fruits and vegetables, but also reveal biological impact on animal and human nutrition. Several components of fruits and vegetables have been isolated, structures elucidated and tested, singularly or in mixtures, to determine their effect on cells and animals.

Metabolomic analysis utilize GC, LC-MS, LC-MS (lipid), and nuclear magnetic resonance (NMR) spectroscopy, for structure elucidation and multivariate, a statistical analysis to identify important relationship among biological conditions. Metabolomics is particularly important in the plant field, because plants produce a huge diversity of metabolites—far more than are produced by animals and microorganisms (Saito and Matsuda, 2010). The manipulation of plants and the search for new plants endowed with high antioxidant potential are conducted in tandem with metabolomic analysis, which is integrated with nutrigenetics and nutrigenomics. These disciplines, which have evolved rapidly in recent years, have increased our knowledge of the interactions between life processes and specific components of our diet. The co-occurrence principle of transcripts and metabolites, particularly transcriptome coexpression network analysis, is powerful for decoding functions of genes, not only in a model plant such as *Arabidopsis* but also in crops and medicinal plants. mQTL analysis, along with scoring of gene expression and agronomical traits, is beneficial for crop-breeding. Although comprehensive coverage of metabolomic analysis is achieved not by a single analytical technology but by multiparallel complementary technologies, it increases the challenge level of the annotation rate of unknown signals.

Proteomics is the study of "the entire protein complement expressed by a genome in a cell or tissue type," and is a major research tool in the postgenomics era (Pandey and Mann, 2004). It provides an essential link between the transcriptome and metabolome, and is becoming an important research platform, along with genomics, as the latter is greatly enhanced through identification of corresponding specific proteins derived from transcripts. The development of proteomics originated from "genomics"; however, it has developed to cover all aspects of protein research in a cell, not only the abundance and changes in time and in association with biological behaviors, but also post-translational modifications (PTMs) related to the biological functions, and protein–protein interactions. Recent and continuing development in proteomic technologies has demonstrated that proteomics is the crucial element of the "omics" approach, and contributes greatly to a new understanding of biological systems, especially for fruit and vegetables (Palma et al., 2011). Systems biology, data-driven by metabolomics and other "omics," will play a key role in understanding plant systems, and developing further biotechnology applications. As the proteome consists of all proteins present in a specific cell type, it may actually provide a better indication of the biological effect of phytochemicals in human nutrition studies.

Combining gene and protein expression profiling in colonic cancer cells, Herzog et al. (2004) identified the flavonoid flavone, present in a variety of fruits and vegetables, as a potent apoptosis inducer in human cancer cells (Herzog et al., 2004). Flavone displayed a broad spectrum of effects on gene and protein expression that related to apoptosis induction and cellular metabolism. The effect(s) of the flavonoid quercetin on normal and malignant prostate cells was evaluated, and possible target(s) of quercetin action was identified. This finding demonstrated that quercetin treatment of prostate cancer cells resulted in decreased cell proliferation and viability. Quercetin promoted cancer cell apoptosis by downregulating the levels of heat shock protein 90 (Aalinkeel et al., 2008).

These studies demonstrate that, indeed, proteomics analysis, particularly when combined with transcriptome analysis, may reveal effects of dietary phytochemicals relevant in cancer prevention. Metabolomics approaches aim to identify changes in relevant physiological processes, based on modifications in the occurrence and concentrations of all end-products of metabolic enzyme activity, in response to exposures or changes in environmental conditions. One of the biggest advantages of this technique is that it can be applied to biological samples like urine, serum, or plasma, making it very suitable for biomonitoring purposes. Although metabolomics analysis can bring comprehensive understanding of biological processes a step forward, studies on the effects of dietary phytochemicals are still very sparse. For instance, although each "omics" technique has its own merits and limitations, combined analysis and interpretation of multiomics data are likely to offer the best opportunities for comprehensive understanding of the biological influences induced by phytochemicals. Data integration of different "omics" techniques will become increasingly important in any system biology study, including postharvest and human nutrition.

ACKNOWLEDGMENT

The author thanks Dr. C. Forney at AAFC for his critical review of this manuscript and his constructive suggestions.

REFERENCES

Aaby, K., D. Ekeberg, and G. Skrede, 2007. Characterization of phenolic compounds in strawberry (*Fragaria × ananassa*) fruits by different HPLC detectors and contribution of individual compounds to total antioxidant capacity. *J. Agric. Food Chem.* 55:4395–4406.

Aaby, K., G. Skrede, and R.E. Wrolstad, 2005. Phenolic composition and antioxidant activities in flesh and achenes of strawberries (Fragaria ananassa). *J. Agric. Food Chem.* 55:4032–4040.

Aalinkeel, R., B. Bindukumar, J. Reynolds, D. Sykes, S. Mahajan, K. Chadha, and S. Schwartz, 2008. The dietary bioflavonoid, quercetin, selectively induces apoptosis of prostate cancer cells by down-regulating the expression of heat shock protein 90. *Prostate* 68:1773–1789.

Aharoni, A., L.C.P. Keizer, H.J. Bouwmeester, Z. Sun, M. Alvarez-Huerta, H.A. Verhoeven, J. Blaas et al. 2000. Identification of the *SAAT* gene involved in strawberry flavor biogenesis by use of DNA microarrays. *Plant Cell* 12:647–661.

Alarcón-Flores, M.I., R. Romero-González, J.L.M. Vidal, F.J.E. González, and A.G. Frenich, 2014. Monitoring of phytochemicals in fresh and fresh-cut vegetables: A comparison. *Food Chem.* 142:392–399.

Alexander, L. and D. Grierson, 2002. Ethylene biosynthesis and action in tomato: A model for climacteric fruit ripening. *J. Exp. Bot.* 53:2039–2055.

Amaro, A.L., J.F. Fundo, A. Oliveira, J.C. Beaulieu, J.P. Fernández-Trujillo, and D.P. Almeida, 2013. 1-Methylcyclopropene effects on temporal changes of aroma volatiles and phytochemicals of fresh-cut cantaloupe. *J. Sci. Food Agric.* 93:828–837.

Atkinson, R.G., K. Gunaseelan, Mindy Y. Wang, Luke Luo, Tianchi Wang, C.L. Norling, S.L. Johnston, R. Maddumage, R. Schroöder, and R.J. Schaffer, 2011. Dissecting the role of climacteric ethylene in kiwifruit (*Actinidia chinensis*) ripening using a 1-aminocyclopropane-1-carboxylic acid oxidase knockdown line. *J. Exp. Bot.* 62:3821–3835.

Baldwin, E.A., A. Plotto, and K. Goodner, 2007. Shelf-life versus flavour-life for fruits and vegetables: How to evaluate this complex trait. *Stewart Postharvest Rev.* 1:1–10.

Beekwilder, J., M. Alvarez-Huerta, E. Neef, F.A. Verstappen, H.J. Bouwmeester, and A. Aharoni, 2004. Functional characterization of enzymes forming volatile esters from strawberry and banana. *Plant Physiol.* 135:1865–1878.

Bengtsson, G.B., 2010. Effect of postharvest conditions and treatments on health-related quality of vegetables and fruits. *Acta Hortic.* 858:113–120.

Bennett, A.B., 2012. Taste: Unraveling tomato flavor. *Curr. Biol.* 22:R443–R444.

Bood, K.G. and I. Zabetakis, 2002. The biosynthesis of strawberry flavor (II): Biosynthetic and molecular biology studies. *J. Food Sci.* 67:2–8.

Boyer, J. and R.H. Liu, 2004. Apple phytochemicals and their health benefits. *Nutr. J.* 3:1–45.

Buttery, R.G., R. Teranishi, R.A. Flath, and L.C. Ling, 1989. Fresh tomato volatiles: Composition and sensory studies, pp. 213–222. In: R. Teranishi, R. Buttery, and F. Shahidi (eds.), *Flavor Chemistry: Trends and Development.* Amer. Chem. Soc., Washington, DC.

Cantín, C.M., M.A. Moreno, and Y. Gogorcena, 2009. Evaluation of the antioxidant capacity, phenolic compounds, and vitamin C content of different peach and nectarine [Prunus persica (L.) batsch] breeding progenies. *J. Agric. Food Chem.* 57:4586–4592.

Cliff, M.A., P.M.A. Toivonen, C.F. Forney, P. Liu, and C. Lu, 2010. Quality of fresh-cut apple slices stored in solid and micro-perforated film packages having contrasting headspace atmospheres. *Postharvest Biol. Technol.* 58:254–261.

Costa, F., L. Cappellin, E. Zini, A. Patocchi, M. Kellerhals, M. Komjanc, C. Gessler, and F. Biasioli, 2013. QTL validation and stability for volatile organic compounds (VOCs) in apple. *Plant Sci.* 211:1–7.

Defilippi, B.G., A.M. Dandekar, and A.A. Kader, 2004. Impact of suppression of ethylene action or biosynthesis on flavor metabolites in apple (Malus domestica Borkh) fruits. *J. Agric. Food Chem.* 52:5694–5701.

Dirinck, P., H. De Pooter, and N. Schamp, 1989. Aroma development in ripening fruits, p. 24–34. In: R. Teranishi, R. Buttery, and F. Shahidi (eds.), *Flavor Chemistry: Trends and Development.* Amer. Chem. Soc, Washington, DC.

Dixon, J. and E.W. Hewett, 2000. Factors affecting apple aroma/flavour volatile concentration: A review. *NZ J. Crop Hort. Sci* 28:155–173.

Dunemann, F., D. Ulrich, L. Malysheva-Otto, W.E. Weber, S. Longhi, R. Velasco, and F. Costa, 2012. Functional allelic diversity of the apple alcohol acyl-transferase gene MdAAT1 associated with fruit ester volatile contents in apple cultivars. *Mol. Breeding* 29:609–625.

Fawbush, F., J.F. Nock, and C.B. Watkins, 2009. Antioxidant contents and activity of 1-methylcyclopropene (1-MCP)-treated "Empire" apples in air and controlled atmosphere storage. *Postharvest Biol. Technol.* 52:30–37.

Fellman, J.K., T.W. Miller, D.S. Mattinson, and J.P. Mattheis, 2000. Factors that influence biosynthesis of volatile flavor compound in apple fruits. *HortScience* 35:1026–1033.

Fournier-Level, A., L. Le Cunff, C. Gomez, A. Doligez, A. Ageorges, C. Roux, Y. Bertrand, J.M. Souquet, V. Cheynier, and P. This, 2009. Quantitative genetic bases of anthocyanin variation in grape (Vitis vinifera L. ssp. sativa) berry: A quantitative trait locus to quantitative trait nucleotide integrated study. *Genetics* 183:1127–1139.

Fredericks, C.H., K.J. Fanning, M.J. Gidley, G. Netzel, D. Zabaras, M. Herrington, and M. Netzel, 2013. High-anthocyanin strawberries through cultivar selection. *J. Sci. Food Agric.* 93:846–852.

Giepmans, B.N., S.R. Adams, M.H. Ellisman, and R.Y. Tsien, 2006. The fluorescent toolbox for assessing protein location and function. *Science* 312:217–223.

Herzog, A., B. Kindermann, F. Doring, H. Daniel, and U. Wenzel, 2004. Pleiotropic molecular effects of the pro-apoptotic dietary constituent flavone in human colon cancer cells identified by protein and mRNA expression profiling. *Proteomics* 4:2455–2466.

Isaacson, T., C.M.B. Damasceno, R.S. Saravanan, Y. He, C. Catala, M. Saladie, and J.K.C. Rose, 2006. Sample extraction techniques for enhanced proteomic analysis of plant tissues. *Nat. Prot.* 1:769–774.

Kader, A.A., 2004. Perspective on postharvest horticulture (1978–2003). *HortScience* 38:759–761.

Khan, S.A., P.Y. Chibon, R.C.H. De Vos, B.A. Schipper, E. Walraven, J. Beekwilder, T. Van Dijk, et al. 2012. Genetic analysis of metabolites in apple fruits indicates an mQTL hotspot for phenolic compounds on linkage group 16. *J. Exp. Bot.* 63:2895–2908.

King, C.J., C. Maliepaard, J.R. Lynn, F.H. Alston, C.E. Durel, K.M. Evans, B. Griffon, et al. 2000. Quantitative genetic analysis and comparison of physical and sensory descriptors relating to fruit flesh firmness in apple (Malus pumila Mill.). *Theor. Appl. Genet.* 102:1074–1084.

Klee, H.J., 2010. Improving the flavor of fresh fruits: Genomics, biochemistry, and biotechnology. *New Phytol.* 187:44–56.

Klee, H.J. and D.M. Tieman, 2013. Genetic challenges of flavor improvement in tomato. *Trends Genet.* 29:257–262.

Lara, I., J. Graell, M.L. López, and G. Echeverría, 2006. Multivariate analysis of modifications in biosynthesis of volatile compounds after CA storage of "Fuji" apples. *Postharvest Biol. Technol.* 39:19–28.

Lavid, N., W. Schwab, E. Kafkas, M. Koch-Dean, E. Bar, O. Larkov, U. Ravid, and E. Lewinsohn, 2002. Aroma biosynthesis in strawberry: *S*-adenosylmethionine:Furaneol *O*-methyltransferase activity in ripening fruits. *J. Agric. Food Chem.* 50:4025–4030.

Li, D., Y. Xu, G. Xu, L. Gu, and H. Shu, 2006. Molecular cloning and expression of a gene encoding alcohol acyltransferase (MdAAT2) from apple (cv. Golden Delicious). *Phytochemistry* 67:658–667.

Li, H., R. Tsao, and Z. Deng, 2012. Factors affecting the antioxidant potential and health benefits of plant foods. *Can. J. Plant Sci.* 92:1101–1111.

Li, H., Z. Deng, R. Liu, J.C. Young, H. Zhu, S. Loewen, and R. Tsao, 2011. Characterization of phytochemicals and antioxidant activities of a purple tomato (Solanum lycopersicum L.). *J. Agric. Food Chem.* 59:11803–11811.

Li, L., J. Song, W. Kalt, C. Forney, R. Tsao, D. Pinto, K. Chisholm, L. Campbell, S. Fillmore, and X. Li, 2013. Quantitative proteomic investigation employing stable isotope labeling by peptide dimethylation on proteins of strawberry fruit at different ripening stages. *J. Proteomics* 94:219–239.

Marinquez, D., I. El-Sharkawy, F.B. Flores, F. El-Yahyaoui, F. Regad, M. Bouzayen, A. Latché, and J.-C. Pech, 2006. Two highly divergent alcohol dehydrogenases of melon exhibit fruit ripening-specific expression and distinct biochemical characteristics. *Plant Mol. Biol.* 61:675–685.

Martínez-García, P.J., D.E. Parfitt, E.A. Ogundiwin, J. Fass, H.M. Chan, R. Ahmad, S. Lurie, A. Dandekar, T.M. Gradziel, and C.H. Crisosto, 2013. High density SNP mapping and QTL analysis for fruit quality characteristics in peach (Prunus persica L.). *Tree Genet. Genomes* 9:19–36.

Mehinagic, E., G. Royer, R. Symoneaux, F. Jourjon, and C. Prost, 2006. Characterization of odor-active volatiles in apples: Influence of cultivars and maturity stage. *J. Agric. Food Chem.* 54:2678–2687.

Morris, C.E. and D.C. Sands, 2006. The breeder's dilemma—yield or nutrition. *Nat. Biotechnol.* 24:1078–1080.

Nath, A., B. Bagchi, L.K. Misra, and B.C. Deka, 2011. Changes in post-harvest phytochemical qualities of broccoli florets during ambient and refrigerated storage. *Food Chem.* 127:1510–1514.

Pérez, A.G., R. Olías, P. Luaces, and C. Sanz, 2002. Biosynthesis of strawberry aroma compounds through amino acid metabolism. *J. Agric. Food Chem.* 50:4037–4042.

Palma, J.M., F.J. Corpas, and L.A. del Río, 2011. Proteomics as an approach to the understanding of the molecular physiology of fruit development and ripening. *J. Proteomics* 74:1230–1243.

Pandey, A. and M. Mann, 2004. Proteomic to study genes and genomes. *Nature* 405:837–846.

Panthee, D.R., J.A. Labate, and L.D. Robertson, 2013. Evaluation of tomato accessions for flavour and flavour-contributing components. *Plant Genet. Resour. Characterisation Utilisation* 11:106–113.

Park, S., N. Sugimoto, M.D. Larson, R. Beaudry, and S. van Nocker, 2006. Identification of genes with potential roles in apple fruit development and biochemistry through large-scale statistical analysis of expressed sequence tags. *Plant Physiol.* 141:811–824.

Perez, A.G., R. Olias, P. Lucaces, and C. Sanz, 2002. Biosynthesis of strawberry aroma compounds through amino acid metabolism. *J. Agric. Food. Chem* 50:4037–4042.

Prior, R.L., X. Wu, and K. Schaich, 2005. Standardized methods for the determination of antioxidant capacity and phenolics in foods and dietary supplements. *J. Agric. Food Chem.* 53:4290–4302.

Raab, T., J.A. López-Ráez, D. Klein, J.L. Caballero, E. Moyano, W. Schwab, and J. Muñoz-Blanco, 2006. FaQR, required for the biosynthesis of the strawberry flavor compound 4-hydroxy-2,5-dimethyl-3(2H)-furanone, encodes an enone oxidoreductase. *Plant Cell* 18:1023–1037.

Rowan, D.D., H.P. Lane, J.M. Allen, S. Fielder, and M.B. Hunt, 1996. Biosynthesis of 2-methylbutyl, 2-methyl-2-butenyl, and 2-methylbutanoate esters in "Red Delicious" and "Granny Smith" apples using deuterium-labelled substrates. *J. Agric. Food Chem.* 44:3276–3285.

Rowan, D.D., J.M. Allen, S. Fielder, and M.B. Hunt, 1999. Biosynthesis of straight-chain ester volatiles in "Red Delicious" and "Granny Smith" apples using deuterium-labelled precursors. *J. Agric. Food Chem.* 47:2553–2562.

Rudell, D.R., D.S. Mattinson, J.P. Mattheis, S.G. Wyllie, and J.K. Fellman, 2002. Investigation of aroma volatile biosynthesis under anoxic conditions and in different tissues of "Redchief Delicious" apple fruit (Malus domestica Borkh.). *J. Agric. Food Chem.* 50:2627–2632.

Saito, K. and F. Matsuda, 2010. Metabolomics for functional genomics, systems biology, and biotechnology. *Annu. Rev. Plant Biol.* 61:463–489.

Saquet, A.A., J. Streif, and F. Bangerth, 2003. Impaired aroma production of CA-stored "Jonagold" apples as affected by adenine and pyridine nucleotide levels and fatty acid concentrations. *J. Hort. Sci. Biotechnol.* 78:695–705.

Seymour, G., J. Taylor, and G. Tucker, 1993. *Biochemistry of Fruit Ripening*, Chapman & Hall. 2–6 Boundary Row, London Se1 8HN London, Galssgow, New York, Tokyo, Melbourne and Madras.

Sharma, M., J.K. Jacob, J. Subramanian, and G. Paliyath, 2010. Hexanal and 1-MCP treatments for enhancing the shelf-life and quality of sweet cherry (*Prunus avium* L.). *Sci. Hortic.* 125:239–247.

Sivakumar, D., F. Van Deventer, L.A. Terry, G.A. Polant, and L. Korsten, 2011. Combination of 1-methylcyclopropene treatment and controlled atmosphere storage retains overall fruit quality and bioactive compounds in mango. *J. Sci. Food Agric.* 92:821–830.

Song, J., 1994. *Einfuß verschiedener Erntzeitpunkte auf die Fruchtreife under besonderer Berücksichtigung der Aromabildung bei Äpfeln, Tomaten und Erdberren*, Verlag Ulrich Grauer, Stuttgart.

Song, J. and C.F. Forney, 2008. Flavour volatile production and regulation in fruit. *Can. J. Plant Sci.* 88:537–550.

Song, J. and F. Bangerth, 1996. The effect of harvest date on aroma compound production from "Golden Delicious" apple fruit and relationship to respiration and ethylene production. *Postharvest Biol. Technol.* 8:259–269.

Song, J. and F. Bangerth, 2003. Fatty acids as precursors for aroma volatile biosynthesis in pre-climacteric and climacteric apple fruit. *Postharvest Biol. Technol.* 30:113–121.

Souleyre, E.J.F., D.R. Greenwood, E.N. Friel, S. Karunairetnam, and R.D. Newcomb, 2005. An alcohol acyl transferase from apple (cv. Royal Gala), MpAAT1, produces esters involved in apple fruit flavor. *FEBS J.* 272:3132–3144.

St-Pierre, B. and V. De Luca, 2000. Evolution of acyltransferase genes: Origin and diversification of the BAHD superfamily of acyltransferases involved in secondary metabolism. *Recent Adv. Phytochem.* 34:282–315.

Syamaladevi, R.M., P.K. Andrews, N.M. Davies, T. Walters, and S.S. Sablani, 2012. Storage effects on anthocyanins, phenolics and antioxidant activity of thermally processed conventional and organic blueberries. *J. Sci. Food Agric.* 92:916–924.

Tieman, D., P. Bliss, L.M. McIntyre, A. Blandon-Ubeda, D. Bies, A.Z. Odabasi, G.R. Rodríguez, et al. 2012. The chemical interactions underlying tomato flavor preferences. *Curr. Biol.* 22:1035–1039.

Tiwari, U. and E. Cummins, 2013. Factors influencing levels of phytochemicals in selected fruit and vegetables during pre- and post-harvest food processing operations. *Food Res. Int.* 50:497–506.

Tressl, R. and F. Drawert, 1973. Biogenesis of banana volatiles. *J. Agric. Food Chem.* 21:560–565.

Ulrich, D., E. Hoberg, A. Rapp, and S. Kecke, 1997. Analysis of strawberry flavour–discrimination of aroma types by quantification of volatile compounds. *Eur. Food Res. Technol.* 205:218–223.

Ulrich, D. and F. Dunemann, 2012. Towards the development of molecular markers for apple volatiles. *Flavour Fragrance J.* 27:286–289.

Vogt, J., D. Schiller, D. Ulrich, W. Schwab, and F. Dunemann, 2013. Identification of lipoxygenase (LOX) genes putatively involved in fruit flavour formation in apple (Malus × domestica). *Tree Genet. Genomes* 9:1493–1511.

Wang, Q.L., S. Khanizadeh, and C. Vigneault, 2007. Preharvest ways of enhancing the phytochemical content of fruits and vegetables. *Stewart Postharvest Rev.* 3:1–8.

Wang, Y., S.G. Wyllie, and D.N. Leach, 1996. Chemical changes during the development and ripening of the fruit Cucumis melo (cv. Makdimon). *J. Agric. Food Chem.* 44:21–216.

Wein, M., N. Lavid, S. Lunkenbein, E. Lewinsohn, W. Schwab, and R. Kaldenhoff, 2002. Isolation, cloning and expression of a multifunctional O-methyltransferase capable of forming 2,5-dimethyl-4-methoxy-3(2H)-furanone, one of the key aroma compounds in strawberry fruits. *Plant J.* 31:755–765.

Wolfe, K.L., X. Kang, X. He, M. Dong, Q. Zhang, and R.H. Liu, 2008. Cellular antioxidant activity of common fruits. *J. Agric. Food Chem.* 56:8418–8426.

Wu, J., Z. Wang, Z. Shi, S. Zhang, R. Ming, S. Zhu, M.A. Khan, et al. 2013. The genome of the pear (Pyrus bretschneideri Rehd.). *Genome Res.* 23:396–408.

Wyllie, S.G., D.N. Leach, Y. Wang, and R. Shewfelt, 1995. Key aroma compounds in melons: Their development and cultivar dependence, p. 248–257. In: R.L. Rouseff and M.M. Leahy, (eds.), *Fruit Flavour Biogenesis, Characterization and Authentication.* Amer. Chem. Soc., Washington, DC.

Xu, K., A. Wang, and S. Brown, 2012. Genetic characterization of the Ma locus with pH and titratable acidity in apple. *Mol. Breeding* 30:899–912.

Zhu, Y., D.R. Rudell, and J.P. Mattheis, 2008. Characterization of cultivar differences in alcohol acyltransferase and 1-aminocyclopropane-1-carboxylate synthase gene expression and volatile ester emission during apple fruit maturation and ripening. *Postharvest Biol. Technol.* 49:330–339.

Zini, E., F. Biasioli, F. Gasperi, D. Mott, E. Aprea, T.D. Mark, A. Patocchi, C. Gessler, and M. Komjanc, 2005. QTL mapping of volatile compounds in ripe apples detected by proton transfer reaction-mass spectrometry. *Euphytica* 145:269–279.

Zorrilla-Fontanesi, Y., J.L. Rambla, A. Cabeza, J.J. Medina, J.F. Sánchez-Sevilla, V. Valpuesta, M.A. Botella, A. Granell, and I. Amaya, 2012. Genetic analysis of strawberry fruit aroma and identification of *O*-methyltransferase FaOMT as the locus controlling natural variation in mesifurane content. *Plant Physiol.* 159:851–870.

13 Metabolomics Tools for Postharvest Quality and Safety of Fresh Produce

Sukhvinder Pal Singh

CONTENTS

13.1 INTRODUCTION

Metabolomics is a novel experimental methodology categorized as an "omics" approach, along with genomics, transcriptomics, and proteomics (Hertog et al., 2011), and is defined as the comprehensive, simultaneous determination of endogenous metabolites at the molecular level and their global and dynamic changes over time in complex multicellular systems as a consequence of biological stimuli or genetic manipulation or both (Hu and Xu, 2013). Metabolomics is often used in combination with the other "omics" approaches for a deeper understanding of biological processes, especially metabolism, through global studies on metabolites and their concentration and dynamics within complex samples. As metabolites are considered the downstream products of cellular regulatory processes, metabolomics data can precisely characterize cells, tissues, or whole organisms by defining specific biochemical phenotypes that are representative of physiological or developmental states. The metabolome is the holistic quantitative set of low-molecular-weight compounds (<1000 Da), including many hundreds or thousands of molecules such as carbohydrates, vitamins, lipids and amino or fatty acids. The number of metabolites in the plant kingdom is considered to be far greater than that in the animal kingdom, and is estimated to exceed 200,000 (Fiehn, 2002). This large number of metabolites is due to the great diversity of metabolic pathways that each plant species has evolved, to survive under varying environmental conditions.

13.2 METABOLOMICS: ANALYTICAL APPROACHES

The complexity and diversity of metabolites in terms of their number, localization, and chemical nature, pose considerable challenges in their analysis, and data interpretations. Therefore, the metabolome analysis is broadly classified into two approaches: targeted and untargeted. Targeted metabolomic analysis involves identification and absolute quantification of a limited number of metabolites under a given set of conditions, while an untargeted approach is aimed at qualitative analysis (identification and relative quantitation) of the maximum possible metabolites in a system (cell, tissue or organ). The targeted metabolomics typically represents a bottom-up approach compared to the top-to-bottom approach in untargeted metabolomics. Both of these approaches require different types of analytical platforms and bioinformatics tools. However, the enhanced capabilities of new analytical systems have enabled the broadening of scope, and increased sensitivity and resolution for targeted as well as untargeted analyses.

There are several analytical options available to find an answer to metabolic inquiry. A typical analytical system will consist of a separation step, followed by detection, and then data processing. Three separation techniques—gas chromatography (GC), liquid chromatography (LC) and capillary electrophoresis (CE)—are most commonly employed in metabolomics studies. The choice of a separation technique depends on the chemical nature and sensitivity of small molecules, and compatibility with the detection system. The separation system is hyphenated to a detection system. The approach based on mass spectrometry (MS) has recently emerged as the technique of choice, considering its advantages over nuclear magnetic resonance (NMR) systems. MS offers distinct advantages of unparalleled sensitivity and specificity, high resolution and wide dynamic range, enabling comprehensive quantitative and qualitative measurement of large-scale metabolites in complex biological samples. On the other hand, NMR is also used for metabolomics studies because of its simplicity, rapidity, high selectivity and nondestructive nature, but relatively lower sensitivity.

Gas chromatography–mass spectrometry (GC–MS) has been regarded as the gold standard for metabolomics as the platform that provides reproducible retention characteristics, high chromatographic resolution, high sensitivity, reproducible production of mass spectra, which aids metabolite identification, and the availability of MS libraries, which are much less common for liquid chromatography–mass spectrometry (LC–MS) (Lisec et al., 2006). Coupling LC to MS enables the detection of multiple metabolite classes in a single analysis, even in a very complex matrix such as rich plant extracts. The technology for liquid-phase separation has also advanced in recent years, with the introduction of ultra-performance LC (UPLC) that enables faster separation with better separation and sensitivity as compared to high-performance LC (HPLC). The analytical instruments typically used in metabolomics experiments differ in their ionization technology, for example, electron ionization (EI), chemical ionization (CI), electrospray ionization (ESI), atmospheric pressure chemical ionization (APCI), matrix-assisted laser desorption/ionization (MALDI), and the type of mass analyzer, for example, quadrupole, triple-quadrupole (QQQ), ion-trap, time-of-flight (TOF), and Fourier-transform ion cyclotron resonance (FT-ICR) mass analyzer. Samples can be directly analyzed by direct infusion technique in MS or resolved primarily by different chromatographic techniques. The EI and ESI sources of ionization have been most commonly used in the GC- and LC-based metabolomics approaches, respectively. Despite the recent extensive development in analytical methods, the ultimate goal of plant metabolomics—to gain a complete overview of the metabolite complement of a plant in one or a small series of analyses—is currently impossible. However, continued technological improvements in the applications of hybrid technologies and combinations will help develop a better understanding of the metabolome. The detailed discussion on analytical methods in metabolomics is beyond the scope of this chapter.

In addition, these analytical platforms generate large data-sets, which require further complex processing to derive meaningful interpretations. The analytical systems supplied by various companies often offer informatic solutions to process the data-sets for various functions such as peak-filtering, alignment, data-reduction, and multivariate statistical analysis (e.g., MarkerLynx from Waters, MarkerView from AB Sciex and MassProfiler from Agilent). In addition, there are several public-domain databases (Metlin, MassBank, XCMS, MZmine etc.) that are available to conduct data-processing and identification of metabolites based on mass spectra and other features such as exact mass, elemental composition, etc. The correct identification of a metabolite with high level of confidence is still a major challenge in global metabolomics studies, which tend to separate and detect a huge range of metabolites simultaneously.

13.3   METABOLOMICS APPLICATIONS IN FRESH-FRUIT QUALITY

Among different "omics" tools, metabolomics is more relevant to postharvest research, unless used in tandem with others in a systems-biology approach. In comparison with genome, transcriptome and proteome, metabolome can be best studied in response to the environmental factors. The epigenetic, posttranscriptional, and posttranslational modifications often limit the scope and robustness of genomics, transcriptomics and proteomics, respectively. Postharvest science involves studying the effects of various external environmental stimuli (such as temperature, storage atmosphere, heat and cold treatments, etc.) on the tolerance and quality of products.

These stimuli directly or indirectly influence the metabolome expressions, which are manifested in different forms in terms of quality and the postharvest life of a commodity. The metabolome of fresh fruits and vegetables not only represents small molecules as intermediary and final products of various metabolic pathways, but these molecules have key roles in determining quality attributes such as color, taste, aroma and nutrition. Postharvest factors, such as treatment, storage, and distribution, act as external stimuli having both positive and negative consequences on quality and safety of fruit and vegetables (Hertog et al., 2011). These can be examined using metabolomic evaluation techniques; however, to date, research employing untargeted metabolic profiling to study postharvest issues, is relatively uncommon.

13.3.1 Breeding for Postharvest Quality

In clinical chemistry, metabolomics is applied in the discovery of biomarkers for diagnosing disease, prognosis, and risk-reduction. However, the expansion of this approach in plant sciences is rapidly discovering new biomarkers for the determination of heredity of biochemical traits, linking traits to specific genes, and screening germplasm for bioactive compounds of interest (Zanor et al., 2009). Untargeted metabolic fingerprinting techniques have been used to distinguish individuals with unique quality characteristics within breeding and wild populations. While targeted metabolite analysis of mapping populations is useful for limited discovery of quality-related metabolic quantitative trait loci (QTL), untargeted biochemical phenotyping of genetically mapped populations can reveal multiple metabolic QTLs that comprise sensory traits (Wahyuni et al., 2013). Recently, a genome-wide metabolomic resource for tomato fruit from *Solanum pennellii* has been developed (Perez-Fons et al., 2014). A multiplatform metabolomic analysis, using NMR, MS, and HPLC, of introgression lines of *Solanum pennellii* with a domesticated line, was conducted to analyze and quantify alleles responsible for metabolic traits. QTL for health-related antioxidant carotenoids and tocopherols, as well as molecular signatures for some 2000 compounds, have been reported (Perez-Fons et al., 2014). Correlation analyses have also revealed intricate interactions in isoprenoid formation in the plastid that can be extrapolated to other crop plants. Identification of metabolite alleles in introgression lines, as well as computational integration of the data-sets, has been carried out to construct metabolic networks for the tomato fruit metabolome and the individual pathway components. These data-sets represent a valuable and timely resource to fully capitalize on the sequencing of tomato genomes.

Metabolomics approaches have also been successfully used to complement association genetics studies to identify QTL for important quality traits in apple fruit. The LC–MS analysis of fruit extracts from a segregating F1 population of a cross between apple cultivars "Prima" and "Fiesta" has been used to integrate metabolite data with the existing reference map for genetic linkage, thus allowing the mapping of metabolite quantitative trait loci (Maliepaard et al., 1998; Khan et al., 2012). The same approach was used successfully to map metabolic QTL (mQTL) for apple fruit volatiles (Dunemann et al., 2009). Apple linkage group 16 was found to contain an mQTL hotspot for various phenylpropanoid metabolites, which correlated with the location of genes putatively involved in the phenylpropanoid pathway (Khan et al.,

2012). This knowledge can now be used in marker-assisted breeding, which should be greatly aided by the available apple genome sequence. Cuthbertson et al. (2012) employed a GC–MS approach to evaluate the use of metabolite patterns to differentiate fruit from six commercially grown apple cultivars in central Washington State in the United States. The principal component analysis (PCA) of apple fruit peel and flesh metabolome data indicated that individual cultivar replicates clustered together and were separated from all other cultivar samples and this observation was further supported by a hierarchical clustering analysis. An evaluation of PCA component loadings revealed specific metabolite classes that contributed the most to each principal component, whereas a correlation analysis demonstrated that specific metabolites correlated directly with quality traits such as antioxidant activity, total phenolics, and total anthocyanins, which are important selection parameters in some breeding programs.

13.3.2 FRUIT-RIPENING

Metabolic analysis has been an integral part of postharvest science for over a century. However, the targeted metabolites quantified for quality analysis were limited in number. The new developments in analytical platforms and informatics tools have empowered postharvest researchers to gain better insights into fruit-ripening and to integrate the data from all "omics" for a systems-biology approach. To date, most of the metabolomics studies have focused on changes in metabolome during the growth, maturation and ripening stages in the tomato fruit (Carrari et al., 2006; Moco et al., 2006; Osorio et al., 2012), strawberry (Zhang et al., 2011), peach (Lombardo et al., 2011) grapes (Toffali et al., 2011; Zamboni et al., 2010), and avocado (Pedreschi et al., 2014).

The fruit-development, -ripening and senescence processes are important determinants for quality and postharvest storage and shelf-life. Therefore, understanding the metabolic events involved in these processes is critical to make harvest decisions, choosing the appropriate postharvest handling and storage procedures and managing product quality for the consumer. The integration of "omics," including metabolomics, provides more information about the regulatory networks involved in controlling vital processes, such as fruit-ripening. Osorio et al. (2012) followed a systems-biology approach to reveal a network of candidate regulators that play important roles in tomato fruit-ripening. The combination of transcriptome, proteome, and targeted metabolite analysis helped to refine the ethylene-regulated transcriptome of tomato fruit and added to the knowledge of the role of ethylene in both protein and metabolite regulation in tomato-ripening. The metabolite abundance of specific compound classes, such as tricarboxylic acid (TCA) cycle organic acids and cell-wall-related metabolites, appears to be strictly controlled, with specific compounds influenced by ethylene, transcriptional control, or both. This systems-biology approach helped identify areas of metabolism that seem to be of high importance to the ripening process, such as hormones and cell-wall metabolism in ethylene perception. Therefore, the integrated analysis has enabled the discovery of additional information for the comprehensive understanding of biological events relevant to metabolic regulation during tomato fruit development. The outcomes of this work will provide potential targets for the engineering of metabolism to facilitate the controlled modulation of ripening in tomato fruit (Osorio et al., 2012).

In peach fruit, Lombardo et al. (2011) showed that the principal role of amino acid metabolism during the early stages of ripening were as substrates for the phenylpropanoid pathway. These results suggest that enzymes involved in amino acid metabolism may be useful biomarkers for the different developmental stages as well as key to determining the final quality of peach fruit. Other metabolic determinants for sucrose levels, invertase activity for early stages of ripening, and sucrose cycling for ripening, were also inferred by Lombardo et al. (2011). They also highlighted the relevance of stone-formation by lignification of the fruit endocarp layer, which affects primary metabolism to fulfill this process. Finally, the relevance of the posttranscriptional regulation of enzyme activity indicated the need for integrated studies of metabolic pathways, combining analyses of metabolites, transcripts, and proteins (Lombardo et al., 2011).

Seven developmental stages of strawberry fruit—to maturation and ripening—were subjected to metabolic profiling of both polar and nonpolar compounds to monitor the alterations in several major groups of compounds (Zhang et al., 2011). The results of metabolic profiling, conducted using GC–MS and HPLC–PDA systems, were consistent with strawberry pigment- and flavor-formation, using mostly materials from primary metabolism. Each stage of fruit development represented its unique metabolic profile, with the most drastic changes occurring at the transition toward the red-ripened stage. Like peaches, amino acid biosynthesis was found to play an important role in generating several classes of compounds related to the quality of the strawberry fruits (Zhang et al., 2011). This information has potential to be translated by the strawberry breeders to detect and monitor key components that are important for these output traits.

The comparison of metabolic profiling during the development of peach (Lombardo et al., 2011), tomato (Osorio et al., 2012), strawberry (Zhang et al., 2011), and grapes (Toffali et al., 2011; Zamboni et al., 2010) indicates that each fruit follows a diverse metabolic program. Furthermore, these fruits are models for studies of ripening behavior where tomato and peach are defined as climacteric fruits, while strawberry and grapes are considered nonclimacteric. Since great differences in terms of morphology, physiology, and biochemistry are found among fruits, it seems reasonable that singular metabolic programs support the differential developmental processes. A recent *in silico* study on the comparative metabolomes of three climacteric (peach and two tomato cultivars) and two nonclimacteric species (strawberry and pepper) was conducted by Klie et al. (2014) with data on the metabolic profiles of these fruits during development and ripening stages. The analysis was based on an extension to principal component analysis, called STATIS, in combination with pathway over-enrichment analysis. STATIS is a novel tool that can be used to identify the metabolic processes that are affected to similar extent during fruit-development and -ripening. These results ultimately provide insights into the pathways that are essential during fruit-development and -ripening across species (Klie et al., 2014). This study is an example of the application of new informatic tools to metabolomics data, to derive an overarching view of the biological processes involved in maturation and ripening.

The application of the metabolomics tool has also been extended to determine the sensorial quality of some apple cultivars ("Golden Delicious," "Liberty," "Santana," and "Topaz") grown in organic (ORG) versus integrated production (IP) systems (Vanzo et al., 2013). Metabolic profiling revealed that significantly higher total phenolic content in ORG fruit was found in "Golden Delicious," whereas differences in other cultivars were not significant. Targeted profiling of multiple classes of phenolics confirmed the impact of the production system on the "Golden Delicious" phenolic profile, with higher levels of 4-hydroxybenzoic acid, neo- and chlorogenic acids, phloridzin, procyanidin B2 + B4, quercetin-3-*O*-glucoside and quercetin-3-*O*-galactoside, kaempferol-3-*O*-rutinoside, and rutin being found in ORG fruit. The results also suggested that scab resistance in some apple cultivars is influenced by the phenolic biosynthesis in relation to the agricultural production system (Vanzo et al., 2013). Although such results are not conclusive enough to define the impact of production systems on metabolic profiles of fruit, similar studies have significant scope for future research to discover biomarkers for organic (and other) produce.

Postharvest ripening behavior of "Hass" avocado is heterogeneous and complex. A fruit biopsy methodology was followed to derive samples of mesocarp tissue, which were subjected to untargeted and targeted metabolomics (Pedreschi et al., 2014). The results showed that while C7 sugars (mannoheptulose and perseitol), dry matter, and total Ca2+ were not correlated with time to reach edible ripeness, untargeted metabolomics profiling of polar and semi-polar compounds (based on GC–MS and LC–MS platforms), revealed several metabolites, mainly amino acids, that were related to ripening heterogeneity. The analysis of fatty acids revealed linoleic acid to be accumulating differentially. In general, the slowest ripening avocados had lower amounts of precursors of metabolites involved in key metabolic pathways (Pedreschi et al., 2014). This study indicated that comprehensive metabolomics may provide new markers for the avocado-ripening stage at harvest, and will provide greater insights into the complex ripening physiology of this fruit.

13.3.3 Physiological Disorders

Metabolomics tools are becoming increasing popular to link the development of postharvest physiological disorders in fruits with some metabolites (Lee et al., 2012; Leisso et al., 2013; Rudell et al., 2009; Rudell and Mattheis, 2009). This section will outline the role of metabolomics to uncover the causes of the physiological disorders: superficial scald in apples, flesh-browning in apples, internal-browning in pears, and puffing disorder in citrus.

13.3.3.1 Superficial Scald in Apples

Superficial scald in apple has been well studied due to its commercial importance for the industry. The development of scald has been largely attributed to the oxidation products of α-farnesene and antioxidant metabolism. The application of metabolomics tools in understanding the chemistry of apple peel in relation to scald has provided great insights into the metabolic pathways, leading towards development of predictive biomarkers for scald. Rudell et al. (2009) employed untargeted metabolic profiling to characterize metabolomic changes associated with superficial scald development in "Granny Smith" apple, following 1-MCP or DPA treatment. Partial least squares models revealed metabolomic differentiation between untreated controls and fruit treated with DPA or 1-MCP within one week following initiation of storage. Metabolic divergence between controls and DPA-treated fruit after four weeks of storage, preceded scald symptom-development by two months. The study demonstrated that extensive metabolomic changes in levels of various classes of triterpenoids, including β-sitosterol associated with scald, precede actual symptom-development (Rudell et al., 2009).

To elaborate the role of phytosterol metabolism in the development of apple scald, Rudell et al. (2011) studied changes in peel-tissue levels of conjugates of β-sitosterol and campesterol, including acylated steryl glycosides (ASG), steryl glycosides (SG) and steryl esters (SE), as well as free sterols (FS) during the period of scald-development in response to antiscald postharvest treatments; diphenylamine (DPA), 1-MCP, and intermittent warming. ASG levels increased and SE levels decreased in untreated control fruit during storage. Removing fruit from cold storage to ambient temperature induced rapid shifts in ASG and SE fatty acyl moieties from unsaturated to saturated. FS and SG levels remained relatively stable during storage but SG levels increased, following a temperature increase after storage. ASG, SE, and SG levels did not increase during six months' cold storage in fruit subjected to intermittent warming treatment. These results clearly showed that apple-peel phytosterol conjugate metabolism is influenced by storage duration, oxidative stress, ethylene action/ripening, and storage temperature (Rudell et al., 2011). Leisso et al. (2013) recently showed that the metabolic profile of control and DPA-treated fruit was divergent after 30 days of cold storage due to differing levels of α-farnesene oxidation products, methyl esters, phytosterols, and other compounds potentially associated with chloroplast integrity and oxidative stress response. Leisso et al. (2013) showed evidence of coregulation within the volatile synthesis pathway, including control of the availability of methyl, propyl, ethyl, acetyl, and butyl alcohol and/or acid moieties for ester biosynthesis.

The researchers at the USDA (United States Department of Agriculture) Tree Fruit Research Laboratory in Washington state, United States, have applied metabolomics tools to discover and validate various risk assessment biomarkers for scald (Leisso et al., 2013; Rudell et al., 2008, 2009, 2011; Rudell and Mattheis, 2009). These biomarkers may be useful as apple storage management tools by indicating when and which fruit are at higher risk to develop scald. By selecting biomarkers for scald using storage conditions typically required for scald to develop, and then validation under storage conditions similar to commercial practice, risk assessment tools are expected to be widely applicable. However, the outcomes of these studies require further validation under commercial conditions in order to develop a diagnostic tool to accurately, and in timely fashion, predict the onset and development of superficial scald in apple.

13.3.3.2 Flesh-Browning in Apples

Flesh-browning is a complex physiological disorder that develops in apple during long-term cold storage and is controlled or aggravated by certain postharvest treatments. A GC–MS-based global metabolomics study has been reported to underpin the reasons why "Empire" apple fruit were susceptible to flesh-browning at 3.3°C if treated with 1-MCP (Lee et al., 2012). Most carbohydrates and organic acids in flesh tissue were not appreciably affected, but the levels of amino acids and volatile metabolites were significantly affected by 1-MCP treatment. In addition, sorbitol and levels of some amino acids were elevated towards the end of storage in 1-MCP-treated fruit. CA storage resulted in lower levels of many volatile components while 1-MCP treatment reduced these levels further. Lee et al. (2012) further showed that multiple metabolites were associated with the development of flesh-browning symptoms. Unlike other volatile compounds, methanol levels gradually increased with storage-duration, regardless of 1-MCP treatment, while 1-MCP decreased ethanol production. The metabolic shifts in response to CA storage and 1-MCP need further investigations to understand the interrelationship among different postharvest factors with regard to the development of flesh-browning. This should also be conducted on other apple varieties and storage conditions.

13.3.3.3 Internal Browning in Pears

Metabolomics- and proteomics-based approaches have been followed to explicate the physiological and biochemical aspects of browning in "Conference" pears (Pedreschi et al., 2007, 2008, 2009). The metabolic profiling was carried out on the flesh tissue of late-harvested "Conference" pears stored in CA containing 1% O_2 + 10% CO_2 at $-1°C$ for up to six months, conditions suitable for inducing browning (Pedreschi et al., 2009). The flesh tissue-browning in pears was mainly related to a disturbed energy metabolism, alteration in concentrations of metabolites dependent on energy metabolism pathways, a collapsed antioxidant system, and cell-wall architecture. The brown tissue exhibited a decrease of malic acid and an increase in fumaric acid and gamma aminobutyric acid (GABA), which indicated a reduced metabolic activity at the level of the Krebs cycle and a putative block of the GABA shunt pathway. Increased gluconic acid concentration might be related to AA degradation due to insufficient reducing equivalents or to an impaired pentose phosphate pathway. The concentrations of other compounds, such as trehalose and putrescine, were also considerably higher in brown (damaged) tissue than in sound tissue, suggesting hypoxic stress. The concentration of some sugars, which are typically found in xyloglucans, also increased during the development of brown tissue, possibly indicating cell-wall breakdown due to enzymatic processes or chemical reactions of hydroxyl radicals (Pedreschi et al., 2009). A proteomics study has also shown that down-regulation of ascorbate peroxidase (APX), glutathione transferase (GT), and monodehydroascorbate reductase (MDHAR) in brown tissue was indicative of total impairment of the ascorbate–glutathione cycle (Pedreschi et al., 2007). These studies indicate that development of core breakdown in "Conference" pears is a consequence of an imbalance between oxidative and reductive processes caused by too-low oxygen or too-high carbon dioxide conditions, which lead to a deficiency of reducing equivalents for defensive mechanisms, cell-damage repair processes and biosynthesis reactions.

13.3.3.4 Puffing Disorder in Citrus

Puffing disorder in citrus has recently been studied using a combination of transcriptomics and metabolomics tools (Ibáñez et al., 2014). Flavedo, albedo and juice sac tissues of normal citrus fruits and fruits showing puffing disorder were studied using metabolomics at three developmental stages. Glycolysis, which is the backbone of primary metabolism, appeared to be severely affected by the disorder based on both transcriptomic and metabolomic results. The concentration of citric acid was significantly lower in puffed fruits (Ibáñez et al., 2014). Transcript analysis further showed that glycolysis and also carbohydrate metabolism were significantly altered in puffed samples in both albedo and flavedo. Gene expression of invertases and sucrose exporters, amylose-starch and starch-maltose converters was higher in puffed fruits, while genes related to gibberellin and cytokinin signaling were down-regulated in symptomatic albedo tissues, suggesting that these hormones play key roles in the disorder. The outcomes of this study may be applied toward the development of early diagnostic methods based on host response genes and metabolites (i.e., citric acid), and toward therapeutics based on hormones (Ibáñez et al., 2014).

13.3.4 Preharvest Treatments

The targeted profiling of various metabolites in response to preharvest treatments have the potential to provide greater insights into the understanding of the basic mechanism of action of these treatments.

Preharvest application of ethephon (2-chloroethylphosphonic acid) on sweet cheery (*Prunus avium* L.) has been reported to induce significant shifts in the metabolome of fruit (Smith et al., 2011). GC–MS-based metabolic profiling was conducted on cherry cultivars "Bing," "Chelan," and "Skeena" fruit mesocarp and exocarp tissue to better understand underlying quality-related metabolism associated with ethephon application. Nearly 200 identified and partially characterized metabolites from mesocarp and exocarp tissue were characterized and evaluated. PCA models revealed changes in the metabolome associated with both natural ripening and ethephon-induced changes, including associations to key color, acid, and sugar components, such as cyanidin 3-glucoside, malic acid, and sugar metabolism. This metabolomics approach successfully deciphered the effects of ethephon treatment on primary and secondary metabolites that are responsible for color and flavor attributes of three important cherry cultivars. The overview of the whole metabolome response has the potential to screen the response of different cultivars to such preharvest treatments aimed at improving fruit quality and harvest practices.

NMR-based global metabolomics approach was followed by Zhang et al. (2012) to compare the metabolite profiles obtained for "Satsuma" mandarin orange juices prepared from fruit harvested in eleven separate orchards, to determine whether preharvest orchard factors such as the application of foliar fertilization or pesticides would significantly alter the metabolite concentration of the juices. The results showed that both foliar fertilization and pesticides lowered the Brix/acid ratio, and caused major changes to amino acid levels, as well as levels of other organic molecules. The change of 1 unit Brix/acid ratio was detected by 68% of subjects in an untrained sensory panel (Zhang et al., 2012). This study demonstrated that metabolomic analysis may be useful to optimize orchard practices such as fertilizer- and pesticide-use to obtain an optimal sensory profile.

In a recent study on improvement of litchi pericarp color through preharvest application of plant growth regulators (PGR), abscisic acid (ABA) and ethephon, Singh et al. (2014) reported that exogenous application of ABA at the color-break stage significantly increased the concentration of total anthocyanins and cyanidin-3-*O*-rutinoside (the major anthocyanin contributing 71%–96% of the total anthocyanins) in litchi pericarp, compared to ethephon. Among the different anthocyanins quantified, the relative contribution of cyanidin-3,5-diglucoside to the total anthocyanins was significantly higher in all PGR-treated fruit compared to control, but the concentration of cyanidin-3-*O*-glucoside was specifically enhanced by ABA. No significant effect on the concentrations of epicatechin, and quercetin-3-*O*-rutinoside was observed in response to PGR treatments (Singh et al., 2014). Ethephon treatment did not significantly increase the anthocyanins levels in the pericarp, but resulted in greater degradation of chlorophyll pigments than in the untreated control fruit. This work is an example of comprehensive profiling of different types of anthocyanins, chlorophylls and other flavonoids in the pericarp peel responsible for the appearance-quality of fruit.

13.3.5 Postharvest Treatments

There is increasing interest in the application of metabolomics alone or in conjunction with transcriptomics and proteomics to understand the effects of postharvest treatments on the biological processes in the fruits. The effects of postharvest treatments with 1-MCP and DPA on apple peel metabolome are discussed in the previous Section 13.3.3.1.

Postharvest heat treatments have been widely studied and used to control a range of postharvest disorders and are commercially used as quarantine treatments. Perotti et al. (2011) reported the changes in proteome and metabolome of "Valencia" orange (*Citrus sinensis* "cv. Valencia" late) subject to a postharvest heat treatment. The heat treatment resulted in numerous changes in the general metabolism of orange fruit in both exocarp and in endocarp, such as alterations in the activities of antioxidant enzymes, induction of key proteins in response to pathogen attack, changes in compounds involved in major metabolic pathways and possibly a cellular-reorganization process. All these changes lead to a lower degree of susceptibility of the fruit against fungal pathogens, while explaining the maintenance of postharvest quality (Perotti et al., 2011).

Yun et al. (2013) also conducted a proteomic and metabolomic study of mandarin peel during storage after a two-min. hot-water treatment (HWT) at 52°C. This treatment successfully suppressed the development of *Penicillium italicum* and reduced chilling-related injury during storage. The metabolite analysis was performed following a hybrid approach using both GC–MS and LC–QTOF–MS platforms. About 62 metabolites were identified, which were grouped into alcohols, amino acids, sugars, organic acids, and fatty acids. Most sugars and fatty acids increased in the peel after the HWT, but the sugar levels returned to control levels during storage (Yun et al., 2013). The levels of organic acids were lower immediately after the HWT but then increased to a higher level than control fruit during storage. Amino acids were also shown to decrease with the HWT and remained lower than control during storage. An exception was ornithine, which was 2.5 times higher than control fruit. Interestingly, H_2O_2 content decreased, while lignin content increased in heat-treated peel compared to the control. This may import increased fruit resistibility in response to external stress. In addition flavonoids, substances that are well-known to be effective in reducing external stress, were shown to be up-regulated in heat-treated pericarp (Yun et al., 2013).

To prevent or ameliorate chilling-related injuries in peaches, heat treatment is often applied prior to cold storage. Lauxmann et al. (2014) conducted metabolic profiling to determine the metabolites' dynamics associated with the induction of acquired tolerance to chilling-related injuries in response to heat-shock. "Dixiland" peach fruits exposed to a heat treatment, cold stored, or after a combined treatment of heat and cold, were compared with untreated control fruit ripened at 20°C. Dramatic changes in the levels of compatible solutes such as galactinol and raffinose were observed, while amino acid precursors of the phenylpropanoid pathway were also modified due to the stress treatments, as was the polyamine putrescine (Lauxmann et al., 2014). The changes in the metabolome of peach during and after heat treatment showed similarity to those changes observed in citrus by Yun et al. (2013). The sugars and alcohol sugars were increased in heat-treated fruit and most remained high when fruit was transferred to 20°C, though fructose and glucose returned to control fruit levels. Galactinol was the sugar most elevated by heat treatment and remained high if the fruit was placed in 0°C, but not when the fruit was maintained at 20°C. Organic acids tended to decrease in heat-treated fruit compared to their levels at harvest. The amino acid levels in heat-treated peach fruit generally increased either during or after the treatment. The identification of such key metabolites, which predispose the fruit to cope with different stress situations, will likely greatly accelerate the design and the improvement of plant-breeding programs and postharvest treatment protocols.

Quarantine regulations in several mango-importing countries, such as Australia, Japan, and New Zealand, require the fresh fruit to undergo a postharvest vapor heat treatment (VHT) in order to be accepted for import. The objective of VHT is to eliminate the risk of the entry of insect pests associated with the fruit into the importing country. VHT involves the use of hot air saturated with water vapor to heat the fruit core to a specified temperature and hold that temperature for a defined period to ensure that all target insect pests are killed. However, VHT is known to cause some deleterious effects on fruit quality. Singh and Saini (2014) reported the influence of postharvest VHT on targeted profiling of aroma volatiles during fruit-ripening in mango (cv. "Chausa") using GC–MS. Reversible inhibition of the emission of aroma volatiles was observed in heat-treated fruit, with a significant alteration in the aroma-volatiles profiles at different stages of fruit-ripening. The heat-induced increase in the rate of fruit-ripening proceeded with a significant lag in the emission of aroma volatiles. The suppression of aroma volatiles at the ripe stage in heat-treated fruit might adversely impact the consumers' acceptance of fruit. The temporal and quantitative variation in the aroma-volatiles profiles of mango fruit in response to VHT were observed through this study, and demonstrate the versatility of these techniques.

13.3.6 POSTHARVEST STORAGE CONDITIONS

The targeted analysis of a few metabolites responsible for specific quality attributes such as ethanol and vitamin C content using GC and HPLC techniques, is common in many postharvest studies and will not be discussed in this section. The emphasis of this discussion will focus on the high-throughput targeted profiling of important metabolites and explore their regulatory roles, using informatics tools that can provide comprehensive information about the fate of these molecules in response to various postharvest storage conditions. Matsumoto and Ikoma (2012) recently reported the application of the LC–MS/MS approach to quantitate sugars, organic acids and amino acids in the juice sacs of "Satsuma" mandarins (*Citrus unshiu* Marc. cv. "Aoshima-unshiu") stored at 5°C, 10°C, 20°C, and 30°C for 14 days. Without any significant change in sugars during storage at different temperatures, organic acids decreased slightly at all temperatures, with the exception of malic acid at 30°C, which increased slightly. Two amino acids, ornithine and glutamine, increased at 5°C, but they did not increase at other temperatures. The content of 11 amino acids (phenylalanine, tryptophan, tyrosine, isoleucine, leucine, valine, threonine, lysine, methionine, histidine, and γ-amino butyric acid) was higher at 20°C and 30°C than at other temperatures. Moreover, amino acids responded to temperature differently: two amino acids were cold-responsive, and 11 were heat-responsive. The best temperature to minimize the postharvest changes in amino acid profiles in the juice sacs was 10°C. This study demonstrates the value of high throughput LC–MS/MS in clearly defining the biochemical changes associated with storage.

13.3.7 PRE- AND POSTHARVEST DISEASES

The concept of biomarker discovery is also evolving towards understanding host–pathogen interactions for disease diagnosis in fruit and vegetables. The selective volatiles compounds from the headspace have been found useful in the identification of fungal diseases in apple fruit (Vikram et al., 2004). This approach can lead to the development of biosensors to monitor the development of disease during postharvest storage. Preharvest diseases are known to influence fruit quality and postharvest behavior. Baldwin et al. (2010) performed the targeted analysis of secondary metabolites to study the effect of Huanglongbing (HLB) on citrus fruit flavor. The results showed that asymptomatic fruit from symptomatic trees were similar to healthy fruit for many of the quality factors measured, but that juice from asymptomatic and, especially, symptomatic fruits, were often higher in the bitter compounds limonin and nomilin (Baldwin et al., 2010). The application of the metabolomics tool has even been explored to differentiate HLB infection from zinc deficiency in citrus trees (Cevallos-Cevallos et al., 2011b). HLB is a devastating disease in many parts of the world, but the symptoms are similar to zinc deficiency, and developing a reliable and robust method to differentiate these symptoms is a priority for industry. Cevallos-Cevallos et al. (2011b) identified six possible biomarkers for HLB, of which four were identified as proline, β-elemene, (-) trans-caryophyllene, and α-humulene were very useful for rapid differentiation of HLB from zinc deficiency. A more recent study has compared the effects of HLB on the citrus metabolome analysing fruit, juice from healthy, HLB-asymptomatic, and HLB-symptomatic Hamlin, as well as from healthy and HLB-symptomatic "Valencia" sweet oranges (Chin et al., 2014). The NMR-based metabolomics approach revealed differences in the concentration of several metabolites, including phenylalanine, histidine, limonin, and synephrine, between control or asymptomatic fruit and symptomatic fruit, regardless of the citrus variety or location. The data indicated that HLB infection presents a strong metabolic response that is observed across different cultivars and regions, suggesting the potential for generation of metabolite-based biomarkers of HLB infection.

13.4 METABOLOMICS APPLICATIONS IN SAFETY OF FRESH PRODUCE

Food safety is an emerging challenge for many horticultural industries. Recent food-borne outbreaks have raised serious public-health concerns and prompted debate over the responsibilities and accountabilities of primary producers, food industries, and regulatory authorities. From the public-health perspective, chemical and microbial hazards are the most common in fresh produce. The chemical hazards related to the presence of agrochemical residues, such as fungicides, herbicides, and insecticides, on fresh produce, are strictly monitored and regulated by the authorities. Residue levels in fresh produce tend to decrease after application during any withholding period and during postharvest washing and in storage/transit time in the supply chain. However, the microbial hazards, which include contamination with human pathogens during preharvest and postharvest stages, tend to exponentially increase under favorable conditions in the supply chain and pose severe risk to the consumers. Fresh horticultural produce have recently been implicated in several food-borne illnesses. Some horticultural crops, such as melons, leafy vegetables, berries, nuts, and tomatoes, are considered high-risk due to their susceptibility to carry and support proliferation of human pathogens. There are several risk factors during on-farm production and postharvest handling of the produce, which predispose horticultural produce to microbial and chemical hazards. To prevent exposure of consumers to these risks, risk assessment is a mandatory practice to meet the regulatory requirements of food-safety authorities. The application of advanced analytical techniques with high resolution, sensitivity and specificity, constitutes the core of the food-safety risk-assessment plans for fresh produce. This section will review the recent developments in application of metabolomics tools to determine the maximum residue limits (MRL) in fresh produce, to develop tools for ensuring microbial safety and to address the traceability concerns.

13.4.1 DETECTION AND QUANTIFICATION OF CHEMICAL RESIDUES

The agrichemical residues are among the most important food-safety concerns for fresh produce. The MRLs on fresh produce have been defined for most pesticides and toxins by different countries. However, the MRLs are usually at very low levels, requiring highly accurate and sensitive analytical platforms for measurements. The increasing pressure on food-safety authorities due to public health and environmental issues is further pushing the MRLs to extremely low levels. MS-based approaches have significantly enhanced the detection and quantification limits for various chemical residues and toxins. These methods are reliant upon the hyphenated techniques involving separation (chromatography) and detection systems (MS). The separation techniques have evolved in the recent past towards high-throughput and analyte-specific column chemistry. For example, ultra-high-performance liquid chromatography (UHPLC) is emerging as the technique of choice for coupling with mass spectrometers. The separation of analytes has also been enhanced through two-dimensional chromatography methods such as 2D–GC and 2D–HPLC. Furthermore, sample preparation techniques have also been improved to achieve better and selective recoveries of compounds of interest. A typical sample-preparation protocol for chemical residues involves three steps: extraction, clean-up and preconcentration. Solvent extraction and solid-phase extraction (SPE) are the procedures most commonly applied. However, the development of the QuEChERS (quick, easy, cheap, effective, rugged, and safe) protocol is increasingly employed for sample-preparation before LC–MS applications. This method basically involves an extraction step with acetonitrile, followed by pH adjustment for phase separation, centrifugation, and SPE clean-up. Other extraction techniques, such as solid-phase microextraction (SPME) and stir bar sorptive extraction (SBSE), have been shown to successfully extract and enrich the compounds of interests from different food matrices. In addition, the automation of SPME and SBSE techniques (e.g., Gerstel's Multipurpose Sampler) has accelerated the adoption and outcomes of these techniques.

The goal of metabolomics applications in chemical residues is to achieve the highest sensitivity, accuracy and repeatability of results in the shortest possible analysis time for multiple classes of pesticides in different food matrices. The high-throughput methods based on GC–MS/MS and LC–MS/MS have been developed and commercially adopted for determination of pesticide residues in fresh produce. Camino-Sánchez et al. (2011) reported a rapid, sensitive, accurate and reliable multiresidue method for quantification and confirmation of 121 common agricultural pesticides in fruits and vegetables by GC–EI–MS/MS (QqQ in SRM mode). This method was accredited according to International Standard UNE-EN ISO/IEC 17025:2005 and was validated and applied to 1463 vegetable and fruit samples collected over one year from extensive greenhouse cultures in Spain. This method separated the pesticides in less than 30 min. and the limit of quantitation was <10 µg kg^{-1}. The results showed that only three pepper samples, one of tomato and one of cucumber had residues above the MRL (0.01 mg kg^{-1}) according to European Union directives (Camino-Sánchez et al., 2011). The development of the multiresidue analytical method based on GC–EI–MS/MS was further extended by Banerjee et al. (2012) when they reported an excellent method for the simultaneous estimation of 375 organic contaminants, including 349 pesticides, 11 polychlorinated biphenyls (PCBs), and 15 polyaromatic hydrocarbons (PAHs), extracted from grape, pomegranate, okra, tomato, and onion matrices. The GC–EI–MS/MS parameters were optimized for analysis of all the 375 compounds within a 40 min run-time with limit of quantification for most of the compounds at <10 µg L^{-1}, which is well below their respective European Union MRLs.

Núñez et al. (2012) developed a method based on LC–MS and LC–MS/MS for the confirmation of a group of 100 pesticides (herbicides, insecticides, and fungicides) in fruits and vegetables. LC–MS/MS—using highly-selective selected reaction monitoring (H-SRM) acquisition mode, monitoring two transitions for each compound—was reported to be the most sensitive methodology. Limits of detection (by acquiring two transitions and with the ion-ratio requirements) ranged between

0.01 and 20 µg kg^{-1} were obtained. This method could meet sensitivity requirements for the MRLs established by the European Union regulation for food monitoring programs. The LC–MS and LC–MS/MS strategies developed were successfully applied for the analysis and confirmation of pesticides in different types of fruit and vegetables samples. In addition to these general pesticides-residue-testing methods, there are some reports on commodity-specific analytical methods for various pesticides. For example, Fleury-Filho et al. (2012) developed a highly selective and sensitive method based on LC–MS/MS for the simultaneous determination of 98 pesticide residues in mango fruit. Similarly, Banerjee et al. (2013) developed a GC–MS method for multiresidue determination of 47 pesticides in grapes with limit of quantifications of each compound in compliance with the EU MRL requirements. This method was successfully applied for analysis of real-world samples for incurred residues.

Methods based on QqQ–MS have been most commonly adopted for multiresidue analysis in fresh produce, as these allow identification of target residue and quantification with an adequate concentration range and reproducibility in complex food matrices at extremely low levels (ng kg^{-1}). On the other hand, the high resolution mass spectrometers (HRMS) systems provide high resolution, accurate mass and high full-scan sensitivity and selectivity, making them attractive for residue analysis in food (Gómez-Ramos et al., 2013). They allow operation for both target and non-target compounds, and options to work in full-scan mode or MS/MS mode, and in sequential or simultaneous mode. This translates into a rapid, effective option for routine laboratories. The drawbacks with the HRMS system, such as accurate mass data obtained above 5 ppm, lower resolving powers than 20,000, saturation effects with certain compounds, and short dynamic range, have been now addressed in the advanced systems to greater extent (Gómez-Ramos et al., 2013). Some studies have shown the quantitative capabilities of HRMS instruments to QqQ systems, and are valuable indicators of the potential of LC–TOF–MS for large-scale quantitative multiresidue analysis. Taylor et al. (2008) evaluated the potential of UPLC–TOF–MS for the quantitative analysis of 100 pesticides in strawberry fruit and the results were compared with those obtained using an UPLC–MS/MS MRM. Residues found in the samples ranged from 0.025 to 0.28 mg kg^{-1} and they were in excellent agreement with results obtained using UPLC–MS/MS. Lacina et al. (2010) investigated the potential of UPLC–TOF–MS in the analysis of 212 pesticides in apple, strawberry, spinach, and tomato. Compared in tandem with mass analyzers such as QqQ, the sensitivity of TOF is lower and the linear dynamic range of the studied instrument is rather narrower. The detailed account of application of metabolomics tools, based on LC–MS/MS, for pesticide residues in fresh produce, can be obtained from the reviews published by Castro-Puyana and Herrero (2013) and Gómez-Ramos et al. (2013).

13.4.2 Detection of Mycotoxins

Mycotoxins are secondary metabolites produced by various fungi on a broad range of food, including fresh or dried horticultural fruit and vegetables. These mycotoxins are capable of causing acute toxic, carcinogenic, mutagenic, teratogenic, immunotoxic, or oestrogenic effects in animals and humans. Considering their potential to cause toxicity, maximum concentrations of mycotoxins permissible in food are defined by regulatory authorities to prevent consumer exposure to them. The most important mycotoxins in food and feed, which are regulated, are aflatoxins (AFB1, AFB2, AFG1, AFG2), ochratoxin A (OTA), type A and B trichothecenes (e.g., HT-2 toxin, T-2 toxin and deoxynivalenol (DON)), fumonisins, and zearalenone. The multitarget analytical methods have replaced the single analyte methods due to their versatility and throughput (Ren et al., 2007; Sulyok et al., 2007; Spanjer et al., 2008). New metabolomics-based methods have been recently developed and validated in order to identify, quantify/semi-quantify the levels of mycotoxins in different foods and nuts. Varga et al. (2013) reported a multitarget method for the determination of 191 fungal metabolites in almonds, hazelnuts, peanuts, and pistachios. The method included all mycotoxins regularly found in food, and the mycotoxins regulated in the European Union. The applicability of the developed method was demonstrated through the analysis of 53 naturally-contaminated nut samples from Austria and Turkey. Varga et al. (2013) showed that of the 40 toxins that were quantified; the most frequently found mycotoxins were beauvericin (79%), enniatin B (62%) and macrosporin (57%). In the most contaminated hazelnut sample, 26 different fungal metabolites were detected. The application of UPLC-MS/MS has also been extended to quantitate 26 mycotoxins in finished grain and nut products (Liao et al., 2013).

Patulin is another very important mycotoxin in apple and apple-derived products such as apple juice, cider and puree. It has also been reported from other fresh fruits and fruit-derived products such as blueberry, cranberry, raspberry syrup, and grape juice. A recent study has shown that patulin can be found on moldy fresh fruits such as tomatoes, bell pepper, and soft red fruits (Van de Perre et al., 2014). Patulin is produced by more than 60 fungal species, of which *Penicillium expansum* (blue mould) is the most important and often involved in postharvest rots of fresh fruits. The Codex Alimentarius recommends levels of patulin in fruits and fruit juices to be lower than 0.05 mg kg^{-1}. In 2006, the European Commission established the following maximum levels of patulin in apple products: 0.05 mg kg^{-1} for fruit juices and other drinks derived from apple or apple juice; 0.025 mg kg^{-1} for solid apple products; and, 0.01 mg kg^{-1} for apple products intended for infants and young children, and baby foods that are not cereals-based products. Therefore, the estimation of patulin at such low concentrations is a challenging task for commercial testing laboratories. Most multitarget analytical methods developed for mycotoxins have not included patulin among the targets. This is primarily due to its high polarity and low molecular mass, which commonly lead to low recoveries and/or low sensitivity, hampering its determination at the regulatory levels.

There are several methods available for the determination of patulin in different food products, including a multitarget method for mycotoxins developed by Varga et al. (2013). Recently, a sensitive, quick and reliable method based on UPLC–MS/MS has been reported to estimate patulin content in fruit samples (Beltrán et al., 2014). In this method, chromatographic separation was achieved in less than 4 min. The comparison of the ion sources showed that the use of ESI caused strong signal-suppression in samples; however, matrix effect was negligible using APCI, allowing quantification with calibration standards prepared in solvent. The method was validated in four different apple matrices (juice, fruit, puree, and compote) at two concentrations at the low μg kg⁻¹ level (Beltrán et al., 2014). In addition, a GC–MS based method with a QuEChERS procedure for determination of patulin in apple juices, has been developed and validated by Kharandi et al. (2013). However, this method still requires a derivatization procedure using N,O-bis-trimethylsilyl trifluoroacetamide, which is time-consuming. There is further scope to develop new techniques to provide analytical solutions for mycotoxins widely differing in their polarity.

13.4.3 DETECTION OF PATHOGENS

Rapid detection of food-borne pathogens is critical to reduce food-related disease outbreaks and product recalls. Most important pathogenic bacteria, such as *Salmonella, Escherichia coli OH157: H7* and *Listeria monocytogenes*, have the potential to contaminate fresh horticultural produce during production and postharvest handling. Therefore, the application of diagnostic tools to detect and identify these pathogens within a short span of time is a challenging task for the food industry. Detection of pathogens using traditional plating technologies takes many days, while rapid technologies based on antibody or polymerase chain reaction (PCR) detection usually yield results in less than 48 h. The genomic technologies such as PCR, real-time PCR, and even next-generation sequencing technologies, have the potential to replace the conventional phenotyping approaches (Lauri and Mariani, 2009). In addition, proteomics tools using MALDI–TOF have also been employed to identify and characterize the pathogenic bacterial strains (Barbuddhe et al., 2008; Dieckmann and Malorny, 2011). The discussion of these technologies in food safety is beyond the scope of this chapter.

With regard to metabolomics, some attempts have been made to apply MS tools to identify the pathogenic bacteria in food, but there are no current reports on such applications in fresh horticultural produce. A metabolomic-based method for rapid detection of *Escherichia coli* O157:H7, *Salmonella hartford, Salmonella typhimurium*, and *Salmonella muenchen* in nonselective media was developed (Cevallos-Cevallos et al., 2011a). PCA discriminated pathogenic microorganisms grown in culture media and the metabolites responsible of PCA classification were dextrose, cadaverine, the aminoacids L-histidine, glycine, and L-tyrosine, as well as the volatiles 1-octanol, 1-propanol, 1 butanol, 2-ethyl-1-hexanol, and 2,5-dimethylpyrazine. Partial least square (PLS) models, based on the overall metabolite profile of each bacterium, were able to detect the presence of *Escherichia coli* O157:H7 and *Salmonella* spp. at levels of approximately 7 ± 2 CFU/25 g of ground beef and chicken within 18 h. This research proposed an alternative approach to rapidly detect two major pathogens based on detecting changes in the metabolite profile during incubation in nonselective culture media (Cevallos-Cevallos et al., 2011a). Another recent study has described a proof-of-concept application of a metabolomic technique for the rapid detection of Listeria, applied to nutrient media and a complex food sample (milk) inoculated with a pathogenic Listeria strain (*L. monocytogenes*). It was found that a profile of intracellular and extracellular metabolites associated with *L. monocytogenes* could be obtained using GC coupled to orthogonal acceleration TOF–MS. Chemometric analysis showed that it is possible to differentiate between the uninoculated samples and samples inoculated with Listeria based on *L. monocytogenes* metabolic activity. This research demonstrates that metabolomics has the potential for rapidly identifying food contaminated with Listeria and could provide a means for enhancing monitoring programs and ensuring food safety (Beale et al., 2014). The outcomes of these studies are not directly transferable to an industrial environment, but these proofs of concept provide evidence of potential application and extension to other food commodities, such as fresh fruit and vegetables.

13.4.4   Traceability, Fraud, and Malpractices

Traceability of fresh produce has now become an essential requirement for the trade of fresh produce. It requires regulation and monitoring of the entire supply chain, allowing the food to be traced through every step of its production back to its origin. Labeling is an integral component of traceability that enables the identification of the produce through complex supply-chain systems. The proper label indicating country/region of origin, and production system (organic or integrated) of the produce not only informs the consumer, but also helps the food industry in fetching premium value of produce linked to some geographical indicators. Deliberate or accidental mislabeling, which can mislead the consumers, constitutes food fraud, as per the food safety and standards regulations of many countries. Furthermore, in the event of food-borne outbreaks and recalls, proper labeling facilitates the consumers as well as enforcement agencies to trace the background of the food during investigations and legal procedures.

Metabolomics tools have the potential to discover biomarkers for fresh produce linked to geographical origin and production systems. The discovery of markers having discriminating powers to authenticate the information on labels seems to be the future of research at this stage. The research in this direction has been quite slow and limited to some commodities such as olive oil, tea, coffee, and herbal products, but this area of application is also gaining momentum. The elemental compositions and stable isotope ratios of different fruits and nuts (Anderson and Smith, 2005; Perez et al., 2006) determined by inductively-coupled plasma MS showed the potential for "ionomics" to play an important role in traceability. A comprehensive review on traceability markers of olive oils has been presented by Montealegre et al. (2010) and highlighted the then state-of-art research using different approaches of genomics, chemometry, and ionomics. The high-resolution-magic-angle spinning–NMR (HRMAS–NMR) spectroscopy approach has been recently found to discriminate potato tubers grown under organic versus conventional management systems (Pacifico et al., 2013). Nitrogen content in organic potatoes decreased by 11%–14% and GABA and lysine accumulated in the organic tubers. For the first time, this study provided the basis for constructing a validated and robust experimental model that can efficiently distinguish potato tubers according to the cultivation system. However the metabolite-based markers linked to different types of production systems and production region have not been robustly tested and validated at large scales to bring them into commercial practice.

Metabolomic tools can also be applied to confirm the use of banned/prohibited substances during production and postharvest handling of fresh produce. Though most agrichemical residues can now be easily detected and quantified using metabolomics tools, there are still some gaps that need attention. For example, the use of calcium carbide for postharvest fruit-ripening is prohibited by law in several countries, including India. However, this compound has been continuously used for ripening mango, banana and papaya fruit, as the regulatory authorities are unable to distinguish the fruit ripened with ethylene and calcium carbide. The method of application of carbide barely leaves any residue on the fruit's surface as the powder, when comes in contact with moisture, liberates acetylene gas, which mimics the action of ethylene to induce fruit-ripening. Metabolomics approaches, using both GC–MS and LC–QTOF–MS techniques, have shown promising results to distinguish the fruit ripened with carbide and ethylene (Singh unpublished).

13.5 CONCLUSIONS

As metabolites are considered the downstream products of cellular regulatory processes, metabolomics data can precisely characterize cells, tissues, or whole organisms, by defining specific biochemical phenotypes that are representative of physiological or developmental states. The metabolomics approach is considered superior to other "omics" tools, which are often compromised with biological enigmas such as epigenetic modifications, posttranscriptional and posttranslational changes. The metabolites are the ultimate signatures of a genome, offering strong correlations with phenotypes and chemotypes. In plant and food systems, metabolomics are beginning to be applied to gain better understanding of various pathways and systems involved in improvement of horticultural produce for better nutrition, processing and postharvest storage and quality. The metabolomics applications have been extended to define and measure comprehensive quality traits in various food crops, to ensure food safety aspects (adulterants, contaminants, allergens, etc.), and to develop quality and processing standards for regulatory authorities. In the modern era, the definition of quality is expanding beyond a few parameters to multiple parameters determined using sophisticated analytical platforms to achieve excellence in developing new product(s) with highest standards of quality and safety.

REFERENCES

Anderson KA, Smith BW. 2005. Use of chemical profiling to differentiate geographic growing origin of raw pistachios. *J Agric Food Chem* 53: 410–418.

Baldwin E, Plotto A, Manthey J, McCollu G. 2010. Effect of liberibacter infection (Huanglongbing disease) of citrus on orange fruit physiology and fruit/fruit juice quality: Chemical and physical analyses. *J Agric Food Chem* 58: 1247–1262.

Banerjee K, Mujawar S, Utture SC, Dasgupta S, Adsule PG. 2013. Optimization of gas chromatography-single quadrupole mass spectrometry conditions for multiresidue analysis of pesticides in grapes in compliance to EU-MRLs. *Food Chem* 138: 600–607.

Banerjee K, Utture S, Dasgupta S, Kandaswamy C, Pradhan S, Kulkarni S, Adsule P. 2012. Multiresidue determination of 375 organic contaminants including pesticides, polychlorinated biphenyls and polyaromatic hydrocarbons in fruits and vegetables by gas chromatography–triple quadrupole mass spectrometry with introduction of semi-quantification approach. *J Chromatogr A* 1270: 283–295.

Barbuddhe SB, Maier T, Schwarz G, Kostrzewa M, Hof H, Domann E, Chakraborty T, Hain T. 2008. Rapid identification and typing of Listeria species by matrix-assisted laser desorption ionization–time of flight mass spectrometry. *Appl Environ Microbiol* 74: 5402–5407.

Beale DJ, Morrison PD, Palombo EA. 2014. Detection of Listeria in milk using non-targeted metabolic profiling of Listeria monocytogenes: A proof-of-concept application. *Food Control* 42: 343–346.

Beltrán E, Ibáñez M, Sancho JV, Hernández F. 2014. Determination of patulin in apple and derived products by UHPLC-MS/MS. Study of matrix effects with atmospheric pressure ionisation sources. *Food Chem* 142: 400–407.

Camino-Sánchez FJ, Zafra-Gómez A, Ruiz-García J, Bermúdez-Peinado R, Ballesteros O, Navalon A, Vílchez JL. 2011. UNE-EN ISO/IEC 17025:2005 accredited method for the determination of 121 pesticide residues in fruits and vegetables by GC-tandem MS. *J Food Compos Anal* 24: 427–440.

Carrari F, Baxter C, Usadel B, Urbanczyk-Wochniak E, Zanor MI, Nunes-Nesi A, Nikiforova V et al. 2006. Integrated analysis of metabolite and transcript levels reveals the metabolic shifts that underlie tomato fruit development and highlight regulatory aspects of metabolic network behavior. *Plant Physiol* 142: 1380–1396.

Castro-Puyana M, Herrero M. 2013. Metabolomics approaches based on mass spectrometry for food safety, quality and traceability. *Trends Anal Chem* 52: 74–87.

Cevallos-Cevallos JM, Danyluk MD, Reyes-De-Corcuera JI. 2011a. GC–MS-based metabolomics for rapid simultaneous detection of *Escherichia coli* O157:H7, *Salmonella typhimurium*, *Salmonella muenchen*, and *Salmonella hartford* in ground beef and chicken. *J Food Sci* 76: M238–M246.

Cevallos-Cevallos JM, García-Torres R, Etxeberria E, Reyes-De-Corcuera JI. 2011b. GC–MS analysis of headspace and liquid extracts for metabolomic differentiation of citrus Huanglongbing and zinc deficiency in leaves of "Valencia" sweet orange from commercial groves. *Phytochem Anal* 22: 226–236.

Chin EL, Mishchuk DO, Breksa AP, Slupsky CM. 2014. Metabolite signature of *Candidatus Liberibacter asiaticus* infection in two citrus varieties. *J Agric Food Chem* 62: 6585–6591.

Cuthbertson D, Andrews PK, Reganold JP, Davies NM, Lange BM. 2012. Utility of metabolomics toward assessing the metabolic basis of quality traits in apple fruit with an emphasis on antioxidants. *J Agric Food Chem* 60: 8552–8560.

Dieckmann R, Malorny B. 2011. Rapid screening of epidemiologically important *Salmonella enterica* subsp. *enterica serovars* by whole cell matrix-assisted laser desorption ionization–time of flight mass spectrometry. *Appl Environ Microbiol* 77: 4136–4146.

Dunemann F, Ulrich D, Boudichevaskaia A, Grafe C, Weber WE, Dunemann F, Ulrich D, Boudichevaskaia A, Grafe C, Weber WE. 2009. QTL mapping of aroma compounds analysed by headspace solid-phase microextraction gas chromatography in the apple progeny "Discovery" × "Prima." *Mol Breed* 23: 501 – 521.

Fiehn O. 2002. Metabolomics: The link between genotypes and phenotypes. *Plant Mol Biol* 48: 155–171.

Fleury-Filho N, Nascimento CA, Faria EO, Crunivel AR, Oliveira JM. 2012. Within laboratory validation of a multiresidue method for the analysis of 98 pesticides in mango by LC tandem MS. *Food Addit Contam* 29: 641–656.

Gómez-Ramos MM, Ferrer C, Malato O, Agüera A, Fernández-Alba AR. 2013. Liquid chromatography–high-resolution mass spectrometry for pesticide residue analysis in fruit and vegetables: Screening and quantitative studies. *J Chromatogr A* 1287: 24–37.

Hertog MLATM, Rudell DR, Pedreschi R, Schaffer RJ, Geeraerd RH, Nicolaï BM, Ferguson I. 2011. Where systems biology meets postharvest. *Postharv Biol Technol* 62: 223–237.

Hu C, Xu G. 2013. Mass-spectrometry-based metabolomics analysis for foodomics. *Trends Anal Chem* 52: 36–46.

Ibáñez AM, Martinelli F, Reagan RL, Uratsu SL, Vo A, Tinoco MA, Phu ML, Chen Y, Rocke DM, Dandekar AM. 2014. Transcriptome and metabolome analysis of Citrus fruit to elucidate puffing disorder. *Plant Sci* 217: 87–98.

Khan SA, Chibon PY, de Vos RC, Schipper BA, Walraven E, Beekwilder J, van Dijk T et al. 2012. Genetic analysis of metabolites in apple fruits indicates an mQTL hotspot for phenolic compounds on linkage group 16. *J Exp Bot* 63: 2895–2908.

Kharandi N, Babri M, Azad J. 2013. A novel method for determination of patulin in apple juices by GC-MS. *Food Chem* 141: 1619–1623.

Klie S, Osorio S, Tohge T, Drincovich MF, Fait A, Giovannoni JJ, Fernie AR, Nikoloski Z. 2014. Conserved changes in the dynamics of metabolic processes during fruit development and ripening across species. *Plant Physiol* 164: 55–68.

Lacina O, Urbanova J, Poustka J, Hajslova J. 2010. Identification/quantification of multiple pesticide residues in food plants by ultra-high-performance liquid chromatography-time-of-flight mass spectrometry. *J Chromatogr A* 1217: 648–659.

Lauri A, Mariani PO. 2009. Potentials and limitations of molecular diagnostic methods in food safety. *Genes Nutr* 4: 1–12.

Lauxmann MA, Borsani J, Osorio S, Lombardo VA, Budde CO, Bustamante CA, Monti LL, Andreo CS, Fernie AR, Drincovich MF, Lara MV. 2014. Deciphering the metabolic pathways influencing heat and cold responses during post-harvest physiology of peach fruit. *Plant Cell Environ* 37: 601–616.

Lee J, Rudell DR, Davies PJ, Watkins CB. 2012. Metabolic changes in 1-methylcyclopropene (1-MCP)-treated "Empire" apple fruit during storage. *Metabolomics* 8: 742–753.

Leisso R, Buchanan DA, Lee J, Mattheis J, Rudell. 2013. Cell wall, cell membrane, and volatile metabolism are altered by antioxidant treatment, temperature shifts, and peel necrosis during apple fruit storage. *J Agric Food Chem* 61: 1373–1387.

Liao CD, Wong JW, Zhang K, Hayward DG, Lee NS, Trucksess MW. 2013. Multi mycotoxin analysis of finished grain and nut products using high-performance liquid chromatography–triple-quadrupole mass spectrometry. *J Agric Food Chem* 61: 4771–4782.

Lisec J, Schauer N, Kopka J, Willmitzer L, Fernie AR. 2006. Gas chromatography mass spectrometry-based metabolite profiling in plants. *Nat Protoc* 1: 387–396.

Lombardo VA, Osorio S, Borsani J, Lauxmann MA, Bustamante CA, Budde CO, Andreo CS, Lara MV, Fernie AR, Drincovich MF. 2011. Metabolic profiling during peach fruit development and ripening reveals the metabolic networks that underpin each developmental stage. *Plant Physiol* 157: 1696–1710.

Maliepaard C, Alston FH, van Arkel G, Brown LM, Chevreau E, Dunemann F, Evans KM et al. 1998. Aligning male and female linkage maps of apple (*Malus pumila* Mill.) using multi-allelic markers. *Theor Appl Genet* 97: 60–73.

Matsumoto H, Ikoma Y. 2012. Effect of different postharvest temperatures on the accumulation of sugars, organic acids, and amino acids in the juice sacs of Satsuma mandarin (*Citrus unshiu* Marc.) fruit. *J Agric Food Chem* 60: 9900–9909.

Moco S, Bino RJ, Vorst O, Verhoeven HA, de Groot J, van Beek TA, Vervoort J, Ric de Vos A CH. 2006. A liquid chromatography–mass spectrometry-based metabolome database for tomato. *Plant Physiol* 141: 1205–1218.

Montealegre C, Alegre MLM, Garcia-Ruiz C. 2010. Traceability markers to the botanical origin in olive oils. *J Agric Food Chem* 58: 28–38.

Núñez O, Gallart-Ayala H, Ferrer I, Moyano E, Galceran MT. 2012. Strategies for the multi-residue analysis of 100 pesticides by LC–triple quadrupole MS. *J Chromatogr A* 1249: 164–180.

Osorio S, Alba R, Nikoloski Z, Kochevenko A, Fernie AR, Giovannoni JJ. 2012. Integrative comparative analyses of transcript and metabolite profiles from pepper and tomato ripening and development stages uncovers species-specific patterns of network regulatory behavior. *Plant Physiol* 159: 1713–1729.

Pacifico D, Casciani L, Ritota M, Mandolino G, Onofri C, Moschella A, Parisi B, Cafiero C, Valentini M. 2013. NMR-based metabolomics for organic farming traceability of early potatoes. *J Agric Food Chem* 61: 11201–11211.

Pedreschi R, Franck C, Lammertyn J, Erban A, Kopka J, Hertog M, Verlinden B, Nicolaï B. 2009. Metabolic profiling of "Conference" pears under low oxygen stress. *Postharv Biol Technol* 51: 123–130.

Pedreschi R, Hertog M, Robben J, Noben JP, Nicolaï B. 2008. Physiological implications of controlled atmosphere storage of "Conference" pears (*Pyrus communis* L.): A proteomic approach. *Postharv Biol Technol* 50: 110–116.

Pedreschi R, Muñoz P, Robledo P, Becerra C, Defilippi BG, van Eekelen H, Mumm R, Westra E, de Vos RCH. 2014. Metabolomics analysis of postharvest ripening heterogeneity of "Hass" avocadoes. *Postharv Biol Technol* 92: 172–179.

Pedreschi R, Vanstreels E, Carpentier S, Hertog M, Lammertyn J, Robben J, Noben JP, Swennen R, Vanderleyden J, Nicolaï B. 2007. Proteomic analysis of core breakdown disorder in "Conference" pears (*Pyrus communis* L.). *Proteomics* 7: 2083–2099.

Perez AL, Smith BW, Anderson KA. 2006. Stable isotope and trace element profiling combined with classification models to differentiate geographic growing origin for three fruits: Effects of subregion and variety. *J Agric Food Chem* 54: 4506–4516.

Perez-Fons L, Wells T, Corol DI, Ward JL, Gerrish C, Beale MH, Seymour GB, Bramley PM, Fraser PD. 2014. A genome-wide metabolomic resource for tomato fruit from *Solanum pennellii*. *Sci Rep* 4: 3859.

Perotti VE, Del Vecchio HA, Sansevich A, Meier G, Bello F, Cocco M, Garrán SM, Anderson C, Vázquez D, Podestá VE. 2011. Proteomic, metabalomic, and biochemical analysis of heat treated Valencia oranges during storage. *Postharv Biol Technol* 62: 97–114.

Ren Y, Zhang Y, Shao S, Cai Z, Feng L, Pan H, Wang Z. 2007. Simultaneous determination of multi–component mycotoxin contaminants in foods and feeds by ultra-performance liquid chromatography tandem mass spectrometry. *J Chromatogr A* 1143: 48–64.

Rudell DR, Buchanan DA, Leisso RS, Whitaker BD, Mattheis JP, Zhu Y, Varanasi V. 2011. Ripening, storage temperature, ethylene action, and oxidative stress alter apple peel phytosterol metabolism. *Phytochemistry* 72: 1328–1340.

Rudell DR, Mattheis JP. 2009. Superficial scald development and related metabolism is modified by postharvest light irradiation. *Postharv Biol Technol* 51: 174–182.

Rudell DR, Mattheis JP, Curry FA. 2008. Prestorage ultraviolet–white light irradiation alters apple peel metabolome. *J Agric Food Chem* 56: 1138–1147.

Rudell DR, Mattheis JP, Hertog MLATM. 2009. Metabolomic change precedes apple superficial scald symptoms. *J Agric Food Chem* 54: 8459–8466.

Singh SP, Saini MK. 2014. Postharvest vapour heat treatment as a phytosanitary measure influences the aroma volatiles profile of mango fruit. *Food Chem* 164: 387–395.

Singh SP, Saini MK, Singh J, Pongener A, Sidhu GS. 2014. Preharvest application of abscisic acid promotes anthocyanins accumulation in pericarp of litchi fruit without adversely affecting postharvest quality. *Postharv Biol Technol* 96: 14–22.

Siracusa L, Patanè C, Avola G, Ruberto G. 2012. Polyphenols as chemotaxonomic markers in Italian "long-storage" tomato genotypes. *J Agric Food Chem* 60: 309–314.

Smith ED, Whiting MD, Rudell DR. 2011. Metabolic profiling of ethephon-treated sweet cherry (*Prunus avium* L.). *Metabolomics* 7: 126–133.

Spanjer MC, Rensen PM, Scholten JM. 2008. LC-MS/MS multi–method for mycotoxins after single extraction, with validation data for peanut, pistachio, wheat, maize, cornflakes, raisins and figs. *Food Addit Contam* 25: 472–489.

Sulyok M, Krska R, Schuhmacher R. 2007. A liquid chromatography/tandem mass spectrometric multi-mycotoxin method for the quantification of 87 analytes and its application to semiquantitative screening of moldy food samples. *Anal Bioanal Chem* 389: 1505–1523.

Taylor MJ, Keenan GA, Reid KB, Fernández DU. 2008. The utility of ultra-performance liquid chromatography/electrospray ionisation time-of-flight mass spectrometry for multi-residue determination of pesticides in strawberry. *Rapid Commun Mass Spectrom* 22: 2731–2746.

Toffali K, Zamboni A, Anesi A, Stocchero M, Pezzotti M, Levi M, Guzzo F. 2011. Novel aspects of grape berry ripening and post-harvest withering revealed by untargeted LC–ESI–MS metabolomics analysis. *Metabolomics* 7: 424–436.

Van de Perre E, Jacxsens L, Van Der Hauwaert W, Haesaert I, De Meulenaer B. 2014. Screening for the presence of patulin in molded fresh produce and evaluation of its stability in the production of tomato products. *J Agric Food Chem* 62: 304–309.

Vanzo A, Jenko M, Vrhovsek U, Stopar M. 2013. Metabolomic profiling and sensorial quality of "Golden Delicious," "Liberty," "Santana," and "Topaz" apples grown using organic and integrated production systems. *J Agric Food Chem* 61: 6580–6587.

Varga E, Glauner T, Berthiller F, Krska R, Schuhmacher R, Sulyok M. 2013. Development and validation of a (semi-)quantitative UHPLC-MS/MS method for the determination of 191 mycotoxins and other fungal metabolites in almonds, hazelnuts, peanuts and pistachios. *Anal Bioanal Chem* 405: 5087–5104.

Vikram A, Prithiviraj B, Kushalappa A. 2004. Use of volatile metabolite profiles to discriminate fungal diseases of "Cortland" and "Empire" apples. *J Plant Pathol* 86: 215–225.

Wahyuni Y, Ballester AR, Tikunov Y, de Vos RCH, Pelgrom KTB, Maharijaya A, Sudarmonowati E, Bino RJ, Bovy AG. 2013. Metabolomics and molecular marker analysis to explore pepper (*Capsicum* sp.) biodiversity. *Metabolomics* 9: 130–144.

Yun Z, Gao H, Liu P, Liu S, Luo T, Jin S, Xu Q, Xu J, Cheng Y, Deng X. 2013. Comparative proteomic and metabolomic profiling of citrus fruit with enhancement of disease resistance by postharvest heat treatment. *BMC Plant Biol* 13: 44.

Zamboni A, Di Carli M, Guzzo F, Stocchero M, Zenoni S, Ferrarini A, Tononi P et al. 2010. Identification of putative stage-specific grapevine berry biomarkers and omics data integration into networks. *Plant Physiol* 154: 1439–1459.

Zanor MI, Rambla JL, Chaib J, Steppa A, Medina A, Granell A, Fernie AR, Causse M. 2009. Metabolic characterization of loci affecting sensory attributes in tomato allows an assessment of the influence of the levels of primary metabolites and volatile organic contents. *J Exp Bot* 60: 2139–2154.

Zhang J, Wang X, Yu O, Tang J, Gu X, Wan X, Fang C. 2011. Metabolic profiling of strawberry (*Fragaria × ananassa* Duch.) during fruit development and maturation. *J Exp Bot* 62: 1112–1118.

Zhang X, Breska AP, Mishchuk DO, Fake CE, O'Mahony MA, Slupsky CM. 2012. Fertilisation and pesticides affect mandarin orange nutrient composition. *Food Chem* 134: 1020–1024.

14 Recent Developments in Proteomic Analysis of Fruits

Jun Song

CONTENTS

14.1 INTRODUCTION

Proteomics is the study of "the entire protein complement expressed by a genome in a cell or tissue type" (Pandey and Mann, 2004). Proteomics now covers all aspects of protein research in a cell, encompassing the abundance and changes in time and biological behavior, as well as posttranslational modifications (PTMs) related to the biological functions and protein–protein interactions. Proteomics has become the crucial element of "omic" approaches or system biology tools, and is contributing to a new understanding of biological systems.

Fruit and vegetable quality is the result of combination of multiple attributes including appearance, flavor (taste and aroma), and nutritional components (Kader, 2004). Significant research efforts on both pre- and postharvest technologies aim to improve and maintain eating quality and nutritional content. Although there have been many successful postharvest applications implemented on many fruit and vegetables, there is little fundamental understanding of the gene, protein, and metabolite networks regulating the quality of fruits and vegetables. Genomic research has recently been applied on fruit and vegetables, and has begun to identify genes related to ripening and quality (Giovannoni, 2004). However, many of the biosynthetic pathways associated with fruit and vegetable quality are not fully understood (Song and Forney, 2008).

A common technique used in proteomics is the examination of differences between pairs of samples, such as produce at different developmental stages, held under different abiotic stress conditions, or between diseased and healthy tissue. Proteomics can be extended to examine the fundamental changes during postharvest handling, and storage. Research platforms employed in proteomic studies generally utilize high-resolution mass spectrometry to directly study the proteins and their modifications (top-down), or study the peptides digested from proteins, using liquid chromatograph and mass spectrometry (LC/MS), equipped with nanospray ionization source (bottom-up).

The early development of protein analysis has been challenging, where the results were incomplete, or were merely qualitative. In addition, it is well-known that changes in protein abundance may not always be coincident with gene expression data, or from DNA microarray, as mRNA abundance poorly correlates with protein abundance (Baginsky et al., 2010). However, with recent developments and advances in genome sequencing, considerable progress has been achieved with MS-based proteomics, and many proteins have been identified. MS-based proteomics have allowed the identification and quantitative profiling of organism proteomes, and the systematic analysis of protein modifications, and protein–protein interactions. All these developments have offered a new range of opportunities for geneticists and network biologists to improve existing models of how phenotypes emerge (Gstaiger and Aebersold, 2009). Ideally, proteomic studies should identify and quantify all the proteins in a cell. With the published data-base of *Arabidopsis thaliana*, more than 227,396 proteins have been listed from ca. 30,000 protein coding genes (NCBI: http://www.ncbi.nlm.nih.gov/, accessed on April 2014). However, the number of proteins that can be investigated in a single proteomic study, remains limited. As a classical proteomic separation platform, two-dimensional electrophoresis (2-DE), along with MS, has been applied to resolve hundreds/thousands of proteins, facilitating peptide composition analysis, sequencing, and polypeptide identification. 2-DE is especially useful for comparative studies between pairs of samples, and has been the dominant technique applied to fruits and vegetables, together with sample preparation, and protocol establishment. Significant improvements in proteomic research platforms have been made from a qualitative platform (2-DE) to techniques of a more quantitative nature (nongel based), especially with model systems such as human tissues and yeast.

In this chapter, only qualitative and quantitative analysis of proteins employing LC/MS on peptides, will be discussed. Readers can find detailed information on top-down proteomic research platforms in other reviews (Parks et al., 2007; Zhou et al., 2012). Data will also be presented on the potential application to fruit such as apples and pears.

14.2   SAMPLE PREPARATION

In a proteomic study, a major challenge is how to isolate as many proteins as possible from a biological sample, and in a form that is suitable for downstream analysis.

Fruits are considered recalcitrant plant tissues for proteomic analysis, due to relatively low protein content, and the presence of numerous interfering substances, such as pigments, carbohydrates, polyphenols, polysaccharides, and starch (Saravanan and Rose, 2004). Significant research has optimized sample preparation procedures from sample handling, protein extraction, and solubilization on young immature vegetative tissues (Gallardo et al., 2002; Jacobs et al., 2005). However, many of these protocols have been shown to be unsuitable and not applicable to mature fruit samples (Song et al., 2006).

As a general procedure, plant tissues are ground to a powder in liquid nitrogen, followed by acetone or acetone containing trichloroacetic acid for precipitation of proteins, which are then solubilized in a suitable buffer (Mechin et al., 2003; Wang et al., 2003). The precipitation agents are very important, where both acetone (with and without TCA) as well as methanol/ammonium acetate (with dithiothreitol, DTT) are widely used precipitation agents. In a study to improve proteomic procedures, Saravanan and Rose found no significant difference between TCA/acetone and phenol-based extraction procedures. However, they reported that the phenol extraction was better in removing the interfering compounds (Saravanan and Rose, 2004).

Another study improved the protein extraction protocol for mature fruit, using a protocol developed specifically for apple and banana fruit, that used a heated extraction buffer containing sodium dodecyl sulfate (SDS), DTT, glycerol, and Tris-HCl buffer (pH 8.5) (Song et al., 2006). This study showed that the SDS was useful in enhancing protein solubility, but then it had to be removed from the sample prior to LC/MS analysis, through acetone precipitation (Song et al., 2006).

It is well-known that phenol can be useful in the protein extraction procedure, and it is especially beneficial for the extraction of low concentrations of protein in vegetative plant tissues (Hurkman and Tanaka, 1986). Saravanan and Rose (2004) showed that a phenol extraction procedure—of protein from tomato and banana fruit—was comparable to protein precipitation with acetone, with an increase in protein spots and glycoproteins (Saravanan and Rose, 2004). However, Carpentier et al. (2005) found that TCA/acetone precipitation, and phenol extraction, gave comparable results on banana, apple, and potato plant tissues. Zheng et al. (2007) further showed that phenol protocol resulted in significantly higher protein yields from apple and strawberry fruit, and, through 2-DE analysis of apple protein extracts, revealed 1422 protein spots associated with the phenol protocol, and 849 spots with the SDS protocol. Zheng et al. (2007) suggested that a phenol-based protein extraction protocol should be used as a standard procedure for most fruit tissues. However, Wang et al. (2006) showed that the TCA/acetone and methanol washes, combined with a phenol/SDS protein extraction procedure, are more efficient than phenol and SDS alone (Wang et al., 2006). While Pedreschi et al. (2007) reported that methanol/ammonium acetate precipitation with DTT, and no PVPP at pH 8.0, would be the best procedure for protein extraction from pear fruit. Subsequent solubilization of the protein extract is very important to the downstream protein analysis. Solubilization requires the proteins to be dissolved and displayed on the gels, to generate discrete protein spots that are suitable for further MS analysis. Among the early stages of fruit proteomic studies, contributors to the development and optimization of fruit protein sample preparation procedures (Carpentier et al., 2005; Rose et al., 2004; Song et al., 2006; Wang et al., 2006; Zheng et al., 2007). The protein extraction methods developed for 2-DE are generally suitable for nongel-based approaches, using LC/MS, but SDS and CHAPS (3-[(3-Cholamidopropyl) dimethylammonio]-1-propanesulfonate) must be avoided, because they interfere with downstream LC/MS analysis. Most nongel-based proteomic technologies have been developed on human tissues, yeast, and bacteria, but these can be directly applied on fruit and plant samples.

14.3   ANALYTICAL PROTEOMIC PLATFORMS

14.3.1   2-DE-Based Comparative Proteomics

A gel-based 2-dimensional electrophoresis (2-DE) platform has been developed to separate proteins by their isoelectric point and size. This system has continually been improved with better resolution, as a result of the development of stable immobilized pH gradients (IPG)-based systems, which have considerable resolving power and reproducibility (Görg et al., 2004). In general, 2-DE based proteomic procedure includes isoelectric focusing (IEF, first dimension), sodium dodecyl sulphate poly-acrylamide gel electrophoresis (SDS-PAGE) separation (second dimension), protein visualization and detection, protein identification, protein quantitation, and bioin-formatics (Wittmann-Liebold et al., 2006). 2-DE-based proteomic techniques have shown to successfully investigate the presence and changes of thousands of proteins. With the development and application of mass spectrometry (MS) techniques to determine the molecular weight of proteins and peptides, and continuing completion of many genomic sequence databases of plants (include some fruit), to match with the masses, it has become possible, using MS platform, to conduct comprehensive and high throughput proteomic experiments on identification and quantitation of plant proteins (Newton et al., 2004; Roberts, 2002). Although 2-DE has limited ability to detect low abundance proteins, as well as membrane proteins, and narrow dynamic range of detection, due to the technical difficulties of gel electrophoresis, 2-DE has been widely used as one of the significant proteomic platforms on many fruit.

The detection and quantitation of proteins is through visualization, by staining. Radioactive labeling, nonfluorescent stains such as Coomassie Brilliant Blue™ (CBB), silver staining, or fluorescent stains, can be used as staining agents to visualize the proteins on 2-D gel, and create digital images, which can then be quantitatively analyzed, using specially designed software. Silver is a very sensitive staining agent and is MS-compatible, but its use has been limited due to its narrow dynamic range, prone to interferences, multistep process, and gel-to-gel variation (White et al., 2004). When different stains were compared, it was reported that both silver and fluorescent stains are more sensitive than CBB, with better dynamic range (Harris et al., 2007). Applications of fluorescent stains have become attractive due to their sensitivity and wider dynamic range of protein concentrations (Nishihara and Champion, 2002; Patton, 2000). Many fluorescent stains, such as Sypro Ruby, Deep Purple, RuTBs (ruthenium II-tris(bathophenan-throline dissulfonate), and Sulforhodamine G fluorescent stains, have been developed, and are commercially available, with good sensitivity (Lamanda et al., 2004; Lanne, 2004; Mackintosh et al., 2003), and no major differences, in terms of protein detention, in most protein samples, have been reported (Wheelock et al., 2006). Harris et al. (2007) found that among the five commercially available fluorescent stains, Sypro Ruby was shown to be superior to others, with high detection of total number of spots, and the lowest signal-to-noise ratio (Harris et al., 2007). First dimension separation of proteins on IEF strips requires that the proteins should be adequately resolved, and separated, as much as possible. In order to improve the resolution to separate thousands of proteins, 18 or 24 cm-long Immobiline™ IEF drystrips, and large format gels (18 × 24 cm), have become the standard size to increase the protein loads and efficiency. On the other hand, narrow pH range gradient strips can also be employed to extend the IEF efficiency (Görg et al., 2004). Under current commercial and physical limitations, most 2-DE gel electrophoresis have been conducted on IEF strips up to 24 cm in length, with a 3–11 linear or nonlinear pH range, and sequential SDS-PAGE at 21 cm long. A large scale gel system with combined multi-pI range, and multiple line SDS/PAGE gels, has been shown to improve the resolution which resulted in resolving of 11,000 proteins spots (Inagaki and Katsuta, 2004). Routine use of this platform type may be difficult, but it shows the possibilities for further technical improvements to 2-DE based technologies (Wu et al., 2008).

To overcome the chemical and physical limitations of protein separation by *pI* and molecular mass, using IEF strips and 2-DE, a third-dimensional electrophoresis can be applied. The use of the difference gel electrophoresis (DIGE) technique has improved quantitation, and minimized the gel-to-gel variation, in comparative proteomic research (Tonge et al., 2001). Before mixing the samples and 2-DE separation, the protein samples are labeled with three fluorophores Cy2, Cy3, and Cy5 stains, with the covalent labeling on lysine residues. After 2-DE separation, the proteins are scanned with a fluorescence scanner at different wavelengths specific to each dye, to reveal the different proteomes. To control the variation, one of the stains is used as internal standard, which then is used in a mix of all the experimental samples (treated, and control). While other two dyes will be used to differentially label the treated and control samples, all three labeled samples are then separated on the same gels (Tonge et al., 2001). The DIGE technique has been applied on several plant proteomic studies, such as *Arapidopsis* (Dunkley et al., 2006), and poplar leaves (Bohler et al., 2007). Further improvement in the use of the DIGE technique is with the use of an Alexa-labeled internal standard (ALIS) spiked into protein samples (Wheelock et al., 2006). The ALIS normalization method is based on incorporation of an identical protein standard into every sample that is spectrally separated from the sample to be analyzed. This makes the method robust, and resistant to the overall changes in protein abundance (Wheelock et al., 2006).

Mass spectrometry has become the central technological platform for proteomic research. Depending on the mass spectrometry for protein spot identification, either peptide mass fingerprint (PMF) from a MALDI-TOF (time of flight MS), or fragment of MS (MS/MS spectra), which are produced in the collision cell within the mass spectrometer to generate the MS/MS spectra can be applied. Both PMF and PFF rely on the protein sequences in order to interpret spectra (Palagi et al., 2006). For the sequences of peptides from MS/MS, without known sequences, *de novo*

sequencing approach can then be applied (Palagi et al., 2006). Using an apple allergen protein (mal d1) as an example, the 2-DE gel-based procedure followed by a LC/
MS analysis of digested proteins (peptides) with MS/MS spectra, has been illustrated in Figure 14.1.

14.3.2 NONGEL-BASED PROTEOMICS

Nongel-based proteomic techniques are based on either label-free or different
labeling technologies, which allow large-scale quantitative proteomic research on
protein populations, and are believed to overcome the shortcomings of gel-based
technologies.

14.3.2.1 Label-Free Approaches

To overcome the technical challenges in 1D and 2D electrophoresis, a higher resolution and higher capacity 2D separation has been achieved, with an in-line system using a biphasic nanocolumn, known as multidimensional protein identification
(MudPIT) (Link et al., 1999). This technique uses a 3–5 cm section of strong cation
exchange resin, upstream from a C_{18} resin, in a nanocolumn, to allow the accumulation of a large amount of protein, that is released onto the C_{18} column, with an elu-
ent with an increasing salt concentration. This technique increases the number of
digested proteins that can be identified, and enhances the detection of low abundance
proteins in the mixture (Delahunty and Yates, 2005; Washburn et al., 2002).

In such a label-free protocol, no label is required to be induced to the samples,
and the protein digest is directly analyzed on a mass spectrometer, coupled with the
nanospray LC separation. In general, the acquisition of LC/MS/MS data is only for
the most abundant peptides peaks at each time point (data dependent acquisition).
Therefore, the label-free technique for protein quantitation consists of two different strategies; either spectra counting, or intensity-based measurement of peptides.
Spectra-counting is widely used to count the number of MS/MS spectra, acquired
for a certain protein, and, thereby, to estimate the quantity of the protein. While
peptide-intensity can be measured by their precursor ions, it is used as a quantitative
measure of the quantity of peptides. Despite the inexpensive and easy nature, the
label-free platform requires a high resolution mass spectrometer, and is more sensitive to technical variations of LC/MS runs. An additional technique called MS^e on
the qTOF MS enables the detection of peptides, precursors, ions, and fragment ions
(MS/MS), using different collision energy from 10 eV to 35 eV (Silva et al., 2006).
A linear relationship between the MS signals, from three most intensive peptides and
proteins, was found (Silva et al., 2006). An excellent review of label-free proteomic
techniques is presented by Levin and Sabine (2010), Matros et al. (2011).

14.3.2.2 Labeling and Targeted Quantitative Proteomic Approaches

The labeled quantitative proteomic platform has been widely applied, and allows
quantitative comparison between/among biological samples, which are differentially
chemical tagged proteins or peptides. Isotope Coded Affinity Tag labeling (iCAT)
uses heavy versus light isotope coded tags, in two comparable samples. This technique allows differential chemical tagging of proteins from different samples, by

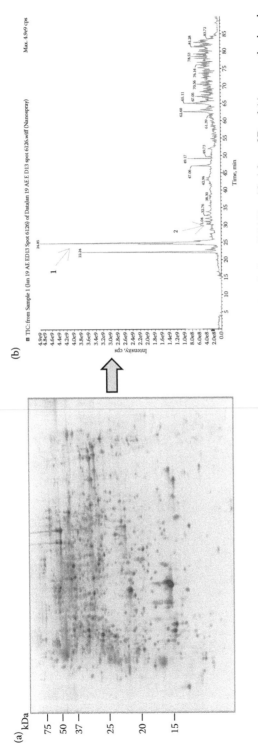

FIGURE 14.1 Illustration of spot identification from a 2D gel of apple fruit employing LC/MS. Spot 6126 identified from 2D gel (a) was excised and digested with trypsine and corresponding peptides were analyzed on a LC/MS system equipped with a nanospray ion source. Total ion chromatography is shown in (b). During LC/MS analysis, two peptides were detected and their MS/MS spectra were collected at retention time of 24.45 and 29.60 min, respectively (c) and (d). The MS/MS ion search employing MASCOT revealed doubly charged GAEILEGNGGPGTIK and IFTGEGSQYGYVK, respectively as peptide 1 and 2. The spot no. 6126 was therefore identified as major allergen mal d1 protein (MW: 17.69 kDa, pI 5.67). More detailed information can be obtained. (From Zheng, Q. et al. 2007. *J. Agric. Food Chem.* 55:1663–1673.)

(*Continued*)

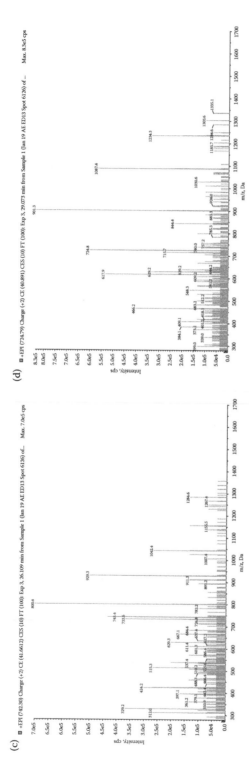

FIGURE 14.1 (Continued) Illustration of spot identification from a 2D gel of apple fruit employing LC/MS. Spot 6126 identified from 2D gel (a) was excised and digested with trypsin and corresponding peptides were analyzed on a LC/MS system equipped with a nanospray ion source. Total ion chromatography is shown in (b). During LC/MS analysis, two peptides were detected and their MS/MS spectra were collected at retention time of 24.45 and 29.60 min, respectively (c) and (d). The MS/MS ion search employing MASCOT revealed doubly charged GAEILEGNGGPGTIK and IFTGEGSQYGYVK, respectively as peptide 1 and 2. The spot no. 6126 was therefore identified as major allergen mal d1 protein (MW: 17,69 kDa, pI 5.67). More detailed information can be obtained. (From Zheng, Q. et al. 2007. *J. Agric. Food Chem.* 55:1663–1673.)

using heavy versus light isotope coded tags in two comparable samples. The relative abundance of proteins can be analyzed as peak-doubling in the same MS spectrum, thus reducing the analytical noise, and increasing the quantitative information of comparable sample data (Gygi et al., 1999). Similarly, stable isotope labeling by amino acids in cell culture (SILAC) allows differential metabolic labeling, followed by mass analysis of cell culture samples (Ong et al., 2002). Each labeling technique has its own advantages and disadvantages. For example, iCAT fails to quantify proteins with no cysteine residues, and it can only compare two samples at a time. Unlike iCAT, SILAC is an *in vivo* labeling method, which requires labeling on the amino acids involved in the growth medium. Among the available labeling approaches, another labeling on peptides, using amine reactive isobaric peptide derivatization reagents, such as iTRAQ (isobaric tags for relative and absolute quantitation), has been developed (Ross et al., 2004). These derivatized peptides are indistinguishable by their MS spectra or MS/MS series, but exhibit intense, low mass MS/MS signature ions (m/z 114–117), also called reporter ions, that permit quantitation of each member of a multiplex set, where four comparable samples can be analyzed in parallel (Ross et al., 2004). This iTRAQ approach can be applied to a set of samples up to 8, with quantitative information on peptides. As a recent development, iTRAQ has been applied on many plant research samples, such as *Arapidopsis* (Dunkley et al., 2004; Zieske, 2006), and fruit proteomes in grapes (Luecker et al., 2009) and tomatoes (Pan et al., 2014).

Another labeling strategy involves methylation of peptide amino groups via reductive amination with isotopically coded formaldehydes (Hsu et al., 2003). This technique was further developed by coupling it with 2D-LC-MS/MS (Boersema et al., 2008; Boersema et al., 2009). In this method, all primary amines (the N-terminus and the side chain of lysine residues) in a peptide mixture are converted to dimethylamines. By using combination of several isotopomers of formaldehyde and cyanoborohyride, peptide triplets can be obtained, that differ in mass by a minimum of 4 Da between different samples. Extracted proteins are digested with protease (trypsine), and the derived peptides are labeled with isotopomeric dimethyl labels, which are then mixed, and simultaneously analyzed, by LC/MS. The quantitation is based on the integration of the entire extracted ion peaks of each of the three m/z values for a peptide triplet, and used to compare the peptide abundance in the difference samples. The modification provides enhanced confidence in protein abundance ratios, statistical evaluation of significance, and discards proteins identified with only a single peptide (Melanson et al., 2006). As a new technique, an isotopic dimethylation labeling technique was applied to apple fruit, to investigate the significantly changed proteins during fruit ripening, and in response to ethylene treatment. This study lead to the identification and quantitation of more than 1000 proteins in apples (Jun Song, unpublished). Use of the technique to quantify apple mal d1 allergen protein during apple fruit ripening, and in response to ethylene treatment, is shown in Figure 14.2.

Direct comparison of iCAT, iTRAQ, DIGE, and 2-DE based techniques showed that DIGE had the same sensitivity as iCAT, but DIGE showed the shortcomings, such as comigration and partial comigration of spots, which may lead to misidentification of spots, and compromise the accuracy of quantitation of proteins (Wu et al., 2006). The iTRAQ technique showed the highest sensitivity, but it can only applied on

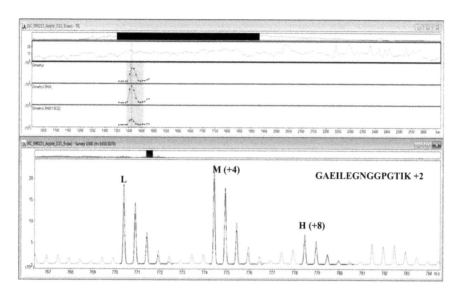

FIGURE 14.2 Example of quantitative proteomic analysis of apple proteins employing dimethylation isotope labeling during fruit ripening and in response to ethylene treatment. Among the identified and quantified 790 proteins, a major allergen, mal d1 (gi|4590376, MW:17.679 kD), was identified and quantified, based on a total of 32 peptides at day 21, after ethylene treatment at 20°C (Tables 14.1 and 14.2). The dimethylation resulted in a mass shift of 0 per label (light) or 4 per label (medium) or 8 per label (heavy). Changes in doubly charged peptide GAEILEGNGGPGTIK +2, one of the 32 quantified peptides, which has two labeling sites (both N-terminus and K), changed significantly among the labels from three different samples, control day 0, labeled as H (heavy, +8), ethylene treated at day 21, labeled as M (intermediate, +4) and control at day 21 as L (light). The total quantitative change of mal d1 protein was summarized by 34 labeled peptides from three different samples. The relative quantitative changes of mal d1 was shown to increase by 3.36 ± 1.94 folds during fruit ripening and further increased by 1.70 ± 1.28 folds to ethylene treatment. (Jun Song, unpublished.)

certain types of mass spectrometry instruments. Nevertheless, all three methods are complementary, and provide improved information for quantitative proteomics (Wu et al., 2006). An investigation on comparison between 2D LC technique and 2D–DIGE was conducted on rice proteome samples (Komatsu et al., 2006), and showed that 2D LC resulted in better detection of low abundance proteins, but it showed

---

TABLE 14.1

Sequence Table of Quantitative Changes in Peptide GAEILEGNGGPGTIK (an Apple Major Allergen, Mal.d1) in Apples Employing Dimethylation Isotope Labeling during Fruit Ripening and in Response to Ethylene Treatment

	L/M	SD	L/H	SD	Fraction	Correlation	Intensity	Modifications
17 GAEILEGNGGPGTIK	0.87	0.11	2.87	0.13	0.67	0.97	3222	

TABLE 14.2

Quantitation Table of Some Allergen Proteins in Apples Employing Dimethylation Isotope Labeling during Fruit Ripening and in Response to Ethylene Treatment

	Accession	Sore	Mass (Da)	L/M	SD	#	L/H	SD	#	Description
1.1	gi|4590368	2296	18,106	0.58	1.61	41	3.73	2.08	41	*Malus domestica*
1.2	gi|2443824	1952	18,059	0.54	1.55	35	3.59	1.95	35	*Malus domestica*
1.3	**gi|4590376**	**1598**	**18,208**	**0.59**	**1.26**	**32**	**3.36**	**1.94**	**32**	*Malus domestica*

poor reproducibility and, therefore, required a large number of repeats, enough for determining statistical significance. It also demonstrated that the 2D-DIGE method is comparable to 2D LC for quantitative comparisons, with very good sensitivity (Komatsu et al., 2006). Techniques such as MudPIT are well suited for the identification of hydrophobic proteins (Link et al., 1999; Washburn et al., 2002), and peptide sequencing by mass spectrometry (Steen and Mann, 2004). On the other hand, various peptide labeling techniques, such as ICAT (Gygi et al., 1999), SILAC (Ong et al., 2003) or iTRAQ (Ross et al., 2004), have become more frequently used for proteomic research.

Proteins undergo a variety of post-translational modifications (PTMs) in a living cell. Phosphorylation is one of the important PTMs of proteins, and plays an important role in proteomic studies, especially involved in cellular signaling pathways in plants and fruit (Peck, 2006). A strategy that combines the application of enrichments of both phosphorylated protein and peptides, with a separation of proteins analysis, was reported for *Arabidopsis* and *M. truncatula* (Laugesen et al., 2006). Affinity chromatography with immobilized metal ions (IMAC) was shown to allow the characterization of a high number of phosphorylation sites in *Arabidopsis* (Nuehse et al., 2007). In addition, three ethylene-enhanced and three ethylene-repressed unique phospopeptides in *Arabidopsis* mutant ein2 were also identified (Li et al., 2009). There have, however, not been any reports on PTMs of proteins in fruit and vegetables.

Another MS-based proteomic approach based on the selection of proteolyitc peptide detection as a targeted proteomic technique, has been developed (Anderson and Hunter, 2006). In this approach, protein identification and quantitation can be achieved by a selected reaction monitoring (MRM/SRM) technique, using triple quadrupole mass spectrometer. Peptide mixtures are separated by liquid chromatography, and then ionized, and detected. In the MRM mode, the mass spectrometer can be programed to only the list of target proteotypic peptides. Only the predefined precursor ions can be selected in the first quadrupole. A second mass filter, in the third quadrupole, allows filtering of the corresponding fragment ions, following collision-induced dissociation in the second quadrupole (Lange et al., 2008). The resulting precursor fragment ion pairs (transitions) are highly specific for a given proteotypic peptide, and are, therefore, unique for a given protein. By including isotope-labeled versions of the proteotypic peptides that are being analyzed, it is possible to simultaneously determine the absolute amounts of the protein. The approximate retention-time information can be used to restrict the time for the detection of a specific transition, and, therefore, allows the detection of multiple peptide ions in one measurement, a process that is referred to as scheduled selected reaction monitoring.

The detection range can be extended to single-digit copies/cell and to proteins undetected by classical methods (Picotti et al., 2010). Recently, a proteomic protocol combining OFFGEL and MRM studies to identify and quantitatively investigate monodehydro-ascorbate reductase, a key enzyme in apple fruit during fruit ripening, was reported (Yang et al., 2014).

14.3.3 INFORMATICS/BIOINFORMATICS

Mass spectrometry is the central technology for proteomic research, but several approaches can be applied for peptide and protein identification. For example, a set of peptide molecular masses from a specific enzyme digested proteins, peptide mass fingerprint (PMF), can be generated and determined, using a matrix-assisted laser desorption ionization (MALDI) MS. The experimental spectrum is compared with theoretical ones computed from protein sequence databases to create "similarity scores" for candidate protein (Henzel et al., 1993; Mann et al., 1993). PMF offers a fast and simple analysis, but it can only be used with a pure protein or a simple mixture, such as a spot from the 2-D gel (Palagi et al., 2006). Using tandem mass spectrometry (MS/MS) analysis, ionized peptides are isolated and fragmented (Pandey and Mann, 2004). During the analysis, the peptide is analyzed at the first stage of the mass analysis, then induced to fragment by collision, to generate a second mass analysis. The collected MS/MS spectra will then be used to match proteins in a database. Another limitation is that a database of protein or nucleic acid sequence is required, and it is therefore not applicable to an unsequenced fruit species.

One of the most applied common strategies in the "proteomic pipeline" is called MS/MS ion search. It uses software to search the uninterpreted MS/MS data from a single peptide, or an LC/MS run, to match the peak list. A peptide sequence is matched to individual MS/MS spectrum by a cross correlation algorithm, to compare experimental MS/MS spectra against predicted peptide sequence from a database (Eng et al., 1994). It should be remembered that it is peptide being identified, not the protein. To ensure the accuracy or minimize the false positive, two levels of control become necessary. First, the match of a peptide to a database must be estimated by determining the false positive rate (FDR). Using identical search parameters, the search can be repeated against a database, in which the sequence have been reversed or shuffled. Another level of control is to exam the identified protein family. Depending on the rules for protein and the protein family, it is usually advisable to use at least two identified peptides to call significant match of a protein.

Statistical analysis in comparative proteomics is one of the most important data analysis processes involving numerous variables, many of which cannot be controlled. Experiment design, biological repeats, and technical replicates, are necessary to ensure the validation of any proteomic study. This implies that data normalization and statistical analysis must be optimized for accurate detection of differentially expressed proteins. A review on design and analysis issues in quantitative proteomics studies was provided by Karp and Lilley (Karp and Lilley, 2007). Clear guidelines for publication of proteomic research data, which included statistical analysis, biological design, and normalization research results, should be followed (Carr et al., 2004; Valledor and Jorrín, 2011; Wilkins et al., 2006).

14.4   PROTEOMIC RESEARCH ON APPLE AND PEAR

Although the genome sequences of most fruits are incomplete, numerous proteomic studies have been conducted to reveal the key proteins in relation to metabolic pathways and specific functions in many fruit (Palma et al., 2011). Several proteomic investigations have been carried out on apple (Buts et al., 2014; Guarino et al., 2007; Qin et al., 2009; Zheng et al., 2007; Zheng et al., 2013) and pears (Pedreschi et al., 2007, 2008). Most proteomic techniques have used 2-DE technology, using IPG strips, IEF/SDS-PAGE, DIGE and MS. Recent publications on apple (*Malus domestica*) and pear (*Pyrus pyrifolia*) fruit proteomic research are summarized in Table 14.3.

Early proteomic work conducted on apples, identified a few proteins in relation to fruit-ripening and as well as some glycoproteins. A proteomic investigation conducted by Guarino et al. (2007) on apple (*Malus domestica* cv. "Annurca") flesh tissue, using 2-DE and MALDI-TOF as well as LC-MS/MS, identified 44 proteins, which were significantly involved in energy production, ripening, and stress responses, but also some proteins as fruit allergens. (Guarino et al., 2007) Another

---

TABLE 14.3
Publications on Proteomic Research with Apples and Pears

Plant Material	Target	Proteomic Technique	No. Proteins Identified	Reference
Apple (*Malus domestica*)	Sample preparation	2-DE, MT[a], and LC-MS/MS	32	Guarino et al. (2007)
	Fruit ripening	2-DE, LC-MS/MS OFFGEL-LC/MS/MS, MRM	3, 13 1709	Guarino et al. (2007), Zheng et al. (2013), Yang et al. (2014)
	Fruit senescence	2-DE, LC-MS/MS	34, 189	Qin et al. (2009), Zheng (2013)
	Fruit senescence	Label free (DID)	287[b]	Buts et al. (2014)
	Allergen	2-DE, LC-MS/MS	5, 4	Song et al. (2006), Wu et al. (2008), Zheng et al. (2007, 2013) Herndl et al. (2007)
Pear (*Pyrus communis*)	Disorder	2-DE, LC-MS/MS	1	Carpentier et al. (2005)
	Fruit ripening	2-DE, MT, and LC-MS/MS	43, 22, 151	Pedreschi et al. (2007, 2009) Pedreschi et al. (2008)

[a] MT: MALDI-TOF.
[b] Protein families with 2528 peptides.

proteomic study on "Red Delicious" apple fruit, using 2-DE, identified three allergens in fruit peel samples (Zheng et al., 2007). In other study from the water-soluble
fraction of an apple extract, four previously classified allergens, Mal d1, Mal d2, Mal
d3, and Mal d4, were identified in Western blots with polyclonal rabbit antibodies
directed toward the four respective allergens. All four known apple allergens were
localized on a 2-DE map, and they were matched with spots recognized by sera of
patients with different allergic patterns, whilst a new, putative allergen, was also
identified using MS (Herndl et al., 2007).

Apple fruit senescence at the proteomic level was also investigated, using comparative proteomic techniques to characterize the dynamic alterations in the mitochondrial proteome during fruit senescence. In purified mitochondria from apple
fruit under various stages of senescence, Qin et al. (2009) identified a set of 22 proteins spots that appeared to change significantly in abundance. These proteins are
mainly involved in metabolism related to TCA cycle, respiratory chain, and carbon
metabolism. They also showed that alteration of energy metabolism contributed
to cellular and organismic senescence (Qin et al., 2009). A proteomic approach,
employing a 2-DE technique with SYPRO Ruby, was performed by Zheng et al.
(2007) to separate the total proteins from apple fruit at different stages of ripening,
and senescence, after ethylene treatment. From 2340 spots, a total of 316 spots, or
approximately 14% of the total protein population, was found to be significantly
changed. Of these, 219 spots were only present at a specific ripening stage, while
97 spots varied throughout ripening, and in response to ethylene treatment. From
the 316 candidate spots, 221 proteins were further identified by liquid chromatography and MS, with protein sequence, and express sequence tag database search.
This indicated that apple fruit ripening is associated with an increased abundance
of proteins, with functions such as ethylene production, antioxidation and redox,
carbohydrate metabolism, oxidative stress, energy, and defense response. Ethylene
treatment increased a number of unique proteins that were not present during normal
fruit ripening, and reduced some proteins involved in primary metabolism, including those of the last few steps of the glycolytic pathway. This study demonstrated the
complexity and dynamic changes of protein profiles of apple fruit during ripening,
and in response to exogenous ethylene treatment.

Using 2-DE and LC-MS/MS, Pedreschi et al. (2007) identified proteins related
to brown core, a physiological disorder in "Conference" pear fruit during storage.
Seventy-eight proteins were identified as being related to the development of this
disorder. After the protein profiles from each group were analyzed, the upregulated
proteins in brown tissues were found mainly involved in energy and defense systems.
In this study, a unique approach, using univariate and multivariate statistical techniques, seems to be an efficient method to elucidate the mechanisms and pathways
leading to the disorder. Another study conducted by Pedreschi et al. (2008) investigated the combination of oxygen and carbon dioxide concentrations, precooling
period, and harvest maturities, on core breakdown, presumably in the protein levels
during controlled atmosphere storage of "Conference" pears. The results demonstrated that impaired respiration is highly related to protein synthesis alterations,
and the activation of defense mechanisms. Triosephosphate isomerase, a key enzyme
of the energy metabolism, was shown to be upregulated under browning-inducing

conditions, in an attempt to use alternative, more efficient, anaplerotic pathways, to cope with the applied stresses. The changes in the accumulation of proteins related to ethylene biosynthesis (ACC oxidase) and allergens (major allergen Pyrc 1) were highly dependent on the oxygen and carbon dioxide concentrations. ACC oxidase and Pyrc 1 were also clearly downregulated under low oxygen, or high carbon dioxide, concentrations (Pedreschi et al., 2008).

In addition, a new proteomic study, applying label-free technique, was developed, using data-independent acquisition to investigate the apple proteome change during the postharvest period. It identified 287 proteins families in "Braeburn" apples that increased significantly after 120 days of storage (Buts et al., 2014).

Despite limited numbers of publications and identified proteins, these examples clearly indicate that comparative proteomics, combined with various MS technologies and bioinformatic tools, are powerful and important tools to reveal numerous physiological and biological processes during ripening and development, as well as to reveal physiological disorders in fruits and vegetables. Data from apple and pear research demonstrated that proteomic tools provide an additional tool, besides genomics, to research fruit allergens.

14.5   FUTURE RESEARCH PERSPECTIVES

Despite the significant advances in technology, proteomic research on fruit and vegetables has been limited mainly to gel-based 2-DE systems, with MS protein identification. In order to improve the coverage and quantitative protein analysis, nongel-based and high throughput techniques should be applied to study complex fruit proteomes. Efficient sample preparation into fractions and cost-effective new labeling techniques also need to be developed, and applied to fruit proteomics research. New analytical concepts, such as targeted proteomics (MRM/SRM), are being explored, to overcome many of the existing limitations, and MS instrumentation will continue to improve the sensitivity and accuracy of current MS measurements. It will be equally important to develop an effective computational framework for the integration of proteomics data with phenomic and functional genomic information, to reconstruct molecular networks and how they interact in cells—an essential step in bridging the genotype–phenotype gap. Given the pivotal role that proteins play in fruit physiology, proteomic techniques, such as high throughput studies on protein abundance and activity, become very useful and necessary. Protein-based microarrays offer the possibility to investigate the specific proteins, using antibodies to conduct high throughput studies, by printing a collection of target proteins on the array surface, and assess their interactions and biochemical activities (Labaer and Ramachandran, 2005). Immunolocalization, immunoblot, and fluorescence imaging techniques, could also help to determine protein expression, localization, activity state, and cell compartmentation (Giepmans et al., 2006).

One important aspect is the lack of proteomic studies on protein PTMs during postharvest ripening and handling. PTMs play an important role in cell regulation and signal transduction. MS is the tool of choice for the analysis of protein PTMs, but without prior knowledge, PTMs can be profiled and quantified at the amino acid level, and used to elucidate a signal transduction pathway.

Another important future research direction of fruit proteomics is to study protein complexes, and their interaction. Investigations on those protein complexes, such as interactions of protein–protein, protein–DNA and protein–RNA, will provide further insights into fruit-ripening initiation, and regulation. As a common practice, the immunoprecipitation techniques have been developed and applied to protein interactions (McHugh et al., 2014; Thelen and Miernyk, 2012). Ethylene-regulated climacteric ripening, as well as the response of fruit to environment stresses, provide long-term, intriguing biological questions for postharvest physiologists who can apply state-of-the-art proteomic research tools to address these challenges. With the further development of proteomic research tools/platforms, including bioinformatics, it is certain that proteomics will provide unique and important insights into the regulation of fruit-ripening, an important sector of systems biology, and future marker selection, and validation for breeding programs.

Although few of the genomes of fruits and vegetables have been completely sequenced so far, many genomic tools available for *Arapidopsis* (Arabidopsis Information Resource), *Solanaceae* (Solanaceae Genomic Network), and *Rosaceae* (Rosaceae Genome Database) can be applied to future research and development in proteomics (Wheeler et al., 2005). In comparison, apple has more than 60,000 protein sequences in NCBI (accessed on December 2014). A high quality draft of the complete genome of pear (*Pyrus × bretschneideri*), reporting 42,812 protein coding genes, and with ca. 28% encoding multiple isomers (Wu et al., 2013), there were 46,518 protein sequences for public use on http://www.ncbi.nlm.nih.gov/protein (accessed on December 2014). Undoubtedly, the ultimate success of proteomic research is based on the completion of the genome sequences of fruits and vegetables.

ACKNOWLEDGMENTS

The author thanks Drs. G. Braun at AAFC, J. Golding, and R. Wills, for their critical reviews and constructive suggestions in preparing this manuscript.

REFERENCES

Anderson, L. and C.L. Hunter, 2006. Quantitative mass spectrometric multiple reaction monitoring assays for major plasma proteins. *Molecular and Cellular Proteomics: MCP.* 5(4):573–588.

Baginsky, S., L. Hennig, P. Zimmermann, and W. Gruissem, 2010. Gene expression analysis, proteomics, and network discovery. *Plant Physiol.* 152:402–410.

Boersema, P.J., R. Raijmakers, S. Lemeer, S. Mohammed, and A.J.R. Heck, 2009. Multiplex peptide stable isotope dimethyl labeling for quantitative proteomics. *Nat. Protoc.* 4:484–494.

Boersema, P.J., T.T. Aye, T.A.B. Van Veen, A.J.R. Heck, and S. Mohammed, 2008. Triplex protein quantification based on stable isotope labeling by peptide dimethylation applied to cell and tissue lysates. *Proteomics* 8:4624–4632.

Bohler, S., M. Bagard, M. Oufir, S. Planchon, L. Hoffmann, Y. Jolivet, J.F. Hausman, P. Dizengremel, and J. Renaut, 2007. A DIGE analysis of developing poplar leaves subjected to ozone reveals major changes in carbon metabolism. *Proteomics* 7:1584–1599.

Buts, K., S. Michielssens, M.L.A.T.M. Hertog, E. Hayakawa, J. Cordewener, A.H.P. America, B.M. Nicolai, and S.C. Carpentier, 2014. Improving the identification rate of data independent label-free quantitative proteomics experiments on non-model crops: A case study on apple fruit. *J. Proteomics* 105:31–45.

Carpentier, S.C., E. Witters, K. Laukens, P. Deckers, R. Swennen, and B. Panis, 2005. Preparation of protein extracts from recalcitrant plant tissues: An evaluation of different methods for two-dimensional gel electrophoresis analysis. *Proteomics* 5:2497–2507.

Carr, S., R. Aebersold, M. Baldwin, A. Burlingame, K. Clauser, and A. Nesvizhskii, 2004. The need for guidelines in publication of peptide and protein identification data. *Mol. Cell. Proteomics* 3:531–533.

Delahunty, C. and J.R. Yates, 2005. Protein identification using 2D-LC-MS/MS. *Methods in Molecular Biology (Clifton, N.J.)* 35:248–255.

Dunkley, T.P.J., P. Dupree, R.B. Watson, and K.S. Lilley, 2004. The use of isotope-coded affinity tags (ICAT) to study organelle proteomes in Arabidopsis thaliana. *Biochem. Soc. Trans.* 32:520–523.

Dunkley, T.P.J., S. Hester, I.P. Shadforth, J. Runions, T. Weimar, S. Hanton, and J.L. Griffin, 2006. Mapping the Arabidopsis organelle proteome. *PNAS* 103:6518–6523.

Eng, J.K., A.L. McMcomack, and J.R. Yates, 1994. An approach to correlate tandem mass spectral data of peptides with amino acid sequences in a protein database *J. Am. Soc. Mass. Spectrom* 5:967–989.

Görg, A., W. Weiss, and M.J. Dunn, 2004. Current two-dimensional electrophoresis technology for proteomics. *Proteomics* 4:3665–3685.

Gallardo, K., C. Job, S.P.C. Groot, M. Puype, H. Demol, J. Vandekerckhove, and D. Job, 2002. Proteomics of Arabidopsis seed germination. A comparative study of wild-type and gibberellin-deficient seeds. *Plant Physiol.* 129:823–837.

Giepmans, B.N., S.R. Adams, M.H. Ellisman, and R.Y. Tsien, 2006. The fluorescent toolbox for assessing protein location and function. *Science* 312:3733–3740.

Giovannoni, J.J., 2004. Genetic regulation of fruit development and ripening. *Plant Cell* 16:S170–S180.

Gstaiger, M. and R. Aebersold, 2009. Applying mass spectrometry-based proteomics to genetics, genomics and network biology. *Nat. Rev. Genet.* 10:617–627.

Guarino, C., S. Arena, L. De Simone, C. D'Ambrosio, S. Santoro, M. Rocco, A. Scaloni, and M. Marra, 2007. Proteomic analysis of the major soluble components in Annurca apple flesh. *Mol. Nutr. Food Res.* 51:255–262.

Gygi, S.P., B. Rist, S.A. Gerber, F. Turecek, M.H. Gelb, and R. Aerbersold, 1999. Quantitative analysis of complex protein mixtures using isotope-coded affinity tags. *Nat. Biotech.* 17:994–999.

Harris, L.R., M.A. Chirchward, H. Butt, and J.R. Coorssen, 2007. Assessing detection methods for gel-based proteomic analysis. *J. Proteome. Res.* 6:1418–1425.

Henzel, W.J., T.M. Billeci, J.T. Stults, S.C. Wong, C. Grimley, and C. Watanabe, 1993. Identifying proteins from two-dimensional gels by molecular mass searching of peptide fragments in protein sequence databases. *Proc. Natl. Acad. Sci. U.S.A.* 90:5011–5015.

Herndl, A., G. Marzban, D. Kolarich, R. Hahn, D. Boscia, W. Hemmer, F. Maghuly, E. Stoyanova, H. Katinger, and M. Laimer, 2007. Mapping of Malus domestica allergens by 2-D electrophoresis and IgE-reactivity. *Electrophoresis* 28:437–448.

Hsu, J.L., S.Y. Huang, N.H. Chow, and S.H. Chen, 2003. Stable-isotope dimethyl labeling for quantitative proteomics. *Anal. Chem.* 75:6843–6852.

Hurkman, W.J. and C.K. Tanaka, 1986. Solubilization of plant membrane proteins for analysis by two-dimensional gel electrophoresis. *Plant Physiol.* 81:802–806.

Inagaki, N. and K. Katsuta, 2004. Large gel two-dimensional electrophoresis: Improving recovery of cellular proteome. *Curr. Proteomics* 1:35–39.

Jacobs, D.I., M. Gaspari, J. Van Der Greef, R. Van Der Heijden, and R. Verpoorte, 2005. Proteome analysis of the medicinal plant Catharanthus roseus. *Planta* 221:690–704.

Kader, A.A., 2004. Perspective on postharvest horticulture (1978–2003). *Hort Science* 38:759–761.

Karp, N.A. and K.S. Lilley, 2007. Design and analysis issues in quantitative proteomic studies. *Practical Proteomics* 1:42–50.

Komatsu, S., X. Zang, and N. Tanaka, 2006. Comparison of two proteomics techniques used to identify proteins regulated say gibberellin in rice. *J. Proteome Res.* 5:270–276.

LaBaer, J. and N. Ramachandran, 2005. Protein microarrays as tools for functional proteomics. *Curr. Opin. Chem. Biol.* 9:14–19.

Lamanda, A., A. Zahn, D. Rodger, and H. Langen, 2004. Improved Ruthenium II tris (bathophenantroline disulfonate) staining and destaining protocol for a better signal-to-background ratio and improved baseline resolution. *Proteomics* 4:599–608.

Lange, V., P. Picotti, B. Domon, and R. Aebersold, 2008. Selected reaction monitoring for quantitative proteomics: A tutorial. *Mol. Syst. Biol.* 4:222.

Lanne, B.a.P., O., 2004. Protein staining influences the quality of mass spectra obtained by peptide mass fingerprinting after separation on 2-Dgels. A comparison of staining with Coomassie Brilliant Blue and Sypro Ruby. *J. Prot. Res.* 4:175–179.

Laugesen, S., E. Messinese, S. Hem, C. Pichereaux, S. Grat, R. Ranjeva, M. Rossignol, and J.J. Bono, 2006. Phosphoproteins analysis in plants: A proteomic approach. *Phytochemistry* 67:2208–2214.

Levin, Y. and B. Sabine, 2010. Quantification of Proteins by Label-Free LC-MS/MS, p. 217–231. In: C.P.R. and J.F. Timms (eds.), *LC-MS/MS in Proteomics, Methods in Moleular Biology.* 658, Springner Sciences + Business Media, LLC, New York, USA.

Li, H., S.W. Wai, L. Zhu, W.G. Hong, J. Ecker, and N. Li, 2009. Phosphoproteomic analysis of ethylene-regulated protein phosphorylation in etiolated seedlings of Arabidopsis mutant ein2 using two-dimensional separations coupled with a hybrid quadrupole time-of-flight mass spectrometer. *Proteomics* 9:1646–1661.

Link, A.J., J. Eng, D.M. Schieltz, E.C. Carmack, G.J. Mize, D.R. Morris, B.M. Garvik, and J.R. Yates, 1999. Direct analysis of protein complex using mass spectrometry. *Nat. Biotechnol.* 17:676–682.

Luecker, J., M. Laszczak, D. Smith, and S.T. Lund, 2009. Generation of a predicted protein database from EST data and application to iTRAQ analysis in grape (*Vitis vinifera* cv. Carernet Sauvignon) berries at ripening initiation *BMC Genomics* 10:1–17.

Mackintosh, J.A.C., H-Y. Bae, S-H. Veal, D.A. Bell, P.J. Ferrari, B.C. Van Dyk, D.D. Verrills, N.M., 2003. A fluorescent natural product for ultrasensitive detection of proteins in one-dimensional and two-dimensional gel electrophoresis. *Proteomics* 3:2273–2288.

Mann, M., P. Hojrup, and P. Rsepstorff, 1993. Use of mass spectrometric molecular weight information to identify proteins in sequence databases. *Biol. Mass Spectrom.* 22:338–345.

Matros, A., S. Kaspar, S. Tenzer, M. Kipping, U. Seiffert, and H.P. Mock, 2011. Label-free liquid chromatography-based quantitative proteomics: Challenges and recent developments, p. 103–135. Proteomics: Methods, Applications and Limitations.

McHugh, C.A., P. Russell, and M. Guttman, 2014. Methods for comprehensive experimental identification of RNA-protein interactions. *Genome Biol.* 15:203.

Mechin, V., L. Consoli, M. Le Guilloux, and C. Damerval, 2003. An efficient solubilization buffer for plant proteins focused in immobilized pH gradients. *Proteomics* 3:1299–1302.

Melanson, J.E., S.L. Avery, and D.M. Pinto, 2006. High-coverage quantitative proteomics using amine-specific isotopic labeling. *Proteomics* 6:4466–4474.

Newton, R.P., A.G. Brenton, C.J. Smith, and E. Dudley, 2004. Plant proteome analysis by mass spectrometry: Principles, problems, pitfalls and recent developments. *Phytochemistry* 65:1449–1485.

Nishihara, J. and K. Champion, 2002. Quantitative evaluation of proteins in one-and two-dimensional polyacrylamide gels using a fluorescence stain. *Electrophoresis* 23:2203–2215.

Nuehse, T.S., A.R. Bottrill, A.M.E. Jones, and S.C. Peck, 2007. Quantitative phosphoproteomic analysis of plasma membrane proteins reveals regulatory mechanisms of plant innate immune responses. *Plant J.* 51:931–940.

Ong, S.E., B. Blagoev, I. Kratchmarova, D.B. Kristensen, H. Steen, A. Pandey, and M. Mann, 2002. Stable isotope labeling by amino acids in cell culture, SILAC, as a simple and accurate approach to expression proteomics. *Molecular and Cellular Protoemics* 1:376–386.

Ong, S.E., I. Kratchmarova, and M. Mann, 2003. Properties of ^{13}C –substituted arginine in stable isotope labelling by amino acids in cell culture (SILAC) *J. Proteome Res* 2:173–181.

Palagi, P., P. Hernanadez, D. Walther, and R.D. Appel, 2006. Proteome information I: Bioinformatics tools for processing experiment data. *Proteomics* 6:5435–5444.

Palma, J.M., F.J. Corpas, and L.A. del Río, 2011. Proteomics as an approach to the understanding of the molecular physiology of fruit development and ripening. *J. Proteomics* 74:1230–1243.

Pan, X., B. Zhu, H. Zhu, Y. Chen, H. Tian, Y. Luo, and D. Fu, 2014. ITRAQ protein profile analysis of tomato green-ripe mutant reveals new aspects critical for fruit ripening. *J. Proteome Res.* 13:1979–1993.

Pandey, A. and M. Mann, 2004. Proteomic to study genes and genomes. *Nature* 405:837–846.

Parks, B.A., L. Jiang, P.M. Thomas, C.D. Wenger, M.J. Roth, M.T.B. II, P.V. Burke, K.E. Kwast, and N.L. Kelleher, 2007. Top-down proteomics on a chromatographic time scale using linear ion trap fourier transform hybrid mass spectrometers. *Anal. Chem.* 79:7984–7991.

Patton, W., 2000. A thousand points of light: The application of fluorescence detection technologies to two-dimensional gel electrophoresis and proteomics. *Electrophoresis* 21:1123–1144.

Peck, S.C., 2006. Analysis of protein phosphorylation: Methods and strategies for studying kinases and substrates. *Plant J.* 45:512–522.

Pedreschi, R., C. Franck, J. Lammertyn, A. Erban, J. Kopka, M. Hertog, B. Verlinden, and B. Nicolaï, 2009. Metabolic profiling of "Conference" pears under low oxygen stress. *Postharvest Biol. Technol.* 51:123–130.

Pedreschi, R., E. Vanstreels, S. Carpentier, M. Hertog, J. Lammertyn, J. Robben, J.P. Noben, R. Swennen, J. Vanderleyden, and B.M. Nicolaï, 2007. Proteomic analysis of core breakdown disorder in Conference pears (*Pyrus communis* L.). *Proteomics* 7:2083–2099.

Pedreschi, R., M. Hertog, J. Robben, J.P. Noben, and B. Nicolaï, 2008. Physiological implications of controlled atmosphere storage of "Conference" pears (*Pyrus communis* L.): A proteomic approach. *Postharvest Biol. Technol.* 50:110–116.

Picotti, P., O. Rinner, R. Stallmach, F. Dautel, T. Farrah, B. Domon, H. Wenschuh, and R. Aebersold, 2010. High-throughput generation of selected reaction-monitoring assays for proteins and proteomes. *Nature Methods* 7:43–46.

Qin, G., Q. Wang, J. Liu, B. Li, and S. Tian, 2009. Proteomic analysis of changes in mitochondrial protein expression during fruit senescence. *Proteomics* 9:4241–4253.

Roberts, J.K.M., 2002. Proteomics and a future generation of plant molecular biologists. *Plant Mol. Biol.* 48:143–154.

Rose, J.K.C., S. Bashir, J. Giovannoni, M.M. Jahn, and S.R. Saravanan, 2004. Tackling the plant proteome: Practical approaches, hurdles and experimental tools. *The Plant J.* 39:715–733.

Ross, P.L., Y.N. Huang, J.N. Marchese, B. Williamson, K. Parker, S. Hattan, N. Khainovski, and S. Pillai, 2004. Multiplexed protein quantitation in Saccharomyces cerevisiae using amine-reactive isobaric tagging reagents. *Mol. Cell Proteomics* 3:1154–1169.

Saravanan, R.S. and J.K.C. Rose, 2004. A critical evaluation of sample extraction techniques for enhanced proteomic analysis of recalcitrant plant tissues. *Proteomics* 4:2522–2532.

Silva, J.C., M.V. Goreenstein, Z.G. Li, P.C. Vissers, and S.J. Geromanos, 2006. Absolute quantificaion of proteins by LC/MS^e. *Mol. Cell Proteomics* 5.1:144–156.

Song, J. and C.F. Forney, 2008. Flavour volatile production and regulation in fruit. *Can. J. Plant Sci.* 88:537–550.

Song, J., G. Braun, E. Bevis, and K. Doncaster, 2006. A simple protocol for protein extraction of recalcitrant fruit tissues suitable for two-dimensional electrophoresis and mass spectrometry analysis. *Electrophoresis* 27:3144–3151.

Steen, H. and M. Mann, 2004. The ABC's (and XYZ's) of peptide sequencing. *Nat. Rev. Mol. Cell. Biol.* 5:699–711.

Thelen, J.J. and J.A. Miernyk, 2012. The proteomic future: Where mass spectrometry should be taking us. *J. Biochem.* 444:169–181.

Tonge, R., J. Shaw, B. Middleton, R. Rowlinson, S. Rayner, J. Young, and F. Pognan, 2001. Validation and development of fluorescence two-dimensional differential gel electrophoresis proteomics technology. *Proteomics* 1:277–396.

Valledor, L. and J. Jorrín, 2011. Back to the basics: Maximizing the information obtained by quantitative two dimensional gel electrophoresis analyses by an appropriate experimental design and statistical analyses. *J. Proteomics* 74:1–18.

Wang, S.B., Q. Hu, M. Sommerfeld, and F. Chen, 2003. An optimized protocol for isolation of soluble proteins from microalgae for two-dimensional gel electrophoresis analysis. *J. Appl. Phycol.* 15:485–496.

Wang, W., R. Vignani, M. Scali, and M. Cresti, 2006. A universal and rapid protocol for protein extraction from recalcitrant plant tissues for proteomic analysis. *Electrophoresis* 27:2782–2786.

Washburn, M.P., R. Ulaszek, C. Deciu, D.M. Schietz, and J.R.I. Yates, 2002. Analysis of quantitative proteomic data generated via multidimensional protein identification technology. *Anal. Chem.* 74:1650–1657.

Wheeler, D.L., B. Smith-White, V. Chetvernin, S. Resenchuk, S.M. Dombrowski, S.W. Pechous, T. Tatusova, and J. Ostell, 2005. Plant genome resources at the National Center for Biotechnology Information. *Plant Physiol.* 138:1280–1288.

Wheelock, A.M., D. Morin, M. Bartosiewicz, and A. Buckpitt, 2006. Use of fluorescence internal protein standard to achieve quantitative two-dimensional gel electrophoresis. *Proteomics* 6:1385–1395.

White, I.R.P., R. Wood, J. Skehel, J. M. Gangadharan, B. Cutler, P., 2004. A statistical comparison of silver and SYPRO Ruby staining for proteomic analysis. *Electrophoresis* 25:3048–3054.

Wilkins, M.R., R.D. Appel, J.E. Van Eyk, M.C.M. Chung, A. Gorg, M. Hecker, H. L.A., H. Lamgen, and A.J. Link, 2006. Guideline for the next 10 years of proteomics. *Proteomics* 6:4–8.

Wittmann-Liebold, B., H.R. Graack, and T. Pohl, 2006. Two-dimensional gel electrophoresis as tool for proteomics studies in combination with protein identification by mass spectrometry. *Proteomics* 6:4688–4703.

Wu, C.Y., J.Y. Song, and S.L. Chen, 2008. Application of expressed sequence tags to study on medicinal plant. *Chinese Traditional and Herbal Drugs* 39:778–782.

Wu, J., Z. Wang, Z. Shi, S. Zhang, R. Ming, S. Zhu, M.A. Khan et al., 2013. The genome of the pear (*Pyrus bretschneideri* Rehd.). *Genome Res.* 23:396–408.

Wu, W.W., G. Wang, S.J. Baek, and R.F. Shen, 2006. Comparative study of three proteomic quantitative methods, DIGE, cICAT, and iTRAQ, using 2D gel-or LC-MALDI TOF/TOF. *J. Prot. Res.* 5:651–658.

Yang, X., L. Li, J. Song, L. Campbell Palmer, X. Li, and Z. Zhang, 2014. Peptide prefractionation is essential for proteomic approaches employing multiple-reaction monitoring of fruit proteomic research. *J. Sep. Sci.* 37:77–84.

Zheng, Q., J. Song, K. Doncaster, E. Rowland, and D.M. Byers, 2007. Qualitative and quantitative evaluation of protein extraction protocols for apple and strawberry fruit suitable for two-dimensional electrophoresis and mass spectrometry analysis. *J. Agric. Food Chem.* 55:1663–1673.

Zheng, Q., J. Song, L. Campbell-Palmer, K. Thompson, L. Li, B. Walker, Y. Cui, and X. Li, 2013. A proteomic investigation of apple fruit during ripening and in response to ethylene treatment. *J. Proteomics* 93:276–294.

Zhou, H., Z. Ning, A.E. Starr, M. Abu-Farha, and D. Figeys, 2012. Advancements in top-down proteomics. *Anal. Chem.* 84:720–734.

Zieske, L.R., 2006. A perspective on the use of iTRAQ reagent technology for protein complex and profiling studies. *J. Exp. Bot* 57:1501–1508.

15 Organic Postharvest Technology

Apiradee Uthairatanakij and Pongphen Jitareerat

CONTENTS

15.1 INTRODUCTION

Organic agriculture has a number of potential environmental benefits and is considered a positive force for green consumerism. These potential benefits have led to increased production and consumption of organic produce; this is one of the major market trends of this decade. To ensure this success, organic fruits and vegetables at harvest are expected to be of high quality in terms of color, shape, texture, and unique flavor, and must be free from decay and injury, as expected of conventionally grown produce. In addition, the postharvest handling system for organically grown produce must follow the guidelines for organic commodities, and have the necessary documentation to demonstrate compliance with these standards from the farm to the consumer.

15.1.1 WHAT ARE ORGANIC FRUITS AND VEGETABLES?

The term "organic" refers to the way that agricultural products are grown, handled, and processed. Organic fruits and vegetables are produced using an organic farming system that emphasizes renewable resources and conservation of soil fertility and water, rich biodiversity, and long-term sustainability (Plotto and Narciso, 2006). This production system relies on crop-rotation, soil-building, and biological pest management (UNCTAD, 2003). Similar production practices are also used for conventionally grown produce, but organic production operates without the use of synthetic fertilizers and pesticides. Postharvest, organically grown fresh fruits and vegetables must be handled without synthetic chemicals, except for those specifically allowed by regulation, and without ionizing irradiation. If these organic practices are followed and certified, the products can be labeled as "organic."

Labeling is essential to allow consumers to recognize organic products and distinguish them from conventional grown commodities (Hemmerling et al., 2013). Organic product labels are useful instruments to establish consumer trust and to quickly allow consumers to identify organic commodities. At the same time, labels help to avoid confusion and increase the profile of organic food, positively influencing the consumer's choice of produce (Padel and Midmore, 2005).

15.2 POSTHARVEST HANDLING SYSTEMS FOR ORGANIC PRODUCE

The use of postharvest technology for organic produce, in order to achieve similar outcomes as conventional crops, is one of the most challenging goals for organic producers, handlers, and processors. An integrated approach for postharvest handling is needed, one that incorporates "layers of prevention" to reduce chemical and microbial contamination of fresh organic produce, but still maintains produce quality during handling, packing, storage, and shipping operations. A basic requirement for certified organic produce is that there should no cross-contamination between organic produce and prohibited substances. A major problem for shippers and handlers is the need to segregate organic products from conventionally grown fruits and vegetables, to avoid comingling. Ingredients and materials used postharvest must also be organic or otherwise approved (Plotto and Narciso, 2006).

15.2.1 HANDLING AT HARVEST

Careful investigation is needed of all the steps in the postharvest chain, to develop and refine a proper handling protocol that ensures that organic products are correctly handled, according to organic protocols. In organic operations, the International

Federation of Organic Agriculture Movements (IFOAM, 2012) requires that handlers shall not comingle organic produce with nonorganic produce throughout the entire postharvest handling process.

The receiving area should be clean and free from prohibited substances such as pest control products and other debris that may result in contamination of organic produce or its comingling with nonorganic produce. Adequate sanitation and water disinfection during postharvest handling are vital components of a postharvest management plan. Food safety issues are of concern with all horticultural crops, and have become increasingly important to the sales and marketing of fresh fruits and vegetables. Several cases of food-borne illnesses, due to contamination with human pathogens, have been traced to improper sanitation during postharvest handling. For example, organic sprouts from a farm in Illinois infected at least 140 people in 26 states in the United States with salmonella in 2012. In 2011, a massive outbreak of a deadly strain of *E. coli* was linked to sprouts from an organic farm in Germany that killed 50 people and sickened more than 4,300. However studies have shown no significant difference in prevalence of *E. coli* and other pathogens between organically and conventionally grown produce (Mukherjee et al., 2004; Winter and Davis, 2006; Leifert et al., 2008; Oliveira et al., 2010). It is also estimated that 30% of fresh produce is lost to microbial spoilage due to nonhuman pathogens and saprophytes from harvest, through postharvest handling, and delivery to consumers. For organic produce, disinfectants and other sanitation products must comply with specific regulations (IFOAM, 2012; OMRI [Organic Materials Review Institute], 2013). If unloading organic produce from bins involves floatation, only approved substances may be used in the floatation water to reduce damage and cross-contamination (Suslow, 2000). Allowed substances that include lignin sulfonates can be used in certified organic handling (OMRI, 2013). For organic produce, packaging materials such as storage containers, bins, and boxes cannot contain synthetic fungicides, preservatives, fumigants, or nano-materials. Polyvinyl chloride (PVC) and aluminum should be avoided.

15.2.2 WAXING

Water loss of organic fresh produce is a primary factor in loss of postharvest quality and of saleable weight. Waxing or coating can be effective in reducing water-loss in produce and can also enhance appearance. According to the National Organic Program (NOP) in the USA, wax for organic produce must not contain any prohibited synthetic substances, preservatives or fungicides, and may not have petroleum-based ingredients. Beeswax, wood rosin (resin from pines), and carnauba (from palm leaves) are allowed (www.ams.usda.gov/standard/). Natural waxes and resins are difficult to apply directly to fruit and vegetables as coating, without first making an emulsion. Emulsification of natural waxes and resins for fruit and vegetable coatings are made by adding fatty acids at high temperatures (Plotto and Narciso, 2006). Commercial carnauba wax-coating often contains a fatty acid (10%–30% of solids), ammonia or morpholine (3%–10% of solids), and antifoam (Hagenmaier and Baker 1997). However, morpholine and ammonia are prohibited for organic produce. Therefore, beeswax, wood rosin, and carnauba are emulsified with vegetable oil, vegetable-based fatty acids, and ethyl alcohol in water (www.ams.usda.gov/standard/). Only beeswax (INS 901) and carnauba (INS 903) are allowed in postharvest handling by IFOAM regulations (IFOAM, 2012).

15.2.3 STORAGE

Temperature and humidity management inside the storage room play a key role in determining the storage-life and shelf-life of organic produce, similar to conventional produce. Proper cooling reduces the respiration rate and maintains visual quality, texture, flavor and nutritional composition. Whatever the cooling method utilized, if both organic and conventional produce are being handled, the organic produce must be cooled at the beginning of daily operation to prevent cross-contamination of organic produce with prohibited substances. An acceptable practice is the injection of ozone into the cooling water to reduce pesticide residues that may remain in the water after cooling nonorganic produce (Suslow, 2000). In addition, the same restrictions on the storage of conventionally produced commodities, especially those from the tropics and subtropics sensitive to chilling injuries, applies to organic produce. Ethylene is allowed for postharvest ripening of tropical fruits and degreening of citrus fruits.

15.2.4 TRANSPORTATION AND DISTRIBUTION

Organic produce must be segregated from nonorganic conventional produce to avoid cross-contamination. Thus, organic produce should never be stacked beneath conventional produce during storage and transport. When organic produce is being shipped in the same truck, it must be packed in separate boxes and on separate pallets to prevent cross-contamination with conventional produce and must be clearly labeled as "organic." A full review of the storage, transportation, and distribution center is essential to identify potential risks for comingling. These reviews need to be carried out on a regular basis and documented.

Upon shipment arrival, the organic produce cartons should be in good condition, without rips and tears or damage. Broken packaging may mean that the organic produce could have been cross-contaminated with nonorganic produce. Also, any organic packages accepted during receiving should be clearly labeled as organic and have a certifying agent's name.

During storage at a distribution center, a barrier is required between packed and unpacked conventional and organic produce, using space buffers. If repacking is necessary, packing materials and storage containers for organic produce must be free from synthetic fungicides and fumigants.

15.3 DISINFECTION OF ORGANIC PRODUCE

Microbial contamination in organic produce is a critical point for both shelf-life and consumer safety. To achieve this goal, many organic and conventional growers have implemented Good Agricultural Practices (GAPs) during production and harvest operations to ensure quality and minimize contamination. Also, many packers or handlers have implemented Best Management Practices (BMPs) to maintain quality and reduce the microbial load on fresh organic produce after harvest.

The postharvest handling operation of organic produce cannot use unapproved chemicals and most synthetic inputs are prohibited. Those synthetic inputs allowed can only be used with restrictions. Additionally, there are no internationally recognized regulatory agencies harmonizing the approval of either biopesticides or chemicals for use in organic commodities, postharvest (Prange et al., 2006). The United States and the European Union recently accepted each other's organic standards as equivalent, with minor exceptions in animal production. For other countries, chemicals used in organic postharvest operations must comply with the organic requirements of each country. Of particular interest are the types of sanitizing agents that can be used.

15.3.1 CHLORINE

Chlorine is the most common disinfectant that can be added to transport flumes, or to produce cooling, or in wash-water. In water, chlorine exists in equilibrium as hypochlorous acid and hypochlorite ion, depending upon the pH. Hypochlorous acid provides the strongest antimicrobial properties. At pH 6.5, 95% of the chlorine is in the hypochlorous form; therefore, maintaining the water pH at the range of pH 6.5–7.5) provides the greatest disinfecting power. Chlorine gas will be released if highly acidic water is used (Suslow, 2000). Products used for adjusting acidity in water must be from natural sources such as citric acid, sodium bicarbonate, or vinegar (Suslow, 2000). Chlorine in the water reacts with organic compounds to produce chlorinated compounds suspected of detrimental effects on humans and wildlife (Suslow, 2006). The most common compounds are trihalomethanes and haloacetic acids.

Organic processors and shippers may use chlorine within specified limits. All liquid sodium hypochlorite, granular calcium hypochlorite, and chlorine dioxide are restricted materials by organic standards. The application of chlorine must conform with the Maximum Residual Disinfectant Limit under the Safe Drinking Water Act at 4 mg/L (ppm) expressed as chlorine (Suslow, 2006; Silva, 2008). Sodium hypochlorite is typically used as a source of chlorine.

The chlorine levels in the wash-tanks and flumes need to be continuously monitored. Chlorine readily binds to soil, debris, and other organic matters in the water, and is no longer available for disinfection (Silva, 2008). A common way is to monitor both the pH and the oxidation-reduction potential of the wash-water. The NOP standard permits a threshold of 4 ppm residual chlorine in the effluent. From handling organic products, the downstream product-wash needs to be monitored to ensure that the effluent water does not exceed this limit. California Certified Organic Farmers (CCOF) has recently modified this threshold to permit 10 ppm residual chlorine measured downstream of the wash. However, the levels of chlorine used to prepare water for sanitation of equipment, tools, product surfaces, or edible products, should be in high enough concentrations to control microbial contaminants. In order to maximize the antimicrobial activity, it is beneficial to prewash the produce arriving from the field, or before loading. This may include a vigorous prewash with brushes or sponges to remove excess debris from the produce, or a clear-water rinse to remove soil and other debris, prior to using the sanitizer solution. Also, cleaning dump tanks and residue screens helps to minimize the presence of soil and debris, and maximize chlorine effectiveness.

15.3.2 Ozone

Ozone is considered to be GRAS (Generally Regarded as Safe) for fresh produce and equipment, and ozonated water is becoming an increasingly popular effective alternative to chlorine for postharvest application, due to it not producing any unacceptable byproducts, and it has a higher antimicrobial activity than chlorine (Kim et al., 2003; Ölmez and Särkka-Tirkkonen, 2008). Ozone provides comparable disinfection power to chlorine, and, in addition, attacks bacterial cell-walls and thick-walled spores of plant pathogens. Ölmez and Kretzschmar (2009) stated that low concentrations (1–5 ppm) of ozone at short exposure times (1–5 min) is effective against many bacteria, yeasts and molds, as well as some viruses, but it has low stability of less than 20 min in clean water and must be generated continuously *in situ* (Silva, 2008). A high level of ozone can lead to produce injury such as damage to the surface of lettuce (Kim et al., 2006), and browning of iceberg lettuce (Koseki and Isobe, 2006). An added advantage of ozone treatment is that its activity is not dependent on pH, as chlorine is. It is therefore not necessary to adjust the wash-water pH (Ölmez and Särkka-Tirkkonen, 2008). However, the efficacy of ozone treatment for disinfection is depending on several factors, including types of microorganism, fresh produce, the level of initial inoculum, the growth stage of microorganism, and the application method of the ozone treatments (Ölmez and Kretzschmar, 2009). In addition, the threshold level of ozone for a safe working environment are: 0.1 for long period (8 h); and and 0.3 ppm for short period (15 min) (Ölmez and Särkka-Tirkkonen, 2008).

15.3.3 Peroxyacetic Acid

Peroxyacetic acid (PAA), also called peracetic acid, is allowed in an organic postharvest system. The disinfection performance of PAA is comparable to chlorine and ozone in eliminating microbial biofilm in dump-tank and flume sanitation. Like ozone, the treatment results in safer byproducts than chlorine. However, a disadvantage of PAA is its higher unit cost. To maximize effectiveness, PAA should be maintained at a level of 80 ppm in the wash-water, and clean water is required for washing after a PAA disinfection treatment (Silva, 2008).

15.3.4 Organic Acids

Many organic acids, such as citric acid, lactic acid, acetic acid, and ascorbic acid, have been applied to minimize microbial growth in fresh and fresh-cut produce due to their low pH of 2.1–2.7 (Arites et al., 2009; Ölmez and Kretzschmar, 2009). Their antimicrobial efficacy is greater on bacteria than yeasts and molds (Arites et al., 2009). The efficacy of these organic acids against microbes is dependent on the type and concentration of organic acid, water quality (such as pH, temperature, turbidity, and organic content) used to dissolve the organic acid, application method (dipping, spraying), exposure time (generally between 5 and 15 min), and target microorganisms, and inoculum level (Gil et al., 2009; Ölmez and Kretzchmar, 2009; Alexopoulos et al., 2013).

15.3.5 ELECTROLYZED WATER

Electrolyzed water (EW) is an alternative disinfectant of fresh and fresh-cut produce, as it is safer and more ecologically friendly than chlorination. EW is generated by electrolysis of small amount of sodium chloride (0.1%–1.0%) in wash-water (Arites et al., 2009; Ramos et al., 2013). Two types of EW have been used as the sanitized water: neutral electrolyzed water (NEW) and acidic electrolyzed water (AEW). Both types strongly inactivate spoilage microorganisms and food-borne pathogens (Ramos et al., 2013). They are more effective than hypochlorite due to their high oxidation–reduction potential (Izumi, 1999).

15.3.6 HYDROGEN PEROXIDE

Hydrogen peroxide (H_2O_2) is used as a disinfecting agent due to its strong oxidizing power, and generates phytotoxic hydroxyl free radicles. Hydrogen peroxide is allowed in food processing and packaging as it does not leave harmful residues and decomposes into water and oxygen through the action of catalase (Arites et al., 2009; Ölmez and Kretzchmar, 2009). The concentration of hydrogen peroxide used in wash-water as the antimicrobial agent ranges between 0.04 and 1.25% (Ölmez and Kretzchmar, 2009). However, Ramos et al. (2013) recommend a higher concentration of 2%–4%, though concentrations of 4%–5% may be phytotoxic to fresh produce.

15.4 POSTHARVEST TREATMENTS FOR ORGANIC PRODUCE

15.4.1 MODIFIED ATMOSPHERE PACKAGING

Modified atmosphere packaging (MAP) is a method of extending the shelf-life of both organic and conventionally grown fresh produce and is used widely for fresh-cut produce. The technology involves the replacement of air inside the package with a beneficial mixture of elevated carbon dioxide, and reduced oxygen, which is achieved by the natural interplay between the respiration of the produce and the transfer of gases through the packing material. The efficacy of MAP on extending shelf-life is dependent on several factors, such as the type of produce, gas mixture, storage temperature, packing material and hygiene during handling (Fonseca et al., 2002; Sandhya, 2010). Since there is no addition of nonnatural materials in the system, MAP is an acceptable technology for organic produce; however, the use of MAP on specific organic produce has similar requirements and performance as for conventional produce.

15.4.2 HEAT TREATMENT

Heat treatment is an acceptable technology for organic produce, as it leaves no chemical residues, and has been approved as an insect quarantine treatment by the United

States Department of Agriculture (USDA) Animal and Plant Health Inspection Service (APHIS) against several pests (USDA, 1993), and is widely used for disease control. It is considered as a safe alternative physical treatment to reduce the quantities of post-harvest chemicals used to inhibit growth of pathogens and insects (Lu et al., 2010).

Hot-water treatments, by themselves, or in combination with other heat and mechanical treatments, can substantially control postharvest diseases of fresh produce. Porat et al. (2000) found that hot-water brushing (HWB) at 56°C for 20 s reduced decay in organically grown citrus cultivars such as "Minneola" tangerines, "Shamouti" oranges, and "Star Ruby" red grapefruit, but did not cause surface damage, and did not affect fruit weight-loss or internal quality. The treatment of peaches and nectarines with a drench of heated water at 55°C or 60°C for 20 s over rotating brushes effectively controlled brown rot (Karabulut et al., 2002). Lydakis and Aked (2003) concluded that vapor-heat treatments at 52.5°C or 55°C for up to 24 min. can be applied to "Sultanina" table grapes to control grey mold disease, without compromising fruit quality. Similarly, Lu et al. (2010) showed that heat treatment of plums and nectarines delayed ripening (higher TA and TSS values), but did not have an effect on lycopene synthesis. Immersion in water at 60°C for 60 s was effective to reduce the incidence of brown rot (Karabulut et al., 2010). In addition, hot water at 55°C for 5 min. significantly reduced postharvest decay in hybrid organic melons, without affecting other qualities of the melons (total soluble solids, fruit firmness or the color of mesocarp tissue) during storage at 13°C, and also showed lowest fresh-weight-loss (Uthairatanakij et al., 2011).

In contrast, disinfestation procedures for mangoes and papaya by hot forced air for 4 h at 50°C, led to faster softening and, occasionally, hard lumps in the flesh after treatment (Shellie and Mangan, 1993). Heating 'Hujin' peaches by moist air at 37°C for 12 h is the most effective treatment to maintaining hardness and reducing water loss, while hot water caused heat injury (Zhou et al., 2002). Fruit sensitivity to heat treatments is modified by preharvest weather conditions, cultivar, time of heating, and subsequent storage conditions, and is related to the level of heat protective proteins at harvest and the postharvest production of heat-shock proteins (Paull and Chen, 2000).

15.4.3 ANTIBROWNING AGENTS FOR FRESH-CUT PRODUCE

There is a growing demand for minimally processed, fresh-cut or ready-to-eat fruit and vegetables. The operational processes used to generate fresh-cut produce cause cell damage that limits the subsequent shelf-life (Soliva-Fortuny and Martin-Belloso, 2003). A major issue is browning on the cut surfaces. Browning is due to the enzymatic oxidation of phenolic compounds by polyphenol oxidase (PPO) to o-quinones and subsequent polymerization to brown tissue (Mayer, 1987). A wide range of antibrowning agents have been developed, but their use on fresh-cut organic fruits and vegetables is restricted to agents of organic origin or from natural sources.

15.4.3.1 Ascorbic Acid

Ascorbic acid (AA), recognized as a GRAS substance, is allowed in organic produce (OMRI, 2013) to prevent browning. Its effectiveness arises from its ability to reduce *o*-quinones back to their phenolic substrates (Hsu et al., 1988). Dips of AA and its derivatives have been widely used in the concentration ranges of 0.5%–4%, and have been applied in various fresh-cut produce (Soliva-Fortuny and Martin-Belloso, 2003). Generally, AA is applied in combination with organic acids such as citric acid (Pizzocaro et al., 1993) and a mixture of 1% AA + 0.2% citric acid inhibits PPO 90%–100% in apple cubes. Moreover, the combination of 1% AA + 0.5% $CaCl_2$ preserves the color of apple cubes under appropriate MAP conditions (Soliva-Fortuny et al., 2001, 2002). However, this treatment is not completely effective in controlling enzymatic browning of fresh-cut fruit, since once the AA is completely oxidized to dehydroascorbic acid, *o*-quinones are no longer reduced, and browning may occur (Nicolas et al., 1994). In addition, AA may cause important oxidative damage in fresh-cut "Fuji" apples (Larrigaudiere et al., 2008).

15.4.3.2 Carboxylic Acids

Carboxylic acids have been widely used commercially due to their antibrowning activity. Citric acid exerts a double inhibitory effect by reducing pH and chelating copper in the active site of PPO and, thus, inactivating the enzyme (Son et al., 2001). Optimum PPO activity is observed at pH 6.0–6.5, while little activity is detected below pH 4.5 (Whitaker, 1995). Pizzocaro et al. (1993) reported more than a 90% inhibition of PPO activity in apple cubes by using a mixture of 1% AA + 0.2% citric acid or 1% AA + 0.5% sodium chloride. Other carboxylic acids, such as oxalic acid and oxaloacetic acid, show higher antibrowning activity than citric acid on fresh-cut apples (Son et al., 2001). Immersion of banana and apple slices in oxalic acid solutions is effective in reducing browning (Son et al., 2001; Yoruk et al., 2004). Although the mechanism of browning inhibition is unknown, oxalic acid seems to inhibit PPO *per se*, by chelating copper from the active site of the enzyme, and has a high affinity to copper ions forming a copper II metal complex (Tong et al., 1995). The extent of inhibition is influenced not only by oxalic acid concentration, but also by pH (Altunkaya and Gokmen, 2008). In addition, PPO enzymes from different sources exhibit different types of inhibition mechanisms (Son et al., 2001; Aydemir and Akkanli, 2006).

15.4.3.3 Ultraviolet Radiation

Ultraviolet (UV) radiation can be divided into four types on the basis of wavelengths: UV-A (400–320 nm), UV-B (320–280 nm), UV-C (280–200 nm), and UV vacuum (Blaustein and Searle, 2013). Of these, UV-C has a high potential to inhibit microbial growth in air, water, container surfaces, and vegetables, by the formation of photoproducts in the DNA, called pyrimidine dimers. Pyrimidine dimer molecules are generated on the same strand of DNA and are able to interfere with DNA transcription and translation, leading to cell malfunction and cell dead (Franz et al., 2009; Bermudez-Aguirre and Barbosa-Canovas, 2013). The maximum antimicrobial activity of UV-C wavelengths is at 254 nm (Chang et al., 1985). UV-C is nonthermal treatment for inactivation of microbial in food industry. The advantage of this technology is safety, and leaving no chemical residue on wash-water and the produce. However, the efficacy of UV-C for water disinfection is dependent on many factors, such as turbidity, suspended solids, and the presence of absorbing compounds (Selma et al., 2008). Graca et al. (2013) showed that UV-C irradiation at 10 kJ/m^2 could be an alternative treatment to washing with hypochlorite solutions for microbial disinfection of minimally processed apples. However, the use of UV-C treatment in the food processing industry is limited, due to UV-C irradiation having multiple effects on human skin, such as the risk of the common skin cancers, malignant melanoma, squamous cell carcinoma, and basal cell carcinoma (Moon et al., 2005).

15.4.3.4   Ultrasound

Decontamination of food products can be achieved with nonthermal technology, such as ultrasound. It has been extensively used to eliminate spoilage and food-borne pathogens in a range of food products (Birmpa et al., 2013). Ultrasound technology is safe, nontoxic, and friendly to the environment (Feng and Yang, 2011). High frequency of ultrasonic wave (≥ 20 kHz) leads to the chemical and physical changes in biological structure (Butz and Tauscher, 2002). Ultrasound causes disruption of microbial cell walls, membranes, and DNA, by free radical production (Scouten and Beuchat, 2002; Hulsmans et al., 2010).

15.5   POSTHARVEST QUALITY OF ORGANIC PRODUCE

Irrespective of the health awareness concerns, the continuing increase in the consumption of fresh organic produce can be attributed to the near absence of synthetic pesticide residues when compared to conventionally grown commodities (Crinnion, 2010; Willer, 2011). In addition, there has been considerable interest in the effect of organic production methods on bioactive compounds that are being positively associated with human health benefits. Plant bioactive metabolites are produced by a plant in response to various stimuli, including pest and environmental stresses (Brandt and MØlgaard, 2001). These substances are an important source of antioxidant compounds or "antioxidants," and can be divided into nonnitrogenous compounds such as phenolic acids, flavonoids, and terpenoids (e.g., carotenoids, xanthophylls), and nitrogen-containing compounds such as alkaloids, amines, nonprotein amino acids, glycosides, and glucosinolates (Ölmez and Särkka-Tirkkonen, 2008).

However, despite numerous studies, there is no consensus on the effect of organic production on levels of bioactives. Table 15.1 summarizes the findings from some of the studies showing the range of effects on antioxidant levels in a range of produce. To this list could be added the positive effect of phytonutrients reported by Worthington (2001), Zhao et al. (2006), Kazimierczak et al. (2008), and Crinnion (2010), while no significant differences or lower levels were reported by Rosen, 2010; Smith-Spangler et al. (2012), Asami et al. (2003), Rossi et al (2008), Kahu et al. (2010) and Bogs et al. (2012). The different responses of nutritional content in organic produce is complicated by interaction with cultivars, farm management, soil quality, weather conditions, and length of time using organic methods (Crinnion, 2010; Aldrich et al., 2011; Camargo et al., 2011).

15.6   CONCLUSIONS

The demand of organic fresh produce continues to grow dramatically (Falguera et al., 2012; Schaack et al., 2013). Organic fruit and vegetables may have higher

TABLE 15.1
Comparison of Nutritional Quality in Fruit and Vegetables Produced from Organic and Conventional Systems

Produce	Results	Reference
Peach	Higher polyphenol in organic	Carbonaro et al. (2002)
Apple	Higher polyphenol in organic	Weibel et al. (2000)
	Similar polyphenol content	Briviba et al. (2007)
	Higher polyphenol in organic	Bogs et al. (2012)
	Similar antioxidant capacity	
Kiwifruit	Higher ascorbic acid and total phenolic in organic	Amodio et al. (2007)
Orange	Higher ascorbic acid in organic	Masamba and Nguyen (2008)
Melon	Higher ascorbic acid in organic, but inconsistency in phenolic content	Salandanan et al. (2009)
Blueberry	Higher total phenolic, anthocyanin and antioxidant activity in organic	Wang et al. (2008)
Strawberry	Higher anthocyanin in organic	Camargo et al. (2011)
Persimmon	Higher β-carotene in organic	Cardoso et al. (2011)
	Lower dehydroascorbate	
	Similar lycopene content	
Tomato	Lower polyphenol in organic	Barrett et al. (2007)
	Higher flavonoid and kaempferol	Mitchell et al. (2007)
	Similar content of nutrient	Pieper and Barrett (2008)
Pak choi	Higher polyphenol in organic	Young et al. (2005)
Lettuce	Similar polyphenol content	Young et al. (2005)
	Similar ascorbic acid	Masamba and Nguyen (2008)
Cabbage	Similar ascorbic acid	Masamba and Nguyen (2008)
Carrot	Similar ascorbic acid	Masamba and Nguyen (2008)
Broccoli	Higher total phenolic content and antioxidant capacity in organic	Aldrich et al. (2011)

levels of antioxidants, although many studies demonstrate no difference between organic and conventional crops (Dangour et al., 2009; Hoefkens et al., 2009; Smith-Spangler et al., 2012).

Postharvest handling practices have not changed dramatically with the expansion of organic fruit and vegetable production over the last decade. Organic packers and handlers are becoming more aware of the special requires of organic commodities and greater emphasis on product quality and safety could be anticipated in the future. There is a need to evaluate fruit and vegetable varieties for better suitability for organic production systems and postharvest quality.

Fraud is another area of concern as the organic market expands. Documentation and traceability, especially in the postharvest logistics chain, is especially important, to avoid mislabeling and a loss of consumer confidence. Though it is a nonbiological aspect to the postharvest handling of organic fruits and vegetables, traceability systems need to be integrated in an effective manner.

ACKNOWLEDGMENT

We thank Dr. Robert E Paul, University of Hawaii at Manoa, for his helpful suggestions and for reviewing this chapter.

REFERENCES

Aldrich HT, Kendall P, Bunning M, Stonaker F, Kulen O, Stushnoff C. 2011. Environmental temperatures influence antioxidant properties and mineral content in broccoli cultivars grown organically and conventionally. *J Agro Crop Sci* 2: 1–10.

Alexopoulos A, Plessas S, Ceciu S, Lazar V, Mantzourani I, Voidarou C, Stavropoulou E. 2013. Evaluation of ozone efficacy on the reduction of microbial population of fresh cut lettuce (*Lactuca sativa*) and green bell pepper (*Capsicum annuum*). *Food Control* 30: 491–496.

Altunkaya A, Gokmen V. 2008. Effect of various inhibitors on enzymatic browning, antioxidant activity and total phenol content of fresh lettuce (*Lactuca sativa*). *Food Chem* 107: 1173–1179.

Amodio ML, Colelli G, Hasey JK, Kader AA. 2007. A comparative study of composition and postharvest performance of organically and conventionally grown kiwifruits. *J Sci Food Agric* 87: 1228–1236.

Arites F, Gomez P, Aguayo E, Escalona V, Artes-Hernandez F. 2009. Sustainable sanitation techniques for keeping quality and safety of fresh-cut plant commodities. *Postharv Biol Technol* 51: 287–296.

Asami DK, Hong YJ, Barrett DM, Mitchell AE. 2003. Comparison of the total phenolic and ascorbic acid content of freeze-dried and air-dried marionberry, strawberry, and corn grown using conventional, organic, and sustainable agricultural practices. *J Agric Food Chem* 51: 1237–1241.

Aydemir T, Akkanli G. 2006. Purification and characterization of polyphenol oxidase from celery root (*Apiumgraveolens*) and the investigation on enzyme activity of some inhibitors. *Internat J Food Sci Technol* 4: 1090–1098.

Barrett DM, Weakley C, Diaz JV, Watnik M. 2007. Qualitative and nutritional differences in processing tomatoes grown under commercial organic and conventional production systems. *J Food Sci* 72: 441–451.

Bermudez-Aguirre D, Barbossa-Canovas VG. 2013. Disinfection of selected vegetables under nonthermal treatments: Chlorine, acid citric, ultraviolet light and ozone. *Food Control* 29: 82–90.

Birmpa A, Sfika V, Vantarakis A. 2013. Ultraviolet light and ultrasound as non-thermal treatments for the inactivation of microorganisms in fresh ready-to-eat foods. *Int J Food Microbiol* 167: 96–102.

Blaustein RA, Searle C. 2013. Ultraviolet radiation. In: *Encyclopedia of Biodiversity* 2nd ed (ed. A Simon), Elsevier, Levin, pp. 296–303.

Bogs J, Bunning M, Stushnoff C. 2012. Influence of biologically enhanced organic production on antioxidant and sensory qualities of (*Malus x domestica* Borkn. cv. Braeburn) apples. *Organic Agric* 2: 117–126.

Brandt K, Mølgaard JP. 2001. Organic agriculture: does it enhance or reduce the nutritional value of plant foods? *J Sci Food Agric* 81: 924–931.

Briviba K, Stracke BA, Rüfer CE, Watzl B, Bub A. 2007. Effect of consumption of organically and conventionally produced apples on antioxidant activity and DNA damage in humans. *J Agric Food Chem* 55: 7716–7721.

Butz P, Tauscher B. 2002. Emerging technologies: Chemical aspects. *Food Res Int* 35: 279–284.

Camargo LKP, de Resende JTV, Tominaga TT, Kurchaidt SM, Camargo CK, Figueiredo AST. 2011. Postharvest quality of strawberry fruits produced in organic and conventional systems. *Hortic Brasileira* 29: 577–583.

Carbonaro M, Mittera M, Nicoli S, Bergoma P, Cappelloni M. 2002. Modulation of antioxidant compounds in organic vs. conventional fruit (peach, *Prunus persica* L. and pear, *Pyrus communis* L. cv. Williams). *Food Chem* 72: 419–424.

Cardoso PC, Tomazini APB, Stringheta PC, Ribeiro SMR, Pinheiro-Sant'Ana HM. 2011. Vitamin C and carotenoids in organic and conventional fruits grown in Brazil. *Food Chem* 126: 411–416.

Chang JCH, Ossoff SF, Lobe DC, Dorfman, MH, Dumais CM, Qualls RG, Johnson JD. 1985. UV inactivation of pathogenic and indicator microorganisms. *Appl Environ Microbiol* 49: 1361–1365.

Crinnion J. 2010. Organic foods contain higher levels of certain nutrients, lower levels of pesticides, and may provide health benefits for the consumer. *Altern Med Rev* 15: 4–12.

Dangour AD, Dodhia SK, Hayter A, Allen E, Lock K, Uauy R. 2009. Nutritional quality of organic foods: A systematic review. *Am J Clin Nutr* 90: 680–685.

Falguera V, Aliguer N, Falguera M. 2012. An integrated approach to current trends in food consumption: Moving toward functional and organic products? *Food Control* 26: 274–281.

Feng H, Yang W. 2011. Ultrasonic process. In: *Handbook on Nonthermal Processing Technologies for Food.* (eds HQ Zhang, GV Barbosa-Canovas, VM Bala-Balasubramaniam, CP Dunne, DF Farkas, JTC Yuan), Wiley-Blackwell, NJ, pp. 135–152.

Fonseca SC, Oliveira FAP, Brecht JK. 2002. Modeling respiration rate of fresh fruits and vegetables for modified atmosphere packages. *J Food Eng* 52: 99–119.

Franz C, Specht I, Cho GS, Graef V, Stahl M. 2009. UV-C inactivation of microorganisms in naturally cloudy apple juice using novel inactivation equipment based on Dean vortex technology. *Food Control* 20: 1103–1107.

Gil IM, Selma VM, Lopez-Galvez F, Allenda A. 2009. Fresh-cut product sanitation and wash water disinfection: Problems and solutions. *Int J Food Microbiol* 134: 37–45.

Graca A, Salazar M, Quintas C, Nunes C. 2013. Low dose UV-C illumination as an eco-innovative disinfection system on minimally processed apples. *Postharv Biol Technol* 85: 1–7.

Hagenmaier RD, Baker RA. 1997. Edible coatings from morpholine-free wax microemulsions. *J Agric Food Chem* 45: 349–352.

Hemmerling S, Obermowe T, Canavari M, Sidali KL, Stolz H,. Spiller A. 2013. Organic food labels as a signal of sensory quality-insights from a cross-cultural consumer survey. *Organic Agric* 3: 57–69.

Hoefkens C, Verbeke W, Aertsens J, Mondelaers K, van Camp J. 2009. The nutritional and toxicological value of organic vegetables: consumer perception versus scientific evidence. *Brit Food J* 111: 1062–1077.

Hsu AF, Shien JJ, Bills DD, White K. 1988. Inhibition of mushroom polyphenol oxidase by ascorbic acid derivatives. *J Food Sci* 53: 765–767.

Hulsmans A, Joris K, Lambert N, Rediers H, Declerck P, Delaedt Y. 2010. Evaluation of process parameters of ultrasonic treatment of bacterial suspensions in a pilot scale water disinfection system. *Ultrasonics Sonochem* 17: 1004–1009.

IFOAM. 2012. IFOAM norms for organic production and processing, www.ifoam.org/sites/default/files/page/files/ifoam_norms_version_august_2012_with_cover.pdf (accessed July 12, 2013).

Izumi H. 1999. Electrolyzed water as a disinfectant for fresh-cut vegetables. *J Food Sci* 64: 536–539.

Kahu K, Klaas L, Kika SA. 2010. Effect of cultivars and different growing technologies on strawberry yield and fruit quality. *Agron Res* 8: 589–593.

Karabulut OA, Cohen L, Wiess B, Daus A, Lurie S, Droby S. 2002. Control of brown rot and blue mold of peach and nectarine by short hot water brushing and yeast antagonists. *Postharv Biol Technol* 24: 103–111.

Karabulut OA, Smilanick JL, Crisosto CH, Palou L. 2010. Control of brown rot of stone fruits by brief heated water immersion treatments. *Crop Prot* 29: 903–906.

Kazimierczak R, Hallmann E, Rusaczonek A, Rembialkowska E. 2008. Anti oxidant content in black currants from organic and conventional cultivation. *Food Sci Technol Res* 2: 57–61.

Kim BS, Kwon JY, Kwon KH, Cha HS, Jeong JW. 2006. Antimicrobial effect of cold ozonated water washing on fresh-cut lettuce. *Acta Hortic* 699: 235–242.

Kim JG, Yousef AE, Khadre MA. 2003. Ozone and its current and future application in the food industry. *Adv Food Nutri Res* 45: 167–218.

Koseki S, Isobe S. 2006. Effect of ozonated water treatment on microbial control and on browning of iceberg lettuce (*Lactuca sativa* L.). *J Food Prot* 69: 154–160.

Larrigaudiere C, Ubach D, Soria Y, Rojas-Grau MA, Martin-Belloso O. 2008. Oxidative behavior of fresh-cut "Fuji" apple treated with stabilizing substances. *J Sci Food Agric* 88: 1170–1176.

Leifert C, Ball K, Volakakis N, Cooper JM. 2008. Control of enteric pathogens in ready-to-eat vegetable crops in organic and 'low input' production systems: a HACCP-based approach. *J Appl Microbiol* 105: 931–950.

Lu J, Charles MT, Vigneault C. 2010. Effect of heat treatment uniformity on tomato ripening and chilling injury. *Postharv Biol Technol* 56: 155–162.

Lydakis D, Aked J. 2003. Vapour heat treatment of Sultana table grapes. II: Effects on postharvest quality. *Postharv Biol Technol* 27: 117–126.

Masamba KG, Nguyen M. 2008. Determination and comparison of vitamin C, calcium and potassium in four selected conventionally and organically grown fruits and vegetables. *Afr J Biotech* 7: 2915–2918.

Mayer AM. 1987. Polyphenol oxidase in plants-recent progress. *Phytochemistry* 26: 11–20.

Mitchell AE, Hong YJ, Koh E, Barrett DM, Bryant DE, Denison RF, Kaffka S. 2007. Ten-year comparison of the influence of organic and conventional crop management practices on the content of flavonoids in tomatoes. *J Agric Food Chem* 55: 6154–6159.

Moon JS, Fryer AA, Strange CR. 2005. Ultraviolet radiation: Effects on risks of prostate cancer and other internal cancers. *Mutation Res Fundamental Molec Mechanisms Mutagen* 571: 206–219.

Mukherjee A, Speh D, Dyck E, Diez-Gonzalez F. 2004. Preharvest evaluation of coliforms, *Escherichia coli*, *Salmonella*, and *Escherichia coli* O157: H7 in organic and conventional produce grown by Minnesota farmers. *J Food Prot* 67: 894–900.

National Organic Program. www.ams.usda.gov/nop (accessed July12, 2013).

Nicolas JJ, Richard-Forget FC, Goupy PM, Amiot MJ, Aubert SY. 1994. Enzymatic browning reaction in apple and products. *Crit Rev Food Sci Nutr* 34: 109–157.

Oliveira M, Usall J, Viñas I, Anguera M, Gatius F, Abadias M. 2010. Microbiological quality of fresh lettuce from organic and conventional production. *Food Microbiol* 27: 679–684.

Ölmez H, Kretzschmar U. 2009. Potential alternative disinfestation methods for organic fresh-cut industry for minimizing water consumption and environmental impact. *LWT Food Sci Technol* 42: 686–693.

Ölmez H, Särkka-Tirkkonen M. 2008. *Case study: Assessment of chlorine replacement strategies for fresh cut vegetables.* Research Institute of Organic Agriculture FiBL, 5070 Frick, Switzerland.

OMRI. 2013. *Organic materials list.* Organic Materials Review Institute. www.omri.org (accessed July 12, 2013).

Padel S, Midmore P. 2005. The development of the European market for organic products: insights from a Delphi study. *Brit Food J* 107: 626–647.

Paull RE, Chen NJ. 2000. Heat treatment and fruit ripening. *Postharv Biol Technol* 21: 21–37.

Pieper JR, Barrett DM. 2009. Effects of organic and conventional production systems on quality and nutritional parameters of processing tomatoes. *J Sci Food Agric* 89: 177–194.

Pizzocaro F, Torregiani D, Gilardi G. 1993. Inhibition of apple polyphenol oxidase (PPO) by ascorbic acid, citric acid and sodium chloride. *J Food Process Pres* 17: 21–30.

Plotto A, Narciso JA. 2006. Guidelines and acceptable postharvest practices for organically grown produce. *HortScience* 41: 287–291.

Porat R, Daus A, Weiss B, Cohen L, Fallik E, Droby S. 2000. Reduction of postharvest decay in organic citrus fruit by a short hot water brushing treatment. *Postharv Biol Technol* 18: 151–157.

Prange RK, Ramin AA, Daniels-Lake B, DeLong JM, Braun PG. 2006. Perspectives on post-harvest biopesticides and storage technologies for organic produce. *HortScience* 41: 301–303.

Ramos B, Miller, FA, Brandoa TRS, Teixeira P, Silva CLM. 2013. Fresh fruits and vegetables: An overview on applied methodologies to improve its quality and safety. *Innovative Food Sci Emerging Technol* 20: 1–15.

Rosen JD. 2010. A Review of the nutrition claims made by proponents of organic food. *Comprehensive Rev Food Sci Food Safety* 9: 270–277.

Rossi F, Godani F, Bertuzzi T, Trevisan M, Ferrari F, Gatti S. 2008. Health-promoting sub-stances and heavy metal content in tomatoes grown with different farming techniques. *Eur J Nutr* 47: 266–272.

Salandanan K, Bunning M, Stonaker F, Kulen O, Kendall P, Stushnoff C. 2009. Comparative analysis of antioxidant properties and fruit quality attributes of organically and conven-tionally grown melons (*Cucumis melo* L.). *HortScience* 44: 1825–1832.

Sandhya. 2010. Modified atmosphere packaging of fresh produce: Current status and future needs: A review. *LWT Food Sci Technol* 43: 381–392.

Schaack D, Lernoud J, Padel S, Willer H. 2013. The Organic Market in Europe 2011—Nine Percent Increase Compared with 2010. In: *The World of Organic Agriculture - Statistics and Emerging Trends 2013.* (eds H Willer, J Lernoud, L Kilcher), Research Institute of Organic Agriculture (FiBL) and International Federation of Organic Agriculture Movements (IFOAM), Frick and Bonn, pp. 224–229.

Scouten AJ, Beuchat LR. 2002. Combined effects of chemical, heat and ultrasound treatments to kill *Salmonella* and *Escherichia coli* O157:H7 on alfalfa seeds. *J Appl Microbiol* 92: 668–674.

Selma VM, Allende A, Lopez-Galvez F, Conesa MA, Gil MI. 2008. Disinfection potential of ozone, ultraviolet-C and their combination in wash water for the fresh-cut vegetable industry. *Food Microbiol* 25: 809–814.

Shellie KC, Mangan RL. 1993. Disinfestation of subtropical commodities with forced air. *Acta Hortic* 343: 367–370.

Silva E. 2008. Approved chemicals for use in organic postharvest systems. In: *Wholesale Success: A Farmer's Guide to Selling, Postharvest Handling and Packing Produce* (Midwest ed). www.familyfarmed.org/wholesale-success (accessed July 12, 2013).

Smith-Spangler C, Brandeau ML, Hunnter GE, Bavinger JC, Pearson M, Eschbach PJ, Liu H et al. 2012. Are organic foods safer or healthier than conventional alternatives?: A sys-tematic review. *Ann Intern Med* 157: 48–366.

Soliva-Fortuny RC, Grigelmo-Miguel N, Odriozola-Serrano I, Gorinstein S, Martın-Belloso O. 2001. Browning evaluation of ready-to-eat apples as affected by modified atmosphere packaging. *J Agric Food Chem* 49: 3685–3690.

Soliva-Fortuny RC, Martin-Belloso O. 2003. New advances in extending the shelf life of fresh-cut fruits: A review. *Trends Food Sci Technol* 14: 341–353.

Soliva-Fortuny RC, Oms-Oliu G, Martin-Belloso O. 2002. Effects of ripeness stages on the storage atmosphere, color and textural properties of minimally processed apple slices. *J Food Sci* 67: 1958–1963.

Son SM, Moon KD, Lee CY. 2001. Inhibitory effects of various anti-browning agents on apple slices. *Food Chem* 73: 23–30.

Suslow T. 2000. *Postharvest handling for organic crops*. Publicn No.7254. Div. Agric Natural Resources, Univ. Calif.

Suslow T. 2006. *Making sense of rules governing chlorine contact in postharvest handling of organic produce*. Publicn No.8198. Div. Agric Natural Resources, Univ. Calif.

Tong CBS, Hicks KB, Osman SF, Hotchkiss AT, Hains RM. 1995. Oxalic acid in commercial pectins inhibits browning of raw apple juice. *J Agric Food Chem* 43: 592–597.

Uthairatanakij A, Jitareerat P, Photchanachai S, Supbua D, Maneenoi A. 2011. Effect of hot water dipping on disease incidence and marketable quality of organic hybrid muskmelons during storage. Proc Internat Soc Organic Agric Res held at 17th IFOAM Organic World Congress Sept 28–Oct 1, Namyangju, Korea.

UNCTAD. 2003. Organic fruit and vegetables from the tropics: Market, certification and production information for producers and international trading companies. A publication of the UN Conf Trade Development, New York http://unctad.org/en/docs/ditccom20032_en.pdf (accessed August 12, 2013)

USDA. 2009. National Organic Program. www.ams.usda.gov (accessed June 29, 2013)

U.S. Department of Agriculture. 1993. *Plant protection and quarantine manual, Section VI-T100*. U.S. Department of Agriculture, Animal and Plant Health Inspection Service, Hyattsville, MD.

Wang SY, Chen CT, Sciarappa W, Wang CY, Camp MJ. 2008. Fruit quality, antioxidant capacity, and flavonoid content of organically and conventionally grown blueberries. *J Agric Food Chem* 56: 5788–5794.

Weibel FP, Bickel R, Leuthold S, Alfoldi T. 2000. Are organically grown apples tastier and healthier? A comparative field study using conventional and alternative methods to measure fruit quality. *Acta Hortic* 7: 417–427.

Whitaker JR. 1995. Polyphenol oxidase. In: *Food Enzymes, Structure and Mechanism*. (ed. DWS Wong), Chapman Hall, New York, pp. 271–307.

Willer H. 2011. Organic Agriculture in Europe: Overview. In: *The World of Organic Agriculture, Statistics & Emerging Trends 2011, FiBL-IFOAM Report*. (eds H Willer, L Kilcher), IFOAM. Bonn and FiBL, Frick.

Winter CK, Davis SF. 2006. Organic foods. *J Food Sci* 71: R117–R124.

Worthington V. 2001. Nutritional quality of organic versus conventional fruits, vegetable, and grains. *J Altern Complem Med* 7: 161–173.

Yoruk R, Yoruk S, Balaban MO, Marshall MR. 2004. Machine vision analysis of antibrowning potency for oxalic acid: a comparative investigation on Banana and apple. *J Food Sci* 69: E281–E289.

Young JE, Zhao X, Carey EE, Welti R, Yang SS, Wang W. 2005. Phytochemical phenolics in organically grown vegetables. *Mol Nutr Food Res* 49: 1136–1142.

Zhao X, Carey EE, Wang W, Rajashekar CB. 2006. Does organic production enhance phytochemical content of fruit and vegetables? Current knowledge and prospects for research. *HortTech* 16: 449–456.

Zhou T, Xu S, Sun D, Wang Z. 2002. Effects of heat treatment on postharvest quality of peaches. *J Food Eng* 54: 17–22.

16 Modeling in Postharvest Horticulture

Maarten L.A.T.M. Hertog and Bart M. Nicolaï

CONTENTS

16.1 INTRODUCTION

16.1.1 POSTHARVEST AND MODELING

Postharvest research is relatively young and has been rapidly increasing since the 1980s (Figure 16.1). Its main objective is to reduce postharvest losses by enlarging insight in the underlying physiological process to understand the responsible mechanisms. As highlighted in the previous chapters, thanks to enhanced insights in the underlying mechanisms, advanced storage conditions, and postharvest treatments are being developed to continuously improve the postharvest handling chain. Given the complex nature of the biochemistry and physiology driving the metabolism of living produce during postharvest, it is not always evident how to interpret the experimental results. Mathematical modeling can help to gain insights in the mechanisms

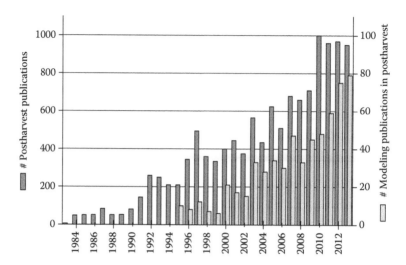

FIGURE 16.1 Number of published items retrieved from Web of Science (All databases) using the keyword "postharvest" either in or not in combination with the keyword "model" in the topic field (status on May 01, 2014).

underlying the observed phenomena by screening hypothetical conceptual models for their fitness to explain the data. The main purpose of modeling in postharvest is often not the model itself, but the process of modeling, in which the researcher is forced to line up concepts to bring order in the chaos. A wide range of mathematical models has found their application in the wider food area (Tijskens et al., 2001) and the importance in the field of postharvest research can be appreciated from the increasing contribution of modeling papers (Figure 16.1).

In biological systems, models are, by definition, a simplification of the real world and, again by definition, will never be true as the only true model is the system itself (Oreskes et al., 1994). In the current context, a mathematical model is defined as an abstract representation expressed in terms of mathematical and/or logical constructs. These can include all sorts of equations (e.g., algebraic, ordinary and partial differential, and integral equations), different types of Boolean expressions (e.g., propositions, predicates, and fuzzy logic), or rule-based expressions (e.g., decision tables, finite state machines and Turing machines) (Nicolai et al., 2005). So far, only few have been applied to postharvest situations. Once a system has been modeled using mathematical constructs, relevant real-life actions affecting the system can be substituted through quantitative reasoning studying their impact. This allows the capture of relations that are impossible to effectively describe otherwise (Staub and Stern, 1997).

When developing mathematical models to describe the behavior of living produce, one needs to find a balance between two extremes. One extreme is to describe all processes in detail. This will inevitably lead to the limitations of current knowledge and will result in a theoretical model that is impossible to parameterize. The other extreme is to leave out too much detail, simplifying the underlying mechanisms to relationships that only hold for the current situation. In the end, the level of complexity of the model chosen should match the objectives of the model.

If the objective of the model is to facilitate sorting of fruit by ripeness, a simple Boolean model could be used, linking fruit color to fruit ripeness, providing a color threshold above (or below) which the fruit can be considered ripe. If the objective is to optimize postharvest climacteric ripening conditions in order to deliver ready-to-eat fruit to the market, a more complex model is needed, quantitatively linking applied ethylene levels and ripening temperatures to the dynamics of postharvest fruit-ripening. If the objective would be to develop new varieties with specific ripening properties, even more detailed mechanistic models might be required, modeling fruit-ripening control at the metabolic, enzymatic, and/or molecular level. In this way, a hierarchy of models with increasing complexity can be constructed. In general, models with the lowest complexity are often more generic, flexible to transfer to similar situations, and, therefore, sufficient for management decisions. With increasing level of complexity, models become more specific, can be less readily extrapolated to other situations and conditions without considerable expert knowledge, but are more useful in generating fundamental insight and understanding of the underlying biology and physics.

In functionalizing the model objective, one will be forced to reduce the available data to the essentials to be covered by the model. In this context, it is important to realize that data reduction invariably leads to loss of information. In many cases, the information filtered out may be not useful (e.g., measurement noise) or not essential (e.g., aroma data when modeling the softening of mango fruit), but in other cases it might decrease the accuracy of the model in representing reality (e.g., by ignoring information on fruit maturity stage when modeling climacteric fruit ripening). To guide this process of data reduction, proper expert knowledge is essential.

16.1.2   INDUCTIVE VERSUS DEDUCTIVE MODELS

The different modeling approaches can be roughly separated into two groups, inductive and deductive modeling techniques. With *inductive* modeling techniques, the model is induced by the data and no explicit expert knowledge is required. These techniques are completely data-driven and might result in some generally valid relationships. Examples of these techniques are the traditional statistical techniques and modern data mining techniques like neural networks. The *deductive* modeling techniques start from the underlying processes, are based on explicit expert knowledge, and are constructed using fundamental laws and generally valid relationships. Starting from these basic building blocks, the model is constructed and calibrated against the experimental data. Examples of these techniques are models of heat and mass transfer, biochemical kinetic models, and models of population dynamics.

Often, one differentiates between fundamental or white-box models, and empirical or black-box models. Whereas the first class is assumed to be built upon well-established laws of physics and first principles, the latter class is based on statistical induction. In reality, many of the so-called mechanistic models are effectively phenomenological models that are based on a simplified assumed mechanism far less complex than the actual underlying process. For example, to describe respiration behavior of fruit and vegetables as a function of the oxygen concentration in the storage atmosphere, often Michaëlis-Menten kinetics, with various modifications to account for the effect of temperature and the inhibition by carbon dioxide, have been developed (Fonseca et al., 2002; Hagger et al., 1992; Helena Gomes et al., 2010; Hertog et al., 1998; Peppelenbos and van't Leven, 1996). However, the Michaëlis-Menten model describes the kinetics of a linear chain of enzyme-mediated reactions with often a single rate-limiting step while, in reality, the metabolic pathways involved in respiration are much more complex and cyclic in nature, as can be retrieved from the kyoto encyclopedia of genes and genomes (KEGG) pathways (Ogata et al., 1999). Although Michaëlis-Menten kinetics has been shown to describe respiration well, it must be considered, at most, as a semi-mechanistic or gray-box model.

Metabolic network models take this approach several steps further by providing a more detailed mechanistic overview by including the detailed stoichiometry of most important biochemical pathways. While such metabolic models can be detailed in their biochemistry, they have their own assumptions—for instance, by assuming no spatial concentration differences of metabolites inside the object of study. While this is a reasonable assumption at the cellular level, it is not necessarily the case at the tissue or whole-fruit level. The obvious examples where this homogeneity assumption does not hold, are the case of compartmentalization within plant cells (Tiessen et al., 2012), the often observed concentration gradients along a tissue differentiation axis (Macklon and DeKock, 1967), and gas gradients related to the microstructure of the fruit (Ho et al., 2011).

This example clearly shows that fundamental knowledge on the underlying mechanisms allow us to move away from purely explorative, statistical approaches by gradually incorporating more of the available knowledge into the model structure. Still, in spite of containing more knowledge, detailed mechanistic models are not always more suitable given computational restrictions, and given the difficulty of properly calibrating all parts of such complex model. However, it is easy to foresee that their future scope in ongoing optimization of postharvest systems is considerable.

16.2 VARIATION IN THE DATA

All experimental data collected in a postharvest context is affected by different sources of variation (De Ketelaere et al., 2006), some willingly imposed, some inevitable, some unknowingly, and some avoidable. When modeling such experimental data, using either statistical or mechanistic modeling approaches, this variation can hamper the interpretation of the data, if not properly accounted for.

16.2.1 TREATMENT VARIATION

Treatment variation is the variability induced by the independent treatment variables controlled by the experimenter. When modeling the data, these variables will be included as part of the model structure to, hopefully, account for the observed differences in the postharvest responses. This implies a certain level of experimental control with the treatment variables being well-defined.

In postharvest literature, one often defines constant set points for the intended levels of the treatment variables that do not necessarily match the actual fluctuating values. If the temporal fluctuations are limited, one might be able to assume these conditions to be constant. However, while 0.5°C difference at room temperature might be acceptable, the same accuracy during cold storage might make the difference between either severe or no incidence of chilling injury. In such situations, it is better to include the actual measured values when modeling the data as compared to taking the set-point values. Such dynamically changing conditions do, of course, impose restrictions on the modeling approach that can be applied, but also creates opportunities, as will be shown later. Similar considerations can be made when spatial variation is involved. When storing packed fruit, conditions in the storage room atmosphere might not resemble the conditions at the individual fruit level. For this reason, one should try to take the actual conditions the fruit is exposed to as an input to analyze the fruit behavior. Alternately, one can extend the model to include the transport phenomena governing the produce environment.

This concept of replacing assumed constant conditions by either actual measurements or by their modeled counterparts can be extended to the various organization levels at which postharvest experimentation takes place (temporal and spatial variation in, e.g., storage room conditions, dosing levels of postharvest chemicals, temperature and gas composition inside a fruit, and substrate levels, pH and gas composition of experimental liquid media). Applying this concept will improve the quality of the subsequent model analyses, as more of the observed variation can be attributed to the experimental variables. Any of the treatment variation left unassigned will result in unexplained variation in the studied postharvest responses.

16.2.2 Technical Variation

Technical variation is the variability due to measurement error. Some of this error is random variation inherent to the methodology applied, but often this error can be related to the design of the experiment. For example, many methodologies depend on calibration of equipment, stock solutions of reagents, multiple instruments of the same type (e.g., pipettes, loggers, balances, and climate chambers), or different brands of consumables that are prone to vary between batches or are sensitive to drifts due to fluctuations in temperature, humidity, atmospheric pressure, and more. For this reason, it is important to be aware and track any such relevant variables to understand changes in the measured responses between, for instance, measurement days. If this information is available, it can be taken into account during the data analysis by including such effects as separate factors. If this is practically unfeasible, the least one should do is to prevent unwanted technical variation from overlapping with the expected treatment variation by randomizing the measurements of samples. For instance, when comparing the effect of two temperatures on fruit quality, one should not analyze samples from one temperature on one day (or using one instrument, or stock solution) while analyzing samples from the other temperature on a second day (or using another instrument, or stock solution). In such cases, one can never be sure if the observed effect was due to temperature or technical variation between the measurement days. In combination with randomization, the inclusion of reference samples and replication of the regular samples is essential to obtain a proper measure of the technical variation.

16.2.3 BIOLOGICAL VARIATION

Biological variation is the variability arising from the fact that no two biological items are identical. Most of the time, postharvest management aims at controlling the average batch behavior and limiting biological variation as much as possible by sorting and grading the product at the different stages in the postharvest chain. Biological variation of tomatoes, for instance, becomes discernible through their initial color at harvest. Assuming all tomatoes go through the same developmental ripening process from fruit set to fruit senescence, the initial color at harvest can be interpreted as a measure of the biological age of an individual tomato (Van de Poel et al., 2012). If all fruit could be harvested at the same maturity, variation at harvest would be negligible and would remain negligible throughout the postharvest period. As fruit are not harvested at such a homogenous stage, variation does exist both at harvest and during postharvest storage (Tijskens and Wilkinson, 1996). Depending on the underlying mechanism, the biological variation at harvest can remain the same or can be transformed during postharvest storage. In general, postharvest research tends to take individual fruit behavior as an unavoidable nuisance and works mainly on the averaged fruit behavior. With such an approach, biological variation is reducing the statistical power of the model analysis (Tijskens et al., 2003). Over the years, alternatives have been developed to integrate biological variation as an explicit part of the modeling approach enhancing the interpretation of the data (as reviewed by Hertog et al., 2007a; Jordan and Loeffen, 2013).

16.2.4 POOLING OF VARIATION

One approach often observed in postharvest research to reduce variation in the data is the pooling of experimental data. This can be done physically by collecting a single measurement per sample that was pooled by combining multiple sub-samples (e.g., ethylene production by a pooled group of multiple fruits, enzyme activity in a tissue sample collected from multiple fruit, or titratable acidity on fruit juice collected from a batch of fruit), or virtually by collecting data as measurements on the individual fruit and then averaging the values. In both cases, the measurement represents the average value for a batch with the biological variation completely removed. When interpreting such pooled data, no information is available to test for a significant effect of the treatments applied. To do so, one needs to check whether the effect induced by the treatment is larger than the variation within a treatment (Cumming et al., 2007). However, this information has been filtered out by the pooling. Even when replicate pooled measurements are collected, these will mainly mirror the technical measurement variation resulting in over-optimistic interpretation of the data.

The golden rule should be not to pool any data at all, as valuable information is discarded, thus inhibiting correct scientific interpretation of the data. The only case in which pooling is inevitable, is when a single fruit would not provide enough material for the planned analyses. However, in that case, one should at least collect replicate measurement (on different pooled samples), and be aware of the limitations of the data when interpreting the results.

16.3 INDUCTIVE MODELING APPROACHES

Inductive modeling techniques are largely data-driven and do not require expert knowledge to build the models. Often, these statistical techniques are applied just because expert knowledge is lacking and one mainly tries to understand which factors might affect the studied response.

16.3.1 CLASSICAL STATISTICS

The most common "model" applied in postharvest is that of a normal distribution. When comparing treatment effects using the commonly used Student's *t*-test, the underlying assumption is that the data follows such normal distribution. When comparing treatments in postharvest research, the number of groups to be compared quickly increases. Analysis of variance (ANOVA) overcomes this issue by generalizing the *t*-test to more than two groups (Payne, 2014). The treatment effects are calculated by partitioning the observed response variance over the treatment factors and comparing the mean of a treatment to the general mean over all treatments. When more than one treatment variable is involved, the model underlying the ANOVA can become more extended with the response variable being interpreted as the combined effect of the various treatments, including possible interaction terms.

In postharvest studies, the treatment factors might be either well-controlled nonrandom factors (so-called fixed effects) or factors still prone to some level of randomness (so-called random effects). Both situations are included in a special class of ANOVA models referred to as mixed effects models (Molenberghs and Verbeke, 2000). If the assumption of normality is violated, generalized mixed models provide the ultimate alternative, allowing the use of other (user-defined) distributions (Payne, 2014). By explicitly taking into account random sources at the treatment level, less variation will remain unexplained when interpreting the dependent postharvest responses.

Besides studying the effect of various postharvest treatments, time plays an important role in postharvest research as well as in trying to explain the treatment effect throughout storage and shelf-life. The (generalized) mixed model approaches can take into account such repeated measurements, typically showing a strong correlation for a single fruit, while being independent between the individual fruit. Given their additional capability to account for the appropriate sources of biological variation, (generalized) mixed model approaches are widely applicable in the area of postharvest operations (Lammertyn et al., 2003a).

A special class of postharvest data that should be treated in a different way is that concerning the incidence of diseases or disorders, often relying on countable data, and measured in a binomial manner as either a "yes" or a "no" (often coded as 1 or 0) response. Similar data is obtained when working on consumer acceptance, resulting in an "accept" or a "reject" response. Such data can be properly analyzed using so-called logistic regression (Lammertyn et al., 2000) in which the dependence of the response on the treatment factors is modeled through probability scores. This approach can be seen as a special case of generalized models.

Once ANOVA-based techniques have been applied to determine which treatment factors significantly contribute to the observed differences between groups, the same underlying model structure can be used as a regression model to predict the postharvest response as a function of the imposed treatments. By developing such regression models, the incorporated treatment-factors do have their proven relevance based on the experimental data. Of course, other regression models can be developed in a complete empirical way by simple curve fitting and might describe the data even better, but such empirical models cannot claim any scientific experimental basis.

Reviewing the usage of these various statistical techniques in the area of postharvest operations, ANOVA was found to be applied in 35% of all articles published in the journal named *Postharvest Biology and Technology* (as on May 01, 2014), while the *t*-test was applied in 11% of all published manuscripts. On inspection it appears that ANOVA is often applied just to obtain the results of the multiple comparison *t*-tests that are commonly available as a posthoc test with ANOVA, without the actual ANOVA results being presented. The application of (generalized) mixed model approaches and logistic regression was limited to less than 2% of all publications in *Postharvest Biology and Technology*. The application of this last group of models is clearly hampered by their conceptual complexity as compared to the basic ANOVA and multiple *t*-tests. With the increased availability of (generalized) mixed model tools in most standard statistical software packages, the door is, however, open for a more intensive application of these advanced statistical modeling techniques.

16.3.2 MULTIVARIATE STATISTICS

Multivariate statistics is often used synonymous with chemometrics, which has its origin in analytical chemistry. Chemometrics was defined as "*How to get chemically relevant information out of measured chemical data, how to represent and display this information, and how to get such information into data*" (Wold, 1995). Multivariate statistical methods are extremely suitable to extract information from large datasets. They can, for instance, be used to profile complex (chemical) produce properties, to find relationships between produce composition and sensory properties, to discriminate between cultivars and species, or to detect adulteration in food products. Especially with the development of data-rich measurement techniques like the various spectroscopic techniques (e.g., nuclear magnetic resonance, UV, visible, and near infrared spectroscopy, Fourier transform infrared spectroscopy, Raman spectroscopy) and the various "omic" techniques generally applied in the framework of systems biology (Hertog et al., 2011), the application of multivariate statistics has a wide potential field of application.

Central to most chemometrics is the concept of reducing the complexity of the data by organizing the original measured variables into groups of highly correlated variables, with each group being represented by a single new synthetic variable (referred to as a principal component or latent variable). These principle components are defined in such a way that they show no mutual correlation, with all redundant information from the original data being removed, and only the unique information being conserved. By this reduction in the complexity of the data, patterns can be more easily recognized. As such, multivariate statistics is less about detecting statistically significant differences, and more about recognizing patterns by examining relationships among multiple variables at the same time.

The two most prominent techniques applied in chemometrics are principal component analysis (PCA) and partial least square regression (PLS). While PCA is an unsupervised explorative technique to detect the underlying correlation structures in the data, PLS provides a supervised approach in which prior knowledge on the underlying latent structure is provided, and where the PLS regression model will try to explain this latent structure from the multivariate data-set.

A more recent multivariate modeling technique that was introduced in chemometrics is that of artificial neural networks (Marini et al., 2008). These are mathematical systems developed in analogy to the human brain, which learns by example. An artificial neural network can be seen as a black-box that receives multiple inputs and produces multiple outputs with the in- and out-puts being interconnected through layers of parallel connected simple arithmetic units (neurons). Parameterization of the neural network takes place by training using representative examples. Artificial neural networks are highly suitable for classification purposes.

Chemometrics has been applied in postharvest research, especially in the field of visual and near infrared spectral data analysis, to link spectral features to the chemical composition of fruit and vegetables to establish nondestructive measurement techniques for produce properties like dry matter content, acidity, starch content, and soluble solids content (as reviewed by Nicolai et al., 2007). The same approach has been applied with various degrees of success to describe derived quality attributes like firmness (Bobelyn et al., 2010), incidence of internal disorders (Magwaza et al., 2012), fruit maturity (Peirs et al., 2005), and sensory evaluation (Parpinello et al., 2013). The application of chemometrics accounts for about 12% of all publications in *Postharvest Biology and Technology*.

With "omic" applications still being under-represented in postharvest research (about 2% of all publications in *Postharvest Biology and Technology*), chemometrics for "omics" data still plays only a minor role (Pedreschi et al., 2007; Rudell et al., 2009; Vandendriessche et al., 2013). Other applications can be found in research in the area of postharvest volatiles (Ciesa et al., 2013; Vandendriessche et al., 2012), where the complex aroma composition can be linked to sensory evaluation (Berna et al., 2005; Obando-Ulloa et al., 2009). Within postharvest research, artificial neural networks are sporadically used as pattern recognition models to classify produce according shelf-life, maturity, or the incidence of various defects (ElMasry et al., 2009; In et al., 2009; Schouten et al., 1997).

16.4 DEDUCTIVE MODELING APPROACHES

As soon as inductive models have identified some leads to how the studied process might function, conceptual models will evolve which can be taken as a starting point for developing deductive models. The basic strategy to develop deductive models is to apply a systematic process of problem decomposition, dissecting the problem into its basic building blocks, subsequently reassembling them while leaving out unnecessary detail. What is essential and what is redundant depends largely on the intended application of the model. Going through this process of modeling, one is forced to think about cause and effect. The modeling process helps to structure ideas and helps to turn them into sound conceptual models. The final mathematical model can be used to test these concepts and check their validity, given the current level of knowledge. The mathematical models can combine expertise from different scientific disciplines and can be targeted to different applications and organizational levels. Examples can be found in the area of economics, operations management, logistics, heat and mass transfer, reaction diffusion models, and biochemical pathway models.

16.4.1   MODELING SUPPLY CHAIN LOGISTICS

Within postharvest systems, there are many processes taking place at various organization levels that can benefit from modeling. This starts at the level of supply chain logistics. In the fresh-food supply chain, one has to consider many factors such as time-to-market, traceability, transport/storage conditions, handling, production/process control, demand variability, seasonal behavior, all of which affect the operation of the supply chain. Fresh-food supply chain model models can be applied to optimize the chain with regard to speed of operation, cost reduction, and optimized shelf-life (Dabbene et al., 2008a,b; Hertog et al., 2014; Rong et al., 2011). Such logistic models focus on the timely distribution of the product to satisfy the needs in the chain and to optimize warehouse management. Common strategies aim at efficient product management across the distribution chain through FIFO (First-In-First-Out) or FEFO (First-Expired-First-Out). FIFO is the more commonly adopted approach while FEFO is more dedicated to perishable products as it will only ship products depending on the potential of their shelf-life in relation to their end-destination (East, 2011). This requires proper insight in product behavior in response to the conditions imposed (De Baerdemaeker et al., 2006).

16.4.2   MODELING THE POSTHARVEST ENVIRONMENT

While, eventually, the main interest is on the produce itself, its postharvest fate is largely defined by the correct functioning of the infrastructures involved during harvesting, transport, storage, cooling, drying, packaging, and so on. For this reason, much effort still goes toward optimization and improved design of this postharvest infrastructure through engineering models (Ambaw et al., 2013). The ultimate aim is to create the best possible environment for the produce under study, which also requires proper knowledge on produce requirements concerning their storage conditions.

To model the produce environment, the main emphasis is on the flow, heat, and mass transfer processes responsible for the temporal and spatial changes in temperature, in the main gas conditions (being the levels of oxygen, carbon dioxide, and water vapor), and in any additional compounds introduced in the storage atmosphere to control, for instance, postharvest ripening (e.g., ethylene and 1-methylcyclopropene), or to suppress tuber-sprouting or the incidence of diseases or fungal and bacterial growth (Ambaw et al., 2012; Delele et al., 2009; Nahor et al., 2005). Generally computational fluid dynamics (CFD) approaches are applied as they take into account both the 3D geometry of storage rooms and produce stacking, and the numerical calculation of fluid flow and heat and mass transfer, based on fundamental laws of physics, thus allowing for the incorporation of processes like gas diffusion, kinetics, and droplet or particle dispersion (Norton and Sun, 2006). While such modeling can be performed starting from the appropriate engineering expertise as such, most added value is obtained when properly embedded within the multidisciplinary context involved in postharvest operations.

16.4.3 MODELING PRODUCE BEHAVIOR

Most published produce quality models are phenotypic models that describe post-harvest-produce behavior in terms of responses measured at the whole-fruit or vegetable level, such as a change in firmness (Schouten et al., 2010), color (Lana et al., 2006), sugar content (Hertog et al., 1997), or other relevant product properties and quality attributes. Some of these changes are part of the physiological process of fruit-ripening while other processes, such as shriveling (Maguire et al., 2000), are driven by merely physics. Often such models are based on semimechanistic concepts formulated through kinetic concepts, but not necessarily one-to-one founded in produce biochemistry. In addition, they assume the produce to be homogeneous and do not take into account any spatial variation of produce properties or quality attributes.

For instance, fruit-softening has been modeled many times, using various approaches with the level of detail often determined by the available experimental data and the purpose in mind. This ranged from purely empirical curve fitting models describing kiwifruit softening with the main emphasis on the industrial application towards batch segregation (Benge et al., 2000; Jabbar et al., 2014), to relatively detailed models based on a detailed conceptual approach but calibrated on firmness data only (Schouten et al., 2010; Tijskens et al., 1999b). Other models were defined closer to the underlying enzyme actions and calibrated on both firmness data and the actual enzyme activities measured (Róth et al., 2008; Tijskens et al., 1998), while a last group of models focused merely on a detailed description of the effect of oxygen and carbon dioxide on the fruit-softening rate to describe the effect of modified atmosphere conditions (Hertog et al., 2001, 2004).

Overall, this type of models has shown itself to be relatively successful in describing the dynamics of fruit and vegetable quality attributes, as they are detailed enough to have some generic value, while they are not hampered by an overly detailed description that would make them hard to calibrate. When developing these semi-mechanistic models, one should always try to respect basic fundamental conservation laws and, at the same time, introduce the required simplifications to deliver valid models that are scientifically useful. One of the more detailed models of this kind was a semimechanistic model describing the dynamics of mealiness in apple (De Smedt et al., 2002). This model described measured textural changes related to mealiness, starting from the underlying cell-wall physiology and the cellular water management, incorporating both simplified biochemical reactions as well as the cellular water transfer processes.

16.4.3.1 Produce Models and Biological Variance

What differentiates the postharvest chain from most other handling chains is that the postharvest chain involves dealing with living tissues that inherently change with time and which show large sources of biological variance. For the sake of simplicity, one generally focuses on understanding the average batch behavior, but from a marketing point of view, one has to deal with the biological variance present. Recently, new impulses have been given to include biological variation as part of postharvest product models (as reviewed by Hertog et al., 2007a). One of the driving forces for this has been the increased availability of nondestructive measuring techniques that allow monitoring of individual objects during time (De Baerdemaeker et al., 2006; Nicolai et al., 2014; Tijskens et al., 1999a; Zerbini et al., 2006). By introducing biological variance into the product models, both the analysis and the subsequent prediction of postharvest batch behavior can be considerably improved as it allows for predicting the propagation of the initial biological variance at harvest throughout the whole postharvest chain taking into account all relevant aspects affecting postharvest fruit behavior (Hertog et al., 2008; Schouten et al., 2004). In recent years, more and more applications have been found where models incorporate biological variation to allow for optimization of the postharvest handling chain based on the whole batch behavior (Eccher Zerbini et al., 2009; Guillard et al., 2012; Gwanpua et al., 2013; Hertog et al., 2007b; Jabbar et al., 2014; Jordan and Loeffen, 2013).

16.4.4 MODELING PRODUCE MORPHOLOGY

When progressing beyond phenotypic models to create more true-to-life models, a first step is to include the overall morphology of the produce under study through its 3D geometry. Such macroscale models have been proven successful to model the combination of respiration, gas diffusion, and gas permeation (Ho et al., 2008; Lammertyn et al., 2003b), either in or not in combination with heat transfer (Benítez et al., 2012). These models allow accounting for spatial differences arising from diffusion processes under the assumption of a homogenous internal structure. However, this ignores the microscale structure of plant tissue arising from tissue differentiation, which contributes greatly to a heterogeneity in heat and mass transfer properties of the tissue, as the various structures have different characteristics, such as porosity, connectivity, and conductance (Herremans et al., 2013; Kuroki et al., 2004). Microstructural models have been developed, starting from various imaging techniques to provide realistic tissue geometries, either using simple 2D reconstructions (Ho et al., 2009), or based on more detailed 3D reconstructions (Herremans et al., 2013). A 3D microscale model of a complete fruit is, however, not yet feasible because of its excessive computational requirements. For this purpose the multiscale modeling paradigm is applied, which combines continuum-type macroscale models with microscale description of features in the regions of interest, in this way creating a hierarchy of models (Ambaw et al., 2013; Ho et al., 2013; Figure 16.2). The models are coupled via multiscale analysis, in which the model results relevant to the macroscale are linked to simulations at the microscale by means of homogenization and localization procedures. Such a multiscale approach has been extensively applied to model gas-exchange of fruit (Ho et al., 2011) and to model mechanical properties of whole fruit and plant tissues (Ghysels et al., 2010; Li et al., 2013). Instead of using real-imaged geometries, there is also the possibility to start from generated virtual geometries mimicking real tissue (Abera et al., 2013; Fanta et al., 2014;

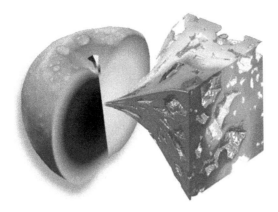

FIGURE 16.2   Artist's impression of the multiscale modeling approach applied to apple. At the microscale model a computational tomographic image of apple fruit tissue was used to generate a 3D geometry of the cortical fruit tissue. Based on simulations at this microscale tissue properties of gas permeability were estimated. These diffusion properties were applied to a homogenized lumped model at the whole-fruit scale under the assumption that all parts of the cortex have the same diffusion properties. Based on this assumption, gas gradients at the whole fruit level were calculated.

Mebatsion et al., 2009; Pieczywek et al., 2011). The added advantage of including realistic geometries obviously lies in accounting for detailed spatial differences induced by the geometry affecting, through the governing physics of heat and mass transfer, the functioning of the plant's metabolism.

16.4.5   MODELING PRODUCE METABOLISM

To quantitatively understand the functioning of fruit metabolism, the phenotypic models have to be extended towards the relevant metabolic pathways involved and their control. This control can be at the metabolite level (e.g., substrate availability and product inhibition), at the enzymatic level (e.g., reaction kinetics, allosteric regulation, enzyme turnover, and posttranslational modification), and at the transcription level (e.g., enzyme induction, switching between specific isoforms and gene silencing). While the main metabolic pathways are often well-defined, the various control systems are less well-known, limiting the development of system-wide regulatory models (de Jong, 2002).

The most essential part of plant metabolism governing postharvest maintenance of fruits and vegetables is respiration (e.g., glycolysis, Krebs cycle, and pentose phosphate pathway). Respiration releases the energy enclosed in carbon compounds for maintenance, and generates carbon precursors for biosynthesis. While at the plant or plant organ level, respiration has been mainly modeled through strongly simplified Michaelis-Menten models (e.g., Hertog et al., 1998), detailed fundamental modeling approaches have been applied to unicellular organisms such as yeast (Herrgård et al., 2008; Teusink et al., 2000). More recently, these efforts have been extended to simple plant systems such as *Arabidopsis thalliana* (Szecowka et al., 2013; Williams et al.,

2010). The difficulty with the application to multicellular plant systems is that: (1) inside plant cells, metabolic reactions are spatially distributed over multiple cellular compartments; and (2) tissue differentiation has resulted in multiple cell types, each having their own metabolic characteristics. These two aspects require the introduction of transport terms between cells and between cellular compartments.

To make such extensive metabolic models work, kinetic information is required on the involved enzymes and transporters. Most literature data on enzyme activity is based on *in vitro* measurements under ideal conditions, concerning factors such as substrate availability, pH, temperature, and cofactors. Such theoretical enzyme activities no longer hold *in vivo* and have to be estimated from the experimental data on *in vivo* fluxes. When dynamic responses are studied, the time profiles of the observed changes are largely dictated by the underlying enzyme specific kinetic mechanisms. As these mechanisms are often not known, steady state approaches are taken. When a system is at steady state, the actual kinetic mechanism is no longer relevant and can be omitted from the model. While these simplified steady state models are easier to derive, their application is inherently limited to steady state situations only; no conclusions can be drawn on, for instance, how fast a fruit will adapt to an applied postharvest condition, only on the eventual steady state itself. Given the dynamic nature of fruit ripening and senescence, the steady state assumption can only be valid when relative short time periods are considered. When evaluating dynamic postharvest conditions, steady state cannot be assumed and the full dynamic changes have to be included. Where the underlying kinetics are not known, they will have to be estimated from experiments designed to excite the system in such a way that it reveals its characteristic dynamic behavior. While the first steps have been taken to develop detailed models on the main respiration pathways, some effort is still needed to extend these models to the level of intact fruits and vegetables.

Of special postharvest interest is the plant hormone ethylene, which is involved in many physiological processes, including the ripening of climacteric fruit. Over the years, various models describing ethylene biosynthesis have been developed, going from schematic representations, including the importance of different enzymatic isoforms (Alexander and Grierson, 2002), to elaborated Michaelis-Menten kinetics to describe the activity of ACO (Sanders and de Wild, 2003). A more mechanistic approach was followed by Génard and Gouble (2005), who developed a theory of ethylene emission by peach, based on a mathematical representation of the respiration process and ethylene pathway. They incorporated the effect of ambient fluctuating temperatures through their effect on respiration. A similar model was applied to apple by East (2007). In all cases, the models were only calibrated and validated based on experimental data concerning respiration and ethylene production rates, while all intermediate metabolites and related enzyme activities remained hypothetical. More recently, a detailed kinetic model of the ethylene biosynthesis pathway was developed, based on actual quantitative data of all the intermediates and enzymes involved in the ethylene biosynthesis in CA-stored apple (Bulens et al., 2012), and ripening tomato fruit (Van de Poel et al., 2014) starting from the level of transcriptomics (Figure 16.3). Over the years, models of the ethylene biosynthesis have progressively included more detail, integrating aspects from the different organization levels but without considering the spatial organization within the cell or tissue. What

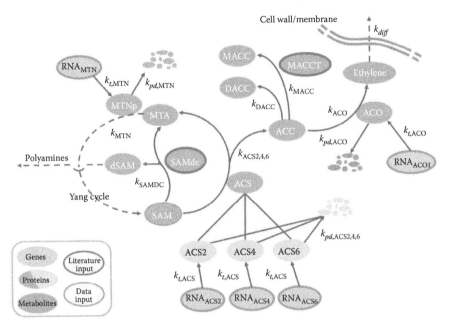

FIGURE 16.3 Visual representation of the mathematical model developed by Van de Poel et al. (2014) describing the ethylene biosynthesis pathway for ripening tomato fruit. Solid arrows represent direct conversions, while the dotted arrows indicate more general pathway interactions containing multiple conversion steps. Ethylene is the end-product of a cascade of enzymatic reactions and diffuses out of the cellular tissue through the cell wall/membrane. The enzyme levels result from both gene expression and protein degradation.

remains is the inclusion of the binding kinetics of ethylene molecules to the ethylene receptors and the downstream signaling cascade leading to the activation of multiple ethylene response factors and the associated physiological responses such as fruit softening (Bennett and Labavitch, 2008).

16.5 PERSPECTIVES OF INTEGRATED MODELS FOR POSTHARVEST OPTIMIZATION

Overall, postharvest modeling allows postharvest systems to be analyzed, predicted and improved, either in a manual way or through an integrated approach of model predictive control, in which postharvest management decisions are made based on model-based prediction of the product behavior (Lukasse and Polderdijk, 2003; Verdijck et al., 1999). The aim of modeling in postharvest is therefore not to develop true models but to develop valid models that are consistent with our current knowledge level and that contain no known or detectable flaws of logic. Also, models should be sufficiently detailed for the intended purpose but, at the same time, simplified to give robust manageable models.

While the previous sections have summarized an already large volume of modeling research, there is still a need and a potential for integrated models to progress scientific concepts and support postharvest optimization. These applications can be found in the area of optimizing the design of packages, shipping containers, and storage rooms, to create the best possible conditions for postharvest storage, either in or not in combination with product models, to understand the effect of the governing conditions on produce metabolism and, eventually, its quality.

Still, the link between what happens at the level of metabolic pathways and what becomes measurable at the intact fruit level, might not always be fully known. For instance, fruit firmness is the resultant of complex cell-wall biochemistry (Johnston et al., 2002). While, over the years, the pathways and the enzyme involvements have been increasingly unraveled, the challenge remains to translate the biochemical changes observed at the level of cell-wall polymers (e.g., degree of methylation and esterification; Brummell, 2006) to the level of the mechanical properties of fruit firmness. This is complicated by the observation that clear differences exist in the genetic regulation of fruit-softening between fruits and even between cultivars of the same fruit (Goulao and Oliveira, 2008). While complicating the transfer of experimental findings, phenomena such as fruit-softening provide a good playing ground for developing an integrative generic model that, depending on the fruit, is activated in a different configuration and that, through such variations, can orchestrate the fruit-specific process of cell-wall breakdown, leading to the softening patterns characteristic for the selected fruit. In this way, modeling can be used to fine-tune and develop our understanding of the whole process through the iterative process of model-based hypothesis development, and the subsequent challenge to these hypotheses through experimental data.

Another important potential application field for modeling in postharvest operations is that of predictive microbiology, as postharvest shelf-life is often limited by microbial growth. In addition, fresh fruit and vegetables are nowadays recognized as providing an important route of entry for zoonotic human pathogens into the food chain (Brandl, 2006). While a large volume of modeling research exists on predictive microbiology (McMeekin et al., 2013), it mainly applies to well-controlled growing conditions, either on artificial growth media or in the framework of fermentation processes, or in relation to the microbial safety of fish and meat products. Some predictive microbiology has been performed in relation to modified atmosphere packaging of fruits and vegetables either with or not with real produce, and under actual logistic-chain conditions (Geysen et al., 2005; Jacxsens et al., 2002; Koseki and Isobe, 2005). The extension of predictive microbiology to postharvest situations is still limited due to the many unknown factors in terms of initial infection levels, the actual microorganisms involved, and the complexity of fruits and vegetables as a bio-based growth medium. With the booming of the fresh-cut produce industry (Nicola et al., 2006), there is an urgent need for quantitative insights in risk assessments throughout the food supply chain. Ultimately, this requires the combination of logistic risk assessment models (Tromp et al., 2010) with numerical tools to include the appropriate sources of variation (Busschaert et al., 2011; Ferrer et al., 2008; Nicolai and Van Impe, 1996; Poschet, 2003).

In summary, modeling in postharvest operations has two important benefits that still can be capitalized more profoundly. The first is from a scientific point of view, in testing different conceptual models on their fitness to explain the data, enhancing the understanding of what mechanisms could be responsible for the observed phenomena. The second benefit is from an applied point of view, where mathematical modeling can be used as an instrument to disseminate scientific knowledge to the wider industry in such a way that it can be readily applied in the form of management support systems.

REFERENCES

Abera MK, Verboven P, Herremans E, Defraeye T, Fanta SW, Ho QT, Carmeliet J, Nicolai BM. 2013. 3D virtual pome fruit tissue generation based on cell growth modeling. *Food Bioprocess Technol* 7: 542–555.

Alexander L, Grierson D. 2002. Ethylene biosynthesis and action in tomato: A model for climacteric fruit ripening. *J Expt Bot* 53: 2039–2055.

Ambaw A, Delele MA, Defraeye T, Ho QT, Opara LU, Nicolaï BM, Verboven P. 2013. The use of CFD to characterize and design post-harvest storage facilities: Past, present and future. *Comput Electron Agric* 93: 184–194.

Ambaw A, Verboven P, Delele MA, Defraeye T, Tijskens E, Schenk A, Nicolai BM. 2012. CFD modelling of the 3D spatial and temporal distribution of 1-methylcyclopropene in a fruit storage container. *Food Bioprocess Technol* 6: 2235–2250.

Benge JR, De Silva HN, Banks NH, Jeffery PB. 2000. Empirical modelling of postharvest changes in the firmness of kiwifruit. *Postharv Biol Technol* 19: 211–220.

Benítez S, Chiumenti M, Sepulcre F, Achaerandio I, Pujolá M. 2012. Modeling the effect of storage temperature on the respiration rate and texture of fresh-cut pineapple. *J Food Eng* 113: 527–533.

Bennett AB, Labavitch JM. 2008. Ethylene and ripening-regulated expression and function of fruit cell wall modifying proteins. *Plant Sci* 175: 130–136.

Berna AZ, Lammertyn J, Buysens S, Di Natale C, Nicolaï BM. 2005. Mapping consumer liking of tomatoes with fast aroma profiling techniques. *Postharv Biol Technol* 38: 115–127.

Bobelyn E, Serban AS, Nicu M, Lammertyn J, Nicolai BM, Saeys W. 2010. Postharvest quality of apple predicted by NIR-spectroscopy: Study of the effect of biological variability on spectra and model performance. *Postharv Biol Technol* 55: 133–143.

Brandl MT. 2006. Fitness of human enteric pathogens on plants and implications for food safety. *Annu Rev Phytopathol* 44: 367–392.

Brummell DA. 2006. Cell wall disassembly in ripening fruit. *Funct Plant Biol* 33: 103.

Bulens I, de Poel B, Geeraerd AH, Nicolai BM, Hertog MLATM. 2012. Ethylene metabolism in "Jonagold" apple during CA storage and shelf-life—A modeling approach. *Acta Hortic* 945: 331–336.

Busschaert P, Geeraerd AH, Uyttendaele M, Van Impe JF. 2011. Hierarchical Bayesian analysis of censored microbiological contamination data for use in risk assessment and mitigation. *Food Microbiol* 28: 712–719.

Ciesa F, Dalla Via J, Wisthaler A, Zanella A, Guerra W, Mikoviny T, Märk TD, Oberhuber M. 2013. Discrimination of four different postharvest treatments of "Red Delicious" apples based on their volatile organic compound (VOC) emissions during shelf-life measured by proton transfer reaction mass spectrometry (PTR-MS). *Postharv Biol Technol* 86: 329–336.

Cumming G, Fidler F, Vaux DL. 2007. Error bars in experimental biology. *J Cell Biol* 177: 7–11.

Dabbene F, Gay P, Sacco N. 2008a. Optimisation of fresh-food supply chains in uncertain environments, Part I: Background and methodology. *Biosyst Eng* 99: 348–359.

Dabbene F, Gay P, Sacco N. 2008b. Optimisation of fresh-food supply chains in uncertain environments, Part II: A case study. *Biosyst Eng* 99: 360–371.

De Baerdemaeker J, Hertog M, Nicolai B, De Ketelaere B. 2006. Innovative measurements and models for predicting shelf-life of fresh foods during postharvest. *Int J Postharv Technol Innov* 1: 32–47.

De Jong H. 2002. Modeling and simulation of genetic regulatory systems: A literature review. *J Comput Biol* 9: 67–103.

De Ketelaere B, Stulens J, Lammertyn J, Cuong NV, De Baerdemaeker J. 2006. A methodological approach for the identification and quantification of sources of biological variance in postharvest research. *Postharv Biol Technol* 39: 1–9.

Delele MA, Schenk A, Ramon H, Nicolaï BM, Verboven P. 2009. Evaluation of a chicory root cold store humidification system using computational fluid dynamics. *J Food Eng* 94: 110–121.

De Smedt V, Barreiro P, Verlinden BE, Veraverbeke EA, De Baerdemaeker J, Nicolai, B.M. 2002. A mathematical model for the development of mealiness in apples. *Postharv Biol Technol* 25: 273–291.

East AR. 2007. The influence of breaks in optimal storage conditions on "Cripps Pink" apple physiology and quality. PhD Thesis, Massey University, Palmerston North, New Zealand.

East AR. 2011. Accelerated libraries to inform batch sale scheduling and reduce postharvest losses of seasonal fresh produce. *Biosyst Eng* 109: 1–9.

Eccher Zerbini P, Vanoli M, Rizzolo A, Jacob S, Torricelli A, Spinelli L, Schouten RE. 2009. Time-resolved Reflectance Spectroscopy as a management tool in the fruit supply chain: An export trial with nectarines. *Biosyst Eng* 102: 360–363.

ElMasry G, Wang N, Vigneault C. 2009. Detecting chilling injury in "Red Delicious" apple using hyperspectral imaging and neural networks. *Postharv Biol Technol* 52: 1–8.

Fanta SW, Abera MK, Aregawi WA, Ho QT, Verboven P, Carmeliet J, Nicolai BM. 2014. Microscale modeling of coupled water transport and mechanical deformation of fruit tissue during dehydration. *J Food Eng* 124: 86–96.

Ferrer J, Prats C, López D 2008 Individual-based modelling: An essential tool for microbiology. *J Biol Phys* 34: 19–37.

Fonseca SC, Oliveira FAR, Brecht JK. 2002. Modelling respiration rate of fresh fruits and vegetables for modified atmosphere packages: a review. *J Food Eng* 52: 99–119.

Génard M, Gouble B. 2005. ETHY. A theory of fruit climacteric ethylene emission. *Plant Physiol* 139: 531–545.

Geysen S, Geeraerd AH, Verlinden BE, Michiels CW, Van Impe JF, Nicolaï BM. 2005. Predictive modelling and validation of Pseudomonas fluorescens growth at superatmospheric oxygen and carbon dioxide concentrations. *Food Microbiol* 22: 149–158.

Ghysels P, Samaey G, Van Liedekerke P, Tijskens E, Ramon H, Roose D. 2010. Multiscale modeling of viscoelastic plant tissue. *Int J Multiscale Comput Eng* 8: 379–396.

Goulao L, Oliveira C. 2008. Cell wall modifications during fruit ripening: When a fruit is not the fruit. *Trends Food Sci Technol* 19: 4–25.

Guillard V, Guillaume C, Destercke S. 2012. Parameter uncertainties and error propagation in modified atmosphere packaging modelling. *Postharv Biol Technol* 67: 154–166.

Gwanpua SG, Verlinden BE, Hertog MLATM, Van Impe J, Nicolai BM, Geeraerd AH. 2013. Towards flexible management of postharvest variation in fruit firmness of three apple cultivars. *Postharv Biol Technol* 85: 18–29.

Hagger PE, Lee DS, Yam K. 1992. Application of an enzyme kinetics based respiration model to closed system experiments for fresh produce. *J Food Process Eng* 15: 143–157.

Helena Gomes M, Beaudry RM, Almeida DPF, Xavier Malcata F. 2010. Modelling respiration of packaged fresh-cut "Rocha" pear as affected by oxygen concentration and temperature. *J Food Eng* 96: 74–79.

Herremans E, Verboven P, Bongaers E, Estrade P, Verlinden BE, Wevers M, Hertog MLATM, Nicolai BM. 2013. Characterisation of "Braeburn" browning disorder by means of X-ray micro-CT. *Postharv Biol Technol* 75: 114–124.

Herrgård MJ, Swainston N, Dobson P, Dunn WB, Arga KY, Arvas M, Blüthgen N et al. 2008. A consensus yeast metabolic network reconstruction obtained from a community approach to systems biology. *Nat Biotechnol* 26: 1155–1160.

Hertog MLATM, Lammertyn J, De Ketelaere B, Scheerlinck N, Nicolaï BM. 2007a. Managing quality variance in the postharvest food chain. *Trends Food Sci Technol* 18: 320–332.

Hertog MLATM, Lammertyn J, Scheerlinck N, Nicolaï BM 2007b. The impact of biological variation on postharvest behaviour: The case of dynamic temperature conditions. *Postharv Biol Technol* 43: 183–192.

Hertog MLATM, Nicholson SE, Banks NH. 2001. The effect of modified atmospheres on the rate of firmness change in "Braeburn" apples. *Postharv Biol Technol* 23: 175–184.

Hertog MLATM, Nicholson SE, Jeffery PB. 2004. The effect of modified atmospheres on the rate of firmness change of "Hayward" kiwifruit. *Postharv Biol Technol* 31: 251–261.

Hertog MLATM, Ochoa-Ascencio S, Nicolai BM. 2008. Modelling the effect of 1-MCP on avocado fruit. *Acta Hortic* 802: 409–414.

Hertog MLATM, Peppelenbos HW, Evelo RG, Tijskens LMM. 1998. A dynamic and generic model of gas exchange of respiring produce: The effects of oxygen, carbon dioxide and temperature. *Postharv Biol Technol* 14: 335–349.

Hertog MLATM, Rudell DR, Pedreschi R, Schaffe, RJ, Geeraerd AH, Nicolaï BM, Ferguson I. 2011. Where systems biology meets postharvest. *Postharv Biol Technol* 62: 223–237.

Hertog MLATM, Tijskens LMM, Hak PS. 1997. The effects of temperature and senescence on the accumulation of reducing sugars during storage of potato (Solanum tuberosum L) tubers: A mathematical model. *Postharv Biol Technol* 10: 67–79.

Hertog MLATM, Uysal, I, Verlinden BM, Nicolaï BM. 2014. Shelf-life modelling for warehouse management. *Phil Trans R Soc A* 372: 20130306.

Ho QT, Carmeliet J, Datta AK, Defraeye T, Delele MA, Herremans E, Opara L et al. 2013. Multiscale modeling in food engineering. *J Food Eng* 114: 279–291.

Ho QT, Verboven P, Mebatsion HK, Verlinden BE, Vandewalle S, Nicolai BM. 2009. Microscale mechanisms of gas exchange in fruit tissue. *New Phytol* 182: 163–174.

Ho QT, Verboven P, Verlinden BE, Herremans E, Wevers M, Carmeliet J, Nicolai BM. 2011. A three-dimensional multiscale model for gas exchange in fruit. *Plant Physiol* 155: 1158–1168.

Ho QT, Verboven P, Verlinden BE, Lammertyn J, Vandewalle S, Nicolai BM. 2008. A continuum model for metabolic gas exchange in pear fruit. *PLoS Comput Biol* 4: e1000023.

In BC, Inamoto K, Doi M. 2009. A neural network technique to develop a vase life prediction model of cut roses. *Postharv Biol Technol* 52: 273–278.

Jabbar A, East AR, Jones G, Tanner DJ, Heyes JA. 2014. Modelling batch variability in softening of "Hayward" kiwifruit from at-harvest maturity measures. *Postharv Biol Technol* 90: 7–14.

Jacxsens L, Devlieghere F, Debevere J. 2002. Predictive modelling for packaging design: equilibrium modified atmosphere packages of fresh-cut vegetables subjected to a simulated distribution chain. *Int J Food Microbiol* 73: 331–41.

Johnston JW, Hewett EW, Hertog MLATM. 2002. Postharv softening of apple (Malus domestica) fruit: A review. *New Zealand J Crop Hortic Sci* 30: 145–160.

Jordan RB, Loeffen MPF. 2013. A new method for modelling biological variation using quantile functions. *Postharv Biol Technol* 86: 387–401.

Koseki S, Isobe S. 2005. Prediction of pathogen growth on iceberg lettuce under real temperature history during distribution from farm to table. *Int J Food Microbiol* 104: 239–248.

Kuroki S, Oshita S, Sotome I, Kawagoe Y, Seo Y. 2004. Visualization of 3-D network of gas-filled intercellular spaces in cucumber fruit after harvest. *Postharv Biol Technol* 33: 255–262.

Lammertyn J, Aerts M, Verlinden BE, Schotsmans W, Nicolai BM. 2000. Logistic regression analysis of factors influencing core breakdown in "Conference" pears. *Postharv Biol Technol* 20: 25–37.

Lammertyn J, De Ketelaere B, Marquenie D, Molenberghs G, Nicolaï BM. 2003a. Mixed models for multicategorical repeated response: Modelling the time effect of physical treatments on strawberry sepal quality. *Postharv Biol Technol* 30: 195–207.

Lammertyn J, Scheerlinck N, Jancsók P, Verlinden B, Nicolaï BM. 2003b. A respiration–diffusion model for "Conference" pears I: Model development and validation. *Postharv Biol Technol* 30: 29–42.

Lana MM, Tijskens LMM, van Kooten O. 2006. Modelling RGB colour aspects and translucency of fresh-cut tomatoes. *Postharv Biol Technol* 40: 15–25.

Li Z, Li P, Yang H, Liu J. 2013. Internal mechanical damage prediction in tomato compression using multiscale finite element models. *J Food Eng* 116: 639–647.

Lukasse LJS, Polderdijk JJ. 2003. Predictive modelling of post-harvest quality evolution in perishables, applied to mushrooms. *J Food Eng* 59: 191–198.

Macklon AES, DeKock PC. 1967. Physiological gradients in the potato tuber. *Physiol Plantarum* 20: 421–429.

Maguire KM, Banks NH, Lang A, Gordon IL. 2000. Harvest date, cultivar, orchard, and tree effects on water vapor permeance in apples. *J Am Soc Hortic Sci* 125: 100–104.

Magwaza LS, Opara UL, Terry LA, Landahl S, Cronje PJ, Nieuwoudt H, Mouazen AM, Saeys W, Nicolai BM. 2012. Prediction of "Nules Clementine" mandarin susceptibility to rind breakdown disorder using Vis/NIR spectroscopy. *Postharv Biol Technol* 74: 1–10.

Marini F, Bucci R, Magrì AL, Magrì AD. 2008. Artificial neural networks in chemometrics: History, examples and perspectives. *Microchem J* 88: 178–185.

McMeekin T, Olley J, Ratkowsky D, Corkrey R, Ross T. 2013. Predictive microbiology theory and application: Is it all about rates? *Food Control* 29: 290–299.

Mebatsion HK, Verboven P, Melese Endalew A, Billen J, Ho QT, Nicolai BM. 2009. A novel method for 3-D microstructure modeling of pome fruit tissue using synchrotron radiation tomography images. *J Food Eng* 93: 141–148.

Molenberghs G, Verbeke G. 2000. *Linear Mixed Models for Longitudinal Data, Springer Series in Statistics*. Springer, New York.

Nahor HB, Hoang ML, Verboven P, Baelmans M, Nicolai BM. 2005. CFD model of the airflow, heat and mass transfer in cool stores. *Int J Refrig* 28: 368–380.

Nicola S, Fontana E, Torassa C, Hoeberechts J. 2006. Fresh-cut Produce: Postharvest critical issues. *Acta Hortic* 712: 223–230.

Nicolai BM, Beullens K, Bobelyn E, Peirs A, Saeys W, Theron KI, Lammertyn J. 2007. Nondestructive measurement of fruit and vegetable quality by means of NIR spectroscopy: A review. *Postharv Biol Technol* 46: 99–118.

Nicolai BM, Defraeye T, De Ketelaere B, Herremans E, Hertog MLATM, Saeys W, Torricelli A, Vandendriessche T, Verboven P. 2014. Nondestructive measurement of fruit and vegetable quality. *Annu Rev Food Sci Technol* 5: 285–312.

Nicolai BM, Hertog MLATM, Lammertyn J, Scheerlinck N, Veraverbeke EA, Verboven P, Verlinden B. 2005. Quality kinetics and modelling. *Acta Hortic* 687: 23–32.

Nicolai BM, Van Impe JF. 1996. Predictive food microbiology: A probabilistic approach. *Math Comput Simulat* 42: 287–292.

Norton T, Sun DW. 2006. Computational fluid dynamics (CFD)—an effective and efficient design and analysis tool for the food industry: A review. *Trends Food Sci Technol* 17: 600–620.

Obando-Ulloa JM, Nicolai BM, Lammertyn J, Bueso MC, Monforte AJ, Fernández-Trujillo JP. 2009. Aroma volatiles associated with the senescence of climacteric or non-climacteric melon fruit. *Postharv Biol Technol* 52: 146–155.

Ogata H, Goto S, Sato K, Fujibuchi W, Bono H, Kanehisa M. 1999. KEGG: Kyoto encyclopedia of genes and genomes. *Nucleic Acids Res* 27: 29–34.

Oreskes N, Shraderfrechette K, Belitz K. 1994. Verification, validation, and confirmation of numerical models in the earth sciences. *Science* 263: 641–646.

Parpinello GP, Nunziatini G, Rombolà AD, Gottardi F, Versari A. 2013. Relationship between sensory and NIR spectroscopy in consumer preference of table grape ("cv Italia"). *Postharv Biol Technol* 83: 47–53.

Payne RW. 2014. Developments from analysis of variance through to generalized linear models and beyond. *Ann Appl Biol* 164: 11–17.

Pedreschi R, Vanstreels E, Carpentier S, Hertog M, Lammertyn J, Robben J, Noben JP, Swennen R, Vanderleyden J, Nicolai BM. 2007. Proteomic analysis of core breakdown disorder in "Conference" pears (Pyrus communis L.). *Proteomics* 7: 2083–2099.

Peirs A, Schenk A, Nicolaï BM. 2005. Effect of natural variability among apples on the accuracy of VIS–NIR calibration models for optimal harvest date predictions. *Postharv Biol Technol* 35: 1–13.

Peppelenbos HW, van't Leven J. 1996. Evaluation of four types of inhibition for modelling the influence of carbon dioxide on oxygen consumption of fruits and vegetables. *Postharv Biol Technol* 7: 27–40.

Pieczywek PM, Zdunek A, Umeda M. 2011. Study on parameterisation of plant tissue microstructure by confocal microscopy for finite elements modelling. *Comput Electron Agric* 78: 98–105.

Poschet F. 2003. Monte Carlo analysis as a tool to incorporate variation on experimental data in predictive microbiology. *Food Microbiol* 20: 285–295.

Rong A, Akkerman R, Grunow M. 2011. An optimization approach for managing fresh food quality throughout the supply chain. *Int J Prod Econ* 131: 421–429.

Róth E, Hertog MLATM, Kovács E, Nicolaï B. 2008. Modelling the enzymatic softening of apples in relation to cultivar, growing system, picking date and season. *Int J Food Sci Technol* 43: 620–628.

Rudell DR, Mattheis JP, Hertog MLATM. 2009. Metabolomic change precedes apple superficial scald symptoms. *J Agric Food Chem* 57: 8459–8466.

Sanders MG, de Wild HPJ. 2003. The relation between *in vivo* ethylene production and oxygen partial pressure. *Postharv Biol Technol* 30: 143–151.

Schouten RE, Jongbloed G, Tijskens LMM, van Kooten O. 2004. Batch variability and cultivar keeping quality of cucumber. *Postharv Biol Technol* 32: 299–310.

Schouten RE, Natalini A, Tijskens LMM, Woltering EJ, van Kooten O. 2010. Modelling the firmness behaviour of cut tomatoes. *Postharv Biol Technol* 57: 44–51.

Schouten RE, Otma EC, vanKooten O, Tijskens LMM. 1997. Keeping quality of cucumber fruits predicted by biological age. *Postharv Biol Technol* 12: 175–181.

Staub FC, Stern E. 1997. Abstract reasoning with mathematical constructs. *Int J Educ Res* 27: 63–75.

Szecowka M, Heise R, Tohge T, Nunes-Nesi A, Vosloh D, Huege J, Feil R, Lunn J, Nikoloski Z, Stitt M, Fernie AR, Arrivault S. 2013. Metabolic fluxes in an illuminated Arabidopsis rosette. *Plant Cell* 25: 694–714.

Teusink B, Passarge J, Reijenga CA, Esgalhado E, van der Weijden CC, Schepper M, Walsh MC, Bakker BM, van Dam K, Westerhoff HV, Snoep JL. 2000. Can yeast glycolysis be understood in terms of *in vitro* kinetics of the constituent enzymes? Testing biochemistry. *Eur J Biochem* 267: 5313–5329.

Tiessen A, Nerlich A, Faix B, Hümmer C, Fox S, Trafford K, Weber H, Weschke W, Geigenberger P. 2012. Subcellular analysis of starch metabolism in developing barley seeds using a non-aqueous fractionation method. *J Expt Bot* 63: 2071–2087.

Tijskens L, Hertog M, Nicolaï B. (Eds.) 2001. *Food Process Modelling*. Woodhead Publishing Ltd, Cambridge, England, 496p.

Tijskens LMM, Hertog MLATM, Van Kooten O, Simčič M. 1999a. Advantages of non-destructive measurements for understanding biological variance and for modelling of the quality of perishable products. In: *Proc 34 Vortragstagung Der DGQ: Zurstörungsfreie Qualitätsanalyse*. Freising Weihenstephan, pp. 13–24.

Tijskens LMM, Konopacki P, Sim M. 2003. Biological variance, burden or benefit? *Postharv Biol Technol* 27: 15–25.

Tijskens LMM, Rodis PS, Hertog MLATM, Kalantzi U, van Dijk C. 1998. Kinetics of polygalacturonase activity and firmness of peaches during storage. *J Food Eng* 35: 111–126.

Tijskens LMM, Van Schaik ACR, Hertog MLATM, de Jager A. 1999b. Modelling the firmness of "Elstar" apples during storage and transport. *Acta Hortic* 485: 363–371.

Tijskens LMM, Wilkinson EC. 1996. Behaviour of biological variability in batches during postharvest storage. *Asp Appl Biol* 46: 267–268.

Tromp SO, Rijgersberg H, Franz E. 2010. Quantitative microbial risk assessment for Escherichia coli O157:H7, Salmonella enterica, and Listeria monocytogenes in leafy green vegetables consumed at salad bars, based on modeling supply chain logistics. *J Food Protect* 73: 1830–1840.

Van de Poel B, Bulens I, Hertog MLATM, Nicolai BM, Geeraerd AH. 2014. A transcriptomics-based kinetic model for ethylene biosynthesis in tomato (Solanum lycopersicum) fruit: Development, validation and exploration of novel regulatory mechanisms. *New Phytol* 202: 952–963.

Van de Poel B, Bulens I, Hertog MLATM, Van Gastel L, De Proft MP, Nicolai BM, Geeraerd AH. 2012. Model-based classification of tomato fruit development and ripening related to physiological maturity. *Postharv Biol Technol* 67: 59–67.

Vandendriessche T, Schaefer H, Verlinden BE, Humpfer E, Hertog MLATM, Nicolai BM, Schäfer H. 2013. High-throughput NMR based metabolic profiling of Braeburn apple in relation to internal browning. *Postharv Biol Technol* 80: 18–24.

Vandendriessche T, Vermeir S, Martinez CM, Lammertyn J, Nicolai BM, Hertog, MLATM, Hendrickx Y. 2012. High-throughput flavor evaluation of strawberry cultivars: Focus on aroma development during ripening. *Acta Hortic* 945: 227–232.

Verdijck GJC, Hertog MLATM, Weiss M, Preisig HA. 1999. Modelling of a potato storage facility for product quality control purposes. *Comput Chem Eng* 23: S911–S914.

Williams TCR, Poolman MG, Howden AJM, Schwarzlander M, Fell DA, Ratcliffe RG, Sweetlove LJ. 2010. A genome-scale metabolic model accurately predicts fluxes in central carbon metabolism under stress conditions. *Plant Physiol* 154: 311–323.

Wold S. 1995. Chemometrics; what do we mean with it, and what do we want from it? *Chemom Intelligent Lab System* 30: 109–115.

Zerbini PE, Vanoli M, Grassi M, Rizzolo A, Fibiani M, Cubeddu R, Pifferi A, Spinelli L, Torricelli A. 2006. A model for the softening of nectarines based on sorting fruit at harvest by time-resolved reflectance spectroscopy. *Postharv Biol Technol* 39: 223–232.

Index